More praise for

*Fash⋯*
*in the N*

"Something is rotten in the ⋮ ⋯atician and physicist Roger Penrose ide⋯ ⋯ ⋯e rot. . . . He is not one to be intimidated by a ⋯ ⋯ajority, no matter how illustrious and vocal it is. He sets out his objections politely and with exemplary patience towards the keepers of physics orthodoxy. . . . Time will tell whether any of his judgments are correct. In the meantime, his critics would do well to remember George Bernard Shaw's warning: 'The minority is sometimes right; the majority is always wrong.'"

— Graham Farmelo, *The Guardian*

"Acclaimed English mathematical physicist Penrose gets to the heart of modern physics' problem with subjectivity in this insightful and provocative pop-sci title. . . . [A] rewarding discussion of scientific stumbles in the search for truth."

— *Publishers Weekly*

"It is always inspiring to read Penrose's uncompromisingly independent perspective on physics."

— Richard Dawid, *Nature*

"The book is replete with phenomenal visual representations of the physics under discussion, a reminder of Penrose's ability to see and describe physics in a unique way. . . . Ultimately, what is most valuable about the book is the excellent example he offers in how to ask questions."

— Chanda Prescod-Weinstein, *Physics World*

"A valuable insight into what one of the most prominent theoretical physicists of recent times makes of reality's relationship to ideas in quantum theory, standard cosmology, and theories that pretend to replace them."

— Richard Webb, *New Scientist*

"The most important thing is not exactly what he writes about string theory, cosmology and quantum mechanics in his latest book . . . but that a book so wide and deep in its erudition could be written at all. If his successors cannot do the same, science will be all the poorer."

— Philip Ball, *Prospect*

"One of the most original thinkers in the field of mathematical physics. . . . I can't recommend [this book] too highly to any mathematician with a serious interest in fundamental questions about physics."

— Peter Woit, *MAA Reviews*

"The strength of this book is how the reader can appreciate science as a human undertaking."

— *Choice*

"This gem of a book is vintage Roger Penrose: eloquently argued and deeply original on every page. His perspective on the present crisis and future promise of physics and cosmology provides an important corrective to fashionable thinking at this crucial moment in science. This book deserves the widest possible hearing among specialists and the public alike."

— Lee Smolin, author of *Time Reborn: From the Crisis in Physics to the Future of the Universe*

# FASHION
## Faith
### and
## FANTASY
### in the New Physics
### of the Universe

# ROGER PENROSE

# FASHION
# Faith
## and
# FANTASY

## in the New Physics
## of the Universe

PRINCETON UNIVERSITY PRESS

PRINCETON AND OXFORD

Requests for permission to reproduce material from this work
should be sent to Permissions, Princeton University Press

Published by Princeton University Press,

41 William Street, Princeton, New Jersey 08540

In the United Kingdom: Princeton University Press,

6 Oxford Street, Woodstock, Oxfordshire OX20 1TR

press.princeton.edu

Cover design by Carmina Alvarez-Gaffin

Third printing, and first paperback printing, 2017
Paperback ISBN: 978-0-691-17853-0
Cloth ISBN: 978-0-691-11979-3 (alk. paper)

Library of Congress Control Number: 2017941739
British Library Cataloguing-in-Publication Data is available

This book has been composed in Times
Typeset by T&T Productions Ltd, London
Printed on acid-free paper ⊗
Printed in the United States of America

3 5 7 9 10 8 6 4

# Contents

# Acknowledgements

This book's somewhat lengthy gestation has faded my memory of the sources of many contributions to its development. To such helpful but anonymous friends and colleagues I offer both my gratitude and apologies. There are of course others whom I clearly owe especial thanks, most particularly my long-time colleague Florence Tsou (Sheung Tsun) for her enormous help (also with her husband Chan Hong-Mo) regarding particle physics. My even longer-time colleague Ted (Ezra) Newman has provided continual insights and support over many years, and I benefited greatly from the knowledge and understandings of Abhay Ashtekar, Krzysztof Meissner, and Andrzej Trautman. Oxford colleagues Paul Tod, Andrew Hodges, Nick Woodhouse, Lionel Mason, and Keith Hannabuss have also greatly influenced my thinking. I have learned much of approaches to quantum gravity from Carlo Rovelli and Lee Smolin. Special thanks go to Shamit Kachru for his careful study of earlier drafts of this book, and although I doubt he will be happy with its expressed sentiments regarding string theory, his criticisms have been greatly helpful in reducing errors and misunderstandings on both sides.

For input of various kinds I am grateful to Fernando Alday, Nima Arkani-Hamed, Michael Atiyah, Harvey Brown, Robert Bryant, Marek Demianski, Mike Eastwood, George Ellis, Jörge Frauendiener, Ivette Fuentes, Pedro Ferreira, Vahe Gurzadyan, Lucien Hardy, Denny Hill, Lane Hughston, Claude LeBrun, Tristan Needham, Sara Jones Nelson, Pawel Nurowski, James Peebles, Oliver Penrose, Simon Saunders, David Skinner, George Sparling, John Statchel, Paul Steinhardt, Lenny Susskind, Neil Turok, Gabriele Veneziano, Richard Ward, Edward Witten, Anton Zeilinger, and Seth Zimmerman.

Richard Lawrence and his daughter Jessica have been invaluable for providing numerous facts. For help in administrative ways, I thank Ruth Preston, Fiona Martin, Petrona Winton, Edyta Mielczarek, and Anne Pearsall. I am supremely grateful to Vickie Kearn of the Princeton University Press for her enormous patience, support, and encouragement, and to her colleagues Carmina Alvarez for her cover design and Karen Fortgang and Dimitri Karetnikov for their guidance with regard to diagrams, and to Jon Wainwright of T&T Productions Ltd for his careful editing. Finally, my wonderful wife Vanessa has kept me going, through difficult times, with her love, critical support, and technical expertise – often magically rescuing me from seemingly hopeless entanglements with my computer. Huge thanks to her and also to our teenage son, Max, whose technical know-how and loving support have been invaluable.

## ILLUSTRATION CREDITS

The author gratefully acknowledges the copyright holders of the following figures:

Figure 1-35: After Carlo Rovelli, personal communication 2013.

Figure 1-38: M. C. Escher's *Circle Limit I* © 2016 The M. C. Escher Company–The Netherlands. All rights reserved. www.mcescher.com

Figure 3-1: M. C. Escher's (a) *Photo of Sphere*, (b) *Symmetry Drawing E45*, and (c) *Circle Limit IV* © 2016 The M. C. Escher Company–The Netherlands. All rights reserved. www.mcescher.com

Figure 3-38 (a) and (b): From "Cosmic Inflation" by Andreas Albrecht, in *Structure Formation in the Universe* (ed. R. Crittenden and N. Turok). Used with permission of Springer Science and Business Media.

Figure 3-38 (c): From "Inflation for Astronomers" by J. V. Narlikar and T. Padmanabhan as modified by Ethan Siegel in "Why we think there's a Multiverse, not just our Universe" (https://medium.com/starts-with-a-bang/why-we-think-theres-a-multiverse-not-just-our-universe-23d5ecd33707#.3iib9ejum). Reproduced with permission of *Annual Review of Astronomy and Astrophysics*, 1 September 1991, Volume 29 © by Annual Reviews, http://www.annualreviews.org.

Figure 3-38 (d): From "Eternal Inflation, Past and Future" by Anthony Aguirre, in *Beyond the Big Bang: Competing Scenarios for an Eternal Universe* (The Frontiers Collection) (ed. Rudy Vaas). Used with permission of Springer Science and Business Media.

Figure 3-43: Copyright of ESA and the Planck Collaboration

All other figures (excepting the computer curves in figures 2-2, 2-5, 2-10, 2-25, 3-6(b), A-1, A37, A-41, A-44, and A-46) were drawn by the author.

# Preface

## ARE FASHION, FAITH, OR FANTASY RELEVANT TO FUNDAMENTAL SCIENCE?

This book has been developed from an account of three lectures I gave at Princeton University in October 2003 at the invitation of Princeton University Press. The title I had proposed to the Press for these lectures – Fashion, Faith, and Fantasy in the New Physics of the Universe – and which remains as the title of this book may well have been a somewhat rash suggestion on my part. Yet it genuinely expressed a certain unease I felt about some of the trends that were part of the thinking of the time concerning the physical laws governing the universe in which we live. Well over a decade has passed since then, but the topics, and much of what I had to say about them, appear to be, for the most part, at least as relevant today as they were then. I gave those talks with some apprehension, I might add, as I was trying to express some points of view that I worried might resonate not too favourably with many of the resident distinguished experts.

Each of the eponymous words "fashion", "faith", and "fantasy" suggests a quality that would seem to be very much at odds with the procedures normally considered appropriate when applied to a search for the deep principles that underlie the behaviour of our universe at its most basic levels. Indeed, ideally, it would be very reasonable to assert that such influences as fashion, or faith, or fantasy ought to be totally absent from the attitude of mind of those seriously dedicated to searching for the foundational underpinnings of our universe. Nature herself, after all, surely has no serious interest in the ephemeral whims of human fashion. Nor should science be thought of as a faith, the dogmas of science being under continual scrutiny and subject to the rigours of experimental examination, to be abandoned the moment that a convincing conflict arises with what we find to be the actuality of nature. And fantasy is surely the province of certain areas of fiction and entertainment, where it is not deemed essential that significant regard be paid to the requirements of consistency with observation, or to strict logic, or even to good common sense. Indeed, if a proposed scientific theory can be revealed as being too much influenced by the enslavement of fashion, by the unquestioning following of an experimentally unsupported faith, or by the romantic temptations of fantasy, then it is our duty to point out such influences, and to steer away any who might, perhaps unwittingly, be subject to influences of this kind.

Nevertheless, I have no desire to be entirely negative with regard to these qualities. For it can be argued that there is something of distinctly positive value in each of these eponymous terms. A fashionable theory, after all, is unlikely to have such a status for purely sociological reasons. There must indeed be many positive qualities to hold multitudes of researchers to a highly fashionable area of study, and it is unlikely to be the mere desire to be part of a crowd that keeps such researchers so fascinated by what is likely to be an extremely difficult field of study – this very difficulty often having roots in the highly competitive nature of fashionable pursuits.

A further point needs to be made here, with regard to research in theoretical physics that may be fashionable, yet far from what is plausible as a description of the world – indeed, as we shall find, often being in fairly blatant contradiction with current observations. Whereas those who work in such areas might well have found huge gratification, had observational facts turned out to be more in accordance with their own pictures of the world, they often seem relatively undisturbed by facts that are found to be less obliging to them than they would have liked. This is not at all unreasonable; for, to a considerable degree, these researches are merely *exploratory*, the viewpoint being that expertise may well be gained from such work, and that this will eventually be useful in the discovery of better theories which agree more closely with the actual functioning of the universe we know.

When it comes to the extreme faith in some scientific dogmas that is often expressed by researchers, this also is likely to have a powerful rationale, even where the faith is in the applicability of such a dogma in circumstances that lie far beyond the original situations where strong observational support initially laid its foundations. The superb physical theories of the past can continue to be trusted to provide enormous precision even when, in certain circumstances, they have become superseded by better theories that extend their precision or the range of their applicability. This was certainly the case when Newton's magnificent gravitational theory was superseded by Einstein's, or when Maxwell's beautiful electromagnetic theory of light became superseded by its quantized version, wherein the particulate aspects of light (photons) could be understood. In each case the earlier theory would retain its trustworthiness, provided that its limitations are kept appropriately under firm consideration.

But what about fantasy? Surely this is the very opposite of what we should be striving for in science. Yet we shall be seeing that there are some key aspects to the nature of our actual universe that are so exceptionally odd (though not always fully recognized as such) that if we do not indulge in what may appear to be outrageous flights of fantasy, we shall have no chance at all of coming to terms with what may well be an extraordinary fantastical-seeming underlying truth.

In the first three chapters, I shall illustrate these three eponymous qualities with three very well-known theories, or families of theory. I have not chosen areas of relatively minor importance in physics, for I shall be concerned with what are big fish indeed in the ocean of current activity in theoretical physics. In chapter 1, I have chosen to address the still highly fashionable string theory (or superstring theory, or its generalizations such as M-theory, or the currently most fashionable aspect of this general line of work, namely the scheme of things referred to as the *ADS/CFT correspondence*). The faith that I shall address in chapter 2 is an even bigger fish, namely that dogma that the procedures of quantum mechanics must be slavishly followed, no matter how large or massive are the physical elements to which it is being applied. And, in some respects, the topic of chapter 3 is the biggest fish of all, for we shall be concerned with the very origin of the universe that we know, where we shall catch a glimpse of some proposals of seeming sheer fantasy that have been put forward in order to address certain of the genuinely disturbing peculiarities that well-established observations of the very early stages of our entire universe have revealed.

Finally, in chapter 4, I bring forward some particular views of my own, in order to point out that there are alternative routes that could well be taken. We shall find, however, that the following of my own suggested paths would appear to involve certain aspects of irony. There is, indeed, an irony of fashion in my own preferred path to the understanding of basic physics – a path that I shall briefly introduce to the reader in §4.1. This is the path staked out by twistor theory, which I have myself been seminally involved with, and which for some forty years had attracted scant attention from the physics community. But we find that twistor theory has itself now begun to acquire some small measure of string-related fashion.

As to an overriding unshakable faith in quantum mechanics, as appears to be held by the considerable majority of the physics community, this has been further endorsed by remarkable experiments, such as those of Serge Haroche and David Wineland, which received well-deserved recognition by the award of the 2012 Nobel Prize in physics. Moreover, the award of the 2013 Nobel Physics Prize to Peter Higgs and François Englert for their part in the prediction of what has come to be known as the *Higgs boson* is a striking confirmation not only of the particular ideas that they (and some others, most particularly Tom Kibble, Gerald Guralnik, Carl Hagen, and Robert Brout) had put forward in relation to the origin of particle masses, but also of many of the foundational aspects of quantum (field) theory itself. Yet, as I point out in §4.2, all such highly refined experiments performed so far still fall considerably short of the level of displacement of mass (as proposed in §2.13) that will be needed before one might seriously anticipate our quantum faith to be significantly challenged. There are other experiments at present under development, however, that are aimed at such a level of mass displacement, which I argue could well help to resolve some profound conflicts between current quantum mechanics and certain other accepted physical principles, namely those of Einstein's general relativity. In §4.2, I point out a serious conflict between current quantum mechanics and Einstein's foundational principle of equivalence between gravitational fields and accelerations. Perhaps the results of such experiments may indeed undermine the unquestioning quantum-mechanical faith that seems to be so commonly held. On the other hand, one may ask why should one have more faith in Einstein's equivalence principle than in the immensely more broadly tested foundational procedures of quantum mechanics? A good question indeed – and it could well be argued that there is at least as much faith involved in accepting Einstein's principle as in accepting those of quantum mechanics. This is an issue that could well be resolved by experiment in the not-too-distant future.

As to the levels of fantasy that current cosmology has been led into, I suggest in §4.3 (as a final irony) that there is a scheme of things that I put forward myself in 2005 – conformal cyclic cosmology, or CCC – that is, in certain respects, even more fantastical than those extraordinary proposals we shall encounter in chapter 3, some of which have now become part of almost all contemporary discussions of the very early stages of the universe. Yet CCC appears to be beginning to reveal itself, in current observational analyses, as having some basis in actual physical fact. It is certainly to be hoped that clear-cut observational evidence will soon be able to convert what may or may not seem to be sheer fantasy, of one kind or another, into a convincing picture of the factual nature of our actual universe. It may be remarked, indeed, that unlike the fashions of string theory, or most theoretical schemes aimed at undermining our total faith in the principles of quantum mechanics, those fantastical proposals that are being put forward for describing the very origin of our universe are already being confronted by detailed observational tests, such as in the comprehensive information provided by space satellites COBE, WMAP, and the Planck space platform, or by the results of the BICEP2 South Pole observations released in March 2014. At the time of writing, there are serious issues of interpretation concerning the latter, but these ought to be resolvable before too long. Perhaps there will soon be much clearer evidence, enabling definitive choices to be made between rival fantastical theories, or some theory not yet thought of.

In attempting to address all these issues in a satisfactory (but not too technical) way, I have had to face up to one particular fundamental hurdle. This is the issue of mathematics and its central role in any physical theory that can seriously purport to describe nature at any real depth. The critical arguments that I shall be making in this book, aimed at establishing that fashion, faith, and fantasy are indeed inappropriately influencing the progress of fundamental science, have to be based, to some meaningful extent, on genuine technical objections, rather than on mere emotional preferences, and this will require us to get involved in a certain amount of significant mathematics. Yet this account is not intended to be a technical discourse, accessible only to experts in mathematics or physics, for it is certainly my intention that it can be read with profit by non-experts. Accordingly, I shall try to keep the technical content to a reasonable minimum. There are, however, some mathematical notions that would be greatly helpful for the full appreciation of various critical issues that I wish to address. I have therefore included eleven rather basic mathematical sections in an appendix, these providing accounts that are not very technical, but which could, where necessary, help non-experts to gain some greater appreciation of many of the main issues.

The first two of these sections (§§A.1 and A.2) involve only very simple ideas, albeit somewhat unfamiliar ones, with no difficult notation. However, they play a special role for many arguments in this book, most particularly with regard to the fashionable proposals discussed in chapter 1. Any reader wishing to understand the central critical issue discussed there should, at some stage, take note of the material of §§A.1 and A.2, which contain the key to my argument against additional spatial dimensions being actually present in our physical universe. Such supra-dimensionality is a central contention of almost all of modern string theory and its major variants. My critical arguments are aimed at the current string-motivated belief that the dimensionality of physical space must be greater than the three that we directly experience. The key issue I raise here is that of functional freedom, and in §A.8 I outline a somewhat fuller argument to clarify the basic point. The mathematical notion under consideration has its roots in the work of the great French mathematician Élie Cartan, basically dating back to the turn of the twentieth century, but seeming to be little appreciated by theoretical physicists of today, although having great relevance to the plausibility of current supra-dimensional physical ideas.

String theory and its modern variants have moved forward in many ways in the years following these Princeton lectures, and have developed considerably in technical detail. I certainly make no claim to any kind of mastery of such developments, although I have looked at a fair amount of this material. My essential issue of concern lies not in any such detail but in whether this work really moves us forward very much towards an understanding of the actual physical world in which we live. Most particularly, I see little (if any) attempt to address the question of excessive functional freedom arising from the assumed spatial supra-dimensionality. Indeed, no work of string theory that I have seen makes any mention of this problem. I find this to be somewhat surprising, not just because this issue was central to the first of my decade-old three Princeton lectures. It had previously featured in a talk I gave at a conference honouring Stephen Hawking's 60th birthday at Cambridge University, in January 2002, to an audience containing several leading string theorists, and written accounts were subsequently provided.

I need to make an important point here. The issue of functional freedom is often rejected by quantum physicists as applying only to classical physics, and the difficulties it presents for supra-dimensional theories tend to be summarily dismissed with an argument aimed at demonstrating the irrelevance of these matters in quantum-mechanical situations. In §1.10 I present my main case against this basic argument, which I particularly encourage the proponents of spatial supra-dimensionality to read. It is my hope that by repeating such arguments here, and by developing them in certain further physical contexts (§§1.10, 1.11, 2.11, and A.11), I might encourage these arguments to be adequately taken into consideration in future work.

The remaining sections of the appendix briefly introduce vector spaces, manifolds, bundles, harmonic analysis, complex numbers, and their geometry. These topics would certainly be well familiar to the experts, but non-experts may find such self-contained background material helpful for fully understanding the more technical parts of this book. In all my descriptions, I have stopped short of providing any significant introduction of the ideas of differential (or integral) calculus, my viewpoint being that while a proper understanding of calculus would be of benefit to readers, those who do not already have this advantage would gain little from a hurried section on this topic. Even so, in §A.11, I have found it helpful just to touch upon the issue of differential operators and differential equations, in order to help explain some matters that have relevance, in various ways, to the thread of the argument throughout the book.

# 1

## Fashion

---

### 1.1. MATHEMATICAL ELEGANCE AS A DRIVING FORCE

As mentioned in the preface, the issues discussed in this book were developed from three lectures given, by invitation of the Princeton University Press, at Princeton University in October 2003. My nervousness, with these lectures, in addressing such a knowledgeable audience as the Princeton scientific community, was perhaps at its greatest when it came to the topic of fashion, because the illustrative area that I had elected to discuss, namely string theory and some of its various descendants, had been developed to its heights in Princeton probably more than anywhere else in the world. Moreover, that subject is a distinctly technical one, and I cannot claim competence over many of its important ingredients, my familiarity with these technicalities being somewhat limited, particularly in view of my status as an outsider. Yet, it seemed to me, I should not allow myself to be too daunted by this shortcoming, for if only the insiders are considered competent to make critical comments about the subject, then the criticisms are likely to be limited to relatively technical issues, some of the broader aspects of criticism being, no doubt, significantly neglected.

Since these lectures were given, there have been three highly critical accounts of string theory: *Not Even Wrong* by Peter Woit, *The Trouble with Physics* by Lee Smolin, and *Farewell to Reality: How Fairytale Physics Betrays the Search for Scientific Truth* by Jim Baggott. Certainly, Woit and Smolin have had more direct experience than I have of the string-theory community and its over-fashionable status. My own criticisms of string theory in *The Road to Reality*, in chapter 31 and parts of chapter 34, have also appeared in the meantime (predating these three works), but my own critical remarks were perhaps somewhat more favourably disposed towards a physical role for string theory than were these others. Most of my comments will indeed be of a general nature, and are relatively insensitive to issues of great technicality.

Let me first make what surely ought to be a general (and perhaps obvious) point. We take note of the fact that the hugely impressive progress that physical theory has indeed made over several centuries has depended upon extremely precise and sophisticated mathematical schemes. It is evident, therefore, that any further significant progress must again depend crucially upon some distinctive mathematical framework. In order that any proposed new physical theory can improve upon what has been achieved up until now, making precise and unambiguous predictions that go beyond what had been possible before, it must also be based on some clear-cut mathematical scheme. Moreover, one would think, to be a proper mathematical theory it surely ought to make mathematical sense – which means, in effect, that it ought to be *mathematically consistent*. From a self-*in*consistent scheme, one could, in principle, deduce any answer one pleased.

Yet, self-consistency is actually a rather strong criterion and it turns out that not many proposals for physical theories – even among the very successful ones of the past – are in fact fully self-consistent. Often some strong elements of physical judgement must be invoked in order that the theory can be appropriately applied in an unambiguous way. Experiments are, of course, also central to physical theory, and the testing of a theory by experiment is very different from checking it for logical consistency. Both are important, but in practice one often finds that physicists do not care so much about achieving full mathematical self-consistency if the theory appears to fit the physical facts. This has been the case, to some considerable degree, even with the extraordinarily successful theory of quantum mechanics, as we shall be seeing in chapter 2 (and §1.3). The very first work in that subject, namely Max Planck's epoch-making proposal to explain the frequency spectrum of electromagnetic radiation in equilibrium with matter at a fixed temperature (the black-body spectrum; see §§2.2 and 2.11) required something of a hybrid picture which was not really fully self-consistent [Pais 2005]. Nor can it be said that the old quantum theory of the atom, as brilliantly proposed by Niels Bohr in 1913, was a fully self-consistent scheme. In the subsequent developments of quantum theory, a mathematical edifice of great sophistication has been constructed, in which a desire for mathematical consistency had been a powerful driving force. Yet, there remain issues of consistency that are still not properly addressed in current theory, as we shall see later, particularly in §2.13. But it is the *experimental* support, over a vast range of different kinds of physical phenomena, which is quantum theory's bedrock. Physicists tend not to be over-worried by detailed matters of mathematical or ontological inconsistency if the theory, when applied with appropriate judgement and careful calculation, continues to provide answers that are in excellent agreement with the results of observation – often with extraordinary precision – through delicate and precise experiment.

The situation with string theory is completely different from this. Here there appear to be *no* results whatever that provide it with experimental support. It is often argued that this is not surprising, since string theory, as it is now formulated as largely a *quantum gravity* theory, is fundamentally concerned with what is called the *Planck scale* of very tiny distances (or at least close to such distances), some $10^{-15}$ or $10^{-16}$ times smaller ($10^{-16}$ meaning, of course, down by a factor of a tenth of a thousandth of a millionth of a millionth) and hence with energies some $10^{15}$ or $10^{16}$ times larger than those that are accessible to current experimentation. (It should be noted that, according to basic principles of relativity, a small distance is essentially equivalent to a small time, via the speed of light, and, according to basic principles of quantum mechanics, a small time is essentially equivalent to a large energy, via Planck's constant; see §§2.2 and 2.11.) One must certainly face the evident fact that, powerful as our present-day particle accelerators may be, their currently foreseeable achievable energies fall enormously short of those that appear to have direct relevance to theories such as modern string theory that attempt to apply the principles of quantum mechanics to gravitational phenomena. Yet this situation can hardly be regarded as satisfactory for a physical theory, as experimental support is the ultimate criterion whereby it stands or falls.

Of course, it might be the case that we are entering a new phase of basic research into fundamental physics, where requirements of mathematical consistency become paramount, and in those situations where such requirements (together with a coherence with previously established principles) prove insufficient, additional criteria of *mathematical elegance* and simplicity must be invoked. While it may seem unscientific to appeal to such aesthetic desiderata in a fully objective search for the physical principles underlying the workings of the universe, it is remarkable how fruitful – indeed essential – such aesthetic judgements seem to have frequently proved to be. We have come across many examples in physics where beautiful mathematical ideas have turned out to underlie fundamental advances in understanding. The great theoretical physicist Paul Dirac [1963] was very explicit about the importance of aesthetic judgement in his discovery of the equation for the electron, and also in his prediction of anti-particles. Certainly, the Dirac equation has turned out to be absolutely fundamental to basic physics, and the aesthetic appeal of this equation is very widely appreciated. This is also the case with the idea of anti-particles, which resulted from Dirac's deep analysis of his own equation for the electron.

However, this role of aesthetic judgement is a very difficult issue to be objective about. It is often the case that some physicist might think that a particular scheme is very beautiful whereas another might emphatically *not* share that view! Elements of fashion can often assume unreasonable proportions when it comes to aesthetic judgements – in the world of theoretical physics, just as in the case of art or the design of clothing.

It should be made clear that the question of aesthetic judgment in physics is more subtle than just what is often referred to as *Occam's razor* – the removal of unnecessary complication. Indeed, a judgement as to which of two opposing theories is actually the "simpler", and perhaps therefore more elegant, need by no means be a straightforward matter. For example, is Einstein's general relativity a simple theory or not? Is it simpler or more complicated than Newton's theory of gravity? Or is Einstein's theory simpler or more complicated than a theory, put forward in 1894 by Aspeth Hall (some 21 years before Einstein proposed his general theory of relativity), which is just like Newton's but where the inverse square law of gravitation is replaced by one in which the gravitational force between a mass $M$ and a mass $m$ is $GmMr^{-2.00000016}$, rather than Newton's $GmMr^{-2}$. Hall's theory was proposed in order to explain the observed slight deviation from the predictions of Newton's theory with regard to the advance of the perihelion of the planet Mercury that had been known since about 1843. (The perihelion is the closest point to the Sun that a planet reaches while tracing its orbit [Roseveare 1982].) This theory also gave a very slightly better agreement with Venus's motion than did Newton's. In a certain sense, Hall's theory is only marginally more complicated than Newton's, although it depends on how much additional "complication" one considers to be involved in replacing the nice simple number "2" by "2.00000016". Undoubtedly, there is a loss of mathematical elegance in this replacement, but as noted above, a strong element of subjectivity comes into such judgements. Perhaps more to the point is that there are certain elegant mathematical properties that follow from the inverse square law (basically, expressing a conservation of "flux lines" of gravitational force, which would not be exactly true in Hall's theory). But again, one might consider this an aesthetic matter whose physical significance should not be overrated.

But what about Einstein's general relativity? There is certainly an enormous increase in the difficulty of applying Einstein's theory to specific physical systems, beyond the difficulty of applying Newton's theory (or even Hall's), when it comes to examining the implications of this theory in detail. The equations, when written out explicitly, are immensely more complicated in Einstein's theory, and they are difficult even to write down in full detail. Moreover, they are immensely harder to solve, and there are many nonlinearities in Einstein's theory which do not appear in Newton's (these tending to invalidate the simple flux-law arguments that must already be abandoned in Hall's theory). (See §§A.4 and A.11 for the meaning of *linearity*, and for its special role in quantum mechanics see §2.4.) Even more serious is the fact that the physical interpretation of Einstein's theory depends upon eliminating spurious coordinate effects that arise from the making of particular choices of coordinates, such choices being supposed to have no physical relevance in Einstein's theory. In practical terms, there is no doubt that Einstein's theory is usually immensely more difficult to handle than is Newton's (or even Hall's) gravitational theory.

Yet, there is still an important sense in which Einstein's theory is actually a very simple one – even possibly simpler (or more "natural") than Newton's. Einstein's theory depends upon the mathematical theory of Riemannian (or, more strictly, as we shall be seeing in §1.7, *pseudo*-Riemannian) geometry, of arbitrarily curved 4-manifolds (see also §A.5). This is not an altogether easy body of mathematical technique to master, for we need to understand what a tensor is and what the purpose of such quantities is, and how to construct the particular tensor object **R**, called the *Riemann curvature tensor*, from the *metric tensor* **g** which defines the geometry. Then by means of a contraction and a trace-reversal we find how to construct the *Einstein tensor* **G**. Nevertheless, the general geometrical ideas behind the formalism are reasonably simple to grasp, and once the ingredients of this type of curved geometry are indeed understood, one finds that there is a very restricted family of possible (or plausible) equations that can be written down, which are consistent with the proposed general physical and geometrical requirements. Among these possibilities, the very simplest gives us Einstein's famous field equation $\mathbf{G} = 8\pi\gamma\mathbf{T}$ of general relativity (where **T** is the mass–energy tensor of matter and $\gamma$ is Newton's gravitational constant – given according to Newton's particular definition, so that even the "$8\pi$" is not really a complication, but merely a matter of how we wish to define $\gamma$).

There is just one minor, and still very simple, modification of the Einstein field equation that can be made, which leaves the essential requirements of the scheme intact, namely the inclusion of a constant number $\Lambda$, referred to as the *cosmological constant* (which Einstein introduced in 1917 for reasons that he later discarded) so that Einstein's equations with $\Lambda$ now become $\mathbf{G} = 8\pi\gamma\mathbf{T} + \Lambda\mathbf{g}$. The quantity $\Lambda$ is now frequently referred to as *dark energy*, presumably to allow for a possibility of generalizing Einstein's theory so that $\Lambda$ might vary. There are, however, strong mathematical constraints obstructing such considerations, and in §§3.1, 3.7, 3.8, and 4.3, where $\Lambda$ will be playing a significant role for us, I shall restrict attention to situations where $\Lambda$ is indeed non-varying. The cosmological constant will have considerable relevance in chapter 3 (and also §1.15). Indeed, relatively recent observations point strongly to the actual physical presence of $\Lambda$ having a tiny (apparently constant) positive value. This evidence for $\Lambda > 0$ – or possibly for some more general form of "dark energy" – is now very impressive, and has been growing since the initial observations of Perlmutter et al. [1999], Riess et al. [1998], and their collaborators, leading to the award of the 2011 Nobel Prize in physics to Saul Perlmutter, Brian P. Schmidt, and Adam G. Riess. This $\Lambda > 0$ has immediate relevance only to the very distant cosmological scales, and observations concerning celestial motions at a more local scale can be adequately treated according to Einstein's original and simpler $\mathbf{G} = 8\pi\gamma\mathbf{T}$. This equation is now found to have an unprecedented precision in modelling the behaviour, under gravity, of celestial bodies, the observed $\Lambda$ value having no significant impact on such local dynamics.

Historically of most importance, in this regard, is the double-neutron-star system PSR1913+16, one component of which is a *pulsar*, sending very precisely timed electromagnetic signals that are received at the Earth. The motion of each star about the other, being very cleanly a purely gravitational effect, is modelled by general relativity to an extraordinary precision that can be argued to be of about one part in $10^{14}$ overall, accumulated over a period of about 40 years. The period 40 years is roughly $10^9$ seconds, so a precision of one in $10^{14}$ means an agreement between observation and theory to about $10^{-5}$ (one hundred thousandth) of a second over that period – which is, very remarkably, indeed just what is found. More recently, other systems [Kramer et al. 2006] involving one or even a pair of pulsars, have the potential to increase this precision considerably, when the systems have been observed for a comparable length of time as has PSR19+16.

To call this figure of $10^{14}$ a measure of the observed precision of general relativity is open to some question, however. Indeed, the particular masses and orbital parameters have to be calculated from the observed motions, rather than being numbers coming from theory or independent observation. Moreover, much of this extraordinary precision is already in Newton's gravitational theory.

Yet, we are concerned here with gravitational theory overall, and Einstein's theory incorporates the deductions from Newton's theory (giving Kepler's elliptical orbits, etc.) as a first approximation, but provides various corrections to the Keplerian orbits (including the perihelion advance), and finally a loss of energy from the system which is precisely in accord with a remarkable prediction from general relativity: that such a massive system in accelerated motion should lose energy through the emission of gravitational waves – ripples in space-time which are the gravitational analogues of electromagnetic waves (i.e. light) that electrically charged bodies emit when they are involved in accelerated motion. As a striking further confirmation of the existence and precise form of such gravitational radiation is the announcement [Abbott et al. 2016] of their direct detection by the LIGO gravitational wave detector, which also provides excellent direct evidence of another of the predictions of general relativity: the existence of black holes, which we shall be coming to in §3.2, and discussed also in later parts of chapter 3, and in §4.3.

It should be emphasized that this precision goes enormously beyond – by an additional factor of about $10^8$ (i.e. one hundred million) or more – that which was observationally available to Einstein when he first formulated his gravitational theory. The observed precision in Newton's gravitational theory could itself be argued to be around one part in $10^7$. Accordingly, the "1 part in $10^{14}$" precision of general relativity was already "out there" in nature, before Einstein formulated his own theory. Yet that additional precision (by a factor of around one hundred million), being unknown to Einstein, can have played no role whatever in his formulating his theory. Thus this new mathematical model of nature was not a man-made construction invented merely in an attempt to find the best theory to fit the facts; the mathematical scheme was, in a clear sense, already there in the works of nature herself. This mathematical simplicity, or elegance, or however one should describe it, is a genuine part of nature's ways, and it is not simply that our minds are attuned to being impressed by such mathematical beauty.

On the other hand, when we try deliberately to use the criterion of mathematical beauty in formulating our theories, we are easily led astray. General relativity is certainly a very beautiful theory, but how does one judge the elegance of physical theories generally? Different people have very different aesthetic judgements. It is not necessarily obvious that one person's view as to what is elegant will be the same as somebody else's, or whether one person's aesthetic judgement will be superior or inferior to another's, in formulating a successful physical theory. Moreover, the inherent beauty in a theory is often not obvious at first, and may be revealed only later when the depths of its mathematical structure become apparent through later technical developments. Newtonian dynamics is a case in point. Much of the undoubted beauty in Newton's framework was revealed only much later, through the magnificent works of such great mathematicians as Euler, Lagrange, Laplace, and Hamilton (as the terms *Euler–Lagrange equations*, the *Laplacian operator*, *Lagrangians*, and *Hamiltonians* – which are key ingredients of modern physical theory – bear witness). The role of Newton's Third Law, for example, which asserts that every action has an equal and opposite reaction, finds a central place in the Lagrangian formulation of modern physics. It would not surprise me to find that the beauty that is frequently asserted to be present in successful modern theories is often to some extent *post hoc*. The very success of a physical theory, both observational and mathematical, may contribute significantly to the aesthetic qualities that it is later perceived to possess. It would follow from all this that judgements of the merits of some proposed physical theory through its claimed aesthetic qualities are likely to be problematic or at least ambiguous. It is unquestionably more reliable to form one's judgements of a new theory on the basis of its agreement with current observation and on its predictive power.

Yet, with regard to experimental support, often the crucial experiments are not available, such as with the utterly prohibitive high energies that single particles might have to attain – absurdly in excess of those available in current particle accelerators (see §1.10) – that are often argued to be required in any proper observational test of any quantum-gravity theory. More modest experimental proposals may also be unavailable, due perhaps to the expense of the experiments or their intrinsic difficulty. Even with very successful experiments, it is quite often the case that the experimenters collect enormous amounts of data, and the problem is of a quite different kind, namely the matter of digging out some key piece of information from that morass of data. This kind of thing is certainly true in particle physics, where powerful accelerators and particle colliders now produce masses of information, and it is now also becoming true in cosmology, where modern observations of the cosmic microwave background (CMB) produce very large amounts of data (see §§3.4, 3.9, and 4.3). Much of this data is considered not to be especially informative, as it simply confirms what is already known, as gleaned from earlier experiments. A great deal of statistical processing is needed in order to extract some tiny residual – which is the new feature that the experimentalists are looking for – which might confirm or refute some suggested theoretical proposal.

A point that should be made here is that this statistical processing is likely to be very specific to current theory, geared to finding out what slight additional effect that theory might predict. It is very possible that some radically different set of ideas, departing significantly from what is currently fashionable, may remain untested even though some definitive answer might actually lie hidden in the existing data, being unrevealed because the statistical procedures that physicists have adopted are too directly tuned to current theory. We shall be seeing what appears to be a striking example of this in §4.3. Even when it is clear how definitive information might be statistically extracted from an existing morass of reliable data, the inordinate amount of computer time that this can require may sometimes constitute a huge barrier to the actual carrying out of the analysis, particularly when more fashionable pursuits may be in direct competition.

Even more to the point is the fact that the experiments themselves are usually enormously expensive and their specific design is likely to be geared to the testing of theories which are within the framework of conventional ideas. Any theoretical scheme which departs too radically from the general consensus may find it hard for sufficient funds to be provided to enable it to be properly tested. A very expensive experimental apparatus, after all, requires many committees of established experts to approve its construction, and such experts are likely to be those who have played their parts in developing the current perspectives.

In relation to this issue we may consider the Large Hadron Collider (LHC) in Geneva, Switzerland, whose construction was completed in 2008. It has a 27 km (17 mile) tunnel running under two countries (France and Switzerland), initially coming into action in 2010. It is now credited with finding the hitherto elusive Higgs particle, of great importance in particle physics, particularly in relation to its role in assigning mass to weakly interacting particles. The 2013 Nobel Prize in physics was awarded to Peter Higgs and François Englert for their part in the ground-breaking work of predicting the existence and properties of this particle.

This is undoubtedly a magnificent achievement, and I have no wish to underrate its undoubted importance. Nevertheless, the LHC appears to provide a case in point. The way in which the very high-energy encounters between particles are analysed requires the presence of extremely expensive detectors, which have been geared to glean information in relation to prevailing particle-physics theory. It may not be at all easy to obtain information of relevance to unconventional ideas concerning the underlying nature of fundamental particles and their interactions. In general, proposals which depart drastically from a prevailing perspective may well find it much harder to have a chance of being adequately funded, and also may find great difficulty in being tested at all, by definitive experiments.

A further important factor is that graduate students, when in search of a problem to work on for a doctorate degree, tend to be highly constrained with regard to appropriate topics of research. Research students working in unfashionable areas, even if leading to successful doctoral degrees, may well find extreme difficulty in obtaining academic jobs afterwards, no matter how talented, knowledgeable, or original they may be. Jobs are limited and research funding hard to come by. Research supervisors are, more likely than not, interested mainly in developing ideas that they themselves had been involved in promoting, and these are likely to be in areas that are already fashionable. Moreover, a supervisor interested in developing an idea that is outside the mainstream may well be reluctant to encourage a potential student to work in such an area, owing to the disadvantage that it may be to the student when it comes to competing, subsequently, in a highly competitive job market where those with an expertise in fashionable areas will have a distinct advantage.

The same issues arise when it comes to the funding of research projects. Proposals in fashionable areas are far more likely to receive approval (see also §1.12). Again, the proposals will be judged by acknowledged experts, and those are overwhelmingly likely to be working in areas that are already fashionable, and to which they themselves may well have been significant contributors. Projects that deviate too much from the currently accepted norms, even if well thought through and highly original, are very likely to be left without support. Moreover, this is not just a matter of limitations on the funds available, as the influence of fashion appears to be particularly relevant in the United States, where the availability of funds for scientific research remains relatively high.

It must be said, of course, that most unfashionable areas of research will be considerably less likely to develop into successful theories than any of those that are already fashionable. A radical new perspective will in the vast majority of cases have little chance of developing into a viable proposal. Needless to say, as with Einstein's general relativity, any such radical perspective must already agree with what has been previously experimentally established, and if not, then an expensive experimental test may well not be needed for the rejection of inappropriate ideas. But for theoretical proposals that are in agreement with all previously performed experiments, and where there is no current prospect for experimental confirmation or refutation – perhaps for reasons such as those just described – it seems that we must fall back on mathematical consistency, general applicability, and aesthetic criteria when we form our judgements of the plausibility and relevance of some proposed physical theory. It is in such circumstances that the role of fashion may begin to attain excessive proportions, so we must be very careful not to allow the fashionable nature of some particular theory to cloud our judgements as to its actual physical plausibility.

## 1.2. SOME FASHIONABLE PHYSICS OF THE PAST

This is particularly important for theories which purport to be probing the very foundations of physical reality, such as modern-day string theory, and we must be very wary of assigning too much plausibility to such a theory on account of

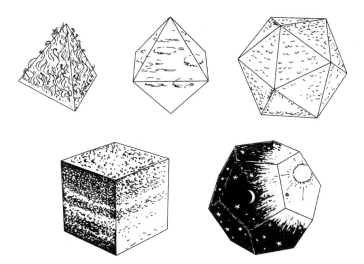

Figure 1-1: The five elements of Ancient Greece: fire (tetrahedron), air (octahedron), water (icosahedron), earth (cube), and aether (dodecahedron).

its fashionable status. Before addressing present-day physical ideas, however, it
will be instructive to mention some fashionable scientific theories of the past that
we do not take seriously today. There are large numbers of them and I am sure
that a good many readers will have little knowledge of most, for the sufficient
reason that if we do not now take these theories seriously, we are fairly unlikely
to learn about them – unless, of course, we are good historians of science; but
most physicists are not. At least let me mention a few of the better-known ones.

In particular, there is the ancient Greek theory that the Platonic solids are to be
associated with what they regarded as the basic elements of material substance, as
represented in figure 1-1. Here, *fire* is represented as the regular tetrahedron, *air*
as the octahedron, *water* as the icosahedron, and *earth* as the cube, and where, in
addition, the celestial *aether* (or *firmament*, or *quintessence*) was later introduced,
of which it was supposed that celestial bodies were composed, and was taken to
be represented as the regular dodecahedron. The ancient Greeks appear to have
formulated this sort of view – or at least many did – and I suppose it could well
have qualified as a fashionable theory of the time.

Initially, they just had the four elements of fire, air, water, and earth, and
this collection of primitive entities seemed to accord well with the four perfectly
regular polyhedral shapes that were known at the time, but when the dodecahedron

was later discovered, the theory needed to be extended to accommodate this additional polyhedron! Accordingly, the celestial substance which composed such supposedly perfect bodies as the Sun, the Moon and planets, and the crystal spheres to which they were supposedly attached was brought into the polyhedral scheme – this substance appearing, to the Greeks, to satisfy very different laws from the ones found on Earth, having a seemingly eternal motion, rather than the universal tendency of familiar substances to slow down and stop. Perhaps there is some lesson here about the way in which even modern sophisticated theories, having been initially presented in a supposedly definitive form, may then become significantly altered and their original doctrines stretched to lengths not previously conceived, in the face of new theoretical or observational evidence. As I understand it, the view that the ancient Greeks held was that somehow the laws governing the motions of stars, planets, Moon, and Sun were indeed quite different from the laws that governed things on Earth. It took Galileo, with his understanding of the relativity of motion, and then Newton, with his theory of universal gravitation – strongly influenced also by Kepler's understanding of planetary orbiting – to appreciate that the same laws actually apply to celestial bodies as to those on Earth.

When I first became acquainted with these ancient Greek ideas, they struck me as sheer romantic fantasy, with no mathematical (let alone physical) rationale. But more recently I learned that there is a bit more of a theory underlying these ideas than I had initially imagined. Some of these polyhedral shapes can be cut up into pieces, and then suitably recombined to make others (as, for example, two cubes can be cut to make two tetrahedra and an octahedron). This might be related to physical behaviour and used as a geometrical model for the basis of transitions which can occur between these different elements. At least here was a bold and imaginative guess as to the nature of material substance, which was not really an unreasonable suggestion at a time when so little had been established as to the actual nature and behaviour of physical material. Here was an early attempt to find a basis for real materials in terms of an elegant mathematical structure – very much in the spirit of what theoretical physicists are still striving to do today – in which theoretical consequences of the model could be tested against actual physical behaviour. Aesthetic criteria were clearly also at work here, and the ideas certainly seem to have appealed to Plato. But, needless to say, the details of the ideas cannot have stood the tests of time too well – otherwise we should surely not have abandoned such a mathematically attractive proposal!

Let us consider a few other things. The Ptolemaic model of planetary motion – in which the Earth was taken to be fixed, located at the centre of the cosmos

– was extremely successful, and remained unchallenged for many centuries. The motions of Sun, Moon, and planets were to be understood in terms of *epicycles*, according to which planetary movements could be explained in terms of the superposition of uniform circular motions upon one another. Though the scheme had to be rather complicated, in order to get good agreement with observation, it was not altogether lacking in mathematical elegance, and it was able to provide a reasonably predictive theory of the future motions of planets. It should be mentioned that epicycles do have a genuine rationale, when one considers external motions relative to a stationary Earth. The motions that we actually directly see from the Earth's perspective involve the composition of the Earth's rotation (so there is a perceived circular motion of the heavens about the Earth's polar axis), which must be composed with the general apparent motions of the Sun, Moon, and planets that are roughly constrained to the ecliptic plane, which appears to us as a fairly closely circular motion about a *different* axis. For good geometrical reasons, we already perceive something of the general nature of epicycles – circular motions upon other circular motions – so it was not so unreasonable to suppose that this idea might be extended more generally in the more detailed motions of the planets.

Moreover, epicycles themselves provide some interesting geometry, and Ptolemy was himself a fine geometer. In his astronomical work, he employed an elegant and powerful geometrical theorem that he possibly discovered, as it now bears his name. (This theorem asserts that the condition for four points in a plane A, B, C, D to lie on a circle – taken in that cyclic order – is that the distances between them satisfy $AB \cdot CD + BC \cdot DA = AC \cdot BD$.) This was the accepted theory of planetary motion for around fourteen centuries, until it was superseded, and eventually completely overturned, through the wonderful work of Copernicus, Galileo, Kepler, and Newton, and it is now regarded as thoroughly incorrect! It must certainly be described as a fashionable theory, however, and it was an extraordinarily successful one, for around fourteen centuries (from the mid second to the mid sixteenth), fairly closely accounting for all observations of planetary motion (when appropriate improvements were introduced from time to time), up until the more precise measurements of Tycho Brahe towards the end of the sixteenth century.

Another famous theory that we do not now believe, though it was very fashionable for over a century between 1667 (when it was put forward by Joshua Becher) and 1778 (when effectively disproved by Antoine Lavoisier), was the phlogiston theory of combustion. According to this theory, any inflammable substance contained an element called *phlogiston*, and the process of burning involved that substance giving up its phlogiston into the atmosphere. The phlogiston theory accounted for most of the facts about burning that were known at the time, such as the fact that when burning took place in a reasonably small sealed container, it would tend to stop before all of the inflammable material was used up, this being explained by the air in the container becoming saturated with phlogiston and unable to absorb any more. Ironically, it was Lavoisier who was responsible for another fashionable but false theory, namely that heat is a material substance, which he referred to as *caloric*. That theory was disproved in 1798 by Count Rumford (Sir Benjamin Thompson).

In each of these two main examples, the success of the theory may be understood by its close relation to the more satisfactory scheme which superseded it. In the case of Ptolemaic dynamics, we can transform to the more satisfactory heliocentric picture of Copernicus by a simple geometric transformation. This involves referring motions to the Sun as centre, rather than the Earth. At first, when everything was described in terms of epicycles, this made little difference – except that the heliocentric picture looked much more systematic, with the more rapid planetary motions being those for planets that were closer to the Sun [Gingerich 2004; Sobel 2011] – and there was a basic equivalence between the two schemes at this stage. But, when Kepler found his three laws of *elliptic* planetary motion, the situation changed completely, since a geocentric description of this kind of motion made no good geometrical sense. Kepler's laws provided the key that opened the way to the extraordinarily precise and broad-ranging Newtonian picture of *universal gravity*. Nevertheless, we might not today regard the geocentric perspective as quite so outrageous as would have been the case in the nineteenth century, in the light of the *general covariance principle* of Einstein's general relativity (see §§1.7, A.5, and 2.13), which allows us to adopt massively inconvenient coordinate descriptions (like a geocentric one in which the Earth's coordinates do not change with time) as nevertheless legitimate. Likewise, the phlogiston theory could be made to correspond closely to the modern perspective on combustion in which the burning of some material is normally taken to involve the taking up of oxygen from the atmosphere, where phlogiston would simply be regarded as "negative oxygen". This provides us with a fairly consistent translation between the phlogiston picture and the now conventional one. But when detailed mass measurements by Lavoisier demonstrated that phlogiston would have to have negative mass, the picture began to lose support. Nevertheless, "negative oxygen" is not such an absurd concept from the perspective of modern particle physics, where every type particle in nature (including a composite one) is supposed to have an anti-particle – an "anti-oxygen atom" is therefore completely in accord with modern theory. It would not, however, have a negative mass!

Sometimes theories that have been out of fashion for some while can come back into consideration in view of later developments. A case in point is an idea that Lord Kelvin (William Thompson) put forward in about 1867, in which atoms (the elementary particles of his day) were to be regarded as being composed of tiny knot-like structures. This idea attracted some considerable attention at the time, and the mathematician J. G. Tait began a systematic study of knots on the basis of this. But the theory did not lead to any clear-cut correspondence with the actual physical behaviour of atoms, so it became largely forgotten. However, more recently, ideas of this general kind have begun to find favour again, partly in view of their connection with string-theoretic notions. The mathematical theory of knots has also encountered a revival, since around 1984, starting with the work of Vaughan Jones, whose seminal ideas had their roots in theoretical considerations within quantum field theory [Jones 1985; Skyrme 1961]. The methods of string theory were subsequently employed by Edward Witten [1989] to obtain a kind of quantum field theory (called a *topological quantum field theory*) which, in a certain sense, encompasses these new developments in the mathematical theory of knots.

As a revival of a far more ancient idea for the nature of the large-scale universe, I might mention – though not entirely seriously – a curious coincidence that occurred at about the time that I was presenting my Princeton lecture on which this particular chapter is based (on 17 October 2003). In that talk, I referred to the ancient Greek idea that the aether was to be associated with the regular dodecahedron. Unbeknown to me, at that time, there were newspaper reports of a proposal, due to Luminet et al. [2003], that the 3-dimensional spatial geometry of the cosmos might actually have a somewhat complicated topology, arising from the identification (with a twist) of opposite faces of a (solid) *regular dodecahedron*. Thus, in a sense, the Platonic idea of a dodecahedral cosmos was also being revived in modern times!

The ambitious idea of a theory of everything, intended to encompass all physical processes, including a description of all the particles of nature and their physical interactions, has been commonly mooted in recent years, especially in connection with string theory. The idea would be to have a complete theory of physical behaviour, based on some notion of primitive particles and/or fields, acting according to some forces or other dynamical principles precisely governing the motions of all constituent elements. This may also be regarded as a revival of an old idea, as we shall see in a moment.

Just as Einstein was producing the final form of his general theory of relativity, towards the end of 1915, the mathematician David Hilbert put forward his own method of deriving the field equations of Einstein's theory,[1] using what is known as a *variational principle*. (This very general type of procedure makes use of the Euler–Lagrange equations, obtained from a Lagrangian, this being a powerful notion, referred to by name in §1.1; see, for example, Penrose [2004, chapter 20]; henceforth this book will be referred to as "TRtR".) Einstein, in his own more direct approach, formulated his equations explicitly in a form which showed how the gravitational field (as described in terms of space-time curvature) would behave, as influenced by its "source", namely the total mass/energy densities of all the particles, or matter fields, etc., collected together in the form of the energy tensor **T** (referred to in §1.1).

Einstein gave no specific prescription for the detailed equations governing how these matter fields were to behave, these being supposed to be taken from some other theory specific to the particular matter fields under consideration. In particular, one such matter field would be the electromagnetic field, whose description would be given according to the wonderful equations of the great Scottish mathematical physicist James Clerk Maxwell in 1864, which fully unified electric and magnetic fields, thereby explaining the nature of light and much of the nature of the forces governing the internal constitution of ordinary materials. This was to be considered matter in this context, and to play its appropriate part in **T**. In addition, other types of field, and all sorts of other kinds of particles could also be involved, being governed by whatever equations as might turn out to be appropriate, would also count as matter and contribute to **T**. The details of this were not important to Einstein's theory, and were left unspecified.

On the other hand, in his own proposal Hilbert was attempting to be more all-embracing. For he put forward what we might now refer to as a *theory of everything*. The gravitational field was to be described in just the same way as in Einstein's proposal, but rather than leaving the source term **T** unspecified, as Einstein had done, Hilbert proposed that this source term should be that of a very specific theory that was fashionable at the time, known as *Mie's theory* [Mie 1908, 1912a,b, 1913]. This involved a nonlinear modification of Maxwell's electromagnetic theory, and it had been proposed by Gustav Mie as a scheme intended to incorporate *all* aspects of matter. Accordingly, Hilbert's all-embracing proposal was supposed to be a complete theory of matter (including electromagnetism) as well as gravity. The strong and weak forces of particle physics were not understood at the time, but Hilbert's proposal could indeed have been viewed as what

---

[1] On the controversial issue of who was first, see Corry et al.'s [1997] commentary.

we now frequently refer to as a theory of everything. Yet, I think it likely that not a great many physicists today will have even heard of the once fashionable Mie's theory, let alone the fact that it was explicitly part of Hilbert's theory-of-everything version of general relativity. That theory plays no part in the modern understanding of matter. Perhaps there is a lesson of caution here for theoreticians of today, intent on proposing their own theories of everything.

## 1.3. PARTICLE-PHYSICS BACKGROUND TO STRING THEORY

One such theoretical proposal is string theory, and many theoretical physicists today do indeed still regard this proposal as providing a definite route to such a theory of everything. String theory originated with some ideas which, when I first heard about them in around 1970 (from Leonard Susskind), I found to be strikingly attractive and of a distinctive compelling nature. But before describing these ideas, I should put them in the appropriate context. We should try to understand why replacing the notion of a point particle by a little loop or curve in space, as was indeed the original idea of string theory, should have any promise as the basis for a physical picture of reality.

In fact there were more reasons than one for the attraction of this idea. Ironically, one of the most specific reasons – having to do with the observational physics of the interactions between hadrons – seems to have become completely left behind by the more modern developments in string theory, and I'm not sure that it has any status in the subject at all now, beyond a historical one. But I ought to discuss it, nevertheless (as I shall, more particularly, in §1.6), as well as some of the other elements of the background of fundamental particle physics which motivated the underlying principles of string theory.

First let me say what a hadron is. We recall that an ordinary atom consists of a positively charged nucleus, and negatively charged electrons orbiting around it. The nucleus is composed of protons and neutrons – collectively called *nucleons* (N) – where each proton has a positive electric charge of one unit (the unit of charge being chosen so that the electron's charge is the negative of one unit) and where each neutron has zero electric charge. The attractive electric force between positive and negative charges is what holds the negatively charged electrons in their orbits around the positively charged nucleus. But if electric forces were the only ones of relevance, then the nucleus itself (apart from that of hydrogen, which has just a single proton) would explode into various constituents, because the protons, all having charges of the same sign, would repel one another. Accordingly, there must be another, stronger, force which holds the nucleus together, and this is what is called the *strong* (nuclear) force. There is, in addition, something called the *weak* (nuclear) force, which has particular relevance in relation to nuclear decay, but this is not the major component in the forces between nucleons. I shall be saying something about the weak force later.

Not all particles are directly affected by the strong force – for example, electrons are not – but those which are so affected are the comparatively massive particles called *hadrons* (from the Greek *hadros*, meaning *bulky*). Accordingly, protons and neutrons are examples of hadrons, but there are many other kinds of hadron now known to exist. Among these others are the cousins of protons and neutrons called *baryons* (from *barys*, meaning *heavy*), which in addition to neutrons and protons themselves include the lambda ($\Lambda$), sigma ($\Sigma$), xi ($\Xi$), delta ($\Delta$), and omega ($\Omega$), most of which come in different versions with different values for the electric charge, and also in a sequence of excited (more rapidly spinning) versions. All these other particles are more massive than the proton and neutron. The reason that we do not find these more exotic particles as parts of ordinary atoms is that they are highly unstable and rapidly decay, ultimately into protons or neutrons, giving up their excessive mass in the form of energy (in accordance with Einstein's famous $E = mc^2$). The proton, in turn, has the mass of about 1836 electrons, and the neutron of around 1839 electrons. Intermediate between the baryons and electrons is another class of hadrons, called *mesons*, the most familiar ones being the pion ($\mu$) and the kayon (K). Each of these comes in a charged version ($\mu^+$ and $\mu^-$, each with a mass of about 273 electrons; $K^+$ and $\bar{K}^-$, each with a mass of about 966 electrons) and an uncharged version ($\mu^0$ has a mass of about 264 electrons; $K^0$ and $\bar{K}^0$, each with a mass of about 974 electrons). The practice here is to use a bar over the particle symbol to denote the anti-particle; we note, however, that the anti-pions are again pions, whereas an anti-kayon differs from a kayon. Again, these particles have many cousins and excited (more highly spinning) versions.

You can begin to see that all this is very complicated – a far cry from the heady days of the early twentieth century when the proton, neutron, and electron (and one or two massless ones such as the photon, the particle of light) had seemed to represent, more or less, the sum total of it all. As the years rolled by, things got more and more complicated, until eventually a unified picture of it all – called the *standard model of particle physics* – took shape [Zee 2010; Thomson 2013] between about 1970 and 1973. According to this scheme, all hadrons are composed of quarks and/or the anti-particles of quarks, known as *anti-quarks*. Each baryon is now taken to be composed of three quarks, and each (ordinary) meson, of a quark and an anti-quark. The quarks come in six different flavours, referred to (rather oddly and unimaginatively) as *up*, *down*, *charm*, *strange*, *top*, and *bottom*, and they have the respective electric charges $\frac{2}{3}$, $-\frac{1}{3}$, $\frac{2}{3}$, $-\frac{1}{3}$, $\frac{2}{3}$, $-\frac{1}{3}$. The fractional charge values seem, at first, to be distinctly odd, but for the observed free particles (such as baryons and mesons), the total electric charge always has a value that is an integer.

The standard model not only systematizes the seeming menagerie of the basic particles of nature, it also provides a good description of the main forces that influence them. Both the strong and the weak force are described in terms of an elegant mathematical procedure – referred to as *gauge theory* – that makes crucial use of the notion of a *bundle*, for which a brief description is given in §A.7, and to which I shall return, particularly in §1.8. The *base space* $\mathcal{M}$ of the bundle (for which notion, see §A.7) is space-time and, in the case of the strong force (which is the more mathematically transparent case), the fibre $\mathcal{F}$ is described in terms of a notion referred to as *colour*, which is assigned to the individual quarks (there being three basic alternative colours available to each quark). The theory of strong-interaction physics is referred to, accordingly, as *quantum chromodynamics* (QCD). I do not want to go into a proper discussion of QCD here because it is difficult to describe properly without using more mathematics than I can provide here [see Tsou and Chan 1993; Zee 2003]. Moreover, it is not "fashionable" in the sense that I mean to use the term here, because the ideas, while sounding exotic and strange, actually work extraordinarily well, not only forming a consistent and tightly knitted mathematical formalism, but finding excellent confirmation in experimental results. The QCD scheme would be studied in any physics research department concerned in a serious way with the theory of strong interactions, but it is not simply fashionable, in the sense intended here, because it is widely studied for very good scientific reasons!

For all its virtues, however, there are also powerful scientific reasons for striving to go beyond the standard model. One of these is that there are some thirty or so numbers in the standard model, for which its theory provides no explanation whatever. These include things like quark and lepton masses, quantities referred to as *fermion mixing parameters* (such as the Cabibbo angle), the Weinberg angle, the theta angle, gauge couplings, and parameters connected with the Higgs mechanism. Related to this issue is another serious drawback, which had already been very much present in other schemes that were around before the emergence of the standard model, and which is only partly resolved by it. This is the disturbing issue of the *infinities* (which are nonsensical answers arising from divergent expressions, like those exhibited in §A.10) that arise in quantum field theory (QFT) – QFT being the form of quantum mechanics that is central not just to QCD and other aspects of the standard model, but to all modern approaches to particle physics, and also to many other aspects of basic physics.

I shall have to say a good deal more about quantum mechanics, generally, in chapter 2. For the moment, let us restrict attention to one very specific but fundamental feature of quantum mechanics, which may be regarded as a root of the problem of the infinities in QFT, and we shall also see how the conventional method of dealing with these infinities precludes any complete answer to the issue of deriving the thirty or so unexplained numbers in the standard model. String theory is largely driven by an ingenious proposal to circumvent the infinities of QFT, as we shall be seeing in §1.6. It therefore appears to offer some hope of providing a route to resolving the mystery of the unexplained numbers.

## 1.4. THE SUPERPOSITION PRINCIPLE IN QFT

A foundation stone of quantum mechanics is the superposition principle, which is a feature common to *all* of quantum theory, not just QFT. In particular, it will be central to the critical discussions of chapter 2. In the present chapter, in order to provide some insight into the source of the problem of the infinities of QFT, I shall need to introduce this principle briefly here, though my main discussion of quantum mechanics will take place in chapter 2 (see, in particular, §§2.5 and 2.7).

To bring out the role of the superposition principle in QFT, let us consider situations of the following kind. Suppose that we have some physical process leading to a particular observed outcome. We shall suppose that this outcome could have arisen via some intermediate action $\Psi$, but there is also another possible intermediate action $\Phi$ which could also have resulted in essentially the same observed outcome. Then, according to the superposition principle, we must consider that, in an appropriate sense, *both* $\Psi$ and $\Phi$ could well have taken place *concurrently* as the intermediate action! This, of course, is very non-intuitive, since at an ordinary macroscopic scale we do not find distinct alternative possibilities taking place at once. Yet, for submicroscopic events, where we do not have the possibility of directly observing whether one intermediate activity has occurred as opposed to another, then we must allow that both could have occurred together, in what is referred to as a *quantum superposition*.

The archetypal example of this sort of thing occurs with the famous two-slit experiment, often used in introductions to quantum mechanics. Here we consider a situation where a beam of quantum particles (say, electrons or photons) is

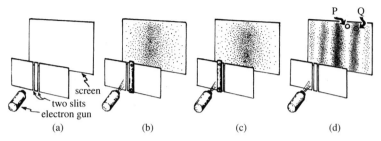

Figure 1-2: The two-slit experiment. Electrons are aimed at a screen through a narrowly separated pair of slits (a). If just one slit is open (b), (c), then a random-looking pattern of impacts is registered at the screen, scattered about the direct route through the slit. However, if both slits are open (d), the pattern acquires a banded appearance, where some places (e.g. P) cannot now be reached by the particle, although they could have been if just one slit were open; moreover, at other places (e.g. Q) there is *four* times the intensity of reception than for a single slit.

directed at a screen, where the beam must pass through a pair of close parallel slits on its way from source to screen (figure 1-2(a)). In the situation that is being considered, upon reaching the screen each particle makes a distinctive dark mark at an individual location at the screen, indicative of the particle's actual *particulate* nature. But after many such particles have passed through, an interference pattern of light and dark bands is built up, the dark bands occurring where many particles reach the screen and the light bands, where relatively few reach it (figure 1-2(d)). A standard careful analysis[2] of the situation leads one to conclude that each individual quantum particle must, in some sense, pass through *both* slits at once, in the manner of a strange kind of superposition of the two alternative possible routes that it might take.

The reasoning behind such an odd conclusion comes from the fact that if either one of the slits is covered up, while the other remains open (figure 1-2(b),(c)), we get no bands, but just a fairly uniform illumination which is darkest at the centre. When both slits are open, however, there are lighter regions at the screen

---

[2] This is what may be regarded as the conventional analysis of the situation. As might be expected, for such a strange-seeming conclusion, there are various other ways of interpreting what happens in this intermediate stage of the particle's existence. The most noteworthy alternative perspective is that of the de Broglie–Bohm theory according to which the particle itself always goes either through one slit or through the other, but there is also an accompanying "carrier wave" which guides the particle and which must *itself* "feel out" the two alternatives which the particle might have adopted [see Bohm and Hiley 1993]. I briefly discuss this viewpoint in §2.12.

situated between darker bands, these lighter regions occurring at places which are perfectly dark when just one of the slits is open. Somehow, when both routes are available to the particle, those lighter places become *inhibited*, whereas the dark places are *enhanced*. If each particle simply either did what it could do when only one of the slits is open or what it could do when only the other is open, then the effects of the routes would just add together, and we wouldn't get these strange interference stripes. This happens only because both of the possible routes are available to the particle, both these alternatives being felt out by the particle to give the ultimate effect. In some sense these routings *coexist* for the particle when it is between source and screen.

This, of course, is very much at odds with our experience of the behaviour of macroscopic bodies. For example, if two rooms are connected to each other by two different doors, and if a cat is observed to have started in one room and is later observed to be in the other room, then we would normally infer that it had passed through one door or through the other door, not that it could, in some strange way, have passed through both doors at the same time. But with an object of the size of a cat, it would be possible, without disturbing its actions significantly, to make continual measurements of its location and thereby ascertain which of the two doors it actually passed through. If we were successful in doing this for a single quantum particle in the two-slit experiment described above, we would have to disturb its behaviour to a degree that would result in the interference pattern at the screen being destroyed. The wave-like behaviour of an individual quantum particle that gives rise to the interference bands of light and dark at the screen depends upon our *not* being able to ascertain which of the two slits it actually went through, thereby allowing for the possibility of this puzzling intermediate superposed state of the particle.

In this two-slit experiment, we can see the extreme strangeness of the behaviour of single quantum particles most particularly by concentrating our attention at a point P of the screen at the middle of a gap between the dark bands, where we find that the particle is simply unable to reach P when both slits are open to it, whereas if only one of the slits is open, the particle could quite readily reach P via the open slit. When both slits are open, the two possibilities that are available to the particle in order to reach P have somehow cancelled each other; yet, at another place on the screen, say Q, where the interference pattern is at its darkest, we find that instead of cancelling, the two possible routes seem to reinforce each other so that when both slits are open, the likelihood of the particle reaching Q is four times as great as it would have been if just one of the two slits were open, not just twice, as would have been the case with an ordinary classical object, rather than a quantum particle. See figure 1-2(d). These strange features are a consequence of what is known as the *Born rule*, which relates the intensities in the superpositions to actual probabilities of occurrence, as we shall come to shortly.

The word *classical*, incidentally, when used in the context of physical theories, models, or situations, simply means *non-quantum*. In particular, Einstein's general theory of relativity is a classical theory, in spite of its having been introduced after many of the seminal ideas of quantum theory (such as the Bohr atom) had come about. Most particularly, classical systems are *not* subject to the curious superpositions of alternative possibilities that we have just encountered above, and which indeed characterizes quantum behaviour, as I come to briefly next.

I shall delay my full discussion of the basis of our present understanding of quantum physics until chapter 2 (see, in particular, §2.3 onwards). For the moment, I recommend that we simply accept the strange mathematical rule whereby modern quantum mechanics describes such intermediate states. The rule turns out to be extraordinarily accurate. But what is this strange rule? The quantum formalism asserts that such a superposed intermediate state, when there are just two alternative intermediate possibilities $\Psi$ and $\Phi$, is to be expressed mathematically as some kind of a sum $\Psi + \Phi$ of the two possibilities or, more generally, as a *linear combination* (see §§A.4 and A.5),

$$w\Psi + z\Phi,$$

where $w$ and $z$ are *complex numbers* (the numbers involving $i = \sqrt{-1}$, as described in §A.9), not both being zero! Moreover, we shall be forced to consider that such complex superpositions of states have to be allowed to *persist* in a quantum system, right up until the time that the system is actually observed, at which point the superposition of alternatives must be replaced by a probability mixture of the alternatives. This is indeed strange, but in §§2.5–2.7 and 2.9 we shall be seeing how to use these complex numbers – sometimes referred to as *amplitudes* – and how they tie in, in remarkable ways, with probabilities, and also with the time evolution of physical systems at the quantum level (Schrödinger's equation); they also relate, fundamentally, to the subtle behaviour of the spin of a quantum particle, and even to the 3-dimensionality of ordinary physical space! Although the precise connections between these amplitudes and probabilities (the *Born rule*) will not be addressed fully in this chapter (since for this we need the notions of *orthogonality* and *normalization* for the $\Psi$ and $\Phi$, which are best left until §2.8), the gist of the Born rule is as follows.

A measurement, geared to ascertaining whether a system is in state $\Psi$ or in state $\Phi$, when presented with the superposed state $w\Psi + z\Phi$, finds:

ratio of probability of $\Psi$ to probability of $\Phi$ = ratio of $|w|^2$ to $|z|^2$.

We note (see §§A.9 and A.10) that the squared modulus $|z|^2$ of a complex number $z$ is the sum of the squares of its real and imaginary parts, this being the squared distance of $z$ from the origin in the Wessel plane (figure A-42 in §A.10). It may also be remarked that the fact that probabilities arise from *squares* of the moduli of these amplitudes accounts for the *fourfold* increase in intensity, as noted earlier, where contributions reinforce each other in the two-slit experiment (see also the end of §2.6).

We must be careful to appreciate that this notion of *plus*, in these superpositions, is quite different from the ordinary notion of *and* (despite a common modern use of *plus* in ordinary conversation simply to mean *and*), or even with *or*. What is meant here is that, in some sense, the two possibilities are actually to be thought of as being *added* together in some abstract mathematical way. Thus, in the case of the two-slit experiment, where $\Psi$ and $\Phi$ represent two distinct transient locations of a single particle, then $\Psi + \Phi$ does *not* represent two particles, one in each location (which would be "one particle in the $\Psi$ position *and* one particle in the $\Phi$ position" – implying *two* particles in total), nor must we think of the two as just being ordinary alternatives, one *or* the other of which actually happened, but where we don't know which. We must indeed think of just a single particle somehow occupying both locations at once, *superposed* according to this strange quantum-mechanical "plus" operation. Of course this looks extremely odd, and the physicists of the early twentieth century would not have been driven to consider such a thing without having some very good reasons to do so. We shall be exploring some of these reasons in chapter 2, but for now I am just asking that the reader simply accept that this formalism indeed works.

It is important to appreciate that, according to standard quantum mechanics, this superposition procedure is taken to be *universal* and, accordingly, applies also if there are more than just two alternatives for the intermediate state. For example, if there are three alternative possibilities, $\Psi$, $\Phi$, and $\Gamma$, then we have to consider triple superpositions like $w\Psi + z\Phi + u\Gamma$ (where $w$, $z$, and $u$ are complex numbers, not all of which are zero). Correspondingly, if there were four alternative intermediate states, we would need to consider quadruple superpositions and so on. Quantum mechanics demands this, and there is excellent experimental support for such behaviour at the submicroscopic level of quantum activity. Strange it is, indeed, but it makes good consistent mathematics. This is, so far, just the mathematics of a *vector space*, with complex-number scalars, as considered in §§A.3, A.4, A.9, and A.10, and we shall be seeing more of the ubiquitous role of quantum superpositions in §2.3 onwards. However, matters are considerably worse in QFT, because we frequently have to consider situations in which there are infinitely many intermediate possibilities. Accordingly, we are led to having to consider infinite sums of alternatives, and then the issue looms large of the possibility that such an infinite sum might provide us with series whose sum may actually *diverge* to infinity (in the sort of way exhibited in §§A.10 and A.11).

## 1.5. THE POWER OF FEYNMAN DIAGRAMS

Let us try to understand in a bit more detail how such divergences actually come about. In particle physics, what we have to consider are situations in which several particles come together to make other particles, where some of them may split apart to make still others, and where pairs of these might join together again, etc., etc., so that they may well be involved in very complicated processes of this kind. The types of situation that particle physicists are frequently concerned with involve some given collection of particles coming together – often at relative speeds close to that of light – and this combination of collisions and separations results in some other collection of particles emerging from it all. The total process would involve a vast quantum superposition of all the possible different kinds of intermediate processes which might take part and are consistent with the given input and output. An example of such a complicated process is illustrated in the *Feynman diagram* of figure 1-3.

We do not go far wrong if we think of a Feynman diagram as a space-time diagram of the particular collection of particle processes involved. I like to represent time as proceeding upwards along the page, being someone who works in relativity theory as opposed to being a professional particle physicist or QFT expert; the professionals usually have time progressing from left to right. Feynman diagrams (or Feynman graphs) are named after the outstanding American physicist Richard Phillips Feynman. Some very basic diagrams of this kind are shown in figure 1-4. Here, figure 1-4(a) shows the splitting of a particle into two and figure 1-4(b) shows the combining of two to make a third.

In figure 1-4(c), we see the exchange of a particle (say a photon, the quantum of electromagnetic field or light, indicated by the wiggly line) between two particles. The use of the term *exchange* for this process, though common among particle

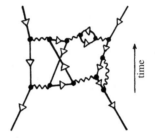

Figure 1-3: A Feynman diagram (drawn with upward time direction) is a schematic space-time picture (with a clear-cut mathematical interpretation) of a particle process often involving creation, annihilation, and exchanges of intermediate particles. Wavy lines indicate photons. Triangular arrows here denote electric charge (positive if the arrow points upwards and negative if downwards).

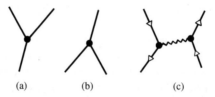

(a)        (b)        (c)

Figure 1-4: Elementary Feynman diagrams: (a) a particle splits into two; (b) two particles combine to make another particle; (c) two oppositely charged particles (e.g. an electron and a positron) "exchange" a photon.

physicists, is perhaps a bit odd here, since a *single* photon simply passes from one external particle to the other – albeit in a way that (deliberately) does not make clear which particle is the emitter and which the receiver. The photon involved in such an exchange is usually what is called *virtual* and its speed is not constrained to be consistent with the requirements of relativity. The usual colloquial use of the term *exchange* might apply more appropriately to the situations depicted in figure 1-5(b), though such processes as shown in figure 1-5 tend to be referred to as the exchange of two photons.

We may think of the general Feynman diagram to be composed of many basic ingredients of this general kind, pieced together in all sorts of combinations. However, the superposition principle tells us not to think of what actually happens in some such particle collision process as being represented by just *one* such Feynman diagram, because there are many alternatives, and the actual physical process is represented as some complicated linear superposition of many different such

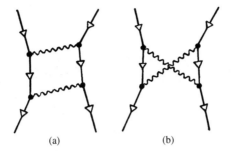

Figure 1-5: Two-photon exchanges.

Feynman diagrams. The magnitude of the contribution to the total superposition from such a diagram – essentially a complex number such as the $w$ or $z$ that we encountered in §1.4 – is what we need to calculate from any particular Feynman diagram, these numbers being called complex *amplitudes* (see §§1.4 and 2.5).

We must bear in mind, however, that the mere arrangement of the connections in the diagram does not tell us the whole story. We also need to know the values of the energies and momenta of all the particles involved. For all the external particles (both incoming and outgoing), we may take these values to be already assigned, but the energies and momenta of the intermediate – or internal – particles could generally take many different values, consistent with a constraint that energy and momentum have to add up appropriately at each vertex, where the momentum of an ordinary particle is its velocity multiplied by its mass; see §§A.4 and A.6. (Momentum has the important property that it is *conserved*, so that in any collision process encountered by particles, the total of the momenta going in – added together in the sense of vector addition – must be equal to the total momentum coming out.) Thus, complicated as our superpositions may appear to be, merely from the elaboration of the succession of increasingly complicated diagrams appearing in the superposition, things are really much *more* complicated than this, because of the generally *infinite* numbers of different possible values that the energies and momenta might take for the internal particles in each diagram (consistent with the given external values).

Thus, even with a single Feynman diagram, with given input and output, we may expect to have to add together an infinite number of such processes. (Technically, this adding together would take the form of a continuous integral, rather than a discrete sum (see §§A.7, A.11, and figure A-44), but the distinctions are not important for us here.) This kind of thing happens with Feynman diagrams containing a *closed loop*, such as occurs in the two examples of figure 1-5. With a

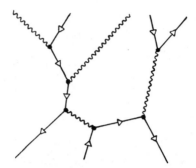

Figure 1-6:  A tree diagram, i.e. containing no closed loops.

*tree diagram*, such as those of figure 1-4 and figure 1-6, where there are no closed loops, the values of the internal energies and momenta turn out to be simply fixed by the external values. But these tree diagrams do not probe the genuinely quantum nature of particle processes; for this we actually need to bring in the closed loops. And the trouble with the closed loops is that there is no limit to the energy-momentum that can, in effect, circulate around the loop, and adding all these up provides us with a *divergence*.

Let us look at this a little more closely. One of the simplest situations where a closed loop occurs is that shown in figure 1-5(a) in which two particles are exchanged. The trouble arises because although at each vertex in the diagram the values of the energy and of the three components of momentum must add up properly (i.e. the sum in equals the sum out), this does not result in enough equations to fix the internal values of these quantities. (For each of the four components of the energy-momentum, separately, there are three independent equations since each of the four vertices provides a conservation equation, but one is redundant, merely re-expressing the overall conservation for the entire process – yet there are four independent unknowns per component, one from each internal line, so there are not enough equations to fix the unknowns, and the redundancy must be summed over.) There is always the freedom to add (or subtract) the same energy-momentum quantity all the way around the loop in the middle. We need to add all these infinitely many possibilities together, involving potentially higher and higher values for the energy-momentum, and this is what leads to the potential divergence.

Thus, we see that the direct application of the quantum rules is indeed likely to give us a divergence. Yet, this does not necessarily mean that the "correct" answer to that quantum field theoretic calculation is actually ∞. It would be useful to keep in mind the divergent series shown in §A.10, where a finite answer can sometimes be assigned the series despite the fact that simply adding the terms up leads to the answer "∞". Although the situation with QFT is not exactly like this, there are some distinct similarities. There are many calculational devices that QFT experts have developed over the years in order to circumvent these infinite answers. Just as with the examples of §A.10, if we are clever about it, we may be able to unearth a "true" finite answer that we do not get simply by "adding up the terms". Accordingly, QFT experts are frequently able to squeeze finite answers out of the wildly divergent expressions that they are presented with, although many of the procedures that are adopted are far less straightforward than simply the method of analytic continuation, referred to in §A.10. (See also §3.8 for some of the curious pitfalls that even the "straightforward" procedures can lead into.)

One key point about the root cause of many of these divergences – those referred to as *ultraviolet divergences* – should be made note of here. The trouble basically arises because, with a closed loop, there is no limit to the scale of energy and momentum that can circulate around it, and the divergence arises from contributions of higher and higher energy (and momentum) having to be added up. Now, according to quantum mechanics, very large values of energy are associated with very tiny times. Basically this comes from Max Planck's famous formula $E = h\nu$, where $E$ is the energy, $\nu$ is the frequency, and $h$ is Planck's constant, so high values of energy correspond to large frequencies and therefore to tiny time intervals between one beat and the next. In the same way, very large values of the momentum correspond to very tiny distances. If we imagine that something strange happens to space-time at very tiny times and distances (as, indeed, most physicists would be inclined to agree would be an implication of quantum-gravity considerations), there might be some kind of effective "cut-off", at the high end of the scale, to the allowed energy-momentum values. Accordingly, some future theory of space-time structure, in which drastic alterations occur at very tiny times or distances, might actually render finite the currently divergent QFT calculations that arise from closed loops in Feynman diagrams. These times and distances would have to be far tinier than those which are relevant to ordinary particle-physics processes, and are frequently taken to be something of the order of those quantities of relevance to quantum-gravity theory, namely the *Planck time* of some $10^{-43}$ s or the *Planck length* of around $10^{-35}$ m (referred to in §1.1), these values being something like $10^{-20}$ of the usual small quantities of direct relevance to particle processes.

It should be mentioned here that there are also divergences in QFT referred to as *infrared divergences*. These occur at the other end of the scale, where energies

Figure 1-7:  Infrared divergences occur when there are indefinitely large numbers of "soft" photons emitted.

and momenta are extremely tiny, so that we are concerned with extraordinarily large times and distances. The problems here are not to do with closed loops, but with Feynman diagrams like those of figure 1-7 in which an unlimited number of *soft photons* (i.e. photons of very tiny energy) might be emitted in a process, and adding all these together again produces a divergence. Infrared divergences tend to be regarded by QFT experts as less serious than the ultraviolet divergences, and there are various ways of sweeping them under the carpet (at least temporarily). However, in recent years their importance is perhaps beginning to be faced up to more seriously. For my own discussions here, I shall not pay too much attention to the infrared problems and concentrate instead on how the problem of the ultraviolet divergences – resulting from the closed loops in the Feynman diagrams – is tackled in standard QFT, and how the ideas of string theory appear to offer hope of providing a resolution of this conundrum.

Of particular note, in this connection, is the standard QFT procedure of renormalization. Let us try to glimpse how this operates. According to various direct QFT calculations we get an infinite scale factor between what would be called the *bare charge* of a particle (such as an electron) and the *dressed charge*, the latter being what would be actually measured in experiments. This comes about because of contributions due to processes like that shown in the Feynman diagram of figure 1-8, which serve to damp down the measured value of the charge. The trouble is that the contribution from figure 1-8 (and many another like it) is "infinity". (It has closed loops.) Accordingly, we find that the bare charge would have to have been infinite in order that we can find a finite value for the observed (dressed) charge. The underlying philosophy of the renormalization procedure is to accept that QFT might not be completely right at very tiny distances, which is where the divergences appear, and some unknown modification of the theory might supply the necessary cut-off that leads to finite answers. The procedure thus involves our giving up on attempting to calculate nature's actual answer for these scale factors (for charge and for other things like mass, etc.), where we instead collect together all such infinite scale factors that QFT burdens us with, and we

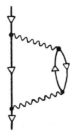

Figure 1-8: Divergent diagrams like this one are dealt with via the procedure of charge renormalization.

effectively bundle these infinite contributions into neat little parcels which we proceed to ignore, by directly adopting the *observed* bare charge (and mass, etc.) values as seen in experiment. Rather remarkably, for appropriate QFTs referred to as *renormalizable*, this can be done in a systematic way, enabling finite answers to be obtained for many other QFT calculations. Numbers such as the dressed charge (and mass, etc.) are taken from observation, rather than calculated from the appropriate QFT, and these values lead to some of the thirty parameters that must be fed into the standard model from their experimentally observed values, as mentioned above.

By adopting procedures such as these, one may often get numbers out of QFT which are extraordinarily precise. For example, there is a now-standard QFT calculation for obtaining the *magnetic moment* of the electron. Most particles behave as little magnets (in addition to sometimes possessing an electric charge), and the magnetic moment of such a particle is a measure of the strength of this magnet. Dirac made an original prediction of the electron's magnetic moment coming directly from his fundamental equation for the electron (referred to briefly in §1.1), and this is almost exactly what is measured by precise experiments. However, it turns out that there are corrections to this value, coming from indirect QFT processes and these must be incorporated into the direct single-electron effect. The QFT calculation finally gives an answer that is 1.001159652... times the "pure" original Dirac value. The observational figure is 1.00115965218073... [Hanneke et al. 2011]. The agreement is pretty incredible – better than determining the distance between New York and Los Angeles to within the width of a human hair, as was pointed out by Richard Feynman [1985]! It provides remarkable support for the *renormalized* QFT theory of electrons and photons (called *quantum electrodynamics* or QED), where the electrons are described by Dirac's theory and the photons by Maxwell's electromagnetic equations (see §1.2), and where their mutual interaction is in accordance with the standard equation of H. A. Lorentz, which describes the response of a charged particle to an electromagnetic field. This latter, in a quantum context, follows from the *gauge procedures* of Hermann Weyl (§1.7). So you see that the theory and observation really do agree to this extraordinary degree, which is telling us that there is something very profoundly true about the theory, although it is not yet strictly consistent as a mathematical scheme.

Renormalization may be regarded as a stopgap, where the hope would be that some improved version of QFT might eventually be discovered in which these infinities would not arise at all, and it might be possible to calculate not only finite values for these scale factors, but also the actual bare values – and hence the experimentally observed values – of the charge and mass, etc., of the various basic particles. No doubt the hope that string theory might provide this improved QFT has provided an important impetus for that theory. But a more modest approach, which has undoubtedly been more successful so far, has been simply to use the applicability of the renormalization procedure within the theory as a criterion for selecting the most promising schemes from within the conventional body of QFTs. As it turns out, only *some* QFTs are amenable to the procedures of renormalization – the *renormalizable* QFTs, as mentioned above – whereas others are not. So, renormalizability is taken as a powerful selection principle for finding the most promising QFTs. In fact, it was discovered (most particularly by Gerardus 't Hooft in 1971 and later ['t Hooft 1971; 't Hooft and Veltman 1972]) that the use of the kind of symmetry that is needed for the gauge theories referred to in §1.3 is extremely helpful in producing renormalizable QFTs, and this fact provided a powerful impetus to the formulation of the standard model.

## 1.6. THE ORIGINAL KEY IDEAS OF STRING THEORY

Let us now try to see how the original ideas of string theory fit into this. Recall from the above discussion that the problems of ultraviolet divergences arise from quantum processes operating at very tiny distances and times. We can think of the source of the problem as being that material objects are viewed as being composed of *particles*, where such basic particles are taken as occupying single *points* in space. Of course, we might regard the point-like character of a bare particle as being a non-realistic approximation, but if such a primitive entity is regarded, alternatively, as being some sort of spread-out distribution, we have a converse problem of how something spread out in this way is to be described,

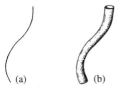

Figure 1-9:  (a) The world-line of an ordinary (point-like) particle is a curve in space-time; (b) in string theory, this becomes a 2-surface world-tube (the string world-sheet).

without having to think of it as somehow being composed of smaller constituents. Moreover, there are always delicate issues, for such models, concerning possible conflicts with relativity (where there is a finite limit to the speed of information propagation), if one requires such a smeared-out entity to behave as one coherent whole.

String theory suggests a different kind of answer to this conundrum. It proposes that the basic ingredient of matter is neither 0-dimensional in spatial extent, like a point particle, nor 3-dimensional, like a smeared-out distribution, but 1-dimensional, like a curved line. Although this may seem like a strange idea, we should bear in mind that, from the 4-dimensional perspective of space-time, even a point particle is not described, classically, as simply a point since it is a (spatial) point which persists in time – so its space-time description is actually a *1-dimensional* manifold (see §A.5), referred to as the *world-line* of the particle (figure 1-9(a)). Accordingly, the way that we should think of the curved line of string theory is as a 2-manifold, or *surface*, in space-time (figure 1-9(b)), referred to as the string *world-sheet*.

To me, one of the particularly attractive features of string theory (at least in its original form) was that these 2-dimensional string histories – or string world-sheets – could, in an appropriate sense, be regarded as *Riemann surfaces* (but see §1.9, especially figure 1-30, in relation to the Wick rotation involved). As is described in more detail in §A.10, a Riemann surface is a *complex* space of 1 dimension (where we bear in mind that 1 complex-number dimension counts for 2 real-number dimensions). Being a complex space, it can benefit from the magic of complex numbers. Riemann surfaces indeed exhibit many aspects of such magic. And the fact that these surfaces (i.e. complex curves) play their roles at a level where the complex-linear rules of quantum mechanics hold sway provides openings for a subtle interplay and perhaps a harmonious unity between two different aspects of the physics of the small.

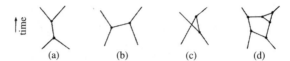

Figure 1-10: (a), (b), (c) The three different tree graphs, where two (unspecified) particles come in and two go out; (d) an example where there are closed loops.

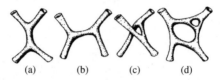

Figure 1-11: String versions of the respective processes shown in figure 1-10.

To be a little more explicit about the role that this fundamental stringy idea is to play, let us return to the Feynman diagrams of §1.5. If we think of the lines of these diagrams as representing the actual world-lines of basic particles, where these particles are being thought of as fundamentally spatially *point-like*, then the vertices of the diagrams represent zero-distance encounters between the particles, and we may think of ultraviolet divergences as arising from the point-like nature of these encounters. If, instead, the basic entities are taken to be tiny loops, then their histories would be narrow tubes in space-time. Now, instead of having to have point-like vertices, as in Feynman diagrams, we can imagine joining these tubes up in a smooth way, as a good plumber might. In figure 1-10(a)–(c), I have drawn some Feynman diagrams (for unspecified particles) without any loops (tree diagrams) and in figure 1-10(d) a more typical one in which there *are* closed loops. In figure 1-11 I have drawn what their stringy counterparts might look like. The point-like encounters are now removed, and the processes are now represented in an entirely smooth-looking way. We can now imagine that the string-history surfaces of figure 1-11, including these junctions, are *Riemann* surfaces, so that we can appeal to their beautiful underlying mathematical theory in order to study basic physical processes. We note, in particular, that the closed loops of standard Feynman theory (which give rise to the ultraviolet divergences) simply provide multiple-connectedness in the topology of the Riemann surfaces. Each closed loop in the Feynman diagram simply provides us with a new "handle" to the topology of our Riemann surface (technically, an increase of the genus, where the *genus* of a Riemann surface is the number of its handles). (See figure 1-44(a) in §1.16 and figure A-11 in §A.5 for examples of topological handles.)

We note, also, that the in- and out-states of the Feynman theory correspond to holes or punctures in our Riemann surfaces, and it is at such places that the information of such things as energy and momentum may be introduced. In some popular accounts of the topology of surfaces, the term *hole* is used for what I am here calling a *handle*. But the non-compact Riemann surfaces that are used in string theory also have the kind of hole (or puncture) according to the terminology that I am using here, so we need to be careful to distinguish these very different notions. We shall be seeing in §1.16 that there are other roles for holes/punctures in the Riemann surfaces.

At this juncture I should describe a particular early motivation for string theory that I hinted at, at the beginning of §1.3, this being concerned with some observed aspects of hadron particle physics that had been puzzling to physicists at the time. In figure 1-10(a)–(c) I have drawn three Feynman diagrams, each exhibiting a low-order process in which two particles – let us say hadrons – go in and two hadrons come out. In figure 1-10(a), the two hadrons come together to form another hadron which almost immediately splits into two others; in figure 1-10(b), the original pair of hadrons exchange a single hadron and they end up as some pair of hadrons. Figure 1-10(c) is similar to figure 1-10(b), except that the two final hadrons are reversed. Now, for a given input and output, we might well find that for each of the three arrangements there are many possibilities for the internal hadron, and we would have to sum over all of these to get the correct answer. This is indeed the case, but to get the full answer, at this order of calculation, it would seem that we should add together all three of these sums (namely those obtained separately for each of the three possibilities of figure 1-10). However, it appears that all three of the sums come out the same, and instead of having to add up all three of these sums, any one of the sums alone gives the required answer!

From the point of view of what has been said above about how Feynman diagrams are to be used, this appears to be very odd, as one would think that all possibilities ought to be added in together, whereas it seems that nature is telling us that each of the processes indicated in any one of the three different-looking diagrams of figure 1-10(a)–(c) would be sufficient, and that to include them would result in some serious "over-counting". In the full formulation of QCD, we can understand this if we think in terms of expressing all these hadronic processes in terms of the basic quarks instead of the hadrons, since the latter are regarded as composite objects, whereas the "counting" of independent states has to be carried out in terms of the elementary quarks. But at the time that string theory was being formulated, the proper formulation of QCD had not been obtained, and it had seemed very appropriate to pursue other ways of addressing this issue (and certain other related issues). The way that the string point of view addresses this is illustrated in figure 1-11(a)–(c), where I have illustrated stringy versions of each of the three respective possibilities in figure 1-10(a)–(c). We note that the stringy versions of all three processes are topologically *identical*. Thus, the string point of view would lead us to conclude that the three processes illustrated in figure 1-10(a)–(c) should indeed not be counted separately, and they are just three ways of looking at what is, deep down, exactly the same basic process.

However, not all the stringy diagrams are the same. Look at the stringy version of figure 1-10(d), namely figure 1-11(d), where the *loops* appearing in this (higher-order) Feynman diagram are represented in terms of *topological handles* on the string histories (see figure 1-44(a),(b) in §1.11 and figure A-11 in §A.5). But again, we find a potentially profound advantage in the string-theoretic approach. Rather than obtaining divergent expressions of the kind we are led to in conventional Feynman-diagram theory when closed loops are present, string-theory presents us with a very elegant way of looking at loops, namely in terms of the 2-dimensional topology that mathematicians are well familiar with in the highly fruitful theory of Riemann surfaces (§A.10).

This kind of reasoning provided an excellent intuitive motive for taking the string idea seriously. A somewhat more technical lead was what guided a number of physicists in this interesting direction. In 1970, Yoichiro Nambu (who won a Nobel Prize in 2008 for a different contribution, namely spontaneous symmetry breaking in subatomic physics) proposed the string idea as a way of explaining a remarkable formula describing such hadron encounters put forward by Gabriele Veneziano about two years earlier. It may be noted that Nambu's strings were rather like rubber bands, in that the force that they exert increases in proportion to the extension of the string (although they differ from ordinary rubber bands in that the force reduces to zero only at the point when the string length is shrunk to zero). We see, from this, that the original strings were really to provide a theory of *strong interactions*, and in this respect it provided a proposal that, for its time, was novel and thoroughly appealing, particularly as QCD had not yet developed into a usable theory. (A key ingredient of QCD known as *asymptotic freedom* was developed only later, in 1973, by David Gross and Frank Wilczek, and independently by David Politzer, eventually winning them the 2004 Nobel Prize in physics.) The string proposal constituted a scheme that seemed, to me as well as to many others, to be well worth developing, but we note that it was the nature of *hadronic* (strong) interactions that motivated the basic original string ideas.

In their attempts to develop a proper quantum theory of such strings, however, the theorists encountered what is referred to as an *anomaly*, and this drove them into some very strange territory. An *anomaly* is something which occurs when a classically described theory – in this case the dynamical theory of string-like basic entities according to ordinary classical (e.g. Newtonian) physics – loses some key property when the rules of quantum mechanics are applied to it, usually a symmetry of some kind. In the case of string theory, this symmetry was an essential invariance under a change of coordinate parameter describing the string. Without this parameter invariance, the mathematical description of the string failed to make proper sense *as* a theory of strings, so the *quantum* version of this classical theory of strings would indeed not make sense as a string theory, owing to this (anomalous) failure of parameter invariance. Yet, in around 1970, it was concluded, very remarkably, that if the number of dimensions of space-time were increased from 4 to 26 (that is, 25 space dimensions and 1 time dimension) – admittedly a very odd idea – then the terms in the theory that gave rise to the anomaly would miraculously cancel out [Goddard and Thorn 1972; see also Greene 1999, §12], so that the quantum version of the theory would work after all!

It appears to be the case that, to many people, there is something of a romantic appeal to the idea that, hidden from direct perception, there might be a world of higher dimensionality and, moreover, that this higher dimensionality could constitute an intimate part of the actual world we inhabit! Yet my own reaction was very different. My immediate response to this news was that no matter how mathematically fascinating this proposal might be, I could not take it seriously as a model relevant to the physics of the universe we know. So, without its being shown that there could perhaps be some other (radically different) way to look at it, all the initial interest and excitement that the string ideas had instilled in the physicist in me was lost. I believe that my reaction was not an uncommon one among physics theorists, though there were some particular reasons that I had for being especially uncomfortable with the great increase in spatial dimensions that was being suggested. I shall be coming to such reasons specifically in §§1.9–1.11, 2.9, 2.11, 4.1, and most explicitly §4.4, but for our current purposes I shall need shortly to explain the attitude of mind adopted by the string theorists that allowed them to be *not* unhappy about the seeming conflict between the manifestly observed 3-dimensionality of physical space (with 1-dimensional physical time) and these postulated 25 spatial dimensions (with 1 time dimension) that string theory was now apparently demanding.

This was for the so-called *bosonic strings*, intended to represent the particles known as *bosons*. We shall be seeing in §1.14 that quantum particles fall into two classes, one consisting of bosons and the other of particles called *fermions*.

Bosons and fermions have characteristically different statistical properties, and differ also in that bosons always spin by an amount that is an *integer* (in absolute units, see §2.11), whereas fermions' spins always differ from an integer by *one half*. These issues will be addressed in §1.14, where they will also be discussed in relation to the proposal of *supersymmetry* which had been put forward as a means of bringing bosons and fermions together into one overall scheme. We shall be seeing that this proposal plays a central role in much of modern string theory. Indeed, it was found by Michael Green and John Schwarz [1984; see also Greene 1999] that, through the incorporation of supersymmetry, the dimensionality of the space-time required by string theory would reduce from 26 to 10 (that is, 9 space dimensions and 1 time dimension). The strings of this theory are referred to as *fermionic* strings, describing fermions, which would be related to the bosons via supersymmetry.

In order that they might feel less unhappy with this enormous and apparently absurd discrepancy between the theory and observational fact, with regard to spatial dimensionality, the string theorists would point to an earlier proposal, put forward in 1921 by the German mathematician Theodor Kaluza[3] and developed by the Swedish physicist Oskar Klein, this scheme being now known as the *Kaluza–Klein theory*, according to which gravity and electromagnetism are simultaneously described by a theory of 5-dimensional space-time. How did Kaluza and Klein envisage that the 5th dimension of the space-time of their theory would not be immediately observable to the inhabitants of their universe? In Kaluza's original scheme, the 5-dimensional space-time would have a metric just as in Einstein's pure gravitational theory, but there would be an exact symmetry along a particular vector field **k** in the 5-dimensional space (see §A.6, figure A-17), with nothing in the geometry changing along the direction of **k**. In the terminology of differential geometry, **k** is what is called a *Killing vector*, which is a vector field that generates such a continuous symmetry (see §A.7, figure A-29). Moreover, any physical object, described within the space-time, would also have a constant description along **k**. Since any such object would have to share in this symmetry, nothing within the space-time could be "aware" of that direction, and the *effective* space-time, with regard to its contents, would be 4-dimensional. Nevertheless, the structure that the 5-metric confers on the 4-dimensional effective space-time would be interpreted within that 4-space as an electromagnetic field satisfying the Maxwell equations, and contributing to Einstein's energy tensor **T**

---

[3] In some accounts, Kaluza is described as being Polish. This is understandable, because the town, Opole (German: Oppeln), in which he was born is now part of Poland.

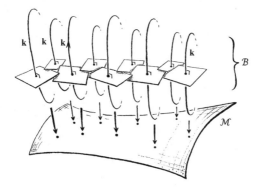

Figure 1-12: Because of the symmetry along the Killing-vector **k**-directions, the Kaluza–Klein 5-space is a bundle $\mathcal{B}$ over our familiar 4-space-time $\mathcal{M}$, with **k** pointing along the $S^1$ fibres (the curves drawn vertically). The Maxwell field is encoded in a "twist" in the fibres, which prevents their orthogonal 4-spaces from knitting together to make consistent 4-space sections that would otherwise have been images of the space-time $\mathcal{M}$.

in exactly the way that it should.[4] This, indeed, was an extremely ingenious idea. Kaluza's 5-space is, in fact, a *bundle*, $\mathcal{B}$, in the sense of §A.7, with 1-dimensional fibre. The base-space is our 4-dimensional space-time $\mathcal{M}$, but $\mathcal{M}$ is not naturally *imbedded* in the 5-space, $\mathcal{B}$, owing to a "twist" in the 4-plane elements orthogonal to the **k** directions, this twist describing the electromagnetic field (see figure 1-12).

Then, in 1926, Klein introduced a different way of looking at Kaluza's 5-space, where the idea was now that this extra dimension in the **k** direction would be "small" in the sense of being curled up in a tiny loop ($S^1$). The picture that is usually presented, in order to provide an intuitive understanding of what is going on, is that of a hosepipe (see figure 1-13). The 4 macroscopic dimensions of ordinary space-time are represented, in this analogy, by the single direction along the length of the pipe, and the extra "small" 5th space-time dimension of Kaluza–Klein theory is represented by the direction of the tiny loop around the pipe, perhaps of the Planck scale of $\sim 10^{-35}$ m (see §1.5). If the hosepipe is viewed on a large scale, then it appears to be only 1-dimensional, and the additional

---

[4] To appreciate the "twisty" differential geometry of Kaluza's 5-space, in technical differential-geometry terms, we first note that the condition on **k** that it be a Killing vector is that the covariant derivative of **k**, when expressed as a *co*vector, be anti-symmetric, and then we ascertain that this 2-form is, in effect, the 4-space Maxwell field.

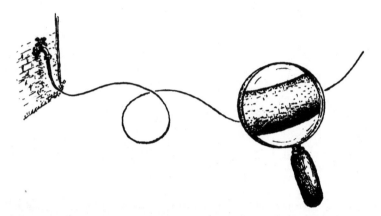

Figure 1-13: A hosepipe provides an intuitive image of Klein's suggestion that the extra dimension(s) should be tiny, of perhaps Planck-length size. When viewed on a large scale, the hosepipe appears 1-dimensional, analogously to the apparent 4-dimensionality of space-time. On a small scale the extra dimension of the hosepipe becomes visible, analogously to the appearance of the hypothesized submicroscopic extra spatial dimension(s).

dimension that gives the hosepipe surface its actual 2-dimensional nature is not directly observed. Accordingly, in this Kaluza–Klein picture, the 5th dimension is analogous to the tiny direction around the hosepipe and taken not to be directly perceived at the scale of ordinary experience.

In a similar way, the string theorists envisaged that the extra 22 spatial dimensions of string theory would be very "tiny" like the single extra Kaluza–Klein 5th dimension and thus, like the single tiny dimension around the hosepipe, unseen when viewed on a very large scale. In this way, they argued, we would not be directly aware of the 22 extra spatial dimensions that string theory seemed to require in order that it be anomaly-free. Indeed, the stringy motivations from hadronic physics, as mentioned at the beginning of this section, would seem to suggest that the hadronic scale of some $10^{-15}$ m might be the appropriate one for the "size" of these extra spatial dimensions – which is indeed very small at the level of ordinary experience, though critically relevant to the sizes of hadronic particles. As we shall be seeing in §1.9, the more modern versions of string theory are generally proposing enormously tinier scales for the extra dimensions, perhaps in the range $10^{-33}$–$10^{-35}$ m.

Does this kind of proposal make sense? I believe that there is a profound problem arising here, namely the issue of *functional freedom* referred to in the preface and discussed in more detail in §A.2 (and §A.8), to which the unfamiliar reader is now directed [see also Cartan 1945; Bryant et al. 1991]. If we are concerned with classical fields, subject to the normal kinds of equation that govern the way that such fields propagate in time, then the number of spatial dimensions involved makes a huge difference, there being vastly greater freedom in the fields, the larger the number of spatial dimensions is taken to be, where there is considered to be just one time dimension in each case. The notation used in §A.2 for the functional freedom of a $c$-component field freely specifiable on a space of $d$ spatial dimensions is

$$\infty^{c\infty^d}.$$

The comparison between this amount of freedom and that for a $C$-component field in a space of a different number $D$ of spatial dimensions is expressed as

$$\infty^{C\infty^D} \gg \infty^{c\infty^d} \quad \text{if } D > d,$$

irrespective of the relative sizes of the respective numbers of components per point, namely $C$ and $c$. The double inequality sign "$\gg$" is used in order to convey the utter unassailable hugeness whereby the functional freedom described by the left-hand side exceeds that described by the right-hand side, when the spatial dimensionality is greater, no matter what the component numbers $C$ and $c$ are (see §§A.2 and A.8). The central point is that for ordinary classical fields, with a finite number of components per point – where we assume the normal kinds of field equations, giving a deterministic time evolution from (effectively) freely specified data on a $d$-dimensional initial space – then the number $d$ is crucial. Such a theory cannot be equivalent to another such theory on which the initial space has a different number $D$ of dimensions. If $D$ is greater than $d$, then the freedom in the $D$-space theory always vastly exceeds that in the $d$-space theory!

Whereas this situation seems to me to be completely clear with regard to classical field theories, the case of quantum (field) theories certainly need not be as clear-cut as this. Nevertheless, quantum theories are usually modelled on classical theories, so that the deviations between a quantum theory and the classical theory on which it is modelled would be expected to be, in the first instance, of the nature of simply providing quantum corrections to the classical theory. In the case of quantum theories of this kind, one needs a very good reason to see why some proposed equivalence between two such quantum theories can possibly hold when there are different numbers of spatial dimensions in each case.

Accordingly, deep questions are raised concerning the physical relevance of quantum theories such as supra-dimensional string theories, for which the number of spatial dimensions is greater than the three we directly perceive. What happens to the floods of excessive degrees of freedom that now become available to the system, by virtue of the huge functional freedom that is potentially available in the extra spatial dimensions? Is it plausible that these vast numbers of degrees of freedom can be kept hidden away and prevented from completely dominating the physics of the world in such schemes?

In a certain sense, even for a classical theory, it *is* plausible but only if these extra degrees of freedom are *not really there* in the first place. This was the situation in Kaluza's original proposal for a 5-dimensional space-time theory, where it was explicitly demanded that there be an *exact* continuous symmetry in the one extra dimension. The symmetry was specified by the Killing-vector nature of **k**, in Kaluza's initial scheme, so that the functional freedom was reduced to that for a conventional, spatially 3-dimensional theory.

Thus, in order to examine the plausibility of the higher-dimensional string-theory ideas, it will be pertinent first to gain some understanding of what Kaluza and Klein were actually attempting to do. This was to provide a geometrical picture of electromagnetism in the spirit of Einstein's general relativity, by exhibiting this force as being somehow a manifestation of the very structure of space-time. We recall from §1.1 that the general theory of relativity, as first fully published in 1916, enabled Einstein to incorporate the fully detailed nature of the gravitational field into the structure of curved 4-dimensional space-time. The basic forces of nature that were known at that time were the gravitational and the electromagnetic fields, and it was natural to think that, with the appropriate point of view, a full description of electromagnetism, incorporating its interrelation with gravity, should also find a full description in terms of some kind of space-time geometry. What Kaluza was indeed able to do, very remarkably, was to achieve this, but at the expense of having to introduce one extra dimension into the space-time continuum.

## 1.7. TIME IN EINSTEIN'S GENERAL RELATIVITY

Before we can look a little more carefully at the 5-dimensional space-time of Kaluza–Klein theory, it will be appropriate for us to examine the method of describing electromagnetic interactions that ultimately became part of standard theory. Our concern here will be particularly with the way that electromagnetic interactions of quantum particles are described (the quantum version of Lorentz's extension of Maxwell's theory which, as was mentioned in §1.5, shows how charged particles respond to an electromagnetic field) and also the generalizations of this to the strong and weak interactions of the standard model. This is the scheme initiated in 1918 by the great German mathematician (and theoretical physicist) Hermann Weyl. (Weyl became one of the mainstays of the Princeton Institute for Advanced Study during the same period, 1933–55, that Einstein was there, although, as with Einstein, his main contributions to physics had been done earlier, in Germany and Switzerland.) Weyl's highly original initial idea was to broaden Einstein's general relativity, so that Maxwell's electromagnetism (the great theory referred to briefly in §§1.2 and 1.6) could be incorporated in a natural way into the geometrical structure of space-time. He did this by introducing the notion of what is now called a *gauge connection*. Ultimately, after some subtle changes were introduced, Weyl's idea became central to the way that interactions are treated generally, in the standard model of particle physics. In mathematical terms (largely through the influence of Andrzej Trautman [1970]), this idea of a gauge connection is now understood in terms of the concept of a *bundle* (§A.7) that we have seen illustrated in figure 1-12 (as was already hinted at in §1.3). It is important that we understand the differences and the similarities that Weyl's original gauge-connection idea had with the slightly later Kaluza–Klein proposal.

In §1.8, I shall describe in a bit more detail how Weyl introduced his geometrical extension of Einstein's general relativity to incorporate Maxwell's theory. We shall see that Weyl's theory does not involve any increase in space-time dimensionality, but it introduces a weakening of the notion of *metric*, on which Einstein's theory is founded. So, as a preliminary, I shall need to address the actual physical role of the *metric tensor* **g** of Einstein's scheme. It is indeed the basic quantity defining the pseudo-Riemannian structure of space-time. Physicists would normally use a notation like $g_{ab}$ (or $g_{ij}$, or $g_{\mu\nu}$, or some such) to denote the *set of components* of this tensor quantity **g**, but I have no intention of entering into the details of these matters here, nor even of explaining what the term *tensor* actually means mathematically. What we really need to know here is just the very direct physical interpretation that can be assigned to **g**.

Suppose that we have a curve $\mathcal{C}$ connecting two points – or *events* – P and Q, in the space-time manifold $\mathcal{M}$, where $\mathcal{C}$ represents the history of some massive particle travelling from the event P to the later event Q. (The term *event* is frequently used for a space-time point.) We call the curve $\mathcal{C}$ the *world-line* of that particle. Then what **g** does, in Einstein's theory, is to determine a "*length*" for the curve $\mathcal{C}$, this length being interpreted physically as the *time* interval (rather than a distance measure) between P and Q that an *ideal clock* carried by the particle would measure (see figure 1-14(a)).

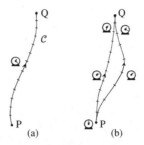

Figure 1-14: (a) The space-time metric **g** assigns a "length" to any segment of a particle's world-line $\mathcal{C}$, this being the time interval measured by an ideal clock following the world-line; (b) if two different such world-lines connect particular events P, Q, these time measures may differ.

We must bear in mind that, according to Einstein's relativity, the "passage of time" is not a given absolute, occurring simultaneously across the universe. We are to think, instead, in fully *space-time* terms. There is no assigned "slicing" of the space-time into 3-dimensional spatial sections each of which would represent the family of events that occur "all at the same time". We are *not* given one absolute "universal clock" which ticks away, whereby for each particular tick of that universal clock there would be given one entire 3-dimensional space of simultaneous events, and at the next tick of the clock another separate instance of 3-dimensional space of simultaneous events, etc., all these 3-spaces fitting together to make space-time (figure 1-15, where we may think of our universal clock as chiming at noon on each day). It is all right to think of space-time in this way provisionally, just so that we can relate this 4-dimensional picture to our everyday experience of a 3-dimensional space within which things "evolve with time", but we are to take the view that there is nothing special, or "God given", about one such slicing of the space-time, as opposed to some other. The *entire* space-time is the absolute notion, but we are not to regard any particular *slicing* of the space-time as a preferred one, according to which there would be a universal notion that we might call "the" time. (This is all part of the principle of general covariance referred to in §1.7 and described more specifically in §A.5, which tells us that particular choices of coordinates – in particular, a "time" coordinate – should not have direct physical relevance.) Instead, each different particle's world-line has its own individual notion of the passage of time, this being determined by the particle's particular world-line and by the metric **g**, as described above. The discrepancies between one particle's time notion and another particle's time notion are very

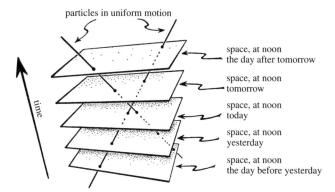

Figure 1-15: The Newtonian picture of a universal time (where, here, a universal clock may be imagined as chiming at noon on each day). This viewpoint is rejected in relativity theory, but we may provisionally think of space-time in this way, as an excellent approximation for objects whose motion is much slower than the speed of light.

small, however – unless the relative speeds between the particles becomes significantly large in relation to the speed of light (or unless we are in some region where the space-time-bending effects of gravity have become enormous) – this smallness being a necessary requirement in order that we do not actually perceive any such discrepancies in our everyday experience of the passage of time.

In Einstein's relativity, if we have two world-lines connecting two particular events P and Q (figure 1-14(b)), then this "length" (i.e. measured elapsed time) may indeed differ in the two cases (an effect that has been repeatedly directly measured, for example, by use of very accurate clocks in fast-moving aeroplanes – or in aeroplanes flying at very different heights from the ground) [Will 1993]. This non-intuitive fact is basically an expression of the familiar (so-called) twin paradox of special relativity, according to which an astronaut travelling at great speed from Earth to a distant star and back may experience a considerably smaller passage of time than does the twin of the astronaut who remains on Earth during the course of the astronaut's journey. The two twins have different world-lines, though they connect the same two events, namely P (when they are together and the astronaut is just at the point of leaving the Earth) and Q (the event of the astronaut's return to Earth).

A space-time depiction of this is shown, for special relativity (with largely uniform motions), in figure 1-16, where, in addition, R is the event of the astronaut's arrival at the distant star. Figure 1-17 similarly illustrates how the metric

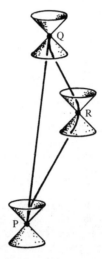

Figure 1-16: The so-called twin paradox of special relativity. The earthbound twin with world-line PQ experiences a longer time than does the space-travelling twin with world-line PRQ (in a curious reversal of the familiar triangle inequality of Euclidean geometry: PR + RQ > PQ). The (double) cones are explained in figure 1-18.

determines the lapse of experienced time, which applies also in the general situation of *general* relativity, where for a (massive) particle's world-line the "length" measure of a stretch of world-line, is determined by **g**, providing the experienced time interval during that period. In each picture, the *null cones* are depicted, these being an important physical manifestation of Einstein's **g**, providing a space-time description of the *speed of light* at each space-time event. We see that at each event along the astronaut's or particle's world-line, the line's direction must be within the (double) null cone at that event, illustrating the important restriction that the speed of light cannot be (locally) exceeded.

Figure 1-18 shows the physical interpretation of the future part of the (double) null cone, as the immediate history of a (hypothetical) flash of light originating at an event X, figure 1-18(a) showing the fully 3-dimensional spatial picture, and figure 1-18(b), the corresponding space-time picture with one spatial dimension suppressed. The past part of the double cone is similarly represented by a (hypothetical) light flash converging on X. Figure 1-18(c) tells us that the null cone is really an *infinitesimal* structure at each event X, existing just locally in, strictly, the *tangent space* at X (see §A.5 and figure A-10).

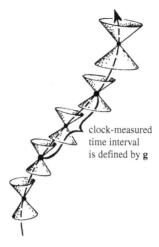

clock-measured
time interval
is defined by **g**

Figure 1-17: In the curved space-time of general relativity, the metric tensor **g** provides the measure of experienced time. This generalizes the flat space-time picture of special relativity shown in figure 1-16.

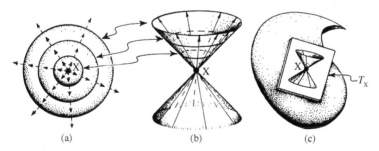

(a)                    (b)                    (c)

Figure 1-18: At each point X of space-time, there is a (double) null cone determined by the metric **g**, consisting of a future null cone and a past null cone, of directions along which the time measure vanishes. The future null cone has a (local) interpretation as the history of a hypothetical flash of light emitted at X: (a) space picture; (b) space-time picture (with one spatial dimension suppressed), where the past null cone would represent the history of a hypothetical light flash converging at X; (c) technically, the null cone is an infinitesimal structure in the neighbourhood of the event X, i.e. lying in the tangent space $T_X$.

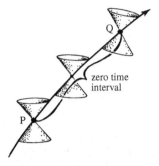

Figure 1-19: Along a light ray (or any null curve) the time measure between any two events P, Q is always zero.

These (double) cones represent the space-time directions along which the "time" measure *vanishes*. This feature comes about because the space-time geometry is, strictly, *pseudo*-Riemannian rather than Riemannian (as noted in §1.1). Frequently, the term *Lorentzian* is used for this particular type of pseudo-Riemannian geometry, where the space-time structure has just 1 time dimension and $(n-1)$ space dimensions, and there will be such a double null cone at every point of the space-time manifold. The null cones provide the most important feature of space-time structure, as they tell us the limits of propagation of information.

How does the *time* measure provided by **g** directly relate to these null cones? Up to this point, the world-lines I have been considering are the histories of ordinary massive particles, and these are deemed to travel more slowly than light, so that their world-lines must lie *within* the null cones. But we must also consider (free) *massless* particles, like photons (the particles of light), and such a particle would travel *at* the speed of light. According to relativity, if a clock were to travel at the speed of light, then it would register no passage of time whatever! Thus, the "length" of the world-line (measured along the curve) of a massless particle is always *zero* between any two events P, Q on the world-line (figure 1-19), no matter how well separated they may be from one another. We call such a world-line a *null* curve. Some null curves are geodesics (see later), and the world-line of a free photon is taken to be a null geodesic.

The family of all null geodesics through a particular point P in space-time sweeps out the complete *light cone* of P (figure 1-20), the null cone at P describing merely the infinitesimal structure at the *vertex* of the light cone of P (see figure 1-18). The null cone tells us the space-time *directions* at P which define

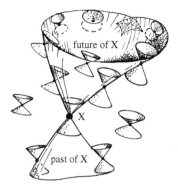

Figure 1-20: The light cone of an event X is a locus in space-time swept out by the null geodesics through X. The tangent structure at its vertex X is the null cone at X.

the speed of light, namely that structure in the *tangent space* at a point P giving the directions of zero "length", according to the metric **g**. (In the literature, the term *light cone* is often also used in the sense that I am here reserving for the term *null cone*.) The light cone (as with the null cone, above) has two parts, one defining the *future null directions* and the other the *past null directions*. The requirement of relativity theory that massive particles are constrained so that they do not exceed the local light speed is expressed explicitly as the fact that the *tangent directions* to the world-lines of massive particles all lie within the null cones at their respective events (figure 1-21). Smooth curves, all of whose tangent directions lie strictly within the null cones, like this, are called *timelike*. Thus, the world-lines of massive particles are indeed timelike curves.

A complementary notion to that of a timelike curve is that of a *spacelike* 3-surface – or spacelike $(n-1)$-surface or spacelike *hypersurface* if we are thinking of an $n$-dimensional space-time. The tangent directions to such a hypersurface are all *external* to the past and future null cones (figure 1-21). In general relativity, this is the appropriate generalization of the idea of "a moment in time" or a "$t =$ constant space", where $t$ is a suitable time coordinate. Clearly, there is a lot of arbitrariness in the choice of such a hypersurface, but it is the kind of thing that is needed if we wish to refer to such issues as *determinism* in dynamical behaviour, where we may ask for "initial data" to be specified on such a hypersurface, where these data are intended (locally) to determine the evolution of the system into the past or future, according to some appropriate equations (normally differential equations; see §A.11).

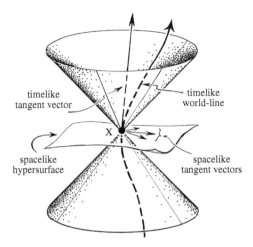

Figure 1-21: Null tangent vectors at X give the null cone, as in figure 1-18, but there are also timelike ones that, if future-pointing, describe tangent vectors (4-velocities) to world-lines of massive particles, and spacelike ones, pointing outside the cone, being the tangents to spacelike surfaces through X.

As another feature of relativity theory, it may be pointed out that if the "length" (in this sense of elapsed measured time) of a world-line $\mathcal{C}$ connecting P to Q is larger than that of *any* other world-line from P to Q, then $\mathcal{C}$ must be what is called a *geodesic*,[5] which is the analogue, in a curved space-time, of a "straight line" (see figure 1-22). Curiously, this maximizing property of "lengths" in space-time is the *opposite* way around from what happens in ordinary Euclidean geometry, where the straight line joining two points P and Q *minimizes* the length of paths connecting P to Q. According to Einstein's theory, the world-line of a particle moving freely under gravity is always a geodesic. The astronaut's journey, in figure 1-16, involves accelerated motion, however, and is not a geodesic.

The flat space-time of special relativity, where there is no gravitational field, is called *Minkowski space* (which I shall tend to denote by the symbol $\mathbb{M}$) after

---

[5] Conversely, *every* world-line $\mathcal{C}$ which happens to be a geodesic has this characteristic property in the *local* sense that for any P on $\mathcal{C}$ there will be a small enough open region $\mathcal{N}$, of $\mathcal{M}$, containing P, such that for every pair of points on $\mathcal{C}$, within $\mathcal{N}$, the maximum length of world-lines connecting them by paths within $\mathcal{N}$ is obtained if we follow the portion of $\mathcal{C}$ that lies within $\mathcal{N}$. (For a pair of points too widely separated along a geodesic $\mathcal{C}$, on the other hand, we may find that $\mathcal{C}$ does not maximize the length owing to the presence of pairs of conjugate points on $\mathcal{C}$ between the points [Penrose 1972; Hawking and Ellis 1973].)

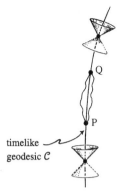

Figure 1-22: A timelike curve which maximizes the time measure between two timelike-separated events P and Q is necessarily a geodesic.

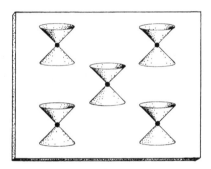

Figure 1-23: Minkowski space is the flat space-time of special relativity. Its null cones are arranged completely uniformly.

the Russian/German mathematician Hermann Minkowski, who first introduced the idea of space-time in 1907. Here the null cones are all uniformly arranged (figure 1-23). Einstein's general relativity follows the same idea, but the null cones may now form a non-uniform arrangement, owing to the presence of a gravitational field (figure 1-24). The metric **g** (10 components per point) defines the null-cone structure, but is not fully defined by it. This null-cone structure is sometimes referred to as the space-time *conformal* structure (9 components per point); see §3.5, particularly. In addition to this Lorentzian conformal structure, **g** determines a *scaling* (1 component per point) and this fixes the rate at which ideal

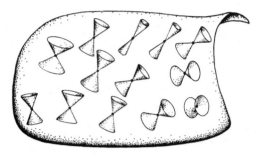

Figure 1-24: In general relativity the null cones may exhibit no particular uniformity.

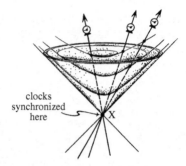

clocks
synchronized
here

Figure 1-25: The metric scaling at an event X would be determined by the rates of ideal clocks passing through X. Here, several identical ideal clocks pass through X, each determining the same metric scaling, where the "ticks" of the various clocks would be related to one another via the bowl-shaped surfaces shown (actually hyperboloidal 3-surfaces).

clocks measure time in Einstein's theory (figure 1-25). For further information on the way that clocks behave in relativity theory, see, for example, Rindler [2001] and Hartle [2003].

## 1.8. WEYL'S GAUGE THEORY OF ELECTROMAGNETISM

Weyl's original 1918 idea for incorporating electromagnetism into general relativity involved weakening the metric structure of space-time to a *conformal structure*, as described above, so that now there is no absolute measure of time rates, although there are still null cones defined [Weyl 1918]. There is, in addition to this, still a

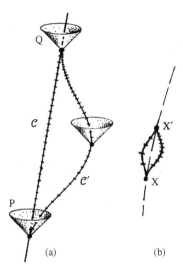

Figure 1-26: (a) Weyl's idea of a gauge connection proposes that the metric scale be not a given thing, but that it can be transferred from a point P to another point Q along a connecting curve $\mathcal{C}$, where a different result may arise with a different curve $\mathcal{C}'$ from P to Q. (b) Weyl's *gauge curvature* arises from the *infinitesimal* version of this discrepancy, and he originally proposed this curvature to be Maxwell's electromagnetic field tensor.

concept of an "ideal clock" in Weyl's theory, so that we can define a measure of "length" to a timelike curve, with respect to any *particular* such clock, although the *rate* at which the clock measures time as passing would depend upon the clock. But there is no *absolute* scale of time in Weyl's theory, because no particular ideal clock is to be preferred over any other. More to the point, we might have two such clocks which tick at exactly identical rates, when at rest with respect to each other at some event P, say, but if they take different space-time routes to a second event Q, we may well find that, upon arrival, the two clock *rates* are in disagreement with each other, i.e. they do not now tick at identical rates when *at rest* with each other at Q. See figure 1-26(a). It is important to note that this is different from – and more extreme than – the "twin paradox" of Einstein's relativity. In that case, whereas the clock *readings* might depend upon their histories, their *rates* would not. Weyl's more general kind of geometry leads to a curious type of space-time "curvature" in the notion of clock rates, which measures this clock-rate discrepancy on an infinitesimal scale (see figure 1-26(b)). It is analogous to the way that

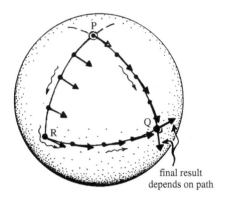

final result
depends on path

Figure 1-27: An affine connection expresses the idea of parallel transport of tangent vectors along curves, where the discrepancy between transporting along different curves provides a measure of the curvature. We can see this explicitly on a sphere, where the transport of a tangent vector by the direct great-circle route from P to Q gives a grossly different answer from that obtained by the route consisting of a great-circle arc from P to R followed by a similar such arc from R to Q.

the curvature of a surface measures a discrepancy in angles, as we shall be seeing shortly (in figure 1-27). Weyl was able to show that the quantity F that describes his type of curvature exactly satisfies the same equations as does the quantity describing the free electromagnetic field in Maxwell's theory! So Weyl proposed that this **F** was to be physically identified with Maxwell's electromagnetic field.

Temporal and spatial measures are essentially equivalent to one another in the neighbourhood of any one event P once we have the concept of the null cone at P, since that fixes the speed of light at P. In particular, in ordinary terms, the speed of light enables space and time measures to be converted one into the other. Thus, for example, the time interval of a year converts to the distance interval of a light year, and of a second to a light second, etc. In fact, in modern measurements, time intervals are far more precisely directly determined than spatial ones, so that the metre is now defined to be *exactly* 1/299792458 of a light second (so that the speed of light is now the exact integer 299792458 in metres per second)! Thus, the terminology *chronometry* for space-time structure (rather than *geometry*), as had been proposed by the distinguished relativity theorist J. L. Synge [1921, 1956], seems particularly apposite.

I have described Weyl's idea in terms of time measures, but Weyl probably had spatial displacements more in mind, and his scheme was referred to as a *gauge*

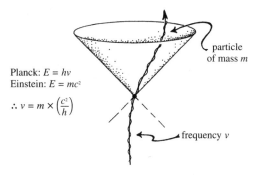

Planck: $E = hv$
Einstein: $E = mc^2$

$$\therefore v = m \times \left(\frac{c^2}{h}\right)$$

Figure 1-28: Any stable massive particle of mass $m$ is a precise quantum-mechanical clock of frequency $v = mc^2/h$.

*theory*, the "gauge" referring to the scale in terms of which physical distances are measured. The point about Weyl's remarkable idea is that a gauge need not be determined globally, all at once for the entire space-time, but if the gauge is specified at one event P, and a curve $\mathcal{C}$ is given connecting P to another event Q, then the gauge can be carried uniquely along $\mathcal{C}$ from P to Q. But if some other curve $\mathcal{C}'$ is also given, connecting P to Q, then carrying the gauge to Q along $\mathcal{C}'$ may give a different result. The mathematical quantity that defines this "gauge-carrying" procedure is called a *gauge connection* and the discrepancies that result from using different paths is a measure of *gauge curvature*. It should be pointed out that Weyl's brilliant idea of a gauge connection would likely have occurred to him from his familiarity with another type of connection that is automatically possessed by any (pseudo-)Riemannian manifold – referred to as an *affine connection* – that is concerned with the parallel transport of tangent vectors along curves, and this is also path-dependent, as strikingly illustrated, on a sphere, in figure 1-27.

When Einstein heard of Weyl's ingenious idea, he was very intrigued, but he pointed out that, from the physical point of view, the scheme had a serious flaw, basically for the physical reason that the *mass* of a particle provides a definite time measure along its world-line. This is given (figure 1-28) by combining Max Planck's quantum relation

$$E = hv$$

with Einstein's own

$$E = mc^2.$$

Here $E$ is the particle's energy (in its own rest frame), $m$ is its (rest) mass, and $\nu$ is the frequency (i.e. a "ticking rate" of the particle) that a particle acquires according to basic quantum mechanic (see §2.2), and where $h$ and $c$ are, respectively, Planck's constant and the speed of light. Thus, combining these according to $h\nu\ (= E) = mc^2$, we see that there is always a precise frequency determined by a single particle, which is directly proportional to its mass:

$$\nu = m \times \frac{c^2}{h},$$

the quantity $c^2/h$ being a universal constant. The mass of any stable particle, therefore, determines a very precise clock rate, as given by this frequency.

In Weyl's proposal, however, any such clock rate would necessarily *not* be a fixed quantity, but something that would depend on the particle's history. Accordingly, the particle's *mass* would have to be dependent upon its history. In particular, in our situation above, if two electrons were to be considered to be *identical* particles (as indeed quantum theory requires) at an event P, then they would be likely to end up at a second event Q with differing masses if they arrived at Q via different routes, in which case they could *not* be identical particles when they arrive at Q! This is, in fact, profoundly inconsistent with the well-established principles of quantum theory, which demand that rules that apply to identical particles are critically different from those applying to non-identical ones (see §1.14).

Thus, it seemed that Weyl's idea had foundered on certain very basic quantum-mechanical principles. However, by an extraordinary turn of events, it was quantum theory itself that then came to the rescue of Weyl's idea, after it had become fully formulated in around 1930 (primarily by Dirac [1930] and von Neumann [1932], as well as by Weyl [1927] himself). As we shall be seeing in chapter 2 (see §§2.5 and 2.6), the quantum description of particles is given in terms of a *complex-number* description (§A.9). We have already seen, in §1.4, this essential role of complex numbers in their appearance as coefficients (the quantities $w$ and $z$) in the superposition principle of quantum mechanics. We shall be finding later (§2.5) that if we multiply all these coefficients by the same complex number $u$ of *unit modulus* (i.e. $|u| = \sqrt{u\bar{u}} = 1$, so that $u$ lies on the *unit circle* in the Wessel plane (see §A.10, figure A-13), then this leaves the physical situation unaltered. We note that the Cotes–De Moivre–Euler formula (see §A.10) shows that such a *unimodular* complex number $u$ can always be written

$$u = e^{i\theta} = \cos\theta + i\sin\theta,$$

where $\theta$ is the angle (measured in radians, in a counter-clockwise sense) that the line joining the origin to $u$ makes with the positive real axis (figure A-13 in §A.10).

In the context of quantum mechanics, a unimodular complex multiplier is frequently referred to as a *phase* (or a phase angle), and it is regarded in the quantum formalism as something that is not directly observable (see §2.5). The subtle change that converts Weyl's ingenious but extraordinary idea into a key ingredient of modern physics is to replace Weyl's real positive scale factor – or *gauge* – into the *complex phase* of quantum mechanics. For these historical reasons, the term *gauge* has stuck, even though it might well have been more appropriate to have referred to Weyl's theory, as modified in this way, as a *phase theory*, and a *gauge connection* as a *phase connection*. However, it might well confuse more people than it would help if we were to change the terminology in this way now.

To be more precise, the phase that appears in Weyl's theory is not exactly the same as the (universal) phase of the quantum formalism, there being a multiplying factor between the two that is given by the *electric charge* of the particle in question. The essential feature that Weyl's theory relies upon is the presence of what is called a *continuous group of symmetries* (see §A.7, final paragraph), which applies at any event P in space-time. In Weyl's original theory, the group of symmetries consists of all the positive real-number factors whereby one might scale the gauge up or down. These possible factors are simply the various *positive real numbers*, the space of which is referred to by mathematicians as $\mathbb{R}^+$, so the symmetry group of relevance here is sometimes referred to as the multiplicative group $\mathbb{R}^+$. In the later, more physically relevant version of Weyl's electromagnetic theory, the elements of the group are the rotations in the Wessel plane (without reflection), called SO(2) or sometimes U(1), and the elements of this group are represented by complex numbers $e^{i\theta}$ of unit modulus, these elements providing the different angles of rotation of the unit circle in the Wessel plane, where I refer to this unit-radius circle simply as $S^1$.

It may be remarked (see also the final paragraph of §A.7 with regard to this notation, and also for the notion of a *group*) that the "O" of "SO(2)" stands for "orthogonal", which means, in effect that we are concerned with a group of *rotations* (i.e. preserving of orthogonality, i.e. of right angles, which in this case is rotations in 2 dimensions, as designated by the "2" in "SO(2)"). The "S" stands for "special", which refers to the fact that *reflections* are here excluded. With regard to "U(1)", the "U" stands for "unitary" (preserving of the unit-norm nature of complex vectors), which refers to a kind of rotation in *complex*-number space that we shall

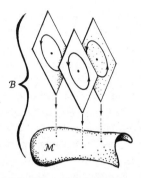

Figure 1-29: Weyl's geometry expresses electromagnetism as a connection on a bundle $\mathcal{B}$ over space-time $\mathcal{M}$. The $S^1$ (circle) fibres are best thought of as unit circles in copies of the complex (Wessel) plane.

be coming to in §§2.5–2.8. Whichever way we refer to it, what we are concerned with are simply the rotations, without reflection, of the ordinary circle $S^1$.

We note that now Weyl's connection is not really a concept that applies simply to the space-time manifold $\mathcal{M}$, since the circle $S^1$ is not really part of space-time at all. The $S^1$ refers, rather, to an abstract space that is specifically to do with quantum mechanics. We can still think of $S^1$ as having a geometrical role, however, namely as the *fibre* of a bundle $\mathcal{B}$ whose base space is the space-time manifold $\mathcal{M}$. This geometry is illustrated in figure 1-29. The fibres are the circles $S^1$, but we see in the picture that these circles are best thought of as the unit circles within copies of the (Wessel) complex plane (§A.10). (The reader is referred to the discussion of §A.7 for the notion of a bundle.) Weyl's concept of a *gauge connection* is indeed a geometrical one, but not one that gives structure to space-time pure and simple; the structure that it supplies is assigned to the bundle $\mathcal{B}$, which is a 5-manifold closely associated with the space-time 4-manifold.

The extensions of Weyl's ideas that express the strong and weak interactions of particle physics are also formulated, in the standard model of §1.3, in terms of a gauge connection, and again the bundle description of §A.7 is appropriate. In each case the base space is 4-dimensional space-time, as before, but the fibre would have to be a space $\mathcal{F}$ of greater dimension than the 1-dimensional $S^1$ which, as noted above, can be used to express electromagnetism. These extensions of Weyl's gauge approach to Maxwell's theory are referred to as *Yang–Mills theories* [Chan and Tsou 1998]. In the case of strong interactions, $\mathcal{F}$ would be a space which has the same symmetry as the space of possible colours that are available to a quark, in accordance with the descriptions given in §1.3. The symmetry group here is the one referred to as SU(3). The case of weak interactions is ostensibly similar where the group, in this case, is that known as SU(2) (or else U(2)), but there is some muddying of the situation in weak-interaction theory because of the fact that the symmetry is taken to be broken, owing to a symmetry-breaking process that is regarded as having occurred in the early stages of the universe's expansion. In fact, there are some issues which I find somewhat worrying in the usual descriptions of this procedure since, strictly speaking, the very idea of a gauge symmetry does not really work unless the symmetry is indeed *exact* [see §A.7 and TRtR, §28.3]. It is fortunate, in my view, that there are reformulations of the usual procedure in which the weak force arises via a mechanism with a somewhat different physical interpretation from the standard one where, in effect, coloured quark-like constituents of leptons are being postulated (analogous to the quark constituents of hadrons) in which the weak-interaction symmetry is regarded as being always exact ['t Hooft 1980b; Chan and Tsou 1980].

## 1.9. FUNCTIONAL FREEDOM IN KALUZA–KLEIN AND STRING MODELS

We now have *two* alternative 5-dimensional spaces, each providing a geometrical procedure for incorporating Maxwell's electromagnetism into a curved-space-time geometry. How does the 5-manifold $\mathcal{B}$, in the $S^1$-bundle representation of Weyl's procedure as described in §1.8, relate to the Kaluza–Klein 5-dimensional space-time picture of electromagnetic interactions referred to in §1.6? They are in fact extremely close to one another, and there is no harm in thinking of them as *identical*! Kaluza's 5-dimensional space-time, as modified by Klein so as to have a tiny circle ($S^1$) as its "extra" dimension, and the bundle $\mathcal{B}$, which we get in Weyl's procedure, are topologically identical, both being (normally) simply the *product space* $\mathcal{M} \times S^1$ of the ordinary 4-dimensional space-time $\mathcal{M}$ with the circle $S^1$ (see figure A-25 in §A.7 and figure 1-29). Moreover, the Kaluza–Klein space automatically has a kind of $S^1$-bundle structure, where to identify the $S^1$ fibres we simply look for geodesics which are *closed* (and belong to the right topological family). There is, however, a bit of a difference between the Weyl and the Kaluza–Klein 5-spaces in the type of structure that is assigned in each case. The Weyl procedure requires that we assign a *gauge connection* (§1.8) to $\mathcal{B}$, considered as a bundle over the 4-dimensional space-time $\mathcal{M}$, whereas, in the Kaluza–Klein theory, the whole 5-manifold is thought of as being "space-time" and a *metric* $\mathbf{g}$ is accordingly assigned to the entire structure. However, Weyl's gauge connection happens to be *already* implicit in the Kaluza construction, as it turns out to be determined simply by the ordinary notion of *affine connection*, discussed in §1.8 (which holds for any Riemannian space, and hence for Kaluza's 5-space), as applied to the directions orthogonal to the $S^1$ fibres. Thus, the Kaluza–Klein 5-space already contains Weyl's gauge connection, and can indeed be *identified* with Weyl's bundle $\mathcal{B}$.

But the Kaluza–Klein space actually gives us something *more*, because it has a *metric* with the property that if it satisfies the appropriate Einstein vacuum field equations $^5\mathbf{G} = \mathbf{0}$ (stating that the energy tensor $^5\mathbf{T}$ of the 5-space is taken to be zero), then not only do we get Weyl's connection but, remarkably, we also get the fact that the Maxwell electromagnetic field $\mathbf{F}$ that emerges from Weyl's connection acts (through its mass/energy density) as a source for the gravitational field – the equations being correctly coupled in this way that is referred to as the *Einstein–Maxwell equations*. This striking fact is *not* something that comes directly out of Weyl's approach.

To be a little more precise about the structure of the Kaluza–Klein 5-space, I must point out that there is actually a proviso to the above assertion, namely that the version of the Kaluza–Klein theory that I am adopting here is the particular one that requires that the length assigned to the $S^1$ loops be the same throughout the 5-space. (Some versions of the theory allow this length to vary, thereby providing scope for an additional scalar field.) I also require that this constant length is chosen so that the constant $8\pi\gamma$ in Einstein's equations (see §1.1) comes out correctly. Most importantly, I insist that when I am referring to the Kaluza–Klein theory, I mean the original version, in which there is an exact symmetry imposed on the entire 5-space, so that it must have complete *rotational symmetry* in the $S^1$ direction (see the essentially similar figure 1-29). In other words, the $\mathbf{k}$ vector is actually a Killing vector, so that the 5-space can be slid over itself along the $S^1$ lines without affecting its metric structure.

Let us now address the issue of *functional freedom* in the Kaluza–Klein theory. If we take that theory in the form just stated, then the extra dimension does not contribute towards an excess of functional freedom. Because of the enforced rotational symmetry along the $S^1$ curves, the freedom is the same as for an ordinary 4-dimensional space-time with a standard type of deterministic evolution from data on an initial *3-space*, in fact the same as for the Einstein–Maxwell equations, to which it is equivalent, namely

$$\infty^{8\infty^3},$$

which is as it should be, for a classical physical theory appropriate to our universe.

A point I wish to stress here is that it is a crucial feature of gauge theories – that class of theories that has been so enormously successful in explaining the basic forces of nature – that there be a (finite-dimensional) *symmetry* possessed by the fibres $\mathcal{F}$ of the bundle to which the gauge theory is being applied. As is firmly pointed out in §A.7, it is the possession of a (continuous) symmetry in our fibre $\mathcal{F}$ that enables the gauge theory to work at all. This symmetry, in the case of Weyl's theory of electromagnetic interactions, is the circle group U(1) (or, equivalently, SO(2)) that must apply, exactly, to the fibres $\mathcal{F}$. (See the end of §A.7 for the meaning of these symbols.) It is also this symmetry that extends, globally, to the entire 5-dimensional manifold $\mathcal{B}$, in Weyl's approach, and which has also been specified in the original Kaluza–Klein procedure. In order to preserve this close relationship between the higher-dimensional space-time approach, as initiated by Kaluza, and the gauge theory approach of Weyl, it appears to be essential that we preserve the fibre symmetry, and do not actually increase the functional freedom (vastly) by treating the fibre spaces $\mathcal{F}$ as though they were properly parts of the space-time with internal degrees of freedom.

But what about string theory? Here, the story appears to be completely different, for it is explicitly demanded that the extra spatial dimension(s) shall take part fully in the dynamical freedom. Such extra spatial dimensions are intended to be playing their role *as* genuine spatial dimensions. This is very much part of the driving philosophy behind string theory as it has been developed, because it is proposed that somehow the "oscillations" that are allowed via these extra dimensions should give explanations for all the complicated forces and parameters that are needed, so as to have the scope to accommodate all the required features of particle physics. In my own view this is a seriously mistaken philosophy – for allowing extra *spatial* dimensions to be freely involved in the dynamics gives us a veritable Pandora's box of unwanted degrees of freedom, with but scant hope of our ever keeping them under control.

Nevertheless, heedless of difficulties of this kind, as would naturally arise from the excessive functional freedom in extra spatial dimensions, the proponents of string theory have chosen this very different course from that of the original Kaluza–Klein scheme. As part of their attempts to resolve anomalies that arose from their requirements of parameter invariance for their quantum theory of strings, they found themselves driven, from around 1970 onwards, to try to adopt (for bosonic strings) a fully dynamical 26-dimensional space-time, where there would be 25 spatial dimensions, with 1 other dimension reserved for time. Then, following the highly influential theoretical development by Michael Green and John Schwarz in 1984, the string theorists managed to reduce this spatial dimensionality to 9 (for fermionic strings) with the aid of what is called *super-symmetry* (see §1.14; but already referred to in §1.6), but this reduction in extra spatial dimensionality (since it does not get the spatial dimensionality down to the directly experienced value of 3) makes little difference to the issues I shall be raising.

In my attempts to get to grips with the various developments in string theory, there has been an additional point of potential confusion for me, particularly when trying to understand the issues of functional freedom. This is that there has often been a shifting of viewpoint about what the space-time dimensionality is actually taken to be. I imagine that many other outsiders must also have similar difficulties in their own attempts to understand the mathematical structure of string theory. The idea of having an ambient space-time of some specific dimension seems to play less of a role in string theory than in conventional physics, and certainly less than the kind of role that I would myself feel comfortable with. It is particularly difficult to assess the functional freedom that is involved in a physical theory unless one has a clear idea of its actual space-time dimensionality.

To be more explicit about this issue, let me return to one of the particularly appealing aspects of the early string ideas, as outlined in §1.6. This was that the string histories could be viewed as being *Riemann surfaces*, i.e. as complex curves (see §A.10), which are particularly elegant structures from the mathematical point of view. The term *world-sheet* is sometimes used for this string history (in analogy with the idea of the world-line of a particle in conventional relativity theory; see §1.7). Now in the early days of string theory, the subject was sometimes viewed from the point of view of a 2-dimensional *conformal field theory* [Francesco et al. 1997; Kaku 2000; Polchinski 1994, chapter 1, 2001, chapter 2], whereby, in rough terms, the analogue of the space-time in the theory would be the 2-dimensional world-sheet *itself*! (Recall the §1.7 notion of *conformal*, in a space-time context.) This would lead us to a picture in which the functional freedom had the form

$$\infty^{a\infty^1}$$

for some positive number $a$. How are we to square this with the far larger functional freedom $\infty^{b\infty^3}$ that is required for ordinary physics?

The answer seems to be that the world-sheet would, in some sense, "feel out" the surrounding space-time and the physics around it in terms of some kind of power-series expansion, where the needed information (effective power series coefficients) would be provided in terms of an infinite number of parameters (actually holomorphic quantities on the world-sheet; see §A.10). Having an infinite number of such parameters is superficially like putting the "$a = \infty$" in the above displayed expression, but this is somewhat unhelpful (for the kind of reason pointed out towards the end of §A.11). The point I am making here is certainly *not* that functional freedom might in some sense be ill defined or irrelevant. The point *is*, however, that for a theory formulated in a way dependent on things like power series coefficients or mode analysis, it may not be at all easy to ascertain what that functional freedom actually is (§A.11). Unfortunately, formulations of this kind seem frequently to be the ones adopted in various approaches to string theory.

It appears that, to some extent, there is a view among string theorists that it is not too important to have a clear view of what the dimensionality of space-time actually is. In some sense, this dimensionality could be supposed to be an energy effect, so that it could become possible that more spatial dimensions become accessible to a system as the energy increases. Accordingly, the view could be taken that there are hidden dimensions, more of which are revealed as the energy gets higher. The lack of clarity in this picture is somewhat disturbing to me, especially with regard to the question of the functional freedom that is intrinsic to the theory.

A case in point arises with what is known at *heterotic* string theory. There are two versions of this proposal, referred to as the *HO theory* and the *HE theory*. The difference between the two is not important for us just now, but I shall say something about this in a moment. The strange feature about heterotic string theory is that it appears to behave simultaneously as a theory in 26 space-time dimensions and in 10 space-time dimensions (the latter with accompanying supersymmetry), depending upon whether we are concerned with left-moving or right-moving excitations of the string – this difference (depending upon an orientation that must be attached to the string) also requiring explanation, which I shall address shortly. This dimensional conflict would seem to cause us problems if we are to try to work out the functional freedom involved (where, for this purpose, each scheme is treated as a classical theory).

This apparent conundrum is officially addressed by regarding the space-time as being 10-dimensional in *both* cases (1 time and 9 space dimensions), but there are 16 extra spatial dimensions that are to be treated in a different way in the two cases. As far as the left-moving excitations are concerned, all the 26 dimensions are to be taken together and regarded as providing the space-time into which the string can wiggle. However, with regard to the right-moving excitations, different directions within the 26 dimensions are interpreted in different ways, 10 of them being regarded as providing directions into which the string can wiggle, the other 16 being taken to be *fibre* directions, so the picture, as far as the string is concerned when in such right-handed modes of vibration, is that of a fibre bundle (see §A.7) with a 10-dimensional base space and a 16-dimensional fibre.

As with fibre bundles generally, there must be a *symmetry group* associated with the fibre, and for the HO theory this is taken to be SO(32) (the group of non-reflective rotations of a sphere in 32 dimensions; see end of §A.7), and for the HE heterotic theory it is the group $E_8 \times E_8$, where $E_8$ is a symmetry group of a particularly mathematically interesting type, referred to as an *exceptional simple continuous group*. No doubt the particular inherent mathematical interest of this exceptional simple group – $E_8$ is the largest and most fascinating of them – provides some encouragement from the direction of aesthetic appeal (see §1.1). The important issue, from the point of view of functional freedom, however, is that whichever group is adopted in the bundle description, this freedom would be of the form $\infty^{a\infty^9}$, appropriate to the fermionic (right-moving) modes of string oscillation, whereas it would appear to be of the form $\infty^{b\infty^{25}}$, as would be appropriate to the bosonic (left-moving) modes. This issue is closely related to the one that we encountered before, when we considered the difference in functional freedom in the original Kaluza–Klein theory (or Weyl circle-bundle theory; see §1.8), with functional freedom of $\infty^{8\infty^3}$, and a fully 5-dimensional space-time theory whose far larger functional theory would be of the form $\infty^{b\infty^4}$. Here, a clear distinction needed to be made between the dimensionality $d + r$ of the total space $\mathcal{B}$ of a bundle (with $r$-dimensional fibre $\mathcal{F}$) and that of the $d$-dimensional base space $\mathcal{M}$. This is described more fully in §A.7.

The above issue concerns the functional freedom possessed by the space-time as a whole, quite independently of what string world-sheet might happen to be present in it. What we are really interested in here, however, is the freedom possessed by the string world-sheets (see §1.6) that lie within this space-time. How can it be that for some kinds of modes of displacement (the fermionic ones) the space-time appears to be 10 dimensional where for other kinds (the bosonic ones) it appears to be 26 dimensional? For the bosonic modes, the picture is reasonably straightforward. The string can wiggle as it may into the ambient space-time, with a functional freedom $\infty^{24\infty^1}$ (the "1" coming from the fact that even though the string world-sheet is a 2-surface, we are considering only the 1-dimensional space of right-moving excitations). But when we come to consider the fermionic modes, we are to be thinking of the string as inhabiting the "space-time" which is 10-dimensional, rather than sitting in the 26-dimensional bundle over it. What this means is that the string itself must *carry with* it those fibres of the bundle that lie above it. It is really quite a different kind of entity from that for the bosonic modes, the string now being itself an 18-dimensional *sub-bundle* of the 26-dimensional total space of the space-time bundle it inhabits. (This fact is not normally appreciated. The effective space-time is a 10-dimensional *factor space* – see figure 1-32 in §1.10 and §A.7 – of the 26-total-dimensional bundle, so the string world-sheet must also be a factor space, now of an 18-dimensional sub-bundle.) The functional freedom in these modes is still of the form $\infty^{a\infty^1}$ (where $a$ depends on the group of the bundle), but the *geometrical picture* is now completely different from that presented for the bosonic modes, where for the bosonic modes the string is to be thought of as a 2-dimensional world tube (as in figure 1-11), but where for the fermionic modes the strings ought, technically, to be sub-bundles of total dimension 18 ( $= 2 + 16$)! I find it very difficult to form a consistent picture of what is going on here, and I have never even seen these geometrical issues properly discussed.

I should, moreover, be more explicit, here, about the geometrical nature of the left- and right-moving modes, quite apart from the issue of how the ambient space-time is to be regarded, since this raises another issue that I have not yet addressed. I have referred to the appealing fact that the string world-sheets may be thought of as *Riemann surfaces*. However, this is not really true for my descriptions above. I have adopted a certain sleight of hand that is very common in discussions of quantum (field) theory and is manifestly being used here, with the employment of a device known as a *Wick rotation*, which has slipped past without explicit mention so far.

What is a Wick rotation? It is a mathematical procedure, originally concerned with converting various problems in quantum field theory in Minkowski space-time $\mathbb{M}$ (the flat space-time of special relativity; see end of §1.7) into frequently more tractable ones in ordinary Euclidean 4-space $\mathbb{E}^4$. The idea comes from the fact that the Lorentzian space-time metric **g** of relativity theory converts to (minus) a Euclidean one if a standard time coordinate $t$ is replaced by i$t$ (where i $= \sqrt{-1}$; see §A.9). This trick is sometimes called *Euclideanization*, and when the problem has been solved in its Euclideanized form, it is to be converted back by a process of *analytic continuation* (see §A.10, and also §3.8) to a solution in the intended Minkowski space-time $\mathbb{M}$. The Wick rotation idea is now so commonly used in quantum field theory that it is often taken as an almost automatic procedure in numerous different kinds of situation, with barely a mention, and its validity is hardly ever questioned. It does, in fact, have a broad applicability, but it is *not* a universally valid procedure. Most particularly, it is highly questionable in the context of the *curved* space-times arising in general relativity, when in

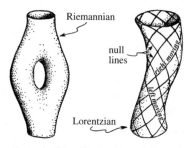

Riemannian

null
lines

right moving

left moving

Lorentzian

Figure 1-30: This figure contrasts two different viewpoints with regard to string world-sheets. In (a) we see the picture of the world-tubes as Riemann surfaces, which may branch and rejoin smoothly in various ways. In (b) we have the more direct way of picturing a (timelike) string history as a Lorentzian 2-manifold, where the left- and right-moving modes of excitation can be depicted, but where branching is not allowed. The two pictures are intended to be related through a Wick rotation, a procedure that is very questionable in a general-relativistic curved-space-time context.

normal circumstances the procedure cannot even be applied, because there is *no natural time coordinate*. In string theory, this is a problem both in the 10-dimensional space-time, in general curved-space situations, and also on the string world-sheet.[6]

It seems to me that this kind of difficulty raises issues that have not really been adequately addressed in string theory. However, let me ignore such general points here and see, instead, what the effect of the Euclideanization of a string *world-sheet* would amount to. We can visualize a string history as a single loop that moves in some way, not exceeding the local speed of light. Then its world-sheet will be a timelike 2-surface, which inherits a Lorentzian 2-metric from the Lorentzian 10-metric of the ambient space-time. This 2-metric will assign a pair of null directions at each point within the world-sheet. As we follow these null directions consistently one way around, or the other, we get a right-handed or left-handed helical null curve on the world-sheet cylinder. Excitations that are constant along one or the other of these families of curves will give us the right- and left-moving modes referred to above (see figure 1-30(b)). However, such cylindrical sheets can never branch and give the kind of pictures that we would need for figure 1-11 because the Lorentzian structure goes wrong at tube

---

[6] An interesting variant of the Wick rotation is used in the Hartle–Hawking approach to space-time quantization [Hartle and Hawking 1983]. However, this is, strictly speaking, a very different procedure, with its own problems.

branching locations. Such topology can arise only for the Euclideanized strings shown in figure 1-30(a), these being Riemann surfaces, possessing a Riemannian-type metric, without null directions, and being interpretable as *complex curves* (see §A.10). The Euclideanized right- and left-moving modes now correspond to *holomorphic* and *anti-holomorphic* functions, respectively, on the Riemann surface (see §A.10).

The issue of *functional freedom*, being my main concern in this section, is not the only directly physical issue that does not appear to be significantly addressed in any string-theoretic considerations I have seen in the standard literature. Indeed, I do not find much that is concerned with the immediate geometrical considerations raised by the seemingly essential, but highly dubious, Wick rotation procedure referred to above. It is my impression that many of the most obvious geometrical and physical issues arising from the perspective of string theory are never really properly discussed at all!

For example, in the case of heterotic string theory, the strings are taken to be necessarily *closed* strings, which means that they do not have holes in them. (See §1.6 and especially §1.16.) If we are trying to think of these strings in a direct physical way – that is to say, before the "trick" of the Wick rotation is brought in – then we must think of a string's world-sheet as being *timelike*, as in figure 1-30(b). If the world-sheet is to be without holes, then it must continue to be a timelike tube extending indefinitely into the future. It is no good thinking of it as wrapping around the "tiny" extra dimensions, since these dimensions are all taken to be spacelike. It can only continue indefinitely into the future, and then it does not really qualify as closed. This is one of many questions that I do not find to be properly addressed in any description of string theory that I have seen.

I find this curious lack of a coherent geometrical picture of how string theory is to be viewed in ordinary physical terms to be very odd. It is especially so for a theory that had not infrequently been referred to in terms such as a *theory of everything*. Moreover, such a lack of a clear geometrical and physical picture in such considerations is in stark contrast with the highly sophisticated geometry and very careful pure-mathematical analysis that enters into the study of the actual 6-manifolds (normally Calabi–Yau spaces; see §§1.13 and 1.14) that are considered to provide the needed Planck-scale curled-up extra 6-space dimensions that are supposed to be required for the consistency of string theory. The acceptance, by a highly knowledgeable section of the physics community, of such a hybrid of great geometrical sophistication on the one hand and a seeming disregard for an overall geometrical coherence on the other is something that I find extremely puzzling!

In the discussion of the issues of functional freedom I shall be giving in the following two sections, I shall phrase things as though the space-time is 10-dimensional, but the arguments are not specific to that particular dimension number. The classical argument of §1.11, that such extra dimensions would be catastrophically unstable, applies to any supra-dimensional theory for which there are at least 2 extra (tiny) space dimensions and subject to the 10-dimensional Einstein ($\Lambda = 0$) vacuum equations $^{10}\mathbf{G} = \mathbf{0}$ (the time-dimensionality remaining at 1). In the standard literature there are arguments that the original 26-dimensional bosonic string theory is indeed catastrophically unstable, but this is not particularly relevant to my arguments here, which apply much more generally.

The argument in §1.10 is of a completely different character from that of §1.9, that being aimed against a frequently expressed quantum-mechanical argument that the extremely tiny extra spatial dimensions would be immune from excitation at any remotely available scales of energy. Here, again the argument is not specific to the number of these additional spatial dimensions, but to be definite I shall phrase it also in terms of the currently popular 10-dimensional theory. In neither case shall I worry about supersymmetry, so that the geometrical notions that can remain reasonably clear-cut. I am assuming that the presence of supersymmetry would not drastically affect the arguments, since these arguments given could be addressing the un-supersymmetric "body" of the geometry (see §1.14).

In all these arguments, I am adopting the viewpoint that seems to be required in string theory, namely that the extra spatial dimensions are being taken to be fully dynamical. Thus, although the similarities between the string-theoretic extra dimensions and the extra dimension introduced by Kaluza and Klein are frequently pointed out by string theorists, I must re-emphasize the huge and essential difference between the original Kaluza–Klein scheme and the kind of proposal that the string theorists have in mind. In all the versions of higher-dimensional string theory that I have seen, except perhaps with regard to the 16 discrepant dimensions in the heterotic models described earlier in this section, there is, indeed, no suggestion of anything analogous to the *rotational symmetry* in the extra dimensions that was adopted in the Kaluza–Klein theory – and indeed, such symmetry is explicitly denied [Greene 1999]. Accordingly, the functional freedom in string theory is likely to be enormously excessive, namely of the form $\infty^{k\infty^9}$ for the now conventional 10-dimensional theory, rather than the $\infty^{k\infty^3}$ that we expect for a realistic physical theory. The key point is that whereas in the Kaluza–Klein theory we do *not* have the freedom to have arbitrary variations in the structure along the extra spatial ($S^1$) dimension (owing to the imposed rotational symmetry), in string theory this freedom is explicitly allowed. It is simply this that is responsible for the excessive functional freedom that comes about in string theory.

This is an issue that, with regard to classical (i.e. non-quantum) considerations, I have never seen seriously addressed by professional string theorists. On the other hand, it has been argued that such considerations are essentially irrelevant to string theory because the problem must be addressed from the point of view of quantum mechanics (or quantum field theory) rather than classical field theory. Indeed, when the matter of the excessive degrees of functional freedom in the extra 6 "small" dimensions is brought up with string theorists, it frequently tends to be dismissed with what would appear to be a somewhat hand-waving quantum-mechanical argument of a general nature that I regard as basically fallacious. I shall discuss this argument in the next section, and then (in §1.11) I put my own case that this argument is not merely thoroughly unconvincing, but that the logical implication of the string theorist's extra spatial dimensions would be a completely unstable universe, in which these extra dimensions would be expected to collapse dynamically, with disastrous consequences for the macroscopic space-time geometry that we are familiar with.

These arguments are primarily concerned with the degrees of freedom in the space-time geometry itself. There is the separate but closely related issue of the excessive functional freedom in other fields defined on higher-dimensional space-time manifolds. I shall discuss such matters briefly, towards the end of §1.10, where a certain relevance to experimental situations has sometimes been suggested. I shall touch upon a related problem in §2.11, and although the points made there are somewhat inconclusive, there are worrying issues involved, not all of which have I seen addressed elsewhere, and which could be worthy of further study.

## 1.10. QUANTUM OBSTRUCTIONS TO FUNCTIONAL FREEDOM?

In this section (and the next) I present an argument that, as it seems to me, provides a very powerful case that we cannot escape the issue of excessive functional freedom in spatially supra-dimensional theories, even in a quantum-mechanical setting. The argument is, in essence, one that I presented in 2002 at the January 2002 Cambridge conference in honour of Stephen Hawking's 60th birthday [Penrose 2003; see also TRtR, §§31.11 and 31.12] but I give it here in a more forceful form. First, in order to understand the relevant quantum issues, as they are usually presented, we need a little more from the procedures of standard quantum theory.

Let us consider a simple quantum system, such as an atom (e.g. of hydrogen) at rest. Basically, what we find is that there will be a number of discretely different *energy levels* for the atom (e.g. the electron's various allowable orbits, in a hydrogen atom). There will be a state of *minimum* energy, referred to as the *ground state*, and it is expected that any other stationary state of the atom, having greater energy, will – provided that the environment in which the atom is placed is not too "hot" (i.e. energetic) – eventually decay into its ground state, via the emission of photons. (In some situations, there may be selection rules forbidding some of these transitions, but this does not affect the general discussion.) Conversely, if sufficient external energy is made available (normally in the form of electromagnetic energy in what is called a *photon bath*, i.e. photons again, in this quantum-mechanical context) and is transferred to the atom, then the atom can be raised from a lower-energy state, e.g. the ground state, to a higher-energy state. In every case the energy $E$ of each photon that is involved will be associated with a particular frequency $\nu$, subject to Planck's famous formula $E = h\nu$ (§§1.5, 1.8, 2.2, and 3.4).

Now let us return to the issue of the supra-dimensional space-times of string theory. There is a complacency that appears to be almost universally expressed by string theorists, when pressed, that somehow the (vast!) functional freedom that resides in the extra spatial dimensions will never come into play in ordinary circumstances. This appears to result from a viewpoint that those degrees of freedom that are involved in *deforming* the geometry of the extra 6 small dimensions would be effectively immune from becoming excited, owing to the hugeness of the energy that would be required in order to activate these degrees of freedom.

In fact, there are some special deformations of the extra spatial dimensions which can be excited *without* the injection of any energy. This applies in the 10-dimensional space-time case when the 6 extra spatial dimensions are taken to be Calabi–Yau spaces; see §§1.13 and 1.14. Such deformations are called *zero-modes*, and they raise problematic issues that are certainly well appreciated by string theorists. These zero-modes do not call upon the excessive functional freedom I am concerned with here, and I shall postpone my discussion of them until §1.16. In the current section and the next I shall be concerned with those deformations which do indeed have access to the full excessive functional freedom, and which require a significant amount of energy to excite them.

To estimate the scale of energy that would be required, we bring in Planck's formula $E = h\nu$ again, and we do not go far wrong if we take for the frequency $\nu$ something of the general order of the reciprocal of the time that it would take for a signal to propagate around one of these extra dimensions. Now the "size" of these small extra dimensions depends upon which version of string theory one is concerned with. In the original 26-dimensional theory, we might well have been thinking of something of the order of $10^{-15}$ m, in which case the energy needed would have been well within the scope of the LHC (see §1.1). In the case of the newer 10-dimensional supersymmetric string theories, on the other hand, the required energy would be far higher, and would be enormously outside the range of the most powerful particle accelerator on Earth (the LHC) or of any other seriously envisaged particle accelerator. In this type of string theory, which attempts to address the issues of quantum *gravity* in a serious way, this energy would be of the rough order of the *Planck energy*, which is the energy associated with the *Planck length*, discussed briefly in §§1.1 and 1.5, and in more detail in §§3.6 and 3.10. Accordingly, it is normally argued that it would need some process involving *individual* particles accelerated to energies of at least this huge amount – which would be something like the energy released in the explosion of a sizeable artillery shell – in order to excite the degrees of freedom in the extra dimensions from their ground state. At least, for versions of string theory with extra dimensions of about this tiny scale, it is argued that these dimensions would be effectively immune from excitation by any means foreseeable at the present time.

In passing, it may be mentioned that there *are* versions of string theory, generally regarded as outside the mainstream, for which some of the extra dimensions may be taken to be as large as something like a millimetre in size. The alleged virtue of these schemes is that they might be amenable to observational test [see Arkani-Hamed et al. 1998]. But, from the point of view of functional freedom, they suffer from the particular difficulty that it should be easy to excite these "large" energies of oscillation, even with current accelerator energies, and it remains particularly unclear to me why the proponents of such schemes do not worry about the enormous excessive functional freedom that ought already to have been made manifest, in accordance with such proposals.

I have to say that, for reasons that I shall describe, I find the argument that the functional freedom in extra spatial dimensions even of *Planck*-scale size should be immune from excitement to be completely unconvincing. Accordingly, I am unable to take seriously the general conclusion that the vast store of degrees of freedom in the extra dimensions should be immune from excitation in the "ordinary" circumstances of the energies available in our current universe. There are various reasons for my serious scepticism. In the first place, we must ask why we should regard the Planck energy as being "large" in this context. I suppose that the intended picture is that the energy is to be injected by the agency of something like a highly energetic particle, as would be experienced in a particle accelerator (this being the analogue of the photon which might raise an atom from its ground state). But we must bear in mind that the picture that the string theorists are presenting is one in which the space-time – at least when the extra dimensions are in their ground state – would be taken as a *product space* $\mathcal{M} \times \mathcal{X}$ (see figure A-25 in §A.7), where $\mathcal{M}$ is something closely resembling our ordinary classical picture of a 4-dimensional space-time, and where $\mathcal{X}$ is the space of extra "small" dimensions. In the 10-dimensional version of string theory, $\mathcal{X}$ is normally taken to be a Calabi–Yau space, which is a particular kind of 6-manifold that we shall come to a little more explicitly in §§1.13 and 1.14. If the extra dimensions themselves were to be excited, the relevant "excited mode" (see §A.11) of the space-time would be exhibited as our higher-dimensional space-time having the form $\mathcal{M} \times \mathcal{X}'$, where $\mathcal{X}'$ is the perturbed (i.e. "excited") system of extra dimensions. (Of course, we have to think of $\mathcal{X}'$ as being, in some sense, a "quantum" space, rather than a classical one, but this does not seriously affect the discussion.) A point that I am making here is that in perturbing $\mathcal{M} \times \mathcal{X}$ to $\mathcal{M} \times \mathcal{X}'$, we have perturbed the *entire universe* (the entire space $\mathcal{M}$ being involved at every point of $\mathcal{X}$) so that when we are thinking of the energy required to effect this mode of perturbation as being "large" we must think of this in the context of the universe as a whole. It seems to me to be quite unreasonable to demand that the injection of this quantum of energy be necessarily effected by some fairly localized high-energy particle.

More to the point would be the consideration of some form of presumably *nonlinear* (cf. §§A.11 and 2.4) instability that affects the dynamics of the (higher-dimensional) universe as a whole. At this point it should be made clear that I do *not* regard the dynamics of the "internal" degrees of freedom, governing the behaviour of the 6 extra spatial dimensions to be *independent* of the dynamics of the "external" ones, governing the behaviour of our familiar 4-dimensional space-time. In order that both may be legitimately regarded as components of an actual overall "space-time", there should be a dynamics that governs both sets of degrees of freedom in one overall scheme (rather than, say, the former being taken as some kind of "bundle" over the latter; see §§A.7 and 1.9). Indeed, some version of Einstein equations is taken to be controlling the evolution of both sets of degrees of freedom all together, which is what I take to be the picture that string theorists have in mind in any case, at least at the classical level, where the evolution of the entire 10-space-time is taken to be well approximated by the 10-dimensional Einstein vacuum equations $^{10}\mathbf{G} = \mathbf{0}$ (see §1.11 below).

I shall come to the issue of such *classical* instabilities in §1.11; the present discussion is relevant to *quantum* issues, the upshot of which will be that we must

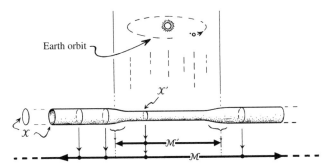

Figure 1-31: A Planck-scale energy, or so, would be needed to excite string theory's tiny compact 6-dimensional extra-dimensional space $X$, yet there is far more than this available in the Earth's motion about the Sun. Here $M$ denotes the ordinary 4-space-time of our experience and $M'$ a comparatively small part of it, encompassing the Earth's orbital motion. Only a very tiny proportion of the disturbance to the space-time from the energy in the Earth's motion would be sufficient to disturb $X$ to a very slightly different space $X'$, this being spread out over the region $M'$.

indeed look to the classical picture in order to understand the question of stability in a serious way. In the context of the dynamics of the entire universe, the Planck energy is not large at all; it is extremely tiny. The Earth's motion about the Sun, for example, involves a kinetic energy that is about a million million million million (i.e. $10^{24}$) times greater! I see no reason why some tiny fraction of this energy, which could easily far far exceed the Planck energy, should not go into disturbing the space $X$ by some very tiny amount, over some spatial region $M'$ of the scale of the Earth – or perhaps somewhat larger, encompassing the entire Earth–Sun system. Being spread over such a relatively large region, the *density* of this energy throughout $M'$ would be extremely tiny (see figure 1-31). Accordingly, the actual *geometry* of these extra spatial dimensions ($X$) would be altered hardly at all over $M'$ by a Planck-energy disturbance, and I see no reason at all why our fairly local space-time geometry $M' \times X$ should not become perturbed to something rather like $M' \times X'$, but smoothly joined to the rest of $M \times X$ outside the region $M'$, where the difference between the *geometries* $X'$ and $X$ could be absurdly tiny, and far far *smaller* than the Planck scale.

The equations governing the entire 10-space would dynamically couple those of $M$ to those of $X$, so a tiny local change in the geometry of $X$ would be an expected consequence of a fairly local ($M'$-neighbourhood) disturbance of the macroscopic space-time geometry $M$. Moreover, this coupling would be mutual.

Accordingly, the unleashing of floods of extra-dimensional degrees of freedom that are potentially there by virtue of the freedom that is in the Planck-scale geometry – involving enormous space-time curvatures, incidentally – could well have devastating effects on the macroscopic dynamics.

Whereas there are arguments presented from supersymmetry that the ground-state $\mathcal{X}$-geometry may be highly restricted (such as that it be necessarily what is called a *Calabi–Yau 6-space*; see §§1.13 and 1.14), that should *not* affect its potential to change away from such a geometry in dynamical situations. For example, whereas the $^{10}\mathbf{G} = \mathbf{0}$ Einstein equations, when applied to geometries that are restricted to have the product form $\mathcal{M} \times \mathcal{X}$, may well imply strong conditions on the geometry of $\mathcal{X}$ itself (as well as on the geometry of $\mathcal{M}$), this very special product form would not be expected to persist in the general dynamical situation – and, indeed, *almost all* the functional freedom would be expressed in solutions that do *not* have this product form (see §A.11). Accordingly, whatever criteria are being used in order to restrict the extra dimensions to have such a particular geometrical structure (e.g. Calabi–Yau) when in their ground state, we cannot expect that this would be maintained in fully dynamical situations.

Some clarification is also needed at this stage, concerning the comparison with atomic quantum transitions that I made earlier, since I have glossed over a technical issue in my discussion at the beginning of this section when I considered an atom at rest. To be exactly at rest, the atom's state (wave function) must, technically, be spread uniformly over the entire universe (since to be at rest, it would have to have zero momentum, from which this uniformity follows; see §§2.13 and 4.2), in a seemingly similar way to $\mathcal{X}$ (or $\mathcal{X}'$) being spread uniformly over the universe in the product $\mathcal{M} \times \mathcal{X}$. Does that in any way invalidate my discussion above? I do not see why it should. The processes involving single atoms must nevertheless be thought of as *localized* events, where a change in the state of an atom would be effected by some local process such as an encounter with some other reasonably localized entity, such as a photon. The fact that a stationary state (or time-independent wave function) of an atom ought, technically, to be thought of as spread out over the entire universe is an irrelevance to the way that calculations are actually performed, as normally one simply considers all spatial considerations to be taken relative to the *mass centre* of the system, and the aforementioned difficulty disappears.

The situation with regard to perturbations of the Planck-scale space $\mathcal{X}$ is completely different, however, because here the ground state of $\mathcal{X}$ is, by its very nature, of necessity *not* localized at any particular place in our ordinary space-time $\mathcal{M}$, being supposed to be omnipresent, permeating the structure of space-time throughout the entire universe. The geometrical quantum state of $\mathcal{X}$ is supposed to influence the detailed physics that is going on in the most remote galaxy, just as much as here on Earth. The string theorist's argument that a Planck-scale energy would be far too great, in relation to what is available, to be able to excite $\mathcal{X}$ seems to me to be inappropriate on various counts. Not only are such energies amply available through non-localized means (e.g. the Earth's motion), but if we were to imagine that $\mathcal{X}$ were actually to be converted to an excited state $\mathcal{X}'$ by such a particle transition (perhaps owing to some advanced technology making a Planck-energy particle accelerator), leading to $\mathcal{M} \times \mathcal{X}'$ for the new state of the universe, this would be clearly absurd, as we could not expect the physics on the Andromeda Galaxy to be instantly changed by such an event here on Earth! We should be thinking more in terms of a much milder event in the vicinity of the Earth propagating outwards with the speed of light. Such things would be described much more plausibly by nonlinear classical equations, rather than abrupt quantum transitions.

In view of such considerations, I should return to a point touched upon earlier and try to see in what way a Planck-scale quantum of energy, spread over some rather large region $\mathcal{M}'$ of $\mathcal{M}$, might be expected to affect the geometry of the space $\mathcal{X}$ over this region. As noted above, $\mathcal{X}$ would be affected very little at all, and the bigger the region $\mathcal{M}'$, the smaller need be the change in $\mathcal{X}$ over this region, where we are thinking of this change as being effected by a spread-out event of Planck energy. Accordingly, if we are to examine changes in the shape or size of $\mathcal{X}$ that are actually significant, thereby changing $\mathcal{X}$ to a space $\mathcal{X}^*$ differing appreciably from $\mathcal{X}$, then we are led to consider energies that are very much larger than Planck scale (such energies, of course, being very abundant in the physical universe that we know, as with the Earth's motion about the Sun). These are not provided by just a *single* "minimal" quantum of Planck-scale energy, but by the injection of a succession of vast numbers of quanta of changes to $\mathcal{X}$. In order to effect a transformation from $\mathcal{X}$ to a significantly different $\mathcal{X}^*$ over some large region, one would indeed need to involve an enormous number of such quanta (perhaps of Planck scale or larger). Now it is usual to assume, when we consider effects requiring such vast numbers of quanta, that such effects are best described purely classically (i.e. without quantum mechanics).

In fact, as we shall be seeing in chapter 2, this issue of how an appearance of classicality might arise from a multiplicity of quantum events actually raises a number of deep questions concerning the way in which the quantum world relates to the classical one. It is, indeed, an interesting (and contentious) question as to whether or not (the appearance of) classicality comes about simply because

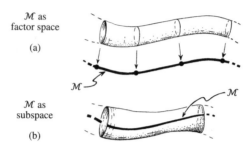

$\mathcal{M}$ as
factor space

(a)

$\mathcal{M}$

$\mathcal{M}$

$\mathcal{M}$

$\mathcal{M}$ as
subspace

(b)

Figure 1-32: The contrast between (a) $\mathcal{M}$ being a factor space and (b) $\mathcal{M}$ being a subspace.

large numbers of quanta are involved or from some other criterion, and I shall have to return to this specific matter in §2.13. For the purposes of the present argument, however, such subtleties seem not to be particularly relevant, and I merely point out that it ought to be regarded as reasonable to use *classical* arguments to address the issue of the perturbations of the space-time $\mathcal{M} \times \mathcal{X}$ which actually do change the space $\mathcal{X}$ in a significant way. This raises its own severe difficulties with the extra spatial dimensions, as we shall see shortly, in the next section.

But before coming to these difficulties – referring specifically to the particular form of spatial supra-dimensionality arising from string theory – I feel that it is helpful to make a comparison with a type of experimental situation that is sometimes bought up as analogous to that considered above. This situation is exemplified by what is known as the *quantum Hall effect* [von Klitzing et al. 1980; von Klitzing 1983], which refers to a well-established 2-space-dimensional quantum phenomenon taking place within ordinary 3-space-dimensional physics. Here there is a large energy barrier constraining the relevant system to a 2-dimensional surface, and the quantum physics of this lower-dimensional world can be apparently oblivious to the extra third space dimension because its contents do not possess the energy sufficient to surmount this energy barrier. Thus, it is sometimes argued that we could take this kind of example as providing an analogy with what is supposed to be happening with the supra-dimensionality of string theory, where our ordinary 3-space-dimensional physics is likewise taken to be oblivious to the 9-space-dimensional ambient world that it is supposedly taking place within, by virtue of a large energy barrier.

This, however, is a completely false analogy. For the above example is more appropriate to the brane-world viewpoint of §1.16, where the lower-dimensional space is a *subspace* of the higher-dimensional one, rather than the *factor space* that is relevant to the standard string theory pictures that I have been discussing (as exemplified by the above expression $\mathcal{M} \times \mathcal{X}$, of which $\mathcal{M}$ is a factor). See §A.7 and figure 1-32. The role of $\mathcal{M}$ as a factor space, as opposed to a subspace, is again pertinent to the argument of the following section. The subspace picture is, however, relevant to the very different brane-world picture that will be described a little more fully in §1.16.

## 1.11. CLASSICAL INSTABILITY OF HIGHER-DIMENSIONAL STRING THEORY

Let us now return to the issue, addressed in §1.10, of the stability of a classical space-time of the form $\mathcal{S} = \mathcal{M} \times \mathcal{X}$, where $\mathcal{X}$ is a compact space of tiny size. Although my arguments are not very specific about the nature of the space $\mathcal{X}$, I shall phrase these arguments in terms of the versions of string theory where $\mathcal{X}$ would be the type of compact 6-dimensional space known as a *Calabi–Yau manifold* that we shall be coming to in a little more detail in §§1.13 and 1.14, $\mathcal{S}$ then being a 10-dimensional space-time. There would be some elements of supersymmetry involved (§1.14), but these will not play any role in the classical discussion that I shall give here (which may be considered to be applying to the "body" part of the system (cf. §1.14)), so I am taking the liberty of ignoring supersymmetry for the time being, postponing our considerations of it until §1.14. In fact, basically all that I shall need about $\mathcal{X}$ is that it should be at least 2-dimensional, which is certainly what is intended in current string theory, although I shall need the space-time $\mathcal{S}$ to satisfy some field equations.

I have mentioned earlier (§1.10) that, according to string theory, there should indeed be field equations satisfied by the metric assigned to this higher-dimensional $\mathcal{S}$. To a first approximation, we can consider this set of equations for $\mathcal{S}$ to be the *Einstein vacuum equations* $^{10}\mathbf{G} = \mathbf{0}$, where $^{10}\mathbf{G}$ is the Einstein tensor constructed from the 10-metric of $\mathcal{S}$. These equations are imposed for the space-time $\mathcal{S}$ in which the strings reside in order to evade an *anomaly* – going beyond the anomaly referred to in §1.6 which led theorists to increase the space-time dimensionality. In fact, the "$^{10}\mathbf{G}$" in $^{10}\mathbf{G} = \mathbf{0}$ is only the first term in a power series in a small quantity $\alpha'$ referred to as the *string constant*, an extremely tiny area parameter normally taken to be only slightly larger than the square of the Planck length (see §1.5):

$$\alpha' \approx 10^{-68} \text{ m}^2.$$

Thus, we have field equations for $\mathcal{S}$ that can be expressed as some kind of power series (§A.10)

$$0 = {}^{10}\mathbf{G} + \alpha'\mathbf{H} + \alpha'^2\mathbf{J} + \alpha'^3\mathbf{K} + \cdots,$$

where $\mathbf{H}$, $\mathbf{J}$, $\mathbf{K}$, etc., would be expressions constructed from the Riemann curvature and its various higher derivatives. Because of the extreme smallness of $\alpha'$, however, the higher-order terms are usually ignored in specific versions of string theory that are put forward (though the validity of doing this is in some doubt, as there is no information concerning the convergence or ultimate behaviour of the series (cf. §§A.10 and A.11)). In particular, the Calabi–Yau spaces referred to above (see §§1.13 and 1.14) can be explicitly taken to satisfy the 6-space equation "${}^6\mathbf{G} = \mathbf{0}$", leading to the corresponding 10-space equation ${}^{10}\mathbf{G} = \mathbf{0}$ holding for the product space $\mathcal{M} \times \mathcal{X}$, provided that the standard Einstein vacuum equation ${}^4\mathbf{G} = \mathbf{0}$ is assumed also to be true for $\mathcal{M}$ (which would be reasonable for the *vacuum* "ground state" of the matter fields).[7] In all this, I am here taking the Einstein equations without any $\Lambda$-term (see §§1.1 and 3.1). The cosmological constant would indeed be completely negligible at the scales under consideration here.

Let us suppose, in accordance with the above, that the vacuum equations ${}^{10}\mathbf{G} = \mathbf{0}$ indeed hold for $\mathcal{S} = \mathcal{M} \times \mathcal{X}$. We are interested in what happens if small perturbations are applied to the (e.g. Calabi–Yau) space of "extra dimensions" $\mathcal{X}$. An important point should be made here about the nature of the perturbation that I shall be considering. There are many discussions, within the string-theory community, of perturbations that deform one example of a Calabi–Yau space into another slightly different one by altering the *moduli* that we shall be coming to in §1.16, these defining the specific shape of a Calabi–Yau manifold within a particular topological class. Among the disturbances that alter the values of such moduli are the *zero-modes* referred to in §1.10. In this section, I am not particularly concerned with deformations of this kind, which do not take us outside the family of Calabi–Yau spaces. In conventional string theory it is normally considered that one must remain within this family, because these spaces are considered to be *stable* because of supersymmetry criteria that restrict the 6 extra spatial dimensions to constitute such a manifold. However, these considerations

---

[7] The Einstein tensor ${}^{m,n}\mathbf{G}$ of the product $\mathcal{M} \times \mathcal{N}$ of two (pseudo-)Riemannian spaces $\mathcal{M}$ and $\mathcal{N}$ (of respective dimensions $m, n$) can be expressed as a *direct sum* ${}^m\mathbf{G} \oplus {}^n\mathbf{G}$ of the respective individual Einstein tensors ${}^m\mathbf{G}$ and ${}^n\mathbf{G}$, where the (pseudo-)metric of $\mathcal{M} \times \mathcal{N}$ is defined to be the corresponding direct sum ${}^{m,n}\mathbf{g} = {}^m\mathbf{g} \oplus {}^n\mathbf{g}$ of the respective individual (pseudo-)metrics ${}^m\mathbf{g}$ and ${}^n\mathbf{g}$ of $\mathcal{M}$ and $\mathcal{N}$. (See Guillemin and Pollack [1974]. For a more in-depth approach, see Besse [1987].) It follows that $\mathcal{M} \times \mathcal{N}$ has vanishing Einstein tensor if and only if *both* of $\mathcal{M}$ and $\mathcal{N}$ have vanishing Einstein tensor.

of "stability" are aimed at showing that these are the only 6-spaces that satisfy the required supersymmetry criteria. The normal idea of stability would be that a small perturbation away from a Calabi–Yau will return it to such a space; but the possibility is not considered that such a perturbation might diverge away from this family, ultimately leading to something *singular*, where no smooth metric exists. In fact, it is this latter possibility of a runaway evolution to such a singular configuration that is the apparent implication of the following arguments.

In order to study this, it will be easiest if we first explicitly take the basic case where $\mathcal{M}$ is not perturbed at all; that is to say $\mathcal{M} = \mathbb{M}$, where $\mathbb{M}$ is the flat Minkowski 4-space of *special* relativity (§1.7). $\mathbb{M}$ being flat, it can be re-expressed as the product space

$$\mathbb{M} = \mathbb{E}^3 \times \mathbb{E}^1$$

(see §A.4, figure A-25; this is basically just grouping the coordinates $x, y, z, t$ first as $(x, y, z)$ and then $t$). The Euclidean 3-space $\mathbb{E}^3$ is ordinary space (coordinates $x, y, z$) and the 1-dimensional Euclidean space $\mathbb{E}^1$ is ordinary time (coordinate $t$), the latter being actually just a copy of the real line $\mathbb{R}$. With $\mathbb{M}$ written this way, the entire (unperturbed) 10-dimensional space-time $\mathcal{S}$ can be expressed ($\mathbb{M}$ and $\mathcal{X}$ being factor spaces of $\mathcal{S}$) as

$$\begin{aligned} \mathcal{S} &= \mathbb{M} \times \mathcal{X} \\ &= \mathbb{E}^3 \times \mathbb{E}^1 \times \mathcal{X} \\ &= \mathbb{E}^3 \times \mathcal{Z}, \end{aligned}$$

simply by regrouping the coordinates, where $\mathcal{Z}$ is the 7-dimensional space-time

$$\mathcal{Z} = \mathbb{E}^1 \times \mathcal{X}$$

(the coordinates for $\mathcal{Z}$ being: $t$ first, and then those for $\mathcal{X}$).

I wish to consider a (small but not infinitesimally small) perturbation of the (e.g. Calabi–Yau) 6-space $\mathcal{X}$ to a new space $\mathcal{X}^*$, at $t = 0$, where we can think of this as propagating in the time-direction provided by $\mathbb{E}^1$ (with time coordinate $t$), to give us an evolved 7-dimensional space-time $\mathcal{Z}^*$. For the moment, I assume that the perturbation applies to $\mathcal{X}$ alone, with the Euclidean external 3-space $\mathbb{E}^3$ remaining unperturbed. This is perfectly consistent with the evolution equations, but since this perturbation will be expected to change the 6-geometry of $\mathcal{X}^*$ in some way as time progresses, we do not expect $\mathcal{Z}^*$ to retain a product form such as $\mathbb{E}^1 \times \mathcal{X}^*$, the exact 7-geometry of $\mathcal{Z}^*$ being governed by the Einstein equations $^7\mathbf{G} = \mathbf{0}$. The entire space-time $\mathcal{S}$ would, however, still retain the product $\mathbb{E}^3 \times \mathcal{Z}^*$

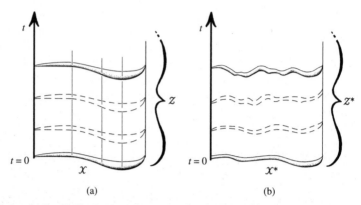

Figure 1-33: (a) The tiny compact space $\mathcal{X}$, when a Calabi–Yau space remains the same upon time evolution, but (b) when slightly perturbed to $\mathcal{X}^*$ it undergoes an evolution to something different.

as it evolves because the full Einstein equations $^{10}\mathbf{G} = \mathbf{0}$ will be satisfied by this product form if $\mathcal{Z}^*$ satisfies $^7\mathbf{G} = \mathbf{0}$, because $^3\mathbf{G} = \mathbf{0}$ certainly holds for the flat space $\mathbb{E}^3$.

The 6-space $\mathcal{X}^*$ is being taken as a $t = 0$ initial value surface for the evolution $\mathcal{Z}^*$ (figure 1-33). The equations $^7\mathbf{G} = \mathbf{0}$ then propagate this perturbation in the future time-direction (given by $t > 0$). There are certain constraint equations that have to be satisfied on $\mathcal{X}$, and it can be a somewhat delicate issue to make sure, in rigorous mathematical terms, that these constraint equations can actually be satisfied everywhere throughout the compact space $\mathcal{X}^*$. Nonetheless, the functional freedom that we expect for such initial perturbations of $\mathcal{X}$ is

$$\infty^{28\infty^6},$$

where the "28" in this expression comes about by putting $n = 7$ into the expression $n(n - 3)$, this being the number of independent initial data components per point on an initial $(n - 1)$-surface, for an $n$-space with vanishing Einstein tensor, the "6" being the dimensionality of the initial 6-surface $\mathcal{X}^*$ [Wald 1984]. This includes both *intrinsic* perturbations of $\mathcal{X}$ itself and *extrinsic* perturbations of how $\mathcal{X}$ is embedded in $\mathcal{Z}$. This classical freedom is, of course, vastly in excess of the functional freedom $\infty^{k\infty^3}$ that we expect for a physical theory appropriate to the kind of activity experienced in our perceived 3-dimensional world.

But matters are far more serious than this, because practically all such perturbations will lead to an evolution for $\mathcal{Z}$ that is *singular* (see figure 1-34). This

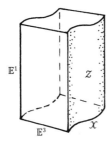

Figure 1-34: The classical instability of the extra dimensions of string theory. The evolution $Z^*$ of the perturbed extra-dimensional 6-space $X^*$ is almost certain to be singular, by a 1970 theorem due to Hawking and the author.

means, in effect, that the extra dimensions must be expected to crumple up to something where curvatures diverge to infinity and further evolution of the classical equations becomes impossible. This conclusion follows from mathematical *singularity theorems* that were proved in the late 1960s – most particularly one that Stephen Hawking and I had established a little before 1970 [Hawking and Penrose 1970] – which showed, among other things, that almost any $n$-dimensional space-time ($n \geqslant 3$) containing a compact spacelike ($n - 1$)-surface (here the initial Calabi–Yau 6-space $X^*$), but which does not contain closed timelike loops, must evolve to a space-time singularity if its Einstein tensor $^n\mathbf{G}$ satisfies an energy (non-negativity) condition called the *strong energy condition* (certainly satisfied here, since $^7\mathbf{G} = \mathbf{0}$ throughout $Z^*$). The provisos "almost" and absence of "closed timelike loops" can be ignored here, since these possible let-outs could occur, if at all, only in exceptional circumstances of much lower functional freedom than would arise in a generic perturbation of $Z$.

A technical point should be made here. This theorem does not actually assert that curvatures must *diverge to infinity*, but merely that, in general, the evolution cannot be extended beyond a certain point. Although there are alternative things that can in principle happen in exceptional cases, it is to be expected that the general reason for the impossibility of continuing the evolution is that curvatures indeed *do* diverge [Clarke 1993]. Another point of relevance here is that the strong energy condition being assumed here, although automatically satisfied by $^7\mathbf{G} = \mathbf{0}$, certainly cannot be guaranteed if we are to consider what happens with the higher-order terms in the power series in $\alpha'$, referred to above. Yet, most current considerations of string theory seem to operate at the level where these higher-order terms in $\alpha'$ are ignored, and $X$ is indeed taken to be a Calabi–Yau space. What this singularity theorem appears to be telling us is that so long as the perturbations of the extra dimensions can be treated *classically* – as indeed appears to be a reasonable thing to do, as a clear conclusion of our earlier considerations in §1.10 – then we must expect a violent instability in the 6 extra spatial dimensions, in which they crumple up and *approach* a singular state. Just prior to this disaster, one might have to take the higher-order terms in $\alpha'$, or further quantum considerations, into serious account. Depending upon the scale of the perturbation, one could well anticipate that this "crumple time" would be likely to occur in some extremely tiny fraction of a second, where one must bear in mind that the *Planck time* (the time that it takes for light to travel over a Planck distance; see §1.5) is of the order of $10^{-43}$ seconds! Whatever the extra dimensions might crumple to, the observed physics is not likely to be other than drastically affected. This is hardly a comfortable picture of the 10-dimensional space-time that string theorists have been proposing for our universe.

There is another issue that needs to be pointed out here, namely that the perturbations considered above are those which affect *only* the extra 6 dimensions, leaving the macroscopic dimensions (here the Euclidean 3-space $\mathbb{E}^3$) untouched. Indeed, there would be *far* more functional freedom (namely $\infty^{70\infty^9}$) in perturbations at the entire spatial 9-space $\mathbb{E}^3 \times \mathcal{X}$ than those that just affect $\mathcal{X}$ alone, these having a functional freedom $\infty^{28\infty^6}$. It appears to be possible to modify the above argument so that the same theorem [Hawking and Penrose 1970] still applies, but in a more complicated way, still leading to the same singular conclusion, but now to the entire space-time [TRtR, note 31.46, p. 932]. Quite apart from this, it is clear that any perturbations of the macroscopic 4-space that are at all comparable with those under consideration with regard to the extra 6 dimensions would be disastrous for ordinary physics, since such tiny curvatures as those involved in $\mathcal{X}$ are simply not experienced in observed phenomena. This does raise an awkward issue that is in any case an unresolved difficulty in modern string theory, namely how it is that such extraordinarily different curvature scales can somehow coexist without significantly affecting one another. This disturbing issue will be raised again in §2.11.

## 1.12. THE FASHIONABLE STATUS OF STRING THEORY

At this stage, the reader may have become puzzled as to why string theory is being taken so seriously by such a huge fraction of the community of extremely able theoretical physicists – particularly by those directly concerned with moving forward to a deeper understanding of the underlying physics of the world in which we actually live. If string theory (and its later developments) indeed leads us to a higher-dimensional space-time picture which appears to be at such odds with the physics that we know, then why does it continue to have such a fashionable status among this extraordinarily large and exceptionably able community of theoretical physicists? Just *how* fashionable it actually still is, I shall come to in a moment. But accepting that it has this status, we must ask why string theorists seem to be so unaffected by arguments against the physical plausibility of higher-dimensional space-time such as those outlined in §§1.10 and 1.11. Why, indeed, does its highly fashionable status appear to be essentially untouched by such arguments against its plausibility?

The arguments that I have been outlining in the previous two sections are essentially ones that I first put forward as the final talk at a workshop part of a meeting in honour of Stephen Hawking's 60th birthday held in Cambridge, England, in January 2002 [Penrose 2003]. There were several leading string theorists present at this lecture, and the next day some of them (particularly Gabriele Veneziano and Michael Green) did raise a few issues with me concerning the arguments that I had presented. Yet since that time there has been extraordinarily little response or counter-argument – and certainly no public refutation of the ideas I had presented. Perhaps the most telling reaction, made to me over lunch on the day following my lecture, was the comment of Leonard Susskind (as recalled by me as closely to verbatim as I can manage):

You are completely right, of course, but totally misguided!

I am not really sure how to interpret this remark, but I take it that what is being expressed is something like the following. While string theorists may be prepared to admit that there are some still unresolved mathematical difficulties appearing to impede the development of their theory – pretty well all of these being already fully appreciated within the string community – such issues are mere technicalities which should not be allowed to obstruct real progress. They would argue that these technicalities must have little genuine importance because string theory is fundamentally on the right lines, and those working in the subject should not be wasting their time on such mathematical niceties nor even bringing such unimportant distractions to light, at this stage of the subject's development, lest they deflect the current or potential string community from its march towards the full realization of all its fundamental goals.

This kind of disregard for overall mathematical consistency seems to me to be especially extraordinary for a theory which is in any case largely mathematically driven (as I shall be explaining shortly). Moreover, the particular objections I have been raising are certainly very far from being the only mathematical obstructions to the coherent development of string theory as a believable physical theory, as we shall be seeing in §1.16. Even the claimed *finiteness* of the string calculations that are supposed to supplant the divergent Feynman diagrams referred to in §1.5 is very far from being mathematically established [Smolin 2006, particularly pp. 278–81]. The apparent lack of genuine interest in clear mathematical demonstration is aptly illustrated by comments such as the following one, which appears to be due to the Nobel laureate David Gross:

> String theory is so obviously finite that if anyone were to present a mathematical proof of this, then I wouldn't be interested in reading it.

Abhay Ashtekar, who provided me with this quote, was not completely certain that Gross was its originator. Curiously, however, when I was giving a lecture in Warsaw in around 2005 on these matters, it was just at the point when I exhibited this quote that David Gross actually entered the room! So I asked him if he was indeed the quote's author. He did not deny it, but then confessed to having now become interested in seeing such a proof.

The hope that string theory should prove to be a *finite* theory, free of the divergences of conventional QFT that arise from the standard analysis in terms of Feynman diagrams (and other mathematical techniques), had certainly provided one of the fundamental drives behind the theory. It is basically the fact that, in the string calculations that supplant those of Feynman diagrams in accordance with figure 1-11 in §1.6, we can make use of the "complex magic" of Riemann surfaces (§§A.10 and 1.6). But even the expected finiteness of an individual amplitude (see §§1.5 and 2.6) arising from a particular string topological arrangement does not, in itself, provide us with a finite theory, since each string topology provides only one term of a *series* of string pictures of increasing topological complication. Unfortunately, even if each individual topological term is indeed finite – as appears to be a basic belief of string theorists, as exemplified by the above quote – the series as a whole is expected to *diverge*, as was shown by Gross himself [see Gross and Periwal 1988]. Mathematically awkward as this may be, its divergence actually tends to be regarded as a good thing by string theorists, merely confirming that the power-series expansion that this provides is being taken "about the wrong point" (see §A.10) thereby illustrating a particular expected feature of string amplitudes. Nevertheless, this awkwardness does seem to invalidate the hope of string theory directly providing us with a finite procedure for calculating the amplitudes of QFT.

So how fashionable *is* string theory? We may get some feeling of its popularity as an approach to quantum gravity (at least in around 1997) from a little survey that was presented as part of a talk given by Carlo Rovelli at the International Congress on General Relativity and Gravitation. This was in Pune, India, in December 1997, and the talk was about the different approaches to quantum gravity that were current at that time. It should be remarked that Rovelli is one of the originators of the rival loop-variable approach to quantum gravity theory [Rovelli 2004; see also TRtR, chapter 32]. He made no pretence at being an unbiased social scientist, and one could, no doubt, query the implications of the survey as a rigorous study, but that's hardly important, so I'm not going to worry about it here. What he did was to examine the Los Angeles archives and find out how many papers on each approach to quantum gravity there were in the previous year. The result of his survey was as follows:

| | |
|---|---|
| String theory: | 69 |
| Loop quantum gravity: | 25 |
| QFT in curved spaces: | 8 |
| Lattice approaches: | 7 |
| Euclidean quantum gravity: | 3 |
| Non-commutative geometry: | 3 |
| Quantum cosmology: | 1 |
| Twistors: | 1 |
| Others: | 6 |

We notice from this that not only did string theory appear to be by far the most popular approach to quantum gravity, but its popularity easily exceeded that of all the others put together.

It turns out that for several subsequent years, Rovelli followed up this survey with a slightly more limited one with regard to topics, but covering all the years from 2000 to 2012, where he traced the relative popularity of just the three quantum-gravity approaches, string theory, loop quantum gravity, and twistor theory (figure 1-35). According to this plot, string theory seems to be very much holding its own as a popular theory – perhaps with a peak at around 2007, but with not a huge fall-off since then. The main change over the years appears to be a steady increase in interest in loop quantum gravity. The noticeable but modest rise in interest in twistor theory from around the beginning of 2004 might be due to factors that I shall refer to in §4.1. However, it may not be appropriate to read too much into these trends.

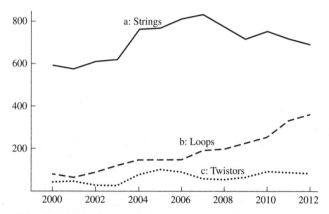

Figure 1-35: Carlo Rovelli's survey, from the Los Angeles archives, of the popularity of three approaches to quantum gravity for the years 2000–12: strings, loops, and twistors.

People assured me after I showed Rovelli's 1997 table at my 2003 Princeton lecture that the proportion of string papers would have been much larger by that date, which I was well able to believe. Indeed we appear to see a significant rise in popularity at round about that time. Also, I suspected that my own baby, namely twistor theory (see §4.1), was rather fortunate to obtain its "−1" in 1997, and that "0" would have been a more likely score around that time. I suspect that, today, non-commutative geometry would score more highly than the "3" that it obtained then, but Rovelli's subsequent survey does not include that topic. Of course, I should make absolutely clear that tables of this kind tell us almost nothing about the probable closeness of any particular proposal to the actualities of nature, giving merely some guide as to how fashionable each of the various schemes might be. Moreover, as I shall explain in chapter 3 and in §4.2, it is my own view that *none* of the current approaches to quantum gravity is likely to provide us with a theory that is fully in accordance with nature's own way of combining together the two great schemes of general relativity and quantum mechanics, for the overriding reason that quantum gravity is, in my view, not really what we should be looking for! That term rather implies that we should be striving for an actual quantum theory that applies to the gravitational field, whereas it is my view that there should be a reaction, of some sort, back on the very structure of quantum mechanics, when gravity is involved. Accordingly, the resulting theory would not be strictly a quantum theory but something that *deviates* from current quantization procedures (see §2.13).

Yet the drive to find the appropriate quantum-gravity theory is a very real one. Many physicists, and especially aspiring young graduate students, have strong desires to make significant headway towards the very creditable goal of uniting those two great revolutions of the twentieth century: the strange but magnificent quantum mechanics and Einstein's extraordinary curved-space-time theory of gravitation. This goal is usually referred to simply as *quantum gravity*, in which the rules of standard quantum (field) theory are applied to gravitational theory (although I shall be presenting my rather different slant on this unification in §§2.13 and 4.2). While it may indeed be reasonably argued that none of the current theories is yet close to this goal, the proponents of string theory seem to have been confident enough to propagate the view expressed in their own conviction that string theory is the only game in town. As a leading string theorist, Joseph Polchinski [1999], has put it:

there are no alternatives ... all good ideas are part of string theory.

On the other hand, one must bear in mind that string theory is a product of one school of thought, with regard to research into theoretical physics. It is a particular culture that has developed from particle physics and the perspective of quantum field theory, where the problematic issues that loom large tend to those of the nature of rendering divergent expressions finite. There is quite a different culture that has been developed by those whose backgrounds have been more directly to do with Einstein's general relativity. Here, particular importance is attached to maintaining general principles, most notably the *principles of equivalence* (between the effects of acceleration and of a gravitational field; see §§3.7 and 4.2) and of *general covariance* (see §§A.5 and 1.7), which are foundational to Einstein's theory. The loop-variable approach to quantum gravity, for example, is foundationally based on the primacy of general covariance, whereas string theory would appear to ignore it almost completely!

I believe that considerations like Rovelli's surveys above give us only the vaguest impression of the dominance of string theory and its descendants (see §§1.13 and 1.15) among theoretical workers trying to probe the foundations of physics. In the vast majority of physics departments and physics institutes right across the globe, the theoreticians among them are likely to include a substantial component primarily concerned either with string theory or with one of its various offshoots. Although this dominance may have tapered off to some degree in recent years, students wishing to do research into foundational physics, such as quantum gravity, are still mainly guided into string theory (or other higher-dimensional relatives), very often at the expense of other approaches with at least as much promise. However, the other approaches are not nearly so well known, and even students with no great desire to do string theory find it hard to pursue such alternative options, particularly because of the paucity of potential relevant supervisors (although loop quantum gravity appears to have gained significant ground, in this respect, in recent years). The very mechanisms for furthering careers in the theoretical physics community (and no doubt in many other research areas also) are strongly biased in favour of the further propagation of already fashionable areas, such factors serving to promulgate string theory's already highly fashionable status.

A strong additional component in the propagation of fashion is funding. Committees set up to judge the relative values of research projects in different areas are highly likely to be strongly influenced by the size of current interest in each area. Indeed, many committee members themselves are likely to have worked in such an area if it is extremely popular – perhaps even being partly responsible for its popularity – and would be consequently more likely to value research in fashionable rather than unfashionable areas. This contributes to an inherent instability, serving to magnify the global interest in areas that are already fashionable, at the expense of those which are not. Moreover, modern electronic communication and jet travel provide great opportunities for the rapid propagation of already fashionable ideas, particularly in a highly competitive world, where the need to get results out rapidly favours those who are quick at building on the results of others and developing an already active area, as opposed to those who wish to break free of the mould, perhaps having to think long and hard to develop ideas that deviate significantly from the mainstream.

I gather, however, that there is now beginning to be a feeling, at least among certain physics departments in the United States, that some kind of saturation point has been reached, and that other topics ought to have more representation among newly recruited faculty members. Is it possible that the string-theory fashion is beginning to taper off? It is my own view that the representation of string theory has for many years been excessive. Undoubtedly, there is enough in the theory which is fascinating and well worth continuing development. This is particularly true with regard to its impact on numerous areas of mathematics, where the effect has certainly been very positive. But its stranglehold on developments in fundamental physics has been stultifying, and has in my view hindered the development of other areas that might have had more promise of ultimate success. I believe that it has provided a remarkable example, perhaps comparable with some of the major misconceptions of the past discussed in §1.2, where the forces of fashion have had an undue influence on the developments of basic physics.

Having said all this, I should make clear that there can, nevertheless, be some genuine merits in pursuing ideas which are fashionable. In a general way, ideas will remain fashionable in science only if they are both mathematically cohesive and well supported by observation. Whether this is true for string theory, however, is debatable at best. Yet, with quantum gravity, it is normally agreed that observational tests lie far beyond the capabilities of any presently feasible experiment, and consequently researchers must rely almost entirely on internal theoretical reasoning, without much guidance from nature herself, where the reason commonly cited for this pessimism is the magnitude of the Planck energy which, with regard to particle interactions, is enormously beyond anything that could be approached with current technology (see §§1.1, 1.5, and 1.10). Quantum-gravity theorists, despairing of ever finding observational confirmation – or refutation – of their respective theories, thus find themselves driven to rely on *mathematical* desiderata, and it is the perception of a mathematical power and elegance that provides the basic criterion for a proposal's substance and plausibility.

Theories of this kind, beyond testing by current experimentation, lie outside the normal scientific criteria of judgement through experiment, and judgement through mathematics (in addition to some basic physical motivation) begins to acquire a considerably greater importance. Of course, the situation would be very different if some such scheme were found, providing not merely a beautifully coherent mathematical structure but also a proposal predictive of new physical phenomena – these being subsequently found to be in accurate accord with observation. Indeed, the type of scheme I shall be promoting in §2.13 (and §4.2), in which the unification of gravitational theory with quantum theory would entail some modifications of the latter, might well be subject to experimental tests that do not lie far beyond the technical possibilities of today. If a properly experimentally testable version of quantum gravity were thereby to emerge, and to be supported by such experiments, one might well expect that a good degree of scientific appreciation would follow, and rightly so. This, however, would then not be a matter of what I am calling fashion but of genuine scientific progress. In string theory, we have not seen anything of this kind.

One might also have expected, in the absence of clear-cut experiments, that quantum-gravity proposals that do not hang together mathematically would be unlikely to survive, so the fashionable status of such a proposal could therefore be taken as indicative of the proposal's worth. Yet, I believe that it is dangerous to attach too much credence to this kind of purely mathematical judgement. Mathematicians tend to be not much concerned with the plausibility – or even the consistency – of a physical theory *as a contribution to our understanding of the physical world* and, instead, will judge such contributions in terms of their value in providing input of new mathematical concepts and powerful techniques for accessing mathematical truth.

This factor has been particularly important for string theory and has no doubt played a distinctive role in maintaining its fashionable status. Indeed, there has been a great deal of very remarkable input from string-theoretic ideas into various areas of pure mathematics. A very striking example is addressed in the following quotation from an email message to me in the early 2000s from the distinguished mathematician Richard Thomas, of Imperial College London, in response to a question I raised with him about the mathematical status of a piece of difficult mathematics that had come out of string-theoretic considerations [see Candelas et al. 1991]:

> I can't emphasise enough how deep some of these dualities are, they constantly surprise us with new predictions. They show up structure never thought possible. Mathematicians confidently predicted several times that these things weren't possible, but people like Candelas, de la Ossa, et al. have shown this to be wrong. Every prediction made, suitably interpreted mathematically, has turned out to be correct. And *not* for any *conceptual* maths reason so far – we have no idea why they're true, we just compute both sides independently and indeed find the same structures, symmetries and answers on both sides. To a mathematician these things cannot be coincidence, they must come from a higher reason. And that reason is the *assumption that this big mathematical theory describes nature...*

The specific type of issue that Thomas was referring to has to do with some profound mathematical ideas that arose from the way that a certain problem confronting string theory's development came to be resolved. This is exhibited in a very remarkable story that we shall come to towards the end of the next section.

## 1.13. M-THEORY

A special virtue that had been claimed for string theory in its early days was that it ought to provide us with a *unique* scheme for the physics of the world. This hope, although having been proclaimed for many years, seemed to have been dashed when it emerged that there were *five* different kinds of string theory, these being referred to as *Type I*, *Type IIA*, *Type IIB*, *Heterotic O(32)*, and *Heterotic* $E_8 \times E_8$ (terms which I shall not even attempt to explain here [Greene 1999] though some discussion of the heterotic models was given in §1.9). This multiplicity of possibilities was disturbing to the string theorists. However, in a highly influential lecture given in 1995 at the University of Southern California, the outstanding theoretician Edward Witten brought together a remarkable family of ideas in which certain transformations referred to as *dualities* would point to a subtle kind of *equivalence* between these different string theories. This was later heralded as ushering in "the second string revolution" (the first having been the work surrounding that of Green and Schwarz referred to in §1.9 in which the introduction of supersymmetry brought the space-time dimensionality down from 26 to 10; see §§1.9 and 1.14). Lying behind all these apparently very different string-theoretic schemes, so the idea went, lay a more profound and supposedly unique theory – though yet unknown as a clear-cut mathematical scheme – that Witten christened *M-theory* (the "M" standing for "master", "matrix", "mystery", "mother", or any of several other different possibilities, according to the researcher's whim).

One of the features of this M-theory is that one needs to consider, in addition to the 1-dimensional strings (with their 2-dimensional space-time histories), that there would be structures of higher dimensions referred to generically as *branes* – generalizing to $p$ spatial dimensions the 2-dimensional notion of a membrane, so that such a $p$-brane has $p + 1$ space-time dimensions. (In fact, such $p$-branes had been studied earlier by others, independently of M-theory [Becker et al. 2006].) The dualities referred to above work only because branes of different dimensions get swapped one into another at the same time as do the various Calabi–Yau spaces that are invoked to play their respective roles as the extra spatial dimensions. This involves an extension of the notion of what is meant by a string theory – and this, of course, also points to why a new name such as M-theory was needed. It may be noted that the elegant original association of strings with complex curves (i.e. the Riemann surfaces referred to in §§1.6 and 1.12, and described in §A.10), which was a key to the original attractiveness and successes of string theory, is abandoned with these higher-dimensional branes. On the other hand, there is undoubtedly a different mathematical elegance in these new ideas – and, indeed, some extraordinary mathematical power (as may be inferred from the remark by Richard Thomas quoted at the end of §1.12) residing in these remarkable dualities.

It will be instructive for us to be a little more explicit, and consider a striking application of the use of one aspect of these dualities, referred to as *mirror symmetry*. This symmetry pairs each Calabi–Yau space with a different Calabi–Yau space, swapping around certain parameters (referred to as *Hodge numbers*) that describe the specific "shape" of each Calabi–Yau space. Calabi–Yau spaces are particular kinds of (real) 6-manifolds which can also be interpreted as complex

3-manifolds – i.e. these 6-manifolds have *complex structures*. Generally, a *complex n-manifold* (see the last part of §A.10) is simply the analogue of an ordinary real *n*-manifold (§A.5), where the system $\mathbb{R}$ of real numbers is replaced by the system $\mathbb{C}$ of complex numbers (see §A.9). We can always reinterpret a complex *n*-manifold as a real 2*n*-manifold endowed with what is known as a *complex structure*. But it is only in favourable circumstances that a real 2*n*-manifold can be assigned such a complex structure so that it can be reinterpreted as a complex *n*-manifold (§A.10). In addition to this, moreover, each Calabi–Yau space also has a different kind of structure referred to as a *symplectic structure* (the very kind of structure possessed by the *phase spaces* referred to in §A.6). Mirror symmetry achieves the very strange mathematical feat of, in effect, *interchanging* the complex structure with the symplectic structure!

The particular application of mirror symmetry that we are concerned with here arose in relation to a problem that some pure mathematicians (algebraic geometers) had been working on for some years earlier. Two Norwegian mathematicians, Geir Ellingstrud and Stein Arilde Strømme, had developed a technique for counting rational curves in a particular type of complex 3-manifold (called a *quintic*, which means defined by a complex polynomial equation of degree 5), which is actually an example of a Calabi–Yau space. Recall (§§1.6 and A.10) that a complex curve is what is called a *Riemann surface*; this complex curve is called *rational* if the topology of the surface is a *sphere*. In algebraic geometry, rational curves can occur in increasingly "twisted-up" forms, where the simplest is just a complex straight line (order 1) and the next simplest a complex conic section (order 2). After that we have rational cubic curves (order 3), quartics (order 4), and so on, where for each successive order there should be a precise, calculable, finite number of rational curves. (The *order* of a curve – often now referred to as its *degree* – lying in a flat ambient *n*-space is the number of points in which it intersects a generically placed $(n-1)$-plane.) What the Norwegians had found, with the aid of complicated computer calculations, were the successive numbers:

$$2875,$$

$$609250,$$

$$2682549425,$$

for orders 1, 2, 3, respectively, but to proceed further than this had proved to be very difficult, owing to the vast complication of the techniques that were available to them.

After hearing about these results, the string-theory expert Philip Candelas and his collaborators set about applying the mirror-symmetry procedures of M-theory, noting that they could perform a different kind of count on the mirrored Calabi–Yau space. On this dual space, instead of counting rational curves, a different calculation can be performed on this second space, of a much simpler kind (in which the system rational curves were "mirrored" into a much more manageable family), and according to mirror symmetry this ought to provide the same numbers as those that Ellingstrud and Strømme were trying to compute. What Candelas and colleagues found was the corresponding sequence of numbers:

2875,

609250,

317206375.

Strikingly, the first two of these agree with what the Norwegians had found, although the third number was, strangely, completely different.

At first, the mathematicians argued that since the mirror-symmetry arguments just came from some kind of physicist's conjecture, with no clear mathematical justification, the agreement at orders 1 and 2 must have been basically a fluke, and there was no reason to have any trust in the numbers for higher order obtained via the mirror-symmetry method. However, it was subsequently found that there was an error in the Norwegians' computer code, and when this was corrected, the answer turned out to be 317206375, just as the mirror-symmetry argument had predicted! Moreover, the mirror-symmetry arguments readily extend, so that the sequence for rational curves of respective orders 4, 5, 6, 7, 8, 9, 10 can be calculated to provide the answers:

242467530000,

229305888887625,

248249742118022000,

295091050570845659250,

375632160937476603550000,

503840510416985243645106250,

704288164978454686113488249750.

This undoubtedly provided a remarkable piece of circumstantial evidence in support of the mirror-symmetry idea – an idea that arose from a desire to demonstrate that two string theories which are apparently completely different can nevertheless be, in some deep sense, the "same" when the two different Calabi–Yau spaces involved are dual to one another in the above sense. Subsequent work by various mathematicians[8] [Givental 1996] have largely demonstrated that what had been merely a physicist's conjecture is actually a firm mathematical truth. But the mathematicians had had no clue previously that such a thing as this mirror symmetry could possibly be true, as with the striking comments by Richard Thomas at the end of §1.13. To a mathematician, perhaps not aware of the tenuousness of the physical underpinnings of these ideas, this seems to be a gift from nature herself, possibly even a little reminiscent of the heady days of the late seventeenth century, when the magic of the differential calculus, having been developed by Newton and others in order to reveal the workings of nature, began also to reveal its tremendous power within mathematics itself.

Of course, there are many of us within the community of theoretical physicists who do indeed believe that the workings of nature depend, with great precision, upon a mathematics that is likely to have a great power and subtlety of structure – as Maxwell's electromagnetism, Einstein's gravity, and the quantum formalism of Schrödinger, Heisenberg, Dirac, and others have impressively revealed. Consequently, we are also likely to be struck by the achievements of mirror symmetry, and to regard these as possibly providing some kind of evidence that the physical theory that gave rise to such powerful and subtle mathematics might also be likely to have some deep validity *as* physics. Yet, we should be very cautious about coming to such conclusions. There are many instances of powerful and impressive mathematical theories where there has been no serious suggestion of any links with the workings of the physical world. A case in point would be the wonderful mathematical achievement of Andrew Wiles, who, basing his work partly on much previous work by others, in 1994 finally established (with some help from Richard Taylor) the more than 350-year-old conjecture referred to as *Fermat's last theorem*. What Wiles established, as the key to his proof, was something that has some similarities to what is achieved by the use of mirror symmetry, namely the establishing that two sequences of numbers, each obtained by a completely different-looking mathematical procedure, are in fact identical. In Wiles's case the identity of the two sequences was an assertion known as the *Taniyama–Shimura conjecture*, and to achieve his proof of Fermat's last theorem, Wiles succeeded in using his methods to establish the needed part of this conjecture (the full conjecture being established a little later, in 1999, by Breuil, Conrad, Diamond, and Taylor, basing their work on the methods developed by Wiles; see Breuil et al. [2001]). There are many other results of this kind in pure mathematics, and it is clear that, for a profound new physical theory, we need a

---

[8] Kontsevich, Givental, Lian, Liu, and Yau.

good deal more than just this kind of mathematics, despite the subtlety, difficulty, and indeed sometimes even genuinely magical qualities that the mathematics may possess. Physical motivation and support from experiment are essential, in order for us to be persuaded that there is likely to be any direct connection with the actual workings of the physical world. These matters are indeed central to the issue that we come to next, which has held a key role in the development of string theory.

## 1.14. SUPERSYMMETRY

Up to this point, I have allowed myself the luxury of ignoring the pivotal issue of supersymmetry, which is what enabled Green and Schwarz to reduce the dimensionality of the string theorist's space-time from 26 to 10, and which has many other foundational roles in modern string theory. In fact, supersymmetry is a notion that has found importance in considerations of physics quite apart from string theory. We may, indeed, regard supersymmetry to be a very fashionable idea in modern physics, and consequently in its own right it deserves some serious consideration in this chapter! While it is true that much of the impetus that this idea has obtained comes from requirements of string theory, its fashionable status is to a considerable extent independent of string theory.

So what is supersymmetry? In order to explain the idea, I shall need to return to the brief discussions of the basic particles of physics in §§1.3 and 1.6. We recall that there were different families of massive particles, such as leptons and hadrons, and also other particles such as the massless photon. In fact there is a more basic classification of particles into just two types, cutting across those which we encountered earlier. This is the classification of particles into fermions and bosons, as briefly indicated in §1.6.

One way of expressing the distinction between fermions and bosons is to think of fermions as being more like the particles we know from classical physics (electrons, protons, neutrons, etc.) and bosons as being more like the carriers of forces between particles (photons being the carriers of electromagnetism and things called W and Z bosons being the carriers of weak interactions, entities referred to as *gluons* being the carriers of strong interactions). This is not a very clear-cut distinction, however, particularly because there are the very particle-like pions, kayons, and other bosons referred to in §1.3. Moreover, some very particle-like atoms can be regarded as bosons to a good approximation, for example, where such composite objects behave, in many respects, like single particles. Bosonic atoms are not all that different from the fermionic ones, both seeming to be rather like classical particles.

But, let us sidestep this issue of composite objects and the question of whether or not they can be legitimately treated as though they are single particles. Insofar as the objects we are concerned with can be taken to be single particles, the distinction between a fermion and a boson comes about because of what is known as *Pauli's exclusion principle*, which applies only to fermions, and which tells us that two fermions cannot be simultaneously in the same state as one another, whereas two bosons can. Roughly speaking, the Pauli principle asserts that two identical fermions cannot remain exactly on top of one another, as they have an effect on each other that may be roughly thought of as pushing them apart when they get too close. On the other hand, bosons have a kind of affinity with other bosons of the same type as themselves, and can indeed remain exactly on top of one another (which is just what happens with the multiple-boson states known as *Bose–Einstein condensates*). For an explanation of these condensates, see Ketterle [2002]. For a more general reference, see Ford [2013].

I shall come back to this rather curious aspect of quantum mechanical particles shortly, and try to clarify this rather vague characterization, which certainly gives us a very incomplete picture of the difference between a boson and a fermion. A somewhat more clear-cut distinction comes from an examination of a particle's rate of spin. Strangely, any (unexcited) quantum particle spins by a definite fixed amount, which is characteristic of that particular type of particle. We should not think of this spin rate as an angular velocity but, rather, as an *angular momentum* – that particular measure of spin possessed by an object, moving freely of external forces, that remains constant throughout the object's motion. Think of a baseball or cricket ball, spinning as it travels through the air, or else of a skater spinning about the point of one skate. In each case, we find that the spin in the sense of angular momentum, maintains itself, and would do so indefinitely in the absence of external (say, frictional) forces.

The skater is perhaps the better example, because we can witness the fact that the angular speed becomes less when the skater's arms are extended and greater when the arms are drawn in. What remains constant throughout this activity is the *angular momentum*, which, for a given angular velocity, would be larger for a mass distribution (e.g. the skater's arms) far from the axis of rotation and smaller when that distribution is closer to the axis (figure 1-36). So the retraction of the arms must be compensated by an increase of the rotation speed in order that the angular momentum may remain constant.

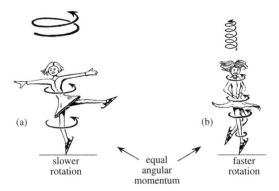

Figure 1-36: Angular momentum is conserved in physical processes. This is illustrated by a spinning skater, whose rotation rate can be increased simply by a retraction of the arms. This is because angular motion at greater distance contributes more to the angular momentum than when closer in.

Thus, we have a notion of angular momentum which applies to all compact isolated bodies. This applies also to individual quantum particles, but the rules at the quantum level turn out to be a bit strange, and take a while to get used to. What we find, for an individual quantum particle, is that for each type of particle the *magnitude* of this angular momentum is always the same fixed number, whatever the situation that the particle may find itself to be in. The direction of the spin axis need not always be the same, in different situations, but the spin direction behaves in a strange essentially quantum-mechanical way, which we shall be exploring in §2.9. For our present purposes, all that we need to know is that if we look to see how much of the particle's spin is distributed about any chosen direction, then for a boson, we find a value that is an integer multiple of $\hbar$ – where $\hbar$ is Dirac's reduced version of Planck's constant $h$; see §2.11, namely

$$\hbar = \frac{h}{2\pi}.$$

Thus, the boson's spin value about any given direction must be one of the values

$$\ldots, \; -2\hbar, \; -\hbar, \; 0, \; \hbar, \; 2\hbar, \; 3\hbar, \; \ldots.$$

In the case of a fermion, however, the spin value about any direction differs from these numbers by $\frac{1}{2}\hbar$, i.e. it has to take one of the values

$$\ldots, \; -\tfrac{3}{2}\hbar, \; -\tfrac{1}{2}\hbar, \; \tfrac{1}{2}\hbar, \; \tfrac{3}{2}\hbar, \; \tfrac{5}{2}\hbar, \; \tfrac{7}{2}\hbar, \; \ldots$$

(so the value is always a half-odd multiple of $\hbar$). We shall be seeing how this curious feature of quantum mechanics works in more detail in §2.9.

There is a famous theorem proved within the framework of QFT, known as the *spin-statistics theorem* [Streater and Wightman 2000], which asserts (in effect) the equivalence of these two notions of the boson/fermion distinction. More precisely, it obtains something much more mathematically far-reaching than just the Pauli exclusion principle referred to above, namely the kind of statistics that bosons and fermions must satisfy. It is hard to explain this satisfactorily, without entering more deeply into the quantum formalism than we have been able to do, so far, in this chapter. But I should at least try to convey something of the essence of what is involved.

We recall the quantum amplitudes referred to in §1.4 (and see §§2.3–2.9), these being the complex numbers that one aims to obtain in QFT calculations (see §1.5), and from which the probabilities of quantum measurement arise (by the Born rule of §2.8). In any quantum process, this amplitude will be a function of all the parameters describing all the quantum particles involved in the process. We can also think of this amplitude as the value of Schrödinger's wave function that we shall be considering in §§2.5–2.7. If $P_1$ and $P_2$ are two identical particles involved in the process, then this amplitude (or wave function) $\psi$ will be a function $\psi(\mathbf{Z}_1, \mathbf{Z}_2)$ of the respective sets of parameters $\mathbf{Z}_1$ and $\mathbf{Z}_2$ for these two particles (where I am using the single bold letter $\mathbf{Z}$ to encompass all these parameters for each particle: position coordinates or momentum coordinates, spin values, etc.). The choice of suffix (1 or 2) refers to the choice of particle. With $n$ particles $P_1, P_2, P_3, \ldots, P_n$ (identical or not), we would have $n$ such sets of parameters $\mathbf{Z}_1, \mathbf{Z}_2, \mathbf{Z}_3, \ldots, \mathbf{Z}_n$. So we have $\psi$ as a function of all these variables:

$$\psi = \psi(\mathbf{Z}_1, \mathbf{Z}_2, \ldots, \mathbf{Z}_n).$$

Now if the type of particle described by $\mathbf{Z}_1$ is the same as that described by $\mathbf{Z}_2$ and is a boson, then we always find the symmetry

$$\psi(\mathbf{Z}_1, \mathbf{Z}_2, \ldots) = \psi(\mathbf{Z}_2, \mathbf{Z}_1, \ldots),$$

so exchanging the particles $P_1$ and $P_2$ makes no difference to the amplitude (or wave function). But if the type of particle (the same for $\mathbf{Z}_1$ as for $\mathbf{Z}_2$) is a fermion, we find

$$\psi(\mathbf{Z}_1, \mathbf{Z}_2, \ldots) = -\psi(\mathbf{Z}_2, \mathbf{Z}_1, \ldots),$$

so exchanging $P_1$ and $P_2$ now changes the sign of the amplitude (or wave function). We may note that if each of the particles $P_1$ and $P_2$ is in the same state as the other one, then $\mathbf{Z}_1 = \mathbf{Z}_2$, whence we must have $\psi = 0$ (since $\psi$ is equal to its negative). From the Born rule (§1.4), we see that $\psi = 0$ implies zero probability. This expresses the Pauli principle that we cannot find two fermions, of identical type, in the same state as one another. If all the $n$ particles are identical, we have, for $n$ bosons, a symmetry that extends to the interchange of any pair:

$$\psi(\dots, \mathbf{Z}_i, \dots, \mathbf{Z}_j, \dots) = \psi(\dots, \mathbf{Z}_j, \dots, \mathbf{Z}_i, \dots);$$

and for $n$ fermions, an antisymmetry in any pair:

$$\psi(\dots, \mathbf{Z}_i, \dots, \mathbf{Z}_j, \dots) = -\psi(\dots, \mathbf{Z}_j, \dots, \mathbf{Z}_i, \dots),$$

The symmetry or antisymmetry that is expressed, respectively, in the above two displayed equations underlies the differing statistics of bosons and fermions. When we "count" the number of different states involving a lot of bosons of the same type, we should not consider that a new state has been arrived at when a couple of the bosons are interchanged. This method of counting gives rise to what is called *Bose–Einstein statistics* (or simply just *Bose statistics*, whence the name *boson*). The same holds for fermions, except for the strange change of sign of the amplitude, and gives rise to *Fermi–Dirac statistics* (or simply *Fermi statistics*, whence the name *fermion*), which has many quantum-mechanical implications, the most evident of which is the Pauli principle. We note that – either in the case of bosons or of fermions – interchanging two of the same kind of particle does nothing to the quantum state (beyond merely changing the sign of the wave function, which does not change the physical state since multiplying by $-1$ is just an example of a phase change: $\times e^{i\theta}$, with $\theta = \pi$; see §1.8). Accordingly, quantum mechanics really demands that two particles of the same kind must actually *be* identical! This illustrates the importance of Einstein's objection to Weyl's original proposal for a gauge theory, in which "gauge" actually referred to a change of scale; see §1.8.

All this is standard quantum mechanics, having numerous consequences excellently supported by observation. Many physicists believe, however, that there ought to be a new kind of symmetry which converts the families of bosons and fermions into one another, rather like those symmetries which relate leptons to one another and give rise to the gauge theory of weak interactions, or the symmetries which relate the different quarks to one another, giving rise to the gauge theory of strong interactions (see §1.3 and the final paragraph of §1.8). This new type of symmetry could not be a symmetry of the ordinary kind, because of the differing statistics that these two families indulge in. Accordingly, various physicists have generalized the usual type of symmetry to a new kind, which has come to be called *supersymmetry* [Kane and Shifman 2000], in which the symmetrical states of the bosons are converted to the antisymmetrical states of the fermions, and vice versa. This involves the introduction of some strange kinds of "number" – called *supersymmetry generators* – which have the property that when you multiply two of them, say $\alpha$ and $\beta$, together, you get *minus* the answer that you get if you multiply them in the opposite order:

$$\alpha\beta = -\beta\alpha.$$

(A non-commutativity of operations, where $\mathbf{AB} \neq \mathbf{BA}$, is actually commonplace in the quantum formalism: see §2.13.) It is this minus sign that enables Bose–Einstein statistics to be converted to Fermi–Dirac statistics, and vice versa.

To be more precise about these non-commuting quantities, I should say a little more about the general formalism of quantum mechanics (and QFT). In §1.4, we encountered the notion of the quantum state of a system, such states ($\mathbf{\Psi}$, $\mathbf{\Phi}$, etc.) being subject to the laws of a complex vector space (§§A.3 and A.9). Several important roles are found in the theory for what are called *linear operators*, as we shall take note of later, particularly in §§1.16, 2.12, 2.13, and 4.1. Such an operator $\mathbf{Q}$, acting on quantum states $\mathbf{\Psi}$, $\mathbf{\Phi}$, etc., is characterized by the fact that it preserves quantum superposition,

$$\mathbf{Q}(w\mathbf{\Psi} + z\mathbf{\Phi}) = w\mathbf{Q}(\mathbf{\Psi}) + z\mathbf{Q}(\mathbf{\Phi}),$$

where $w$ and $z$ are (constant) complex numbers. Examples of quantum operators are the position and momentum operators $\mathbf{x}$ and $\mathbf{p}$, and the energy operator $\mathbf{E}$, which we shall encounter in §2.13, and also the spin operator of §2.12. In standard quantum mechanics *measurements* are usually expressed in terms of linear operators. This will be explained in §2.8.

In the case of supersymmetry generators such as $\alpha$ and $\beta$, these also are linear operators, but their job in QFT is to act upon other linear operators, called *creation* and *annihilation* operators, which are central to the algebraic structure of QFT. A symbol such as $\mathbf{a}$ might well be used for an annihilation operator, where $\mathbf{a}^{\dagger}$ would stand for the corresponding creation operator. If we have a particular quantum state $\mathbf{\Psi}$, then $\mathbf{a}^{\dagger}\mathbf{\Psi}$ would be that state obtained from $\mathbf{\Psi}$ by the insertion of the particular particle state represented by $\mathbf{a}^{\dagger}$; similarly, $\mathbf{a}\mathbf{\Psi}$ would be the state $\mathbf{\Psi}$ but with this particular particle state removed from $\mathbf{\Psi}$ (assuming such removal is a possible operation; if not, we would just get $\mathbf{a}\mathbf{\Psi} = \mathbf{0}$). A supersymmetry generator such as $\alpha$ would then act on a creation (or annihilation) operator for a boson to convert it into the corresponding operator for a fermion, and conversely.

Note that by choosing $\beta = \alpha$, in the relation $\alpha\beta = -\beta\alpha$, we infer that $\alpha^2 = 0$ (since $\alpha^2$ is equal to its negative). According to this, we never get any supersymmetry generator raised to a power ($> 1$). This has the curious effect, assuming we have just a finite number $N$ of supersymmetry generators $\alpha, \beta, \ldots, \omega$ altogether, that any algebraic expression $X$ can be written without powers of these quantities

$$X = X_0 + \alpha X_1 + \beta X_2 + \cdots + \omega X_N + \alpha\beta X_{12} + \cdots$$
$$+ \alpha\omega X_{1N} + \cdots + \alpha\beta \cdots \omega X_{12\cdots N},$$

so there would be $2^N$ terms in the sum altogether (there being one term present for each possible collection of distinct members of the set of supersymmetry generators). This expression explicitly demonstrates the only kind of dependence upon the supersymmetry generators that can occur – although some of the $X$s on the right might be zero. The first term $X_0$ is sometimes referred to as the *body*, and the rest $\alpha X_1 + \cdots + \alpha\beta \cdots \omega X_{12\cdots N}$, where at least one supersymmetry generator is present, is then referred to as the *soul*. Notice that once a part of an expression gets into the soul part, multiplying it by other such expressions never brings it back into the body part. Hence the body part of any algebraic calculation stands on its own, providing us with a perfectly legitimate *classical* calculation, where we just forget about the soul part altogether. This provides a justification of a role for algebraic and geometrical considerations, such as those in §1.11, in which supersymmetry is simply ignored.

The requirement of supersymmetry provides a certain guidance for choosing a physical theory. It is indeed a powerful restriction that a proposed theory be supersymmetric. This restriction endows the theory with a certain balance between its bosonic and fermionic parts, each being related to the other by means of a supersymmetry operation (i.e. an operation constructed with the aid of the supersymmetry generators, like $X$ above). This is considered a valuable asset in the construction of a QFT, intended to model nature in a plausible way, so that it not be plagued by uncontrollable divergences. Requiring it to be supersymmetric very much improves its chance of being renormalizable (see §1.5) and greatly enhances the possibility that it be capable of providing finite answers to physical questions of importance. With supersymmetry, divergences from the bosonic and fermionic parts of the theory, in effect, cancel each other out.

This seems to be one of the main reasons (apart from string theory) for the popularity of supersymmetry in particle physics. However, if nature were actually exactly supersymmetric (with, say, *one* supersymmetry generator), any basic particle would be accompanied by another – called a *supersymmetry partner* –

having the same mass as the original particle, so each pair of supersymmetry partners would consist of a boson and a fermion of the same mass. There would have to be a *selectron*, which would be the boson accompanying the electron, and a bosonic *squark* to accompany each type of quark. There should also be a massless *photino* and *gravitino*, which would be fermions to accompany the photon and graviton respectively, and also further fermions like the *wino* and *zino*, which ought to accompany the W and Z bosons referred to above. In fact the full situation is more alarming than this relatively simple case of just one supersymmetry generator. If there were $N$ supersymmetry generators, with $N > 1$, the basic particles would occur not merely paired up in this way, but there would be $2^N$ in each supersymmetric grouping (multiplet) of mutual partners, half bosons and half fermions, all of the same mass.

Given the alarming nature of this proliferation of basic particles (and, perhaps, the seeming absurdity of the proposed terminology), the reader may be relieved to hear that no such supersymmetric sets of particles have been yet observed! This observational fact has not convincingly deterred the proponents of supersymmetry, however, as it is normally argued that there must be some supersymmetry-breaking mechanism which leads to a serious departure from exact supersymmetry holding for the particles actually observed in nature, where the masses within any multiplet might actually differ very greatly. All these supersymmetry partners (partners to the *single* so-far observed member of each group) would, accordingly, have to have masses that are beyond the range of particle accelerators that have been put into operation so far!

Of course, there is still the possibility that all these particles predicted by supersymmetry actually do exist, remaining unobserved simply because of their enormously large masses. It might have been hoped that the LHC, when it became fully operational again, at increased power, would have provided clear evidence for or against supersymmetry. However, there are very many different proposals for supersymmetric theories, with no consensus as to the level and nature of the required supersymmetry-breaking mechanisms. The situation existing at the time of writing shows no evidence whatever for any supersymmetry partners, yet this still seems to be somewhat far from that scientific ideal, towards which most theorists purport to strive, whereby a proposed theoretical scheme, to be genuinely scientific (at least according to a well-known criterion of the scientific philosopher Karl Popper [1963]), ought to be *falsifiable*. One is left with the uneasy feeling that even if supersymmetry is actually *false*, as a feature of nature, and that accordingly *no* supersymmetry partners are ever found by the LHC or by any later more powerful accelerator, then the conclusion that some supersymmetry proponents might come to would *not* be that supersymmetry is false for the actual particles of nature, but merely that the level of supersymmetry breaking must be greater even than the level reached at that moment, and that a *new* even more powerful machine would be required to observe it!

In fact, the situation is probably not that bad, with regard to scientific refutability. The latest results from the LHC, which include the remarkable discovery of the long sought-for Higgs boson, not only fail to find any evidence for a supersymmetry partner of any known particle, but actually do rule out the most direct and previously hoped-for supersymmetric models. The theoretical and observational constraints may well prove too great for any reasonable version of a supersymmetry theory of the kind so far suggested, and may guide theorists to newer and more promising ideas as to how boson and fermion families might interrelate with one another. It should be pointed out, also, that models in which there is *more than one* supersymmetry generator – such as the 4-generator theory, very popular among theoreticians, referred to as $N = 4$ *supersymmetric Yang–Mills theory* – are very much further from any kind of observational agreement than those with just a single supersymmetry generator.

Nevertheless, supersymmetry remains highly popular among theorists, and it is also a key component of current string theory, as we have seen. Indeed, the very choice of a Calabi–Yau space, as the preferred manifold $\mathcal{X}$ describing the extra spatial dimensions (see §§1.10 and 1.11), depends upon its possessing supersymmetric properties. Another way of expressing this requirement is to say that, on $\mathcal{X}$, there is what is called a (non-zero) *spinor field*, which remains constant throughout $\mathcal{X}$. The terminology *spinor field* refers to one of the most basic kinds of physical field (in the sense of §§A.2 and A.7) – not usually a constant one – which could be used to describe a fermion's wave function. (Compare §§2.5 and 2.6; for more information on spinor fields, see, for example, Penrose and Rindler [1984] and, in higher dimensions, the appendix of Penrose and Rindler [1986].)

This constant spinor field can be used, in effect, to play the role of a supersymmetry generator, and by its use, the supersymmetric nature of the entire higher-dimensional space-time can be expressed. It turns out that this supersymmetry requirement ensures that the overall energy in the space-time has to come out as zero. This zero energy state is taken to be the ground state of the entire universe, and it is argued that this state must be stable by virtue of this supersymmetric character. The idea behind this argument appears to be that perturbations of this zero-energy ground state would have to increase its energy, and so the space-time structure of this slightly perturbed universe would simply settle back into this supersymmetric ground state by the re-emission of that energy.

However, I have to say that I have a great deal of difficulty with this type of argument. As pointed out in §1.10, and taking into account the point made earlier in this section that the body part of any supersymmetric geometry can be extracted as a classical geometry, it seems to be very appropriate to regard such perturbations as providing us with *classical* disturbances and, because of the conclusions of §1.11, we must accept that the vast majority of such classical disturbances are likely to lead to *space-time singularities* in a tiny fraction of a second! (At least, the disturbances in the extra spatial dimensions would rapidly become so great as to be effectively singular, before any higher-order terms in the string constant $\alpha'$ might be able to come into play.) According to this picture, rather than settling gently back into such a stable supersymmetric ground state, the space-time would crumple into a singularity! I see no rational reason to hope that such a catastrophe is to be averted, irrespective of that state's supersymmetric nature.

## 1.15. AdS/CFT

Although I am not aware that many (or even any) professional string theorists have allowed themselves to be diverted from their main goals by arguments like those given above – that is, by the arguments of §§1.10 and 1.11, that at the end of §1.14, and (generally) by the functional-freedom issues of §§A.2, A.8, and A.11 – they have, in fairly recent years, found themselves to be led into areas somewhat different from those I have been describing so far. Nevertheless, the issues of excessive functional freedom remain very relevant, and it seems appropriate to end this chapter by addressing the most significant of these developments. In §1.16, I shall very briefly describe some of the strange territory that the trajectory of string theory appears have led us into, namely what are referred to as *brane worlds*, the *landscape*, and *swamplands*. Of much greater mathematical interest, and with tantalizing connections with various other areas of physics, is what is referred to as the *AdS/CFT correspondence* – or the *holographic conjecture* or *Maldacena duality*.

This *AdS/CFT correspondence* [see Ramallo 2013; Zaffaroni 2000; Susskind and Witten 1998] is what is often referred to as the *holographic principle*. I should first make clear that this is not an established principle but a collection of interesting notions that do have some empirical mathematical support, but which at first sight appear to run counter to certain serious issues of functional freedom. In broad terms, the idea of the holographic principle is that two very different types of physical theory, one of which (a form of string theory, as it turns out) is defined on some $(n + 1)$-dimensional space-time region, referred to as the *bulk*, and the other of which (a more conventional type of quantum field theory) is defined on the $n$-dimensional *boundary* of this region. An initial impression would be that such a correspondence seems unlikely, from the point of view of functional freedom, because the bulk theory appears to have a functional freedom $\infty^{A\infty^n}$ for some $A$, whereas that on the boundary would *seem* to have a far smaller freedom $\infty^{B\infty^{n-1}}$ for some $B$. This would be the case if both are space-time theories of a more-or-less usual type. In order to understand better the underlying reasons for this suspected correspondence, and possible difficulties with the suggested proposal, it will be helpful to look first at some of the background behind it.

One of the early inputs into this idea came from a well-established feature of black-hole thermodynamics which will, in fact, underlie much of the reasoning of chapter 3. This is the fundamental Bekenstein–Hawking formula for the entropy of a black hole that we shall encounter more explicitly in §3.6. What this formula tells us is that the entropy in a black hole is in proportion to the surface area of the black hole. Now, in rough terms, the entropy of an object, if it is in a completely random ("thermalized") state, is basically a count of the total number of degrees of freedom in that object. (This is made more precise in a powerful general formula, due to Boltzmann, that we shall come to more explicitly in §3.3.) What seems peculiar about this black-hole entropy formula is that if we had some ordinary classical body, made out of some substance with a large number of tiny molecules, or other basic localized constituents, then the number of degrees of freedom potentially available to the body would be in proportion to the volume of that body, so we would expect that when it is in a completely thermal (i.e. maximum-entropy) state, its entropy would be a quantity that is in proportion to its volume, not its surface area. Thus the viewpoint has grown up that, with a black hole, we have something going on inside it that is kept track of by its 2-dimensional surface – and this surface information is, in some sense, *equivalent to* all that is carrying on in its 3-dimensional interior. Thus, so the argument goes, we have some kind of instance of a holographic principle with the information in the degrees of freedom inside a black hole being somehow encoded in those on the black-hole's boundary (i.e. horizon).

This general type of argument carried on from some early work in string theory [Strominger and Vafa 1996], where attempts were made to give the Bekenstein–Hawking formula a Boltzmann-type foundation by counting string degrees of freedom in an interior region to a spherical surface, where first the gravitational constant was taken to be small, so the surface did not represent a black-hole boundary, and then "cranking up" the gravitational constant until the point is reached when the bounding surface would indeed become an event horizon to the black hole. At that time, this result was taken by the string theorists as a great step forward in the understanding of black-hole entropy, because there had not previously been any direct connection made between the Boltzmann and Bekenstein–Hawking formulae. There were, however, many objections made to this argument (which was limited, and unrealistic in several ways), and a rival approach was initiated by the proponents of the loop-variable approach to quantum gravity [Ashtekar et al. 1998, 2000]. This, in turn, encountered its own (seemingly more minor) difficulties, and I believe that it is fair to say that there is still no totally convincing and unambiguous procedure for obtaining the Bekenstein–Hawking black-hole formula from the general Boltzmann entropy definition. Nevertheless, the arguments for the correctness of the black-hole entropy formula are very well established and convincing, by other means, and do not really require a *direct* Boltzmannian foundation.

From my own perspective, it is an inappropriate viewpoint to assign a "bulk" to a black hole's interior and to consider that there are "degrees of freedom" maintaining their existence in the interior of the hole (see §3.5). Such a picture is inconsistent with the causality behaviour within a black hole. There is an internal singularity that has to be regarded as capable of destroying information, so a balance that is sought for in the string theoretic approach is, in my opinion, misconceived. The arguments presented in the loop-variable procedure are in my view much better founded than the earlier string-theoretic arguments, but the obtaining of a convincing numerical agreement with the black-hole entropy formula remains elusive.

Let us now address the proposed AdS/CFT version of a holographic principle. As it now stands, it is still an unproven hypothesis (first suggested by Juan Maldacena in 1997 [Maldacena 1998], with a strong endorsement from Edward Witten [1998]) and not an established mathematical principle, yet there is argued to be a good deal of mathematical evidence in support of an actual precise mathematical correspondence relating two very different-looking proposals for physical models. The idea is that one might be able to demonstrate that some theory one hopes to understand better (here a string theory), which is defined on an $(n + 1)$-dimensional bulk region $\mathcal{D}$, is actually *equivalent* to a much better understood theory (here a more conventional type of QFT) on the $n$-dimensional *boundary* $\partial \mathcal{D}$ of this region. Although the origin of the idea relates to deep issues of black-hole physics, as described above, the name *holographic* has its origins in the now-familiar notion of a *hologram*. For it tends to be argued that this seeming inconsistency of dimensional information is not implausible because something of this general nature occurs already with a hologram, whereby information on what is effectively just a 2-dimensional surface encodes a 3-dimensional image, hence the terms *holographic conjecture* and *holographic principle*. However, an actual hologram is not really an example of this principle, because the perceived 3-dimensional image is only of the 2-dimensional *surface* that bounds a body in 3-space, the *bulk* of this body being not perceived at all. Such surfaces have a functional freedom of $\infty^{3\infty^2}$, not the much greater $\infty^{a\infty^3}$ that would be required according to a supposed genuine "holographic principle", in which the bulk, also, would need to be encoded in the image. An ordinary hologram is more like a stereographic image as perceived by our two eyes, giving us the impression of depth but not of bodies' interiors. For the original motivations behind the holographic principle, see 't Hooft [1993] and Susskind [1994].

In the version of this principle that is specifically referred to as the *AdS/CFT correspondence*, the region $\mathcal{D}$ is a 5-dimensional space-time, referred to as the *anti–de Sitter cosmology* $\mathcal{A}^5$. We shall be seeing, in §§3.1, 3.7, and 3.9, how this cosmological model belongs to a broad class of models, generally called FLRW models, some others of which appear to have a good chance of closely modelling the space-time geometry of our actual 4-dimensional universe. Moreover, *de Sitter space* is the model that, according to current observation and theory, would well approximate the remote future of our actual universe (see §§3.1, 3.7, and 4.3). On the other hand, the 4-dimensional *anti*–de Sitter space $\mathcal{A}^4$ is not really a plausible model for the universe, having a sign for the cosmological constant $\Lambda$ that is *opposite* to the one observed (see §§1.1, 3.1, and 3.6). This observational fact seems not to have deterred the string theorists from pinning a great deal of their faith on the usefulness of $\mathcal{A}^5$ for analysing the nature of our universe.

As emphasized in the preface to this book, physical models are often studied for the insights that they may provide in improving our general understanding, and it is not necessary for such purposes that they be physically realistic, although in this case there appeared to have been some genuine hope that $\Lambda$ might actually have turned out to be negative. Juan Maldacena first put forward his AdS/CFT proposal in 1997, which was just before the observations (by Perlmutter et al. [1998] and Riess et al. [1998]) presented persuasive evidence of a positive $\Lambda$, rather than Maldacena's required negative $\Lambda$. Even by 2003, when I recall having conversations with Edward Witten on the matter, there seemed to have been some hope that the observations might still have allowed for a negative $\Lambda$.

The AdS/CFT conjecture proposes that an appropriate string theory on $\mathcal{A}^5$ is, in some suitable sense, completely equivalent to a more conventional type of gauge

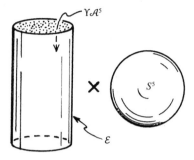

Figure 1-37: The AdS/CFT conjecture refers to the Lorentzian 10-manifold which is the product space $\mathcal{A}^5 \times S^5$ of anti–de Sitter 5-space $\mathcal{A}^5$ with a (spacelike) 5-sphere $S^5$. Here $\Upsilon\mathcal{A}^5$ refers to the "unwrapped" version of $\mathcal{A}^5$ and $\mathcal{E}$ (Einstein's static universe) is the unwrapped version of compactified Minkowski space.

theory (see §§1.3 and 1.8) on the 4-dimensional conformal boundary $\partial\mathcal{A}^5$ of $\mathcal{A}^5$. However, as remarked earlier (§1.9), the current ideas of string theory require that the space-time manifold be 10-dimensional, not 5-dimensional like $\mathcal{A}^5$. The way that this issue is dealt with is to consider that the string theory is to apply not to the 5-space $\mathcal{A}^5$, but actually to the 10-dimensional space-time manifold:

$$\mathcal{A}^5 \times S^5$$

(see figure A-25 in §A.7, or §1.9, for the notion of "×" where $S^5$ is a 5-dimensional sphere of a cosmological-scale radius; see figure 1-37). (The relevant string theory is Type IIB, but I shall not go into the distinctions between the various types of string theory here.)

It is significant that $S^5$, being of a cosmological size (so that quantum considerations are irrelevant), must have a functional freedom available within this $S^5$ factor that certainly completely swamps that of any dynamics within the $\mathcal{A}^5$, *if* that dynamics is to match the conventional 3-space dynamics that AdS/CFT proposes for the boundary $\partial\mathcal{A}^5$. There is no need to appeal to the arguments of §1.10 aimed at possible quantum obstructions to exciting this excessive functional freedom. In the $S^5$, there is no prospect of these vast degrees of freedom being suppressed, and this emphatically tells us that the AdS/CFT model does not in any direct way represent the universe we live in.

In the AdS/CFT picture, the $S^5$ is simply transferred over to $\mathcal{A}^5$'s conformal boundary $\partial\mathcal{A}^5$ so as to provide a sort of boundary

$$\partial\mathcal{A}^5 \times S^5$$

to $\mathcal{A}^5 \times S^5$, but this is very far from actually being a *conformal* boundary to $\mathcal{A}^5 \times S^5$. To explain this, I need to provide some idea of what a conformal boundary actually is, and for that I refer the reader to figure 1-38(a), in which the entire hyperbolic plane (a notion I shall return to in §3.5) is represented in a conformally accurate way, the conformal boundary here being just the surrounding circle. This is a beautiful and well-known woodcut by the Dutch artist M. C. Escher and it accurately depicts a conformal representation of the hyperbolic plane (originally due to Eugenio Beltrami in 1868, but commonly referred to as the *Poincaré disc*) the straight lines of this geometry being represented as circular arcs meeting the bounding circle at right angles (figure 1-38(b)). In this *non-Euclidean* plane geometry, many lines ("parallels") through a point P do not meet a line $a$, and any triangle's angles $\alpha$, $\beta$, $\gamma$ add up to less than $\pi$ ($=180°$). There are also higher-dimensional versions of figure 1-38, such as where 3-dimensional hyperbolic space is represented conformally as the interior of an ordinary sphere $S^2$. "Conformal" basically means that all very small shapes – e.g. the shapes of the fish's fins – are very accurately represented in this depiction, the smaller the shape the better the accuracy, though different instances of the same small shape can vary in size (and the fish's eyes remain circular right up to the edge). Some of the powerful ideas of conformal geometry are alluded to in §A.10 and, in the space-time context, at the end of §1.7 and the beginning of §1.8. (We shall return to the notion of conformal boundary in §§3.5 and 4.3.) It turns out that, in the case of $\mathcal{A}^5$, its conformal boundary $\partial\mathcal{A}^5$ can be interpreted as being essentially a conformal copy of ordinary Minkowski space-time $\mathbb{M}$ (§§1.7 and 1.11), although "compactified" in a certain way that we shall be coming to shortly. The idea of AdS/CFT is that, in this particular case of a string theory on the space-time $\mathcal{A}^5$, the mysteries of string theory's mathematical nature might be resolved through this conjecture, since gauge theories on Minkowski space are pretty well understood.

There is also the issue of the "$\times S^5$" factor, which ought to be supplying a major part of the functional freedom. With regard to the general ideas of holography, the $S^5$ is largely ignored. As mentioned above, $\partial\mathcal{A}^5 \times S^5$ is certainly not a conformal boundary of $\mathcal{A}^5 \times S^5$ because the "squashing down" of the infinite regions of $\mathcal{A}^5$ in order to "reach" $\partial\mathcal{A}^5$ does not apply to $S^5$, whereas for a conformal squashing this would have to apply to all dimensions equally. The way that the information on the $S^5$ is handled is simply to resort to a *mode analysis*, in other words, to encode it all in a sequence of numbers (referred to as a *tower* in AdS/CFT considerations). As pointed out at the end of §A.11, this is a good way of obscuring the issues of functional freedom!

(a)

Figure 1-38: (a) M. C. Escher's *Circle Limit I* uses Beltrami's conformal representation of the hyperbolic plane, whose infinity becomes a circular boundary.

Readers of this book who have followed me so far may well appreciate possible alarm bells that are likely to be set off with a proposal such as AdS/CFT, because if the theory defined on the 4-dimensional boundary $\partial\mathcal{A}^5$ seems to be an ordinary type of 4-dimensional field theory, it would be constructed from quantities with functional freedom $\infty^{A\infty^3}$ (for some positive integer $A$), whereas that for its 5-dimensional interior $\mathcal{A}^5$, we would expect the enormously larger functional freedom $\infty^{B\infty^4}$ (for whatever $B$) if we were able to take the theory in the interior to be also a field theory of the normal type (see §§A.2 and A.8). This would be likely to present us with considerable difficulties in taking seriously such a proposed

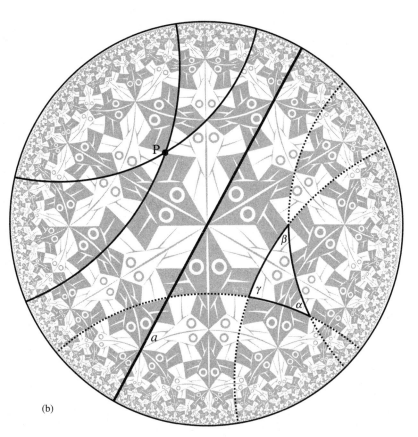

Figure 1-38: (*Cont.*) (b) The straight lines of this geometry are represented as circular arcs meeting the boundary at right angles. There are many "parallels" through P not meeting the line $a$; a triangle's angles $\alpha$, $\beta$, and $\gamma$ add up to less than $\pi$ (=180°).

equivalence between the two theories. However, there are several complicating issues here, which need to be taken into consideration.

The first thing to consider is that the theory in the interior is meant to be a string theory rather than an ordinary QFT. An immediate reaction to such a suggestion could quite reasonably be that the functional freedom ought actually to be much *greater* for a string theory than for a theory in which the basic ingredients were points since, with regard to *classical* functional freedom, there are far more loops of string than there are individual points. However, this gives a very misleading estimate of the amount of functional freedom in a string theory. It is better to think of string theory as simply being another way of addressing normal physics (which, after all, is part of its aims), so we are led to a functional freedom of the same general form (in some sort of classical limit) as one would get for normal classical field theory in the $(n + 1)$-dimensional space-time, i.e. of the general form $\infty^{B\infty^4}$, as considered earlier. (Here I am now ignoring the huge freedom in the $S^5$.) The form $\infty^{B\infty^4}$ would certainly be the answer for the functional freedom of classical Einstein gravity in the bulk – and therefore presumably what we ought to be getting from the appropriate classical limit of string theory in the bulk.

There is, however, another issue of what should be meant by a classical limit. Such matters will be considered from a completely different perspective in §2.13 (and §4.2). Perhaps there is a connection, but I have not attempted to follow this up here. There is the issue, however, that for the bulk and boundary we may well be looking at different limits, and this introduces a complicating factor which may be related to the puzzle of how the holographic principle can be satisfied, when the functional freedom in the boundary appears to be only of the form $\infty^{A\infty^3}$, which is enormously smaller than the $\infty^{B\infty^4}$ that we expect to find in the bulk. Of course, one possibility could certainly be that the AdS/CFT conjecture is not actually true, despite the seemingly strong partial evidence, already unearthed, of a great many close relationships between the bulk and boundary theories. It might, for example, be that every solution of the boundary equations indeed arises from a solution of the bulk equations, but that there are vastly many bulk solutions which do not correspond to a boundary solution. This sort of thing would arise if we consider some spacelike 3-sphere $S^3$ on $\partial \mathcal{A}^5$ which spans a spacelike 4-ball $D^4$ within $\mathcal{A}^5$ and we contemplate equations that are, respectively, the 3-dimensional and 4-dimensional Laplace equation. Each solution on $S^3$ arises from a unique solution on $D^4$ (see §A.11) but many solutions on $D^4$ give *non*-solutions on $S^3$ ($= \partial D^4$). At a much more sophisticated level, we find that certain solutions of the actual equations referred to as *BPS* (Bogomol′nyi–Prasad–Sommerfield) *states*, which are characterized by certain symmetry and supersymmetry properties, turn out to exhibit a surprising exact correspondence between the BPS states of the boundary theory and those relevant to the interior. But, we may ask, to what extent do such specific states illuminate the general case, when the full functional freedom is involved?

Another point to consider (as noted in §A.8) is that our concerns of functional freedom are essentially *local*, so that the above problems confronting a classical version of Ads/CFT might not apply globally. Global restrictions can indeed

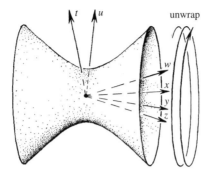

Figure 1-39: Anti–de Sitter space $\mathscr{A}^5$ contains closed timelike curves. These can be removed by "unwrapping" $\mathscr{A}^5$ while rotating it around the $(t, u)$-plane, thereby forming its *universal covering space* $\Upsilon\mathscr{A}^5$.

sometimes drastically reduce the number of solutions to classical field equations. To address this matter for the AdS/CFT correspondence requires us to confront what appears to be a bit of a confusion in the literature as to what "global" actually means in this correspondence. In fact, there are two different versions of the geometry involved. In each case, we have a perfectly valid (though no doubt confusing) piece of conformal geometry – where *conformal*, in the context of space-times, refers to the *family of null cones* (see §§1.7 and 1.8). I shall distinguish these two geometries as the *wrapped* and *unwrapped* versions of $\mathscr{A}^5$, and of its conformal boundary $\partial\mathscr{A}^5$. The symbols $\mathscr{A}^5$ and $\partial\mathscr{A}^5$ will here refer to the *wrapped* versions, and $\Upsilon\mathscr{A}^5$ and $\Upsilon\partial\mathscr{A}^5$ as the *un*wrapped versions. Technically, $\Upsilon\partial\mathscr{A}^5$ is what would be called the *universal cover* of $\mathscr{A}^5$. The concepts involved are (I hope adequately) explained in figure 1-39. The wrapped $\mathscr{A}^5$ is the one most readily arrived at through the appropriate algebraic equations,[9] and it has the topology $S^1 \times \mathbb{R}^4$. The unwrapped version has topology $\mathbb{R}^5 (= \mathbb{R} \times \mathbb{R}^4)$, in which each $S^1$ circle in $S^1 \times \mathbb{R}^4$ is unwrapped (by going around it an unlimited number of times) into a straight line ($\mathbb{R}$). The physical reason for wanting this "unwrapping" is that these circles are *closed timelike world-lines*, normally considered to be unacceptable in any model space-time that is intended to be realistic (because of the possibility of paradoxical actions for observers having these curves as their world-lines, in view of the potential for such observers to make use of free will to alter

---

[9] $\mathscr{A}^5$ is the 5-quadric $t^2 + u^2 - w^2 - x^2 - y^2 - z^2 = R^2$ in the 6-space $\mathbb{R}^6$ of real coordinates $(t, u, w, x, y, z)$ and metric $ds^2 = dt^2 + du^2 - dw^2 - dx^2 - dy^2 - dz^2$. The unwrapped versions $\Upsilon\mathscr{A}^5$ and $\Upsilon M^{\#}$ are the *universal covering spaces* of $\mathscr{A}^5$ and $M^{\#}$, respectively; see Alexakis [2012].

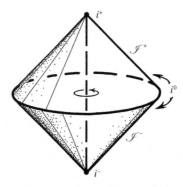

Figure 1-40: Minkowski space represented with its conformal boundary, consisting of two null hypersurface 3-spaces, $\mathscr{I}^+$ (future null infinity) and $\mathscr{I}^-$ (past null infinity), and three points, $i^+$ (future timelike infinity), $i^0$ (spacelike infinity), and $i^-$ (past timelike infinity).

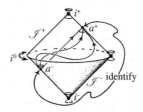

Figure 1-41: To form the compactified Minkowski 4-space with topology $S^1 \times S^3$, we identify $\mathscr{I}^+$ with $\mathscr{I}^-$ (of figure 1-40) pointwise in the way indicated, so that $a^-$ on $\mathscr{I}^-$ is identified with $a^+$ on $\mathscr{I}^+$, where any null geodesic in $\mathbb{M}$ with past endpoint $a^-$ on $\mathscr{I}^-$ acquires the future endpoint $a^+$ on $\mathscr{I}^+$. In addition, all three points $i^-$, $i^0$, and $i^+$ must be identified.

events which had been regarded as having definitely taken place in the past!). The unwrapping process thus makes this model have a better chance of being realistic.

The conformal boundary of $\mathscr{A}^5$ is what is called *compactified Minkowski 4-space* $\mathbb{M}^\#$. We can think of this boundary (conformally) as the ordinary Minkowski 4-space $\mathbb{M}$ of special relativity (see §1.7; figure 1-23) with its own conformal boundary $\mathscr{I}$ adjoined (as indicated in figure 1-40) but where we "wrap it up" by appropriately identifying the future part $\mathscr{I}^+$ of its conformal boundary with the past part $\mathscr{I}^-$, via the identification of the infinite future end $a^+$ of any light ray (null geodesic) in $\mathbb{M}$ with its infinite past end $a^-$ (figure 1-41). The

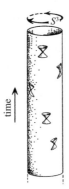

Figure 1-42:  The Einstein static model of the universe $\mathcal{E}$ is a spatial 3-sphere that is constant in time: topologically $\mathbb{R} \times S^3$.

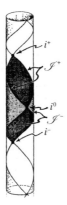

Figure 1-43:  This picture (although only 2-dimensional) indicates the way in which Minkowski space with its conformal boundary can be regarded as a closed portion of the Einstein static model $\mathcal{E}$, and this makes it clearer how $i^0$ can be simply a single point.

unwrapped boundary space $\gamma\mathbb{M}^{\#}$ (the universal covering space of $\gamma\mathbb{M}^{\#}$) turns out to be conformally equivalent to the Einstein static universe $\mathcal{E}$ (figure 1-42); see also §3.5 (figure 3-23), which is a spatial 3-sphere which is unchanging with time: $\mathbb{R} \times S^3$. The portion of this space which is conformal to Minkowski space, with its conformal boundary $\mathscr{I}$ adjoined, is indicated (in the 2-dimensional case) in figure 1-43 (and see also figure 3-23).

The unwrapped versions of these spaces do not appear to impose much of a global constraint on the classical solutions of field equations. Basically, all we should need to worry about would be such things as the vanishing of the total charge, in the case of Maxwell's equations with sources arising from the compactification of the spatial directions (which gives the $S^3$ of the Einstein universe above). I cannot see that there would be any further topological restrictions on the classical fields on the unwrapped $\Upsilon \mathcal{A}^5$ and $\Upsilon \mathbb{M}^\#$, since there are no further constraints on the time evolution. But the compactification of the temporal direction provided by wrapping up the open time direction in $\Upsilon \mathcal{A}^5$ and $\Upsilon \mathbb{M}^\#$ to give $\mathcal{A}^5$ and $\mathbb{M}^\#$ could certainly lead to a drastic reduction in the number of classical solutions, because only those with a periodicity consistent with the compactification would survive the wrapping-up procedure [Jackiw and Rebbi 1976].

I assume, therefore, that it is really the *unwrapped* versions, $\Upsilon \mathcal{A}^5$ and $\Upsilon \partial \mathcal{A}^5$, that are intended to be the spaces of relevance to the AdS/CFT proposal. How then are we to escape the apparent inconsistency in the functional freedoms in the two theories? Very possibly an answer may lie in one feature of the correspondence that I have, as yet, not addressed. This is the fact that the Yang–Mills field theory on the boundary is not really quite a standard field theory (even apart from its 4 supersymmetry generators), because its gauge symmetry group has to be considered in the limit where the dimension of this group goes to infinity. From the point of view of functional freedom, it is a bit like looking at the "towers" of harmonics that go to define what is going on in the $S^5$ spaces. The extra functional freedom can lie "hidden" in the infinitude of these harmonics. In a similar way, the fact that the gauge group's size has been taken to *infinity* for the AdS/CFT correspondence to work could easily resolve the apparent conflict in the functional freedom.

To sum up, it does seem clear that the AdS/CFT correspondence has opened a huge new area of research, which has related many active areas of theoretical physics, making unexpected connections between such disparate fields as condensed matter physics, black holes, and particle physics. On the other hand, there is a strange contrast between this great versatility and wealth of ideas, and the unreality of the immediate picture of the world that it projects. It depends upon the wrong sign for the cosmological constant; it requires 4 generators of supersymmetry, whereas none has been observed; it requires a gauge symmetry group acting on infinitely many parameters instead of the 3 that particle physics requires; and its bulk space-time has 1 too many dimensions! It will be most fascinating to see where all of this leads.

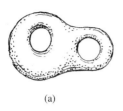 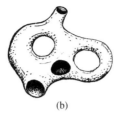

(a)                                    (b)

Figure 1-44: Pictures of Riemann surfaces: (a) exhibiting handles, and (b) exhibiting both handles and holes. (Note: in the literature, what I am here calling *handles* are sometimes confusingly referred to as *holes*.)

## 1.16. BRANE-WORLDS AND THE LANDSCAPE

Let us next come to the issue of brane-worlds. In §1.13, I mentioned that entities referred to as *p*-branes – the higher-dimensional versions of strings – that are needed, in addition to strings, for the various dualities characterizing M-theory. Strings themselves (1-branes) can come in two essentially different forms: the *closed* strings, which can be described as ordinary compact Riemann surfaces (see figure 1-44(a) and §A.10), and the *open* strings for which the Riemann surfaces have holes in them (figure 1-44(b)). In addition to these, there are supposed to be structures known as *D-branes*, which play a different role in string theory. D-branes are regarded as classical structures in the (higher-dimensional) space-time, being presumed to arise from huge conglomerations of elementary string and *p*-brane constituents, but being characterized as classical solutions of supergravity equations through symmetry and supersymmetry requirements. An important role that D-branes are taken to have is as places where the "ends" of the open strings (i.e. the holes) must reside. See figure 1-45.

The concept of a *brane-world* represents a serious (though not so frequently admitted) departure from the original string-theory viewpoint described in §1.6, in which the higher-dimensional space-time was to be thought of (locally) as a product space $\mathcal{M} \times \mathcal{X}$ where the space-time that we directly experience is the 4-space $\mathcal{M}$, and the 6-space $\mathcal{X}$ provides the tiny unseen extra dimensions. According to this original view, the observed 4-dimensional space-time is an example of what is mathematically described as a *factor space*, where the space $\mathcal{M}$ is obtained if we map each instance of $\mathcal{X}$ in $\mathcal{M} \times \mathcal{X}$ down to a point:

$$\mathcal{M} \times \mathcal{X} \to \mathcal{M}.$$

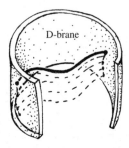

Figure 1-45: A cartoon of a D-brane. This is a (classical) region on which the ends ("holes") of a string would reside.

See figure 1-32(a) (in §1.10). But according to the brane-world point of view, on the other hand, things are quite the opposite, as the observed 4-dimensional universe is regarded as a *subspace* of the ambient 10-dimensional space-time $\mathcal{S}$, being identified with a particular 4-dimensional D-brane $\mathcal{M}$ *within* it:

$$\mathcal{M} \hookrightarrow \mathcal{S}.$$

See figure 1-32(b). To me, this is a very odd idea, since most of the space-time $\mathcal{S}$ appears now to be completely irrelevant to our experiences. Nevertheless, one might regard this as some kind of improvement, since the functional freedom is likely to be far less excessive than before. A disadvantage, however, is that the normal propagation of fields deterministically into the future that we are used to from conventional physics is liable to be lost completely, through information continually leaking away from the subspace $\mathcal{M}$ into the ambient higher-dimensional space-time. It should be pointed out that this is completely at variance with the normal deterministic evolution of classical fields in the space-time we experience. In the brane-world picture, the functional freedom for ordinary classical fields that we would directly experience would now be of the form $\infty^{B\infty^4}$, rather than the much smaller $\infty^{A\infty^3}$ that we actually do experience. This is still far too much. In fact, I find this kind of picture harder to take seriously even than the original one of §1.6.

Finally, we come to the issue of the *landscape*, and also *swamplands*, which, unlike the other problems I have mentioned up to this point, actually *do* seem to worry some string theorists! In §1.10, I referred to the zero-modes of the Calabi–Yau spaces, which do not require any energy in order for them to become excited. These modes of excitation do not tap the excessive functional freedom that is inherent in the extra spatial dimensions, but they do involve finite-dimensional changes in the parameters – referred to as *moduli* – which characterize the *shape* of the Calabi–Yau spaces that are being used to provide the extra spatial dimensions. Deforming these moduli lead us to a vast number of alternative string theories, given by what are called *alternative vacua*.

The notion of a *vacuum* in quantum field theory is an important one, which I have not previously discussed in this book. The fact is that in the specification of a QFT one needs *two* ingredients. One is the *algebra* of the operators of the theory – such as creation and annihilation operators, referred to in §1.14 – and the other is the choice of vacuum, upon which such operators would ultimately act, by building up states with more and more particles, by the use of the particle creation operators. What is frequently found in QFT is that there may well be different "inequivalent" vacua for the same algebra of operators, so that starting from one particular choice of vacuum, one cannot get to the other by means of legitimate operations within the algebra. That is to say, the theory built up starting from one choice of vacuum describes a completely different universe from the theory built up from another inequivalent vacuum, and one is not allowed quantum superpositions between states in the two different QFTs. (This fact will also play a significant role for us in §§3.9, 3.11, and 4.2.) What happens in string theory is that we get an enormous multitude of inequivalent string (or M-) theories in this way.

This contrasts completely with the early aim of string theory that it should provide a *unique* theory of physics. We recall the success claimed for M-theory: the uniting into one, of what had seemed to be five different possible types of string theory. It seems that this apparent success is now completely overwhelmed by this proliferation of different string (or M-) theories (the exact number being presently unknown, but values such as $\sim 10^{500}$ have been quoted [Douglas 2003; Ashok and Douglas 2004]) these arising from the huge numbers of inequivalent vacua that seem to be allowed! In an attempt to deal with this problem, a point of view has arisen according to which different universes all coexist, providing a "landscape" of all these different possibilities. In amongst this vast array of apparently unrealized mathematical possibilities there are also many that *seem* to be possibilities, but actually turn out to be mathematically *inconsistent*. These constitute what has become known as the *swampland*. The idea now seems to be that if we want to explain nature's apparent "choice" of values for the different moduli that determine the nature of the actual universe that we experience, we are to argue as follows: it is possible to find ourselves only in a universe in which the moduli have values leading to constants of nature compatible with the chemistry, physics, and cosmology necessary for the evolving of intelligent life. This is an instance of what is referred to as the *anthropic principle* that we shall come to in §3.10. In my view, this is a very sorry place for such a grand theory to have finally stranded us. The anthropic principle has a role to play in perhaps explaining some apparently coincidental relationships between certain fundamental constants of nature, but generally its explanatory power is extremely limited. I shall return to this issue in §3.10.

What moral is to be gained from these final sections, concerning the grand ambitions of those initial, attractively compelling, string-theoretic ideas? AdS/CFT has indeed led to many intriguing and often unexpected correspondences between different current areas of genuine physical interest (such as a relationship between black holes and solid state physics; see also §3.3 and Cubrovic et al. [2009]). Such correspondences can indeed be fascinating, especially with regard to the mathematics – but the whole drift has been far from the initial aspirations of string theory, where it had been fully intended that the subject would lead to a much closer understanding of nature's deep secrets. What about the notion of brane worlds? To me, there is an air of desperation about it, where existence itself clings to a tiny low-dimensional cliff, while abandoning hope of comprehending those vast reaches of higher-dimensional unfathomable activity. And the landscape is even worse, offering no reasonable expectation of ever even locating such a relatively safe cliff for our very existence!

# 2

## Faith

### 2.1. THE QUANTUM REVELATION

*Faith*, according to my *Concise Oxford Dictionary*, is belief founded on authority. We are used to authority having a strong influence on our thinking, whether it be the authority of our parents when we were young, or of our teachers during our schooling, or of people in respectable professions like doctors, lawyers, scientists, television presenters, or representatives of government or international organizations – or, indeed, senior figures of religious institutions. In one way or another, authorities influence our opinions and the information we receive in this way frequently leads us to having beliefs that we do not seriously question. Indeed, it is often the case that doubt never even enters our minds about the validity of much of the information acquired this way. Moreover, such influences of authority will frequently affect our own behaviour and positions in society, and any authority that we ourselves might come to possess can, in turn, enhance the status of our own opinions as they then affect the beliefs of others.

In many cases such behavioural influences are merely matters of culture, and it may be just a question of good manners that we fit in with these influences so as to avoid unnecessary friction. But it is a more serious issue when it comes to our concerns about what is actually *true*. It is, indeed, one of the ideals of science that we should *not* simply take things on trust, and that our beliefs should, at least from time to time, be tested against the realities of the world. Of course, we are not likely to have the opportunity or the facilities to subject very many of our beliefs to such a test. But at least we should attempt to have open minds. It may well be that, often, what we have at our disposal is just our reason, judgement, objectivity, and common sense. Yet, these qualities should not be underrated. In accordance with them it is rational to suppose that the claims of science are not likely to be the result of a tightly knitted conspiracy of falsehoods. There are, for example, enough magical-seeming gadgets available to us now – such as television sets, mobile telephones, iPads, and global positioning system (GPS) devices, not to mention jet planes and life-saving drugs, and so on – to reassure us that there is something profoundly true in most of the proclamations arising from scientific understanding and the rigorous methods of scientific testing. Accordingly, although there is certainly a new authority that has arisen from the culture of science, it is an authority that – in principle at least – is subject to continual review. Thus, the faith that we have in this scientific authority is not a blind faith, and we must always be prepared for the possibility of an unexpected shift in the views that are being expressed by scientific authority. Moreover, we should not be surprised that some scientific views are liable to be the subject of serious controversy.

The word *faith* is, of course, more commonly used in relation to *religious* doctrine. In that context – although discussion on basic points may sometimes be welcomed, and some details of the official doctrine may be expected to encounter subtle shifts over the years so as to remain in keeping with the changing of circumstance – there tends to be a body of firm doctrinal belief that, at least in the case of the great religions of today, will extend back for time periods that may well amount to thousands of years. In each case, the origins of the body of beliefs that underlie that faith would be taken to date back to an individual (or individuals) of outstanding stature in respect of moral values, strength of character, wisdom, and persuasive powers. Although one can expect that the mists of time may well have introduced subtle alterations in the interpretation and details of the original communications, the contention would be that the central message would indeed have survived essentially unscathed.

All this seems very different from the way that scientific knowledge has developed. Yet it is all too easy for scientists to become complacent and to take the firm proclamations of science to be immutable. We have, in fact, witnessed various significant changes in scientific belief that have, at least in part, overthrown those that had previously been firmly held. However, these changes have come about only with reluctance on the part of those holding to the previous views, and normally come only in the face of very impressive new observational evidence. Kepler's elliptical planetary motion is a case in point, overthrowing the previous ideas of circles and of circles upon circles. The experiments of Faraday and the equations of Maxwell were to herald in another great change in our scientific viewpoint concerning the nature of material substance, showing that the discrete individual particles of Newtonian theory needed to be supplemented by continuous electromagnetic fields. Even more startling were the two great revolutions of twentieth-century physics, namely relativity and quantum mechanics. I have discussed some of the remarkable ideas of special and general relativity in §§1.1 and 1.2, and especially §1.7. But even these revolutions, impressive as they were, pale almost into insignificance when compared with the startling revelations of the quantum theory. It is, indeed, this *quantum revolution* that is the subject of this chapter.

In § 1.4, we have already witnessed one of the strangest features of quantum mechanics: as a consequence of the quantum superposition principle a particle can occupy two distinct locations at the same time! This certainly represents a departure from our cosy Newtonian picture of individual particles, where each one occupies a clear-cut single location. It is evident that such a crazy-seeming description of reality, as appears to be provided by quantum theory, would not have been taken seriously by respectable scientists were it not for the fact that there is a great deal of observational evidence to support such things. Not only this, but once one has become accustomed to the quantum formalism and mastered many of the subtle mathematical procedures that are involved in it, explanations of a vast array of observed physical phenomena begin to present themselves, many of these phenomena having seemed previously to be totally mysterious.

Quantum theory explains the phenomenon of chemical bonding, the colours and physical properties of metals and other substances, the detailed nature of the discrete frequencies of light that particular elements and their compounds emit when heated (spectral lines), the stability of atoms (where classical theory would predict a catastrophic collapse with the emission of radiation as electrons spiral rapidly into their atomic nuclei), superconductors, superfluids, Bose–Einstein condensates; in biology, it explains the puzzling discreteness of inherited characteristics (as first unearthed by Gregor Mendel, in around 1860, and basically explained by Schrödinger, in 1943, in his ground-breaking book *What Is Life?* [see Schrödinger 2012] even before DNA had properly entered the scene); in cosmology, the microwave background radiation that permeates our whole universe (and which will be central to our discussions in §§3.4, 3.9, and 4.3) has a black-body spectrum (see §2.2) whose exact form comes directly from the earliest considerations of an essentially quantum process. Many modern physical devices depend crucially on quantum phenomena, and their construction requires serious understanding of the underlying quantum mechanics. Lasers, CD and DVD players, and laptop computers all crucially involve such quantum ingredients, as do the superconducting magnets that propel particles at practically the speed of light around the 27 km long tunnels of the LHC in Geneva. The list seems almost endless. So we must, indeed, take quantum theory seriously, and accept that it provides a compelling description of physical reality that goes greatly beyond the classical pictures that had been held to so firmly in the centuries of scientific understanding that had preceded the advent of this remarkable theory.

When we combine quantum theory with special relativity we get quantum field theory, which is essential for, in particular, modern particle physics. Recall from §1.5 that quantum field theory explains the value of the magnetic moment of the electron correctly to 10 or 11 significant figures, once the appropriate renormalization procedure for dealing with divergences has been properly incorporated. There are several other examples, and these provide powerful evidence for the extraordinary accuracy that is inherent in quantum field theory when it is appropriately applied.

Quantum theory is commonly regarded as a deeper theory than the classical scheme of particles and forces that had preceded it. Whereas, on the whole, quantum mechanics has been used to describe relatively small things, such as atoms and the particles that compose them, and the molecules which those atoms compose, the theory is not restricted to a description of such elemental constituents of matter. Collections of very large numbers of electrons are often involved in the strange and very quantum-mechanical nature of superconductors, for example, and of hydrogen atoms (around $10^9$ of them) in Bose–Einstein condensates [Greytak et al. 2000]. Moreover, quantum-entanglement effects are now observed to stretch over 143 km [Xiao et al. 2012], where pairs of photons that are separated to that distance must nevertheless be treated as single photon-pair quantum objects. There are also observations that provide measurements of the diameters of distant stars, which depend upon the fact that pairs of photons emitted from opposite sides of the star are automatically entangled with each other because of the Bose–Einstein statistics referred to in §1.14. This effect was impressively established and explained by Robert Hanbury Brown and Richard Q. Twiss (the *Hanbury Brown–Twiss effect*), in 1956, when they correctly measured the diameter of Sirius as being some 2.4 million kilometres, thereby establishing this type of quantum entanglement over such distances [see Hanbury Brown and Twiss 1954, 1956a,b]! Thus it appears that quantum influences are by no means restricted to small distances, and there seem to be no reasons to expect that there is any limit to the distances over which such effects can stretch. Moreover, in a clearly accepted sense, there are *no* observations whatever, to date, which *contradict* the expectations of the theory.

The dogma of quantum mechanics is thus seen to be very well founded indeed, as it rests on an enormous amount of extremely hard evidence. With systems that are simple enough that detailed calculation can be carried out and sufficiently accurate experiments performed, we find an almost incredible precision in the agreement between the theoretical and observational results that are obtained. Moreover, the procedures of quantum mechanics have been successfully applied over a huge range of scales; for, as remarked above, we find that their effects stretch from the scale of elementary particles to atoms, to molecules, to entanglements of over distances of some 150 km, or of the millions of kilometres across the body of a star, and there are precise quantum effects relevant even to the scale of the entire universe (see §3.4).

This dogma does not come from the pronouncements of one single historic individual, but from the agonized deliberations of many dedicated theoretical scientists, each of outstanding ability and insight: Planck, Einstein, de Broglie, Bose, Bohr, Heisenberg, Schrödinger, Born, Pauli, Dirac, Jordan, Fermi, Wigner, Bethe, Feynman, and many others found themselves driven to their mathematical formulations from the results of experiments performed by an even larger number of highly skilled experimenters. It is striking that, in this respect, the origins of quantum mechanics differed very greatly from those of general relativity, the latter coming almost entirely from the theoretical considerations of Albert Einstein alone,[1] and without significant input from observation beyond those of Newtonian theory. (It appears that Einstein was well aware of the previously observed slight anomaly in the motion of the planet Mercury,[2] and this could certainly have had an influence on his early thinking, but there is no direct evidence of this.) Perhaps the multitude of theoreticians involved in the formulation of quantum mechanics is a manifestation of the totally non-intuitive nature of that theory. Yet, as a mathematical structure, there is a remarkable elegance; and the deep coherence between the mathematics and physical behaviour is often as stunning as it is unexpected.

Given all this, it is perhaps not at all surprising that, for all its strangeness, the dogma of quantum mechanics is often taken as absolute truth, so that any phenomenon of nature is regarded as necessarily having to conform to it. Quantum mechanics provides, indeed, an overarching framework that would appear to apply to any physical process, at no matter what scale. There is perhaps no puzzle, therefore, in the fact that a profound *faith* has arisen among physicists, that *all* the phenomena of nature must adhere to it. The extreme oddities that seem to arise when this particular principle of faith is applied to everyday experience must, accordingly, be just things that we have to come to terms with, to get used to, and somehow to understand.

---

[1] With regard to the necessary mathematical formulation of this theory, however, Einstein needed to enlist the assistance of his colleague Marcel Grossmann. It should also be mentioned that special relativity, on the other hand, must be regarded as a several-person theory, as Voigt, FitzGerald, Lorentz, Larmor, Poincaré, Minkowski, and others in addition to Einstein made significant contributions to its birth [see Pais 2005].

[2] General relativity was put forward in 1915; Einstein mentioned the perihelion of Mercury in a letter of 1907 (see footnote 6 on p. 90 in J. Renn and T. Sauer's chapter in Goenner [1999]).

Most particularly, there is the quantum-mechanical conclusion, pointed out to the reader in §1.4, that a quantum particle can exist in a state where it simultaneously occupies two separated locations. Despite the fact that the theory asserts that the same could apply to any macroscopic body – even to the cat referred to in §1.4, which would then seem to be able to go through two separate doors at the *same time* – this is not the kind of occurrence that we ever experience, nor do we have reason to believe that such separated coexistence could ever actually occur on a macroscopic scale, even when our cat is completely out of our sight at the time of her passing through the door(s). The issue of Schrödinger's cat (but where Schrödinger's original version had the cat in a superposed state of life and death [Schrödinger 1935]) will indeed loom large in our deliberations in later sections of this chapter (§§2.5, 2.7, and 2.13), and we shall find, despite our currently presumed quantum faith, that these matters *cannot* be dismissed at all lightly, and that there should, indeed, be a profound *limit* to our quantum faith.

## 2.2. MAX PLANCK'S $E = h\nu$

I shall need to be a good deal more explicit about the structure of quantum mechanics from now on. Let us start by looking at the main initial reason for believing that something beyond classical physics is needed. We shall consider the original circumstance that led the very distinguished German scientist Max Planck, in 1900, to postulate something that was, from the perspective of all the physics that was known at the time, totally outrageous [Planck 1901] – though how outrageous it actually was seems not to have been properly appreciated by Planck, nor by any of his contemporaries at the time. What Planck was concerned with was a situation where matter and electromagnetic radiation, confined in a non-reflective cavity, are in equilibrium with the heated material of the cavity (figure 2-1) and kept at a given temperature. He found that the emission and absorption of electromagnetic radiation by that material had to take place in discrete bundles of energy according to the now famous formula

$$E = h\nu.$$

Here $E$ is the energy bundle just referred to, $\nu$ is this radiation's frequency, and $h$ is a fundamental constant of nature now referred to as *Planck's constant*. Planck's formula was later recognized as providing a basic connection between energy and frequency that holds *universally*, according to quantum mechanics.

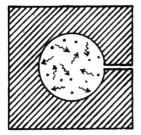

Figure 2-1:  A black-body cavity, with black internal surface, contains matter and electromagnetic radiation in equilibrium with its heated cavity boundary.

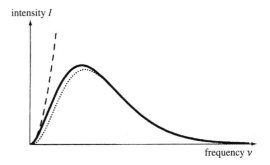

Figure 2-2:  The solid line denotes the observed relation between the intensity $I$ of black-body radiation and the frequency $\nu$ of radiation, which is in precise accord with Max Planck's famous formula. The broken line depicts the relation given by Rayleigh–Jeans formula and the dotted one that given by the Wien formula.

Planck had been trying to explain an experimentally observed functional relationship between the intensity and the frequency of radiation, shown as the continuous curve in figure 2-2, in the situations just described. This relationship is referred to as the *black-body spectrum*. It arises in situations in which radiation and matter, interacting with each other, are together in equilibrium.

There had been previous suggestions for what this relationship should be. One of these, referred to as the *Rayleigh–Jeans formula*, gave the intensity $I$ as a function of the frequency $\nu$, as provided by the expression[3]

$$I = 8\pi k c^{-3} T \nu^2,$$

where $T$ is the temperature, $c$ is the speed of light, and $k$ is a physical constant known as *Boltzmann's constant*, which will play important roles for us later, especially in chapter 3. This formula (plotted as the broken line in figure 2-2) was based on a purely classical (Maxwell field) interpretation of the electromagnetic field. Another suggestion, *Wien's law* (which can be derived from a picture that electromagnetic radiation consists of randomly moving classical massless particles), was

$$I = 8\pi h c^{-3} \nu^3 e^{-h\nu/kT}$$

(the dotted line in figure 2-2). After much effort, Planck found that he was able to derive a very accurate formula for this intensity–frequency relationship, which happened to fit well with the Wien formula at large frequencies $\nu$ and with that of Rayleigh–Jeans when $\nu$ is small, namely

$$\frac{8\pi h c^{-3} \nu^3}{e^{h\nu/kT} - 1}$$

but in order to do this, he needed to make this very strange assumption that the emission and absorption of radiation by the material was indeed always in discrete bundles, in accordance with the above $E = h\nu$.

It does appear that Planck seems not at first to have properly appreciated the very *revolutionary* nature of his assumption, and the matter had to wait some five years before Einstein [see Pais 2005; Stachel 1995] clearly realized that electromagnetic radiation itself had to be made up from individual bundles of energy in accordance with the above Planck formula, these bundles being later called *photons*. In fact, there is a very basic reason (not then actually explicitly employed either by Planck or by Einstein), implicit in the equilibrium nature of black-body radiation and an assumed particulate nature of material bodies, that electromagnetic radiation (i.e. light) must also have a particulate nature, this coming from a principle known as *equipartition of energy*. This principle asserts that as a finite system approaches equilibrium, the energy ultimately becomes equally distributed, on average, among all its degrees of freedom.

This deduction from the equipartition principle can be thought of as yet another exercise in the issue of functional freedom (§§A.2, A.5, 1.9, 1.10, and 2.11). Let

---

[3] In the various formulae given in this section, there is an initial "$8\pi$" that is sometimes written as "2". This is a purely conventional matter concerning the meaning of the term *intensity* in these expressions.

us suppose that our system consists of $N$ individual particles in equilibrium with a continuous electromagnetic field (where energy can be transferred between the two due to the presence of electric charges on some of the particles). The functional freedom in the particles would be $\infty^{6N}$, assuming for simplicity that we are dealing with classical point particles, so the "6" refers to each particle's 3 position degrees of freedom and 3 momentum degrees of freedom (*momentum* being essentially mass times velocity; see §§1.5, A.4, and A.6). Otherwise we might need some more parameters to describe its *internal* degrees of freedom, and replace the "6" by some larger integer. For example, a classically spinning *irregular*-shaped "particle" would have 6 more degrees of freedom: 3 for its spatial orientation and 3 for the direction and magnitude of its angular momentum (cf. §1.14), which in total would give 12 per particle, resulting in $\infty^{12N}$ for the functional freedom of the whole classical $N$-particle system. We shall be seeing in §2.9 that, in quantum mechanics, these numbers are somewhat different, but we would still get a functional freedom of the form $\infty^{kN}$ for some integer $k$. However, in the case of the continuous electromagnetic field, we have the enormously larger functional freedom $\infty^{4\infty^3}$, as follows directly from the discussion of §A.2 (applied to the magnetic and electric fields separately).

What, then, would equipartition of energy tell us about a classical system in which particulate matter and continuous electromagnetic field are together in equilibrium? It would tell us that, in approach to equilibrium, more and more of the energy would have to find its way into those enormously greater number of degrees of freedom that reside in the field, finally draining this energy completely from the degrees of freedom in the particles of the matter. This is what Einstein's colleague, the physicist Paul Ehrenfest, would later call the *ultraviolet catastrophe*, as it is in the high-frequency end (i.e. the ultraviolet end) of the spectrum where the catastrophic draining of degrees of freedom into the electromagnetic field would ultimately occur. The infinite upward rise of the broken-line (Rayleigh–Jeans) curve in figure 2-2 is an illustration of this difficulty. But when a particulate nature is also assigned to the field, where this structure becomes more and more relevant when the frequencies get higher, the catastrophe is averted and consistent equilibrium states can exist. (I shall return to this matter in §2.11, and try to address the functional-freedom issue arising here a little more fully.)

We see from this general argument, moreover, that this is not merely a feature of the electromagnetic field. Any system consisting of continuous fields in interaction with discrete particles would be subject to the same difficulties, whenever an approach to equilibrium is contemplated. Accordingly, it is not an unreasonable expectation that Planck's saving relationship $E = h\nu$ might well also hold for other fields. One is tempted to believe, indeed, that it should be a *universal* feature of physical systems. In fact, this turns out to be the case for quantum mechanics, completely generally.

The profound significance of Planck's work in this area went virtually unnoticed, however, until Einstein, in 1905, published a (now-famous) paper [Stachel 1995, p. 177] in which he put forward the extraordinary idea that, in appropriate circumstances, an electromagnetic field should be treated as though it actually *consisted* of a system of particles, rather than as a continuous field. This was particularly shocking to the physics community because the viewpoint had apparently become well established that electromagnetic fields are thoroughly described by James Clerk Maxwell's beautiful system of equations (see §1.2). Most satisfyingly, Maxwell's equations had provided what appeared to be a complete description of light as self-propagating waves of electromagnetism, where this wave interpretation had explained many detailed properties of light, such as polarization and interference effects, and had led to the prediction of types of light other than the kind that we directly see (i.e. other than the visible spectrum), such as radio waves (when the frequency is much lower) and X-rays (when the frequency is much higher). To suggest that, after all, light should be treated as *particles* – apparently in accordance with the seventeenth-century proposal of Newton, a proposal that had eventually been (seemingly) convincingly refuted by Thomas Young in the early nineteenth century – was fairly outrageous, to say the least. Perhaps even *more* remarkable was the fact that in the very same year of 1905 (Einstein's "miraculous year" [Stachel 1995, pp. 161–64 for the $E = mc^2$ paper and pp. 99–122 for relativity]), Einstein had himself, a little later in the year in fact, based his even more famous two papers introducing special relativity (the second containing his "$E = mc^2$" equation) upon the firm validity of Maxwell's equations!

To cap it all, even in the *very* paper in which Einstein proposed his particle ideas of light, he wrote of Maxwell's theory that it "will probably never be replaced by another theory" [Stachel 1995, p. 177]. Though this would seem to be a contradiction to the very purpose of his paper, we can see, with the hindsight of our more modern treatment of quantum fields, that Einstein's particulate perspective on electromagnetism is, in some deep sense, *not* inconsistent with Maxwell's field, since the modern quantum theory of the electromagnetic field arises through the general process of field quantization when this procedure is applied to, indeed, *Maxwell's theory*! It may also be remarked that Newton himself had realized that his particles of light needed to have some characteristically *wave-like* properties [see Newton 1730]. Yet, there were, indeed, even in Newton's day, some deep reasons for having sympathy with a particle-like picture of light propagation. In my opinion, these would undoubtedly have struck a chord with Newton's thinking [Penrose 1987b, pp. 17–49], and I believe that Newton actually had some very good reasons for his particulate/wave-like view of light, these being of clear relevance even today.

It should be made clear that this particle perspective on physical fields (so long as the particles are actually bosons – and thence satisfy the behaviour needed for the Hanbury Brown–Twiss effect, referred to in §2.1) does not invalidate their thoroughly field-like behaviour at comparatively low frequencies (i.e. long wavelengths), as is in fact directly observed. The quantum constituents of nature are, in some sense, entities which cannot be regarded completely accurately as either particles or fields but as some more mysterious intermediate kind of entity (say, a wave–particle) which exhibits aspects of both. The elementary components, the quanta, are all subject to Planck's $E = h\nu$ law. In a general way, when the frequency, and therefore the energy per quantum, is very large, so the wavelength is small, the particle-like aspect of a collection of such entities becomes more dominant, so we tend to get a good picture if we think of it as composed of particles. But when the frequency (and hence the individual particle energy) is small, so we are thinking of relatively large wavelengths (and extremely vast numbers of low-energy particles), then the classical field picture tends to work well.

At least, that is the case for bosons (see §1.14). With fermions, the long-wavelength limit does not really resemble a classical field, because the very large number of particles that would need to be involved would start to "get in the way" of one another owing to the Pauli principle (see §1.14). However, in certain circumstances, such as with a superconductor, electrons (which are fermions) can pair up to form what are known as *Cooper pairs*, these pairs behaving very much like individual bosons. These bosons together give rise to the supercurrent of a superconductor, which can maintain itself indefinitely without input from outside, and which has some coherent attributes of a classical field (though being perhaps more like the Bose–Einstein condensate referred to briefly in §2.1).

This universal character of Planck's $E = h\nu$ suggests that the particle-like aspect of fields should have a converse in which those entities that we naturally think of as ordinary particles ought also to have a field-like (or wave-like) aspect to them. Accordingly, $E = h\nu$ ought also to apply to such ordinary particles in some sense, where we now assign a wave-like attribution to individual particles, whose frequency $\nu$ would be determined by its energy, in accordance with $\nu = E/h$. This turns out to be the case, and such a proposal was specifically put forward

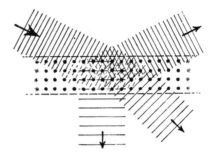

Figure 2-3: Evidence for the wave-like nature of electrons is seen in the Davisson–Germer experiment, where electrons are fired at a crystalline material. It is found that scattering or reflection occurs when the crystalline structure matches the de Broglie wavelength of the electrons.

by Louis de Broglie in 1923. The requirements of relativity tell us that a particle of mass $m$, in its own rest frame, should have an energy $E = mc^2$ (Einstein's famous formula), so by Planck's relation, de Broglie assigns it a natural frequency $v = mc^2/h$, as already noted in §1.8. But when the particle is in motion, it also acquires a momentum $p$, and the requirements of relativity also tell us that this ought, correspondingly, to be inversely proportional to a naturally associated wavelength $\lambda$, where

$$\lambda = \frac{h}{p}.$$

This de Broglie formula is now amply confirmed in innumerable experiments, in the sense that a particle with this momentum value $p$ will experience interference effects as though it were a wave with that wavelength $\lambda$. One of the clearest early examples was the Davisson–Germer experiment, performed in 1927, in which electrons are fired at a crystalline material, and scattering or reflection occurs when the crystalline structure matches the de Broglie wavelength of the electrons (figure 2-3). Conversely, Einstein's earlier proposal of a particle-like character for light had already explained the 1902 observations of Philipp Lenard, concerning the *photoelectric effect*, in which high-frequency light aimed at a metal dislodges electrons, these emerging individually with a specific energy which depends on the wavelength of the light but, surprisingly, does not depend on the light's intensity. These results, very puzzling at the time, were just what Einstein's proposal explains (and it was this which eventually won him the 1921 Nobel Prize) [Pais 2005]. A more direct crucial confirmation of Einstein's proposal was later found by Arthur Compton in 1923, where X-ray quanta fired at charged particles were indeed found to respond just as massless particles – now called *photons* – would, in accordance with standard relativistic dynamics. This makes use of the same formula as de Broglie's, but now read the other way around, assigning a momentum to individual photons of wavelength $\lambda$ according to $p = h/\lambda$.

## 2.3. THE WAVE–PARTICLE PARADOX

So far, we have not got very far with the actual structure of quantum mechanics. Somehow, we have to bring together the wave and particle aspects of our basic quantum ingredients into something a bit more explicit. To understand the issues more clearly, let us consider two different (idealized) experiments, both very similar, but where one of them brings out the particle-like aspect and the other the wave-like aspect of a quantum wave–particle entity. For convenience, I shall simply refer to this entity as a *particle*, or more explicitly as a *photon*, since this type of experiment is in fact most readily carried out in practice with photons. But we should bear in mind that the same kind of thing would apply if it were an electron, a neutron, or any other kind of wave–particle. In my descriptions I shall ignore all sorts of technical difficulties that there might be in performing the experiments in practice.

In each experiment, using an appropriate kind of laser – situated at the point L in figure 2-4(d) – we fire a single photon at what we may think of as a half-silvered mirror, located at M, although, in actual experiments, this mirror would be unlikely to involve the kind of silvering that an ordinary hand mirror usually employs. (Better mirrors, in this kind of quantum-optics experiment might well deliberately take advantage of the wave-like aspect of our photon, via interference effects, but this is not important for our discussion here.) Technically, such a thing would be called a *beam splitter*. What I require of the beam splitter, in these experiments is that it be oriented at 45° to the laser beam, and exactly half of the light that impinges upon it is reflected (by a right angle) and exactly half is transmitted directly through it.

In the first experiment, experiment 1, which is illustrated in figure 2-4(a), there are two detectors: one at A, which is in the transmitted beam (so that LMA is a straight line), and the other at B, which is in the reflected beam (so that LMB is a right angle). I am supposing (for ease of discussion) that each detector is 100%

(a)                                    (b)

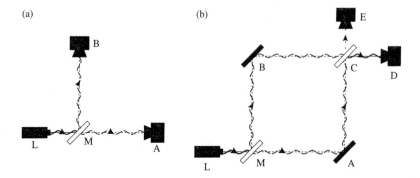

Figure 2-4: The wave–particle aspects of a photon, fired by a laser at L, directed towards the beam splitter at M. (a) Experiment 1: the photon's particle-like behaviour is illustrated by detectors at A and B, exactly one of which detects a photon at each emission. (b) Experiment 2: illustrates wave-like behaviour of a photon (Mach–Zhender interferometer) with mirrors at A and B, detectors at D and E, and a second beam splitter at C; now only D receives an emerging photon.

perfect, in the sense that it registers if and only if it receives the photon; moreover, I assume that the rest of the experimental set-up is also perfect, so the photon is not lost by absorption, misdirection, or any other kind of malfunction. I am also assuming that at each emission of a photon, the laser has a means to keep track of the fact that an emission has taken place.

Now, what happens with experiment 1 is that at each emission of a photon either the detector at A registers reception of a photon or the detector at B registers reception of a photon, but not both. There is a 50% probability of either result. This is exhibiting the *particle*-like aspect of the photon. The photon has either gone one way or the other, the results of this experiment being consistent with a particle-like premise that a decision is made, just as the photon encounters the beam splitter, that it should either be transmitted or reflected, there being a 50% probability for each choice.

Let us next turn to the second experiment, experiment 2, illustrated in figure 2-4(b). Here we replace the detectors at A and B with (fully silvered) mirrors, tilted at 45° to their respective beams so that in each case the beam that the mirror encounters is aimed at a second beam splitter, at C, of identical type to the first one, this being again oriented at 45° to the beams (so all mirrors and beam splitters are parallel to one another). See figure 2-4(b). The two detectors are now placed at locations D and E (where CD is parallel to LMA and CE is parallel to MB), so that MACB is a rectangle (here shown as a square). This arrangement is known as a *Mach–Zhender interferometer*.

Now, what happens when the laser emits a photon? It would seem that, in accordance with experiment 1, the photon would leave the beam splitter at M with a 50% chance of taking the route MA and hence being reflected along AC, and a 50% chance of taking the route MB and hence being reflected along BC. Accordingly, the beam splitter at C would find a 50% probability of encountering a photon coming along AC, which it would dispatch with equal probabilities to the detector at D or at E, and it would find a 50% probability of encountering a photon coming along BC, which it would accordingly dispatch with equal probabilities to the detector at D or at E. This would result in the detectors at D and E each finding a 50% (= 25% + 25%) chance of receiving the photon.

However, this is *not* what happens! Innumerable experiments of this and similar types have been performed, and they all tell us is that there is 100% probability of the detector at D receiving the photon and 0% probability of E receiving it! This is completely incompatible with the particle description of the photon's action, as I have described it in relation to experiment 1. On the other hand, what is found is much more the kind of thing that we would expect if we think of the photon as a little wave. Now, we may picture what is happening at the beam splitter M, with the set-up described (and with the assumption of a perfect experiment), to be that the wave splits into two lesser disturbances, one travelling along MA and the other along MB. These are reflected at, respectively, the mirrors A and B, so that the second beam splitter C encounters two wave-like disturbances simultaneously arriving from each of the two directions AC and BC. Each of these is split by the beam splitter C into a component along CD and a component along CE, but to see how these combine with one another we must look carefully at the phase relations between the crests and troughs of the two superposed waves. What we find (with the equal arm lengths being assumed here) is that the two emergent components coming together out along CE completely *cancel* one another out, since the crests of one coincide with the troughs of the other, whereas the two components emerging along CD *reinforce* one another, with the crests coinciding and the troughs coinciding. Hence the whole emergent wave, coming from the second beam splitter will be along the direction CD, with nothing at all aimed along CE, all this being consistent with 100% detection at D and 0% detection at E, consistently with what would be actually observed in practice.

This rather suggests that a wave–particle might best be thought of as what is referred to as a *wave-packet*, which would be a small burst of oscillatory wave-like activity, but constrained within a small region, so that on a large scale it provides a small localized disturbance resembling a particle. (See figure 2-11 in §2.5, for a picture of a wave-packet within the standard quantum formalism.) However, such a picture has only a very limited explanatory value in quantum mechanics, for all sorts of reasons. In the first place, the wave forms that are frequently used in this type of experiment do not at all resemble such wave-packets, as a single photon wave might well have a wavelength that spreads to much longer than the dimensions of the entire apparatus. Much more important is the fact that such pictures do not come close to explaining what is going on in experiment 1, for which we now return to figure 2-4(a). Our wave-packet picture would (to be consistent with the results of experiment 2) have our photon wave being split into two smaller wave-packets at the beam splitter at M, one of which would be aimed at A and the other aimed at B. In order to reproduce the results of experiment 1, we appear to have to take the view that the detector at A, upon receiving this smaller-sized wave-packet, would have to find that it has a probability of 50% of being activated by this, in accordance with its registering the reception of the photon. The same would apply to the detector at B, so it also would have to find that it has a 50% probability of registering the reception of a photon. This is all very well, but it does not agree with what actually happens, because we now find that whereas this model does give equal probabilities of detections at A and at B, half of these are responses are ones which in fact *never occur* – for this proposal incorrectly predicts that there would be 25% chance that *both* detectors register reception of a photon and a 25% chance that *neither* detector receives one! These two types of combined detector responses do not occur because the photon is neither lost nor duplicated in this experiment. This type of wave-packet description of a single photon simply does not work.

The quantum behaviour of things is indeed much more subtle than this. The wave that describes a quantum particle is not like a water wave or a sound wave, which would describe some kind of *local* disturbance in an ambient medium, so that the effect that one part of the wave might have on a detector in one region is independent of the effect that another part of the wave might have on a detector in some distant region. We see from experiment 1 that the wave picture of a single photon, after it has been "split" into two simultaneous separated beams by a beam splitter, still represents just a *single* particle, despite this separation. The wave appears to describe some kind of *probability distribution* for finding the particle in various different places. This is getting somewhat closer to a description of what the wave is actually doing and, indeed, sometimes people refer to such a wave as a *probability wave*. This, however, is still not a satisfactory picture,

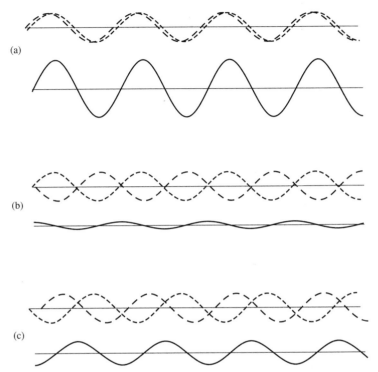

Figure 2-5: The sum of two wave modes of equal amplitude and frequency (shown here with broken curves) can (a) reinforce, (b) cancel, or (c) something intermediate between the two, depending on the phase relation between the modes.

because probabilities, being always positive quantities (or zero) cannot cancel one another out, as would be needed for an explanation of the absence of any responses at E, in experiment 2.

Sometimes people attempt a probability-wave type of explanation of this nature by allowing the probabilities somehow to become *negative* in places, so that cancellation can then take place. However, this is not really how quantum theory operates (see figure 2-5). Instead, it goes one step further than this by allowing the wave amplitudes to be *complex numbers*! (See §A.9.) In fact, we have already encountered some mention of these complex amplitudes in §§1.4, 1.5, and 1.13. These complex numbers are crucial to the whole structure of quantum mechanics. They do indeed have some close relationship to probabilities, but they are *not* probabilities (which of course they cannot be because probabilities are real numbers). But the role of complex numbers in the quantum formalism is much more all-embracing than just that, as we shall be seeing shortly.

## 2.4. QUANTUM AND CLASSICAL LEVELS: **C**, **U**, AND **R**

In 2002 I was invited to give a talk in Odense, Denmark, by the Hans Christian Andersen Academy. Andersen, who was born in Odense in 1805, was approaching his bicentenary. I suppose that the reason I was asked was that I had written a book with the title *The Emperor's New Mind*, which had been inspired by Andersen's *The Emperor's New Clothes*. I thought that I should talk on something different, however, and I wondered whether there might be another Hans Christian Andersen story I could use to illustrate some aspect of the ideas I had recently been concerned with, these having been largely to do with the foundations of quantum mechanics. After some thought, it occurred to me that the story of the *Little Mermaid* could be used in various ways to illustrate issues I intended to discuss.

Have a look at figure 2-6, where I have depicted the mermaid sitting on a rock, with one half of her body under water and the other in the open air. The lower part of the picture depicts what is happening beneath the sea, as an entangled mess of activity involving many odd-looking creatures and alien entities, but with perhaps a kind of special beauty of its own. This represents the strange and unfamiliar world of quantum-level processes. The upper part of the picture depicts the kind of world that we are more familiar with, where different objects are well separated and constitute things that behave as independent objects. This represents the classical world, acting according to laws we have become accustomed to, and had understood – prior to the advent of quantum mechanics – as precisely governing the behaviour of things. The mermaid herself straddles the two, being half fish and half person. She represents the link between the two mutually alien worlds. See figure 2-7. She is also mysterious and apparently magical, as her ability to form her link between these worlds seems to defy the laws of each. Moreover, she brings, from her experiences of the world beneath, a different perspective on our world above, appearing to look down upon it from a great height from her vantage point on the rock.

It is the normal belief of physicists – and it is my own belief also – that there should *not* be fundamentally different laws governing different regimes of physical phenomena, but just *one* overriding system of fundamental laws (or general principles) governing *all* physical processes. On the other hand, some

Figure 2-6:  Inspired by Hans Christian Andersen's *The Little Mermaid*, to illustrate the magic and mystery of quantum mechanics.

philosophers – and no doubt even a fairly significant number of physicists also – take the view that there could well be different levels of phenomena at which fundamentally different physical laws apply, with no necessity for an overriding consistent scheme encompassing them all [see, for example, Cartwright 1997]. Of course, when the external circumstances do become very different from our normal experience, it may be expected that different laws from those we are used to – or, rather, unusual *aspects* of the overreaching fundamental laws – may acquire a previously unfamiliar importance. In practice, it may well be possible, in such circumstances, even to ignore some of the laws that would have been of especial significance in previously familiar situations. Indeed, in any particular circumstance of immediate concern, we should certainly pay most of our attention to those laws that have the greatest relevance, and we may well get away with ignoring the others. Nevertheless, any *fundamental* laws that we are ignoring,

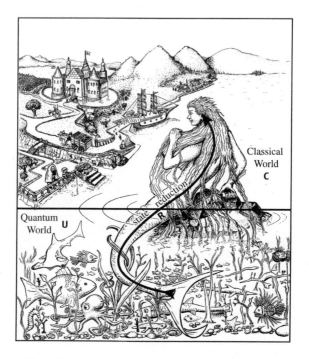

Figure 2-7: The upper half of the picture represents the familiar classical world **C** of separate constituents, and the lower part the alien entangled quantum world **U**. The mermaid straddles the two, representing the mysterious **R**-process that allows quantum entities to enter the classical world.

owing to the smallness of their effects, might still have indirect influence. This, at least, is the generally accepted belief of physicists. We expect that – indeed we may a *faith* that – physics as a whole must be a *unity*, and even when there may be no immediate role for some particular physical principle, this principle ought still to have its underlying role within the entire picture, and may well contribute importantly to the consistency of that picture as a whole.

Accordingly, the worlds depicted in figure 2-7 should not be thought of as really being alien to one another, but with our present understanding of quantum theory and its relation to the macroscopic world, it is merely a convenience for us to treat things as though they inhabit different worlds that obey different laws. In practice, it is certainly the case that we *do* tend to use one set of laws for what I shall refer to as the *quantum level* and another set for the *classical level*. The borderline between these two levels is never made very clear, and there is the common view that classical physics is, in any case, merely a convenient approximation to the "true" quantum physics that is taken to be satisfied *exactly* by its basic constituents. It would be taken that the classical approximation normally works extremely well when there are vast numbers of quantum particles involved. I shall be arguing later (in §§2.13 and 4.2 in particular) that there are, nevertheless, certain severe difficulties involved in holding too strongly to this kind of convenience viewpoint. But for the moment, let us try to go along with it.

Thus, in a general way, we shall think of quantum-level physics as applying precisely to "small" things but that the more readily understood classical-level physics holds, very closely, with "large" things. Yet, we must be careful about the use of the words "small" and "large" in this context because, as has been remarked in §2.1 above, quantum effects can in certain circumstances be observed to stretch over great distances (certainly over 143 km). Later, in §§2.13 and 4.2, I shall be presenting a viewpoint that argues for a different criterion from that of mere distance to be the relevant one to characterize the onset of classical-like behaviour, but for the moment it will not be too important for us to have any such specific criterion in mind.

Accordingly, let us accede to this consensus view for the time being, whereby we may regard the division into classical and quantum worlds as a mere convenience, and that things that are, in some unspecified sense, "small" are to be treated by the dynamical equations of quantum theory, and those that are "large" are treated as behaving extremely closely in accordance with those of classical dynamical theory. In any case, this is certainly a viewpoint which is almost always adopted in practice, and it will be helpful for us in our understandings of how quantum theory is actually used. The classical world seems indeed to be extremely closely governed by the classical laws of Newton, augmented by those of Maxwell to describe continuous electromagnetic fields, and those being further augmented by the Lorentz force law which describes how individual charged particles respond to electromagnetic fields; see §§1.5 and 1.7. If we consider matter in very rapid motion, we need to bring in the laws of special relativity, and when we involve sufficiently significant gravitational potentials, we need also to take into account Einstein's general relativity. These laws combine into a coherent whole, in which behaviour is governed – via differential equations; see §A.11 – in a precise *deterministic* and *local* way. The space-time behaviour is derivable from data that can be specified at any one particular time (where in general relativity we interpret "at one particular time" as "on an appropriate initial spacelike surface"; see §1.7). In this book, I have refrained from entering into any details of calculus, but all we need to know here is that these differential equations govern the future (or past) behaviour of the system precisely in terms of its state (and state of motion) at any one time. I use the letter **C** to denote all this classical evolution.

The quantum world, on the other hand, has a time evolution – which I denote by the letter **U**, standing for *unitary* evolution – described by a different equation, called the *Schrödinger equation*. This is still a deterministic and local time evolution (governed by a differential equation; see §A.11), and it applies to a mathematical entity called the *quantum state*, introduced into quantum theory to describe a system at any one moment. This determinism is very similar to what we have in classical theory, but there are various key differences from the classical evolution process **C**. In fact, some of these differences, particularly some consequences of the *linearity* that we shall be coming to in §2.7, have implications that are so alien to our experiences of the actual behaviour of the world that it becomes totally unreasonable to try to continue to use **U** for our descriptions of reality after macroscopically discernable alternatives become involved. Instead, in standard quantum theory, we adopt a third procedure, called *quantum measurement*, that I shall denote by the letter **R** (standing for the *reduction* of the quantum state). This is where the mermaid herself plays her vital role, performing this necessary link connecting the quantum world with the classical world of our experiences. The **R** procedure (see §2.8) is completely different from the deterministic evolutions of either **C** or **U**, being a *probabilistic* action, and (as we shall be seeing in §2.10) it exhibits curiously non-local features that defy all understanding in terms of the classical laws that we are familiar with.

To get some feeling for the role of **R** let us consider the action of a *Geiger counter*, which is the familiar device for the detection of energetic (charged) particles resulting from radioactivity. Any individual such particle would be considered as a quantum object, subject to the quantum-level laws of **U**. But the Geiger counter, being a classical measuring device, has the effect of magnifying the achievements of our little particle from the quantum to the classical level, the device's detection of this particle resulting in an audible click. Since this click is something that we can directly experience, we treat it as an inhabitant of our familiar classical world, providing it with a classical existence in terms of the vibrations of wave motions in the air that can be very adequately described by the classical (Newtonian) equations of fluid (aerodynamic) motion. In short, the effect of **R** is to replace the continuous evolution that **U** has provided us with, by a sudden *jump* to one of several possible classical **C**-descriptions. In the Geiger-counter example, these alternatives would, prior to detection, all come

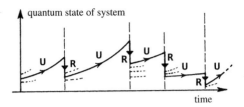

Figure 2-8: The way that the quantum-theoretic world appears to behave, with stretches of deterministic **U**-evolution, punctuated by moments of probabilistic **R**-action, each of which restores some element of classicality.

from various contributions to the particle's **U**-evolved quantum state, in which the particle might be in one place or another, moving this way or that, all taken together in the kind of quantum superposition considered in §1.4. But when the Geiger counter gets involved, these quantum-superposed alternatives become various possible classical outcomes, whereby a *click* takes place at one time or another, these different times being assigned probability values. This is all that remains of the U-evolution, in our calculational procedure.

The actual procedure that we normally adopt for such situations is very much in accordance with the approach of Niels Bohr and his Copenhagen school. And despite the philosophical underpinnings that Bohr tried to attach to his *Copenhagen interpretation* of quantum mechanics, it really provides a very pragmatic viewpoint with regard to its treatment of quantum measurement **R**. A point of view of the "reality" underlying this pragmatism might be, roughly speaking, that the measuring device (such as the Geiger counter just considered), and its random environment, could be viewed as constituting such a large or complicated system that it would be unreasonable to try to provide an exact treatment according to the rules of **U** for it, so instead we would treat the device (and its environment) as being an effectively *classical* system, whose classical actions are presumed to represent some very close approximation to the "correct" quantum behaviour, so that the observed activity following a quantum measurement can be very accurately described by the subsequent classical rules **C**. Nevertheless, the transition from **U** to **C** cannot in general be achieved *without* the introduction of probabilities, so that the determinism involved in the equations of **U** (and of **C**) is broken (figure 2-8), and a "jump" in the quantum description would normally have to occur, in accordance with the operation of **R**. It is taken that the complication involved in any "actual" physics underlying the measurement process can be presumed to be so vast that a supposedly correct description according to **U** would indeed be totally impractical, and the best that one might hope for would be some kind of approximate treatment resulting in probabilistic behaviour rather than determinism. Thus, one might well anticipate that, in a quantum setting, the rules of **R** provide an adequate treatment of this. However, as we shall be seeing in §§2.12 and 2.13 most particularly, there are some profoundly mysterious puzzles involved in the adoption of such a point of view, and it becomes very hard to accept that the strange **U**-rules, when directly applied to macroscopic bodies, can in themselves result in **R**-like or **C**-like behaviour. It is here that the serious difficulties in "interpreting" quantum mechanics begin to loom large.

Accordingly, Bohr's Copenhagen interpretation does not, after all, ask for a "reality" to be attached to the quantum level and, instead, the procedures of **U** and **R** are regarded as providing us with merely a body of *calculational procedure* giving an evolving mathematical description which, at the time of measurement, allows us to use **R** to calculate the probabilities for the different possible outcomes of a measurement. The **U**-actions of the quantum world, being not taken to be physically real, then appear to be regarded as being somehow "all in the mind", to be used only in this pragmatic calculational way, so that only the needed probabilities may be then obtained by means of **R**. Moreover, the *jumping* that the quantum state normally undergoes, when **R** is applied, is not taken to be an actual physical process, but represents merely the "jump" that the physicist's *knowledge* undergoes, upon the reception of the additional information that the result of the measurement provides.

The original virtue of this Copenhagen view, it seems to me, was that it enabled physicists to work with quantum mechanics in a pragmatic way, thereby obtaining numerous wonderful results and, to good measure, it relieved them of the burden of having to understand, at a deeper level, what is "really going on" in the quantum world and its relation to the classical world that we appear to experience directly. However, this will not be sufficient for us here. Indeed, there are various more recent ideas and experiments which allow us actually to probe the nature of the quantum world, and these confirm, to some considerable degree, a genuine measure of *reality* for the strange descriptions that quantum theory actually provides. The strength of this quantum "reality" is something that we shall certainly have to appreciate if we are to explore the possible limits to the dogma of quantum mechanics.

In the next several sections (§§2.5–2.10) we shall be seeing the basic ingredients of the magnificent mathematical framework of quantum mechanics. It is a scheme of things that agrees with the working of the natural world with enormous precision. Many of its implications are very counterintuitive and some blatantly contradict the expectations that our experiences with the classical-scale world have led us to believe in. Yet, all experiments to date, which successfully probe these contradictory expectations, have confirmed the predictions of the quantum formalism, rather than those "common-sense" expectations of our classical experience. Moreover, some of these experiments demonstrate that, contrary to our hard-won classical intuitions, the reach of the quantum world is not at all restricted to tiny submicroscopic distances, but can stretch to vast distances (a current terrestrial record being 143 km). The widespread scientific *faith* in the formalism of quantum mechanics has, indeed, a remarkable basis in observed scientific fact!

Finally, in §§2.12 and 2.13, I present arguments in support of my case that a total faith in this formalism is misplaced. While it is acknowledged that quantum field theory (QFT) has its difficulty with divergences – one of the initial motivations for string theory (see §1.6) – the arguments I shall present are concerned purely with the more basic and overriding rules of quantum mechanics itself. Indeed, in this book, I have not entered into detailed issues of QFT, beyond those touched upon in §§1.3, 1.5, 1.14, and 1.15. Yet, in §2.13, and again in §4.2, I put forward my own ideas concerning the kind of modification that I regard as essential, and I argue that there is a clear need to break free of that frequently expressed total faith in the dogma of the standard quantum formalism.

## 2.5. WAVE FUNCTION OF A POINT-LIKE PARTICLE

What, then, *is* this standard quantum formalism? Recall that, in §1.4, I referred to what is called the *superposition principle*, which is to apply quite generally to quantum systems. We considered there a situation where particle-like entities are successively fired through two closely separated parallel slits towards a sensitive screen behind (figure 1-2(d), in §1.4). A pattern consisting of a large number of tiny dark spots appears on the screen, this being consistent with the picture of individual localized impacts of a great many particles coming from the source, thereby providing support for the actual particle-like (or point-like) character of the quantum entities emitted by the source. Nevertheless, the overall pattern of impacts at the screen concentrates itself along a system of parallel bands, providing clear evidence of interference. This is the kind of interference pattern that would occur between two coherent sources of a wave-like entity emerging simultaneously from the two slits. But the impacts at the screen do seem to be caused by individual entities, and this can be made particularly clear by turning down the intensity of the source so that the time-interval between the emission of each of these entities and the next is greater than its transit time to the screen. Consequently, we are indeed dealing with entities arriving one at a time, but each such wave–particle entity demonstrates an interference that occurs between the different possible trajectories of its motion.

This is very similar to what happens with experiment 2, of §2.3 (figure 2-4(b)), where each wave–particle emerges from the beam splitter (at M) in the form of two spatially separated components, which subsequently come together to interfere at the second beam splitter (at C). We see again that a *single* wave–particle entity can consist of two separated parts, where these can lead to interference effects when the two parts are later brought together. Thus, for each such wave–particle, there is no requirement that it be a localized object; nevertheless it still behaves as a coherent whole. It continues to behave as a single quantum, no matter how far separated are its component parts and whatever number of such separated parts there may be.

How are we to describe such a strange wave–particle entity? Even if its nature seems unfamiliar, we are fortunate that a very elegant mathematical description can be given, so we may perhaps rest content (at least for a while, in accordance with the so-called Copenhagen viewpoint) that we can accurately describe mathematical laws that such an entity adheres to. The key mathematical property is that, like a wave of electromagnetism, we can *add* two of these wave–particle states together (as hinted at in §1.4) and, moreover, that the evolution of the sum is the *same* as the sum of the evolutions – a characterization of the term *linear* that I shall be making more precise in §2.7 (and see also §A.11). This addition notion is what we refer to as *quantum superposition*. When we have a wave–particle consisting of two separated parts, the entire wave–particle is simply the superposition of the two parts. Each of the parts would behave like a wave–particle on its own, but the *entire* wave–particle would be the sum of the two.

Moreover, we can superpose two such wave–particles in different ways, depending upon the phase relations between the wave–particles. What are these different ways? We shall find that they come about in our mathematical formalism from the use of complex numbers in the superpositions (see §A.9, §1.4, and recall the closing remarks of §2.3). Thus if $\alpha$ represents the state of one of these wave–particle entities and $\beta$ represents the state of another one, then we can form different combinations of $\alpha$ and $\beta$ of the form

$$w\alpha + z\beta,$$

where the individual weighting factors $w$ and $z$ are complex numbers (just as we considered in §1.4) which are called, perhaps confusingly, complex *amplitudes*[4] assigned to the respective alternative possibilities, $\alpha$ and $\beta$. The rule will be, assuming that the amplitudes $z$ and $w$ are not both zero, that this combination gives us another possible state of our wave–particle. In fact, the family of quantum states, in any quantum situation, are to form a *complex vector space*, in the sense of §A.3, and we shall explore this aspect of quantum states more fully in §2.8. We need these weighting factors $w$ and $z$ to be complex (rather than, say, the non-negative real numbers that would arise if we were merely concerned with probability-weighted alternatives) so that we can express the *phase relations* between the two components $\alpha$ and $\beta$ that are essential for the *interference* effects that can arise between these components.

The interference effects, arising from the phase relation between the constituents $\alpha$ and $\beta$, come about in both the two-slit experiment of §1.4 and the Mach–Zhender interferometer of §2.3 for the reason that the individual states each have temporal *oscillatory* natures, with specific frequencies, so that they can cancel or reinforce each other (out of phase or in phase respectively), in different circumstances (figure 2-5 in §2.3). These depend upon the relations between the states $\alpha$ and $\beta$, and also between the amplitudes $w$ and $z$ assigned to each. In fact, all we need, concerning the amplitudes $w$ and $z$, is just their *ratio* $w : z$. (The notation $a : b$ here means simply $a/b$, i.e. $a \div b$, but where we allow that $b$ might be zero – in which case we could, if we like, assign the "numerical" value "$\infty$" to this ratio. However, we must be careful that $a$ and $b$ are not *both* zero!)

In terms of the Wessel (i.e. complex) plane (§A.10), as shown in figure A-42, this ratio $w : z$ has an argument (assuming that neither $w$ nor $z$ is zero) which is the angle $\theta$ between the two directions out from the origin, given by 0, to the respective points $w$ and $z$. In the discussion of §A.10, we see that this is the angle $\theta$ in the polar representation of $z/w$:

$$z/w = r e^{i\theta} = r(\cos\theta + i\sin\theta);$$

see figure A-42, but $z/w$ now replaces the "$z$" of that diagram. It is $\theta$ that governs the phase displacements between the states $\alpha$ and $\beta$ and, therefore, whether – or, more appropriately, *where* – they cancel or reinforce one another. Note that $\theta$ is the angle between the lines out from the origin 0 to the respective points $z$

---

[4] In the literature, what I refer to in §A.10 as the *argument* of a complex number (i.e. the "$\theta$" in its polar representation $re^{i\theta}$) is sometimes called its *amplitude*. On the other hand, the *intensity* of a wave (effectively the $r$ in this expression) is often also called its *amplitude*! I avoid both of these confusingly contradictory terminologies here, my (quantum-mechanically *standard*) terminology encompassing both!

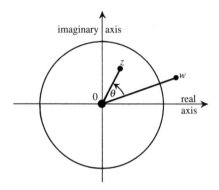

Figure 2-9: In the Wessel (complex) plane, the argument of the ratio $z/w$ (i.e. $z:w$) between two non-zero complex numbers $w$ and $z$ is the angle $\theta$ between the lines out from the origin to these two points.

and $w$, in the Wessel plane, taken in the anticlockwise sense (figure 2-9). The remaining information in the ratio $w:z$ is in the ratio of the distances of $w$ and $z$ from 0, i.e. the "$r$" in the aforementioned polar representation of $z/w$, and this governs the relative intensities of the two components $\boldsymbol{\alpha}$ and $\boldsymbol{\beta}$ of the superposition. We shall be seeing in §2.8 that such ratios of intensities (squared, in fact) play an important role with regard to probabilities, when a measurement **R** is made on a quantum system. Strictly, this probabilistic interpretation applies only when the states $\boldsymbol{\alpha}$ and $\boldsymbol{\beta}$ are "orthogonal", but we shall come to this concept also in §2.8.

Note that the quantities of interest in the previous paragraph with regard to the weighting factors $w$ and $z$ (their phase difference and relative intensity) both have to do with their *ratio* $w:z$. To understand better why it is such ratios that have particular significance, we need to address an important feature of the quantum formalism. This is that not all such combinations $w\boldsymbol{\alpha} + z\boldsymbol{\beta}$ are considered to be physically distinct, the physical nature of the combination indeed depending only on the amplitude ratio $w:z$. This is because of a general principle that applies to the mathematical description – the *state vector* (e.g. $\boldsymbol{\alpha}$) – of *any* quantum system whatever, not simply to a wave–particle entity. This is that the quantum state of the system is considered to be *physically unchanged* if the state vector is multiplied by any non-zero complex number, so the state vector $u\boldsymbol{\alpha}$ represents the *same* wave–particle state (or general quantum state) as does $\boldsymbol{\alpha}$, for any non-zero complex number $u$. Accordingly, this applies also to our combination $w\boldsymbol{\alpha} + z\boldsymbol{\beta}$.

Any multiple of this state vector by a non-zero complex number $u$,

$$u(w\boldsymbol{\alpha} + z\boldsymbol{\beta}) = uw\boldsymbol{\alpha} + uz\boldsymbol{\beta},$$

describes the *same* physical entity as does $w\boldsymbol{\alpha} + z\boldsymbol{\beta}$. We note that the ratio $uz : uw$ is the same as $z : w$, so it is indeed just this ratio $z : w$ that we are really concerned with.

Thus far, we have been considering the superpositions that can arise from just *two* states $\boldsymbol{\alpha}$ and $\boldsymbol{\beta}$. With three states $\boldsymbol{\alpha}$, $\boldsymbol{\beta}$, and $\boldsymbol{\gamma}$, we can form superpositions

$$w\boldsymbol{\alpha} + z\boldsymbol{\beta} + v\boldsymbol{\gamma},$$

where the amplitudes $w$, $z$, and $v$ are complex numbers, not all zero, and where we consider that the physical state is unaltered if we multiply through by any non-zero complex number $u$ (to get $uw\boldsymbol{\alpha} + uz\boldsymbol{\beta} + uv\boldsymbol{\gamma}$), so that again it is only the set of ratios $w : z : v$ that physically distinguish these superpositions from one another. This extends to any finite number of states $\boldsymbol{\alpha}, \boldsymbol{\beta}, \ldots, \boldsymbol{\phi}$ to give superpositions $v\boldsymbol{\alpha} + w\boldsymbol{\beta} + \cdots + z\boldsymbol{\phi}$, for which the various ratios $v : w : \cdots : z$ physically distinguish these states.

In fact, we must be prepared for such superpositions to extend even to an *infinite* number of individual states. Then we have to be careful about such matters as continuity and convergence, etc. (see §A.10). These matters introduce awkward mathematical issues with which I do not wish to concern the reader unduly. Whereas some mathematical physicists might very reasonably take the view that the undoubted difficulties that confront quantum theory (and more particularly, quantum field theory; see §§1.4 and 1.6) require a careful attention to these mathematical niceties, I am proposing to be rather cavalier about these matters here. It is not that I believe that such mathematical subtleties are unimportant – far from it, as I regard mathematical consistency as an essential requirement – it is more that I regard the seeming inconsistencies in quantum theory that we shall be coming to (particularly in §§2.12 and 2.13) as being of a much more basic nature, having little to do with matters of mathematical rigour as such.

Going along with this mathematically relaxed viewpoint, let us try to consider the general state of a single point particle, or *scalar* particle (with no directional qualities, so no direction of spin, i.e. spin 0). The most primitive basic state – a *position* state – would be characterized by its just being in some particular spatial location A, specified by a position vector **a** with respect to some given origin O (see §§A.3 and A.4). This is a very "idealized" kind of state, and it is usual to think of it as given (up to proportionality) by $\delta(\mathbf{x} - \mathbf{a})$, where the "$\delta$"

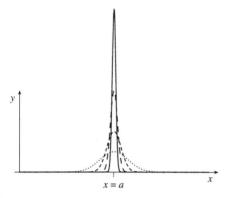

Figure 2-10:  The Dirac delta function $\delta(x)$ is here illustrated by thinking of $\delta(x-a)$ as the limit of a sequence of smooth positive-valued functions, each of which has unit area bounded by the curve and the $x$-axis, concentrating more and more at $x = a$.

being used here denotes Dirac's "delta function" that we shall come to shortly. It is also not a very sensible kind of state for an actual physical particle, because its Schrödinger evolution would cause the state to spread immediately outwards – an effect that can be thought of as an implication of the *Heisenberg uncertainty principle* (that we shall come to briefly at the end of §2.13): absolute precision in the particle's position demands complete uncertainty in its momentum, so the high momentum contributions would cause the state to spread out instantly. In my discussion here, however, I shall not be concerned with how such a state might evolve in the future. It will be appropriate to think merely in terms of how quantum states can behave at one particular time, say $t = t_0$.

A delta function is not actually an ordinary function but a limiting case of such things, where (for a real number $x$) $\delta(x)$ *vanishes* for all $x$ for which $x \neq 0$, but where we are to think of $\delta(0) = \infty$ in such a way that the area under the curve of this function is unity. Correspondingly, $\delta(x - a)$ would be the same thing but displaced so that it vanishes everywhere except at $x = a$, where it becomes infinite, the area under the curve remaining unity, $a$ being some given real number. See figure 2-10 for an impression of this limiting process. (For more mathematical understanding, see, for example, Lighthill [1958] and Stein and Shakarchi [2003].) It is not appropriate for me to go into the mathematical way that one makes sense of such things here, but it will be useful to make use of the idea and the notation. (In fact, technically, such position states are, in any case, not part of the standard Hilbert-space formalism of quantum mechanics that

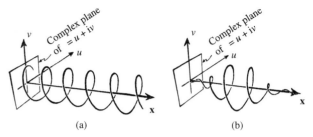

Figure 2-11: (a) Wave function of a momentum state $e^{-i\mathbf{p}\cdot\mathbf{x}}$ for given 3-momentum **p**. (b) Wave function of a wave-packet.

we shall come to in §2.8, strict position states being not physically realizable. Nevertheless, they are discursively very useful.)

In accordance with the above, we can consider a delta function of a *3-vector* **x**, where we can write $\delta(\mathbf{x}) = \delta(x_1)\delta(x_2)\delta(x_3)$, the three Cartesian components of **x** being $x_1, x_2, x_3$. Then $\delta(\mathbf{x}) = 0$ for all *non*-zero values of a 3-vector **y**, but we must again think of $\delta(\mathbf{0})$ as being extremely large, giving rise to a *unit* 3-volume. Then we can think of $\delta(\mathbf{x} - \mathbf{a})$ as a function that gives a zero amplitude for the particle being anywhere *other* than at the point X, with position vector **x** (i.e. $\delta(\mathbf{x} - \mathbf{a})$ is non-zero only when $\mathbf{x} = \mathbf{a}$), and assigns a very large (infinite) amplitude for the particle actually being *at* the point A (i.e. when $\mathbf{x} = \mathbf{a}$). We may next contemplate *continuous* superpositions of such special position states, such superposition assigning a complex amplitude to every spatial point X. This complex amplitude (now simply an ordinary complex number) is therefore just some complex function of the position 3-vector **x** of the variable point X. This function is frequently denoted by the Greek letter $\psi$ ("psi"), and the function $\psi(\mathbf{x})$ is referred to as the particle's (Schrödinger) *wave function*.

Thus the complex number $\psi(\mathbf{x})$ that the function $\psi$ assigns to any individual spatial point X is the amplitude for the particle to be situated exactly at the point X. Again, the physical situation is considered to be unchanged if the amplitude at every point is multiplied by the same non-zero complex number $u$. That is to say, the wave function $w\psi(\mathbf{x})$ represents the same physical situation as $\psi(\mathbf{x})$ if $w$ is any non-zero (constant) complex number.

An important example of a wave function would be an oscillatory plane wave, with a definite frequency and direction. Such a state, called a *momentum state*, would be given by the expression $\psi = e^{-i\mathbf{p}\cdot\mathbf{x}}$, where **p** is a (constant) 3-vector describing the momentum of a particle; see figure 2-11(a), the vertical $(u, v)$-plane in the picture denoting the Wessel plane of $\psi = u + iv$. Momentum states, in the case of photons, will have importance for us in §§2.6 and 2.13. In figure 2-11(b) a wave-packet is illustrated, such things having been discussed in §2.3.

At this stage, it may be helpful to point out that it is usual in quantum mechanics to normalize the description of the quantum state, in terms of a measure of "size" that can be assigned to a wave function $\psi$, which is a positive real number, called its *norm*, that[5] we can write (as in §A.3)

$$\|\psi\|$$

(where we have $\|\psi\| = 0$ if, and only if, $\psi$ is the zero function – not actually an allowable wave function), the norm having the scaling property

$$\|w\psi\| = |w|^2 \|\psi\|$$

for any complex number $w$ ($|w|$ being its modulus; see §A.10). A *normalized* wave function is one of *unit* norm,

$$\|\psi\| = 1,$$

and if $\psi$ were not originally normalized, we could always make it so by replacing $\psi$ by $u\psi$, where $u = \|\psi\|^{-1/2}$. Normalization removes some of the freedom in this replacement $\psi \mapsto w\psi$, and it has the advantage, for a scalar wave function, that we can then regard the squared modulus $|\psi(\mathbf{x})|^2$ of the wave function $\psi(\mathbf{x})$ as providing the *probability density* of finding the particle at the point X.

However, normalization does not remove all of this scaling freedom, since any multiplication of $\psi$ by a *pure phase*, i.e. a complex number of unit modulus $e^{i\theta}$ ($\theta$ being real and constant)

$$\psi \mapsto e^{i\theta}\psi,$$

does not affect the normalization. (This is basically the phase freedom that Weyl was eventually led to consider in his electromagnetic theory described in §1.8.) Although genuine quantum states always have norms, and can therefore be normalized, certain idealizations of quantum states that are in common use do not, such as the position states $\delta(\mathbf{x} - \mathbf{a})$ that we considered earlier, or the momentum states $e^{-i\mathbf{p}\cdot\mathbf{x}}$ just considered. We shall come to momentum states again in §§2.6 and 2.13, for the case of photons. Ignoring this issue is part of the relaxed attitude to mathematical niceties that I am allowing myself. We shall be seeing in §2.8 how this notion of *norm* fits in with the broader framework of quantum mechanics.

---

[5] In the case of a scalar wave function, as considered here, this would be an *integral* (see §A.11) over the whole of 3-space of the form $\int \psi(\mathbf{x})\bar{\psi}(\mathbf{x}) \, d^3\mathbf{x}$. Thus, for a normalized wave function, the interpretation of $|\psi(\mathbf{x})|^2$ as probability density is consistent with this total probability being unity.

## 2.6. WAVE FUNCTION OF A PHOTON

This complex-valued function $\psi$ gives us Schrödinger's picture of a single scalar wave–particle state. But this, so far, is just a particle without structure, so without any directional characteristics (spin 0), but in fact we can also get a pretty good picture of what goes on with an individual photon wave function. It is not a scalar particle, as the photon has spin value $\hbar$, i.e. "spin 1" in the normal Dirac units (see §1.14). Then the wave function has a vector character and it turns out that we can picture it as an *electromagnetic wave*, where we imagine that if the wave were to be extremely weak it could provide us with some kind of image of what a single photon is like. Since a wave function is a *complex*-valued function, we can take it in the form

$$\psi = \mathbf{E} + i\mathbf{B},$$

where $\mathbf{E}$ is the 3-vector of the electric field (cf. §§A.2 and §A.3), and $\mathbf{B}$ the 3-vector of the magnetic field. (Strictly speaking, to get a genuine free photon wave function, we would need to take what is called the *positive frequency part* of $\mathbf{E} + i\mathbf{B}$ – which is an issue to do with the Fourier decomposition, as indicated in §A.11 – and to add this to the positive frequency part of $\mathbf{E} - i\mathbf{B}$ [see TRtR, §24.3], but such matters are much more thoroughly addressed in Streater and Wightman [2000]. However, these technical issues will not concern us here, and do not substantially affect the following discussion.)

A key aspect of this electromagnetic-wave picture is that we need to consider the polarization of the wave. This notion of *polarization* needs explaining. An electromagnetic wave moving in a definite direction in free space, with a given frequency and intensity – i.e. a *monochromatic* wave – may be *plane-polarized*. Then it will have what is called a *polarization plane*, which is a plane that contains the direction of the wave's motion, and it is the plane within which the electric field composing the wave oscillates back and forth; see figure 2-12(a). Accompanying this oscillatory electric field is a magnetic field, which itself oscillates back and forth, with the same frequency and phase as the electric field, but in a plane perpendicular to the polarization plane, though still containing the direction of motion. (We can also think of figure 2-12(a) as depicting the temporal behaviour of the wave, with the arrow pointing in the negative time direction.) We can have plane-polarized monochromatic electromagnetic waves with any chosen plane containing the direction of motion for its polarization plane. Moreover, *any* monochromatic electromagnetic wave, with some given direction of motion $\mathbf{k}$, can be decomposed as a sum of plane-polarized waves, with polarization planes

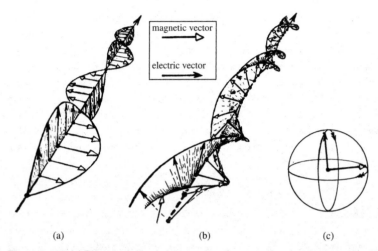

(a)                              (b)                              (c)

Figure 2-12:  (a) Plane-polarized electromagnetic wave, showing the mutual oscillation of the electric and magnetic field vectors; the motion can be viewed either
spatially or temporally. (b) Circularly polarized electromagnetic wave, similarly
depicted. (c) By combining plane- and circularly polarized motions we can get
various degrees of elliptical polarization.

perpendicular to one another. Polaroid sunglasses have the property that they
allow the component with **E** in the vertical direction to be transmitted through
while causing the component with **E** horizontal to be absorbed. Light from low in
the sky, and reflected light from the sea, are both largely polarized in the direction
perpendicular to this (i.e. horizontally), so there is a considerable reduction in the
amount that gets through to the eye when protected by sunglasses of this type.

The polarized glasses that are now commonly used for the viewing of 3-dimen-
sional films are a little different. To understand these, we need a further concept of
polarization, namely *circular* polarization (figure 2-12(b)), twisting, as the wave
passes, in an either right- or left-handed sense, this handedness being referred to
as the *helicity* of the circularly polarized wave.[6] These glasses have the ingenious
behaviour that the semi-transparent material transmitting the light to one eye or
the other allows only *right*-circularly polarized light through, for one eye, and

---

[6] It appears that the fields of particle physics and quantum optics employ opposite conventions for the
sign of helicity. This is covered in Jackson [1999, p. 206].

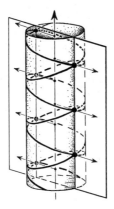

Figure 2-13: By adding equal measures of left- and right-circularly polarized waves we can obtain a plane-polarized one, with peaks and troughs indicated by the horizontal arrows. Changing the phase relation between the circularly polarized components rotates the plane of polarization.

only *left*-circularly polarized light through, for the other – though, curiously, the light in each case emerges from the material towards the eye in a plane-polarized state.

In addition to these polarization states, there are states of elliptical polarization (see figure 2-12(c)) which combine the twist of circular polarization with a certain amount of plane polarization, in some direction. All these states of polarization can come about as combinations of just *two* such states. These two can be the right-handed and the left-handed circularly polarized waves or they can be the horizontally and the vertically plane-polarized waves, or any of many other pairs of possibilities. Let us try to see how this works.

Perhaps easiest to explore is the case of an equal-intensity superposition of a right-handed and a left-handed *circularly* polarized wave. The various phase differences simply give all possible directions of plane polarization. To see how this works, think of each wave as represented as a *helix*, which is a curve drawn on a circular cylinder making some fixed angle (not 0° or 90°) with the axis; see figure 2-13. This represents the track of the electric vector as we move along the axis (the axis representing the direction of the wave). We have a right- and left-handed helix (of equal and opposite pitch) to represent our two waves. Drawing them both on the same cylinder, we find that the intersection points of the two helices all lie in a plane, this being the plane of polarization of the superposition of the two waves that they represent. As we slide one of the helices along the cylinder, with the other fixed – giving us the various phase differences – we get all the possible planes of polarization for the superposed wave. If, now, we allow the contribution of one wave to be increased, in relation to the other, we get all the possible states of elliptical polarization.

Our complex representation of the electric and magnetic vectors, in the form $\mathbf{E} + i\mathbf{B}$ given above, allows us to get a rather direct route towards appreciating what a photon's wave function is like. Let us use the Greek letter $\boldsymbol{\alpha}$ to represent (say) the state of right-handed polarization, in the situation described in the previous paragraph, and the Greek letter $\boldsymbol{\beta}$ the state of left-handed polarization (of equal intensity and frequency). Then we can represent the various states of plane polarization obtained by *equal*-intensity superpositions of the two (up to proportionality) as

$$\boldsymbol{\alpha} + z\boldsymbol{\beta}, \quad \text{where } z = e^{i\theta}.$$

As $\theta$ is increased from 0 to $2\pi$, so that $e^{i\theta} (= \cos\theta + i\sin\theta)$ moves once around the unit circle in the Wessel plane (see §§A.10 and 1.8, and figure 2-9), the polarization plane also rotates about the direction of motion. It is of some significance to compare this plane's rate of rotation with that of $z$, as $z$ goes once around the unit circle.

The state vector $z\boldsymbol{\beta}$ is represented by the left-handed helix in figure 2-13. As $z$ goes continuously once around, this helix can be taken to rotate bodily about its axis continuously through $2\pi$, in a right-handed (anti-clockwise) sense. As we see in figure 2-13, this bodily rotation is equivalent to moving the left-handed helix continuously upwards while leaving the right-handed helix alone, until the left-handed helix reaches its original configuration. We see that this takes the intersection points of the two helices which were originally at the front of the picture around to the back, while restoring the polarization plane to its original position. Thus, whereas $z$ will have gone once round the unit circle, through an angle $2\pi$ (i.e. 360°) around the unit circle, the polarization plane will have returned to its original location by rotating through an angle of only $\pi$, rather than $2\pi$ (i.e. 180° not 360°). If $\varphi$ denotes the angle rotated by the polarization plane at any one stage, we thus find that $\theta = 2\varphi$, where, as above, $\theta$ is the angle in the Wessel plane that the amplitude $z$ makes with the real axis. (There are certain issues of convention here, but for our present discussion I am taking our Wessel plane's orientation to be given as if viewed from above, in figure 2-13.) In terms of points in the Wessel plane, we can use a complex number $q$ (or its negative $-q$) to represent the angle made by the polarization plane, where $q = e^{i\varphi}$, so the relation $\theta = 2\varphi$ becomes

$$z = q^2.$$

We shall see later (figure 2-20 in §2.9) how this $q$ extends so as to describe general states of elliptical polarization also.

It should be made clear, however, that the photon states just considered are merely a very particular type of example, these being referred to as *momentum states*, which carry their energy in one very definite direction. It is clear from the universality of the superposition principle that there are many other possibilities for single-photon states. For example, we could consider just a superposition of two such momentum states aimed in different directions. Such situations again provide electromagnetic waves that are solutions of Maxwell's equations (since Maxwell's equations are just like Schrödinger's in being linear; see §§2.4, A.11, and 2.7). Moreover, by combining many such waves together, travelling in only very slightly different directions, and with closely the same very high frequency, we can construct solutions of Maxwell's equations that are very concentrated at a single location, and travelling with the speed of light in the general direction of the superposed waves. Such solutions are referred to as *wave-packets* and were referred to in §2.3 as possible candidates for the quantum wave–particle entities needed to explain the results of experiment 2, as shown in figure 2-4(b). However, as we saw in §2.3, such a classical description of a single photon will not explain the results of experiment 1, as illustrated in figure 2-4(a). Moreover, such classical wave-packet solutions of Maxwell's equations do not remain particle-like for all time, and will disperse after some short while. This may be contrasted with the behaviour of individual photons coming from very great distances, such as from remote galaxies.

The particle-like aspects of photons do not come about from their wave functions having such a highly localized nature. They come, instead, from the fact that the measurement being performed happens to be one that is geared to observing their particle-like characteristics, such as with a photographic plate or (in the case of charged particles) a Geiger counter. The resulting *particulate* nature of the quantum entity being observed is a feature of the operation of **R** under such circumstances, where the detector is particle-responsive. The wave function of a photon from a very distant galaxy, for example, would be spread over utterly vast regions of space, and its detection at a particular spot on a photographic plate would be the result of an exceedingly tiny probability that this particular measurement would find it in this **R**-process. We would be hardly likely to see such a photon at all, were it not for the utterly vast numbers of photons that would have been emitted from the region being observed in that very distant galaxy, this compensating for the extremely tiny probability for any particular photon!

The example of the polarized photon, just considered, also illustrates how complex numbers can sometimes be used when forming linear superpositions of classical fields. In fact there is a very close connection between the aforementioned procedure of complex superposition of classical (electromagnetic) fields and the quantum superposition of particle states – here in the case of photons. Indeed, the Schrödinger equation for a single free photon turns out to be simply a rewriting of Maxwell's free-field equations, but for a *complex-number*-valued electromagnetic field.

One difference should be pointed out here, however, namely that if we multiply the description of a single photon state by a non-zero complex number, we do not change the state, whereas it turns out that for a *classical* electromagnetic field the intensity (i.e. energy density) of that state would be scaled up by the *square* of the field strength involved. In order to scale up the energy content of a *quantum* state, on the other hand, we would have to increase the *number* of photons, each individual photon having to have its energy constrained by Planck's formula, $E = h\nu$. Thus, in a quantum-mechanical situation, it would be the photon *number* that would be scaled up by the square of the modulus of the complex number, when we consider increasing the intensity of an electromagnetic field. We shall be seeing in §2.8 how this relates to the probability law, for **R**, referred to as the *Born rule* (see also §§1.4 and 2.8).

## 2.7. QUANTUM LINEARITY

Before coming to this, however, we should catch some glimpse of the staggering universal scope of quantum linearity. A major feature of the formalism of quantum mechanics is, indeed, the *linearity* of **U**. As pointed out in §A.11, this particular simplifying quality of **U** is *not* shared by most classical types of evolution – though Maxwell's equations are, in fact, also linear. We shall try to understand what this linearity means.

As pointed out previously (early in §2.5), the meaning of *linearity*, as applied to a time-evolution process such as **U**, depends upon there being a notion of addition or, more specifically, of *linear combination* that can be applied to the states of the system, and then the time evolution is said to be *linear* if it preserves this. In quantum mechanics this is the *superposition* principle of quantum states, where if $\alpha$ is an allowed state of the system, and so also is $\beta$, then the linear combination

$$w\alpha + z\beta,$$

where we take the fixed complex numbers $w$ and $z$ not to be both zero, is again a legitimate quantum state. The property of *linearity*, possessed by $\mathbf{U}$, is simply that if some quantum state $\alpha_0$ would evolve according to $\mathbf{U}$ to a state $\alpha_t$ after a specific time period $t$,

$$\alpha_0 \rightsquigarrow \alpha_t,$$

and if another quantum state $\beta_0$ were to evolve to $\beta_t$ after the same period $t$,

$$\beta_0 \rightsquigarrow \beta_t,$$

then any superposition $w\alpha_0 + z\beta_0$ were to evolve, after this time period $t$, to $w\alpha_t + z\beta_t$,

$$w\alpha_0 + z\beta_0 \rightsquigarrow w\alpha_t + z\beta_t,$$

the complex numbers $w$ and $z$ remaining unchanged over time. This, indeed, characterizes linearity (see also §A.11). In brief terms, we can say that linearity is encapsulated in the aphorism

the evolution of the sum is the sum of the evolutions

though we should really think of "sum" as encompassing "linear combination".

Thus far, as described in §§2.5 and 2.6, we have been considering the linear superpositions of states as applied only to *single* wave–particle entities, but the linearity of Schrödinger's evolution applies to quantum systems completely generally, no matter how many particles might be simultaneously involved. Accordingly, we need to see how this principle applies to systems involving more than one particle. For example, we might have a state consisting of two scalar particles (of different kinds), the first of which is thought of as a wave–particle concentrated within a small region around one particular spatial location, P, and the other concentrated within a quite separate small spatial location, Q. Our state $\alpha$ could represent this *pair* of particle locations. The second state $\beta$ might have the first particle concentrated at a completely different location P′, with the second one concentrated at some other location Q′. The four locations need have no relation to each other. Suppose, now, we consider a superposition like $\alpha + \beta$. How are we to interpret this? I should first make clear that the interpretation can be nothing like finding some kind of average of the positions involved (such as the first particle being located at the mid-point of P and P′, and the second at the mid-point of Q and Q′). This kind of thing would be very far from what quantum linearity actually tells us – and we may recall that even for a *single* wave–particle no localized interpretation analogous to this comes about (i.e. a linear superposition of the state where a particle is at P with another state where that same particle is located at P′ is a state where the particle shares its existence between these two locations, and this is certainly not equivalent to the state where that particle is at some third point). No, the superposition $\alpha + \beta$ involves *all* the four locations P, P′, Q, Q′, just where they are, with the two alternative pairs of locations (P, Q) and (P′, Q′) for the individual particles somehow *coexisting*! We shall find that there are curious subtleties about the kind of state $\alpha + \beta$ that arises here, referred to as an *entangled* state, of where neither particle has a separate state on its own, independently of the other.

This concept of *entanglement* was introduced by Schrödinger, in a letter to Einstein, in which he used the German word *Verschränkung* and translated it into English as *entanglement*, publishing it shortly afterwards [Schrödinger and Born 1935]. The quantum strangeness and importance of this phenomenon was encapsulated by Schrödinger in his comment:

> I would not call [entanglement] *one* but rather *the* characteristic trait of quantum mechanics, the one that enforces its entire departure from classical lines of thought.

Later, in §2.10, we shall be seeing some of the very strange and essentially quantum features that such entangled states can exhibit – these going under the general names of Einstein–Podolsky–Rosen (EPR) effects [see Einstein et al. 1935], which stimulated Schrödinger to appreciate the entanglement concept. Entanglements are now seen to reveal their actual presence through such things as Bell-inequality violations, which we shall come to in §2.10.

It may be helpful, for understanding quantum entanglement, to refer the reader back to the $d$-function notation, briefly mentioned in §2.5. Let us use position vectors **p**, **q**, **p′**, **q′**, for the respective points P, Q, P′, Q′. For simplicity, I shall not bother to put in any complex amplitudes, nor worry about normalizing any of our states. First, let us consider just a single particle. We can write the state for that particle to be at P as $\delta(\mathbf{x} - \mathbf{p})$ and for it to be at Q as $\delta(\mathbf{x} - \mathbf{q})$. The sum of those two states would be

$$\delta(\mathbf{x} - \mathbf{p}) + \delta(\mathbf{x} - \mathbf{q}),$$

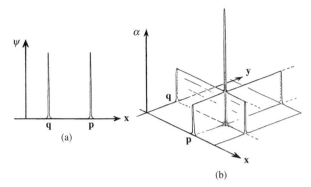

Figure 2-14: (a) The sum of delta functions $\psi(\mathbf{x}) = \delta(\mathbf{x} - \mathbf{p}) + \delta(\mathbf{x} - \mathbf{q})$ provides a wave function for a scalar particle in a superposed state of being at P and Q simultaneously. (b) The product of two delta functions $\alpha(\mathbf{x}, \mathbf{y}) = \delta(\mathbf{x} - \mathbf{p})\delta(\mathbf{y} - \mathbf{q})$ provides the state where there are two different particles, one at P and the other at Q. Note that two variables $\mathbf{x}$ and $\mathbf{y}$ are needed, denoting the locations of each separate particle.

representing a superposition of the particle in these two locations simultaneously (which is completely different from the $\delta(\mathbf{x} - \frac{1}{2}(\mathbf{p} + \mathbf{q}))$) that would represent the particle at the mid-point location; see figure 2-14(a)). (The reader should be reminded of the point made in §2.5 that such an idealized wave function can have this delta-function form only initially, at time $t = t_0$, say. The Schrödinger evolution would demand that such states spread out at the next instant. This matter will not be of significance for the discussion here, however.) When it comes to representing the quantum state of a pair of particles, the first at P and the second at Q, we might try to write this as $\delta(\mathbf{x} - \mathbf{p})\delta(\mathbf{x} - \mathbf{q})$, but this would be wrong for all sorts of reasons. The main objection is that just taking products of wave functions like this – i.e. taking two particle wave functions $\psi(\mathbf{x})$ and $\phi(\mathbf{x})$ and trying to represent the particle pair by just taking their product $\psi(\mathbf{x})\phi(\mathbf{x})$ – would certainly be wrong, as it would ruin the linearity of Schrödinger's evolution. However, the correct answer is not so different (assuming, for now, that the two particles are not of the same type, so we don't need to consider the issues of §1.4, relating to their fermion/boson character). We can use the position vector $\mathbf{x}$ to represent the location of the first particle, as before, but now we use a different vector $\mathbf{y}$ to represent that of the second particle. Then the wave function $\psi(\mathbf{x})\phi(\mathbf{y})$ would indeed describe a genuine state, namely that for which the spatial array of amplitudes for the first particle is given by $\psi(\mathbf{x})$ and that of the second by $\phi(\mathbf{y})$, these being completely *independent* of each other. When they are *not* independent they are what is called *entangled*, and then the state does not have the form of a simple product like $\psi(\mathbf{x})\phi(\mathbf{y})$, but would have a more general form of a function, $\Psi(\mathbf{x}, \mathbf{y})$, of (here) the *two* position-vector variables $\mathbf{x}, \mathbf{y}$. This would apply more generally, where several particles are involved, with position vectors $\mathbf{x}, \mathbf{y}, \ldots, \mathbf{z}$, where their general (entangled) quantum state would be described by a wave function $\Psi(\mathbf{x}, \mathbf{y}, \ldots, \mathbf{z})$, and where a completely unentangled one would have the special form $\psi(\mathbf{x})\phi(\mathbf{y}) \cdots \chi(\mathbf{z})$.

Let us return to the example considered above, where we initially considered the state in which the first particle is located at P and the second at Q. The wave function would then be the unentangled one,

$$\alpha = \delta(\mathbf{x} - \mathbf{p})\delta(\mathbf{y} - \mathbf{q});$$

see figure 2-14(b). In our example, this is to be superposed with $\beta = \delta(\mathbf{x} - \mathbf{p}')\delta(\mathbf{y} - \mathbf{q}')$, where the first particle is at P' and the second at Q', to give (ignoring amplitudes)

$$\alpha + \beta = \delta(\mathbf{x} - \mathbf{p})\delta(\mathbf{y} - \mathbf{q}) + \delta(\mathbf{x} - \mathbf{p}')\delta(\mathbf{y} - \mathbf{q}').$$

This is a simple example of an entangled state. If the first particle's position were to be measured and found to be at P, then the second is automatically found to be at Q, whereas if the first is found to be at P', then the second becomes automatically located at Q'. (The issue of quantum measurement is treated more fully in §2.8, and that of quantum entanglement more extensively in §2.10.)

Such entangled states of pairs of particles are undoubtedly very odd and unfamiliar, but this is only the beginning. We have seen that this feature of entanglement arises as one aspect of the quantum principle of linear superposition. But this principle has considerably broader scope than just applying to pairs of particles. When applied to triplets of particles we can get entangled triplets, and we can also get mutually entangled quadruplets of particles, or of any number of particles, there being no limit to the numbers that can be involved in a quantum superposition.

We have seen that this principle of quantum superposition of states is fundamental to the linearity of quantum evolution, which is itself central to the **U** time evolution of the quantum state (Schrödinger equation). Standard quantum mechanics puts no limit on the scale at which **U** applies to a physical system. For example, recall the cat of §1.4. Let us suppose that there is a room, connected to the outside by two separate doors A and B, and that there is a hungry cat outside, the room itself containing some tempting food, both doors being initially

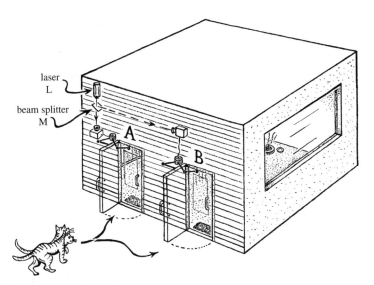

Figure 2-15:   A high-energy photon is emitted by the laser L, aimed at the beam splitter M. The two (superposed) alternative resulting photon beams reach detectors, one of which would open door A upon reception of the photon, the other, door B. Since these photon states are in quantum linear superposition, so also must be the opening of the doors, as follows from the linearity of the quantum (**U**-)formalism. Moreover, according to **U**, so also must the cat's motion be a *superposition* of the two paths to the food within the room!

closed. Suppose there is a high-energy photon detector connected to each door – at respective locations A and B – which automatically opens the particular door to which it is connected, when it receives such a photon from a (50%) beam splitter at some location M. These are to be the two alternative outcomes of a high-energy photon aimed at M by a laser at L (figure 2-15). The situation is just like that of experiment 1 in §2.3 (figure 2-4(a)). In any actual realization of this arrangement, the cat would experience one or other of the doors opening, and she would therefore pass through just one door or the other (with, here, a 50% probability of each outcome). Yet, if we are to follow the detailed evolution of the system, supposed to act in accordance with **U** with its implied linearity, as applied to all the ingredients that are relevant – the laser, the emitted photon, the material in the beam splitter, the detectors, the doors, the cat herself, and the air in the room, etc. – then the superposed state which starts with the photon leaving the beam splitter in a superposition of being reflected and transmitted must evolve to a superposition of two states each with just one of the doors open, and eventually to a superposed cat's motion through the two different doors at once!

This is just a particular instance of the above implication of linearity. We take the evolution $\alpha_0 \rightsquigarrow \alpha_t$ to begin with the photon leaving the beam splitter at M being aimed at the detector at A, with the cat outside, and ending with the cat eating the food inside the room, having passed through the door A. The evolution $\beta_0 \rightsquigarrow \beta_t$ is to be similar, but with the photon aimed at the detector at B when it leaves the beam splitter at M, and consequently the cat reaches the food by passing through the door B. But the *total* state, as the photon leaves the beam splitter, has to be a superposition, $\alpha_0 + \beta_0$ say, and this will evolve $\alpha_0 + \beta_0 \rightsquigarrow \alpha_t + \beta_t$, according to **U**, so the cat's motion between the two rooms would accordingly be a superposition of passing through both doors at the same time – which is certainly *not* what is experienced! This is an example of *Schrödinger's cat paradox*, and we shall need to return to it in §2.13. The way that such matters are handled in standard quantum mechanics would be in accordance with the Copenhagen perspective, whereby the quantum state is somehow *not* regarded as describing physical reality but merely as providing a means whereby the various probabilities of the alternative outcomes of an observation made on the system can be calculated. This is the action of the process **R**, referred to in §2.4, which we come to next.

## 2.8. QUANTUM MEASUREMENT

In order to understand how quantum mechanics deals with such seemingly blatant discrepancies between experienced reality and the **U**-evolution process, we shall need first to understand how the **R** procedure actually operates in quantum theory. This is the issue of *quantum measurement*. Quantum theory allows that only a restricted amount of information can be extracted from the quantum state of a system, and directly ascertaining by measurement what that quantum state actually *is*, is not considered to be achievable. Instead, any particular measuring device can only distinguish a certain limited set of alternatives for the state. If the state prior to the measurement does not happen to *be* one of these alternatives, then – according to the strange procedure that **R** demands we adopt – the state instantaneously *jumps* to one of those allowed alternatives, with a probability that the theory provides (in fact, calculated by the Born rule alluded to in §2.6 above – and also in §1.4 – and described in more detail below).

This quantum jumping is one of the oddest features of quantum mechanics, and many theorists deeply question the actual physical reality of this procedure. Even

Erwin Schrödinger himself is reported (by Werner Heisenberg [1971, pp. 73–76]) to have said: "If all this damned quantum jumping were really here to stay then I should be sorry I ever got involved in quantum theory." In response to Schrödinger's despairing remark, Bohr commented [Pais 1991, p. 299]: "But the rest of us are extremely grateful that you did; your wave mechanics... represents a gigantic advance over all previous forms of quantum mechanics." Nevertheless, this is the procedure which, when adopted, gives us results of quantum mechanics that are, as things stand, in full agreement with observation!

It will be necessary, here, to be a good deal more specific about the **R** procedure. In conventional texts, the issue of quantum measurement is addressed in relation to the properties of certain types of *linear operator* (as referred to in §1.14). However, although I shall return briefly to this connection with operators at the end of this section, it will be simpler to describe the action of **R** in a more direct way.

First, we must take note of the fact (already mentioned in §2.5) that the family of state vectors for some quantum system always forms a *complex vector space* in the sense of §A.3. This requires that we must also include a special element **0**, the *zero vector* which does not correspond to any physical state. I shall refer to this vector space as $\mathcal{H}$ (to stand for *Hilbert space*, a notion that we shall come to more specifically, shortly). I shall initially discuss the general, so-called non-degenerate quantum measurement, but there are also what are called *degenerate measurements*, which are unable to distinguish between certain of the alternatives. These will be discussed briefly at the end of this section, and also in §2.12.

In the case of a non-degenerate measurement, we can describe the above limited set of alternatives (the possible results of the measurement) as constituting an *orthogonal basis* for $\mathcal{H}$, in the sense of §A.4. Accordingly, the basis elements $\varepsilon_1, \varepsilon_2, \varepsilon_3, \dots$ are all orthogonal to one another, in a sense indicated below, and must constitute a basis, so they span the space $\mathcal{H}$. This latter condition (described more fully in §A.4) means that every element of $\mathcal{H}$ can be expressed as a superposition of $\varepsilon_1, \varepsilon_2, \varepsilon_3, \dots$. Moreover, this expression of a given state in $\mathcal{H}$ in terms of $\varepsilon_1, \varepsilon_2, \varepsilon_3, \dots$ is *unique*, in accordance with the states $\varepsilon_1, \varepsilon_2, \varepsilon_3, \dots$ actually forming a *basis* for the family. The term *orthogonal* refers to pairs of states that are, in a particular sense, *independent* of one another.

This notion of *independence* that is involved in quantum orthogonality is not easy to understand in classical terms. Perhaps the closest classical notion is that which occurs in modes of vibration, such as when a bell or a drum is struck and oscillates in different ways, each with its characteristic frequency. The different "pure" modes of vibration can be regarded as being independent or orthogonal to one another (an example being the modes of the vibrating violin string discussed in §A.11) but this analogy does not carry us very far. Quantum theory demands something much more specific (and subtle), and it may be helpful to give some actual examples. Two wave–particle states that do not overlap at all (such as the two routes MAC and MBC around the Mach–Zhender interferometer of §2.3; see figure 2-4(b)) would be orthogonal, but this is not a necessary condition, there being many other ways that orthogonality can arise between wave–particle states. Two infinite wave trains of differing frequency, for example, would also have to count as orthogonal. Moreover, photons (of equal frequency and direction) with polarization planes that are at right angles to one another would be orthogonal, though not if the planes were separated by some other angle. (We recall the classical descriptions of electromagnetic plane waves considered in §2.6, noting the fact that we can legitimately think of these classical descriptions as good models for individual photon states.) A photon state of circular polarization would not be orthogonal to any (otherwise identical) photon state of plane polarization, but states of left-handed and right-handed circular polarization are orthogonal to each other. We should bear in mind that the polarization direction provides only a small part of the complete state of the photon, however, and irrespective of their states of polarization, two photon momentum states (see §§2.6 and 2.13) of differing frequency or direction would be orthogonal.

The geometrical connotation of the term *orthogonal* would be *at right angles* or *perpendicular*, and although this meaning of the term in quantum theory usually has no clear relevance to ordinary spatial geometry, this notion of perpendicularity is indeed appropriate to the geometry of the complex vector space $\mathcal{H}$ of quantum states (together with $\mathbf{0}$). This kind of vector space is referred to as a *Hilbert space*, after the very distinguished twentieth-century mathematician David Hilbert, who introduced this notion in the early twentieth century,[7] in a different context. The relevant concept of *at right angles* refers, indeed, to the geometry of Hilbert space.

What is a Hilbert space? Mathematically, it is a vector space (as explained in §A.3), which can be either finite dimensional or infinite dimensional. The scalars are taken to be complex numbers (elements of $\mathbb{C}$; see §A.9) and there is an inner product $\langle \cdots | \cdots \rangle$ (see §A.3), which is what is called *Hermitian*:

$$\langle \beta | \alpha \rangle = \overline{\langle \alpha | \beta \rangle};$$

---

[7] Hilbert published the bulk of his work on this between 1904 and 1906 (six papers bundled in Hilbert [1912]). The first important paper in the area was actually by Erik Ivar Fredholm [1903], whose work on the general concept of a "Hilbert space" preceded that of Hilbert. The above notes are adapted from Dieudonné [1981].

(the overbar denoting complex conjugate; §A.10) and *positive definite*:

$$\langle \alpha | \alpha \rangle \geqslant 0,$$

where

$$\langle \alpha | \alpha \rangle = 0 \quad \text{only when } \alpha = \mathbf{0}.$$

The notion of orthogonality referred to above is simply that defined by this inner product, so we have orthogonality between state vectors $\alpha$ and $\beta$ (non-zero elements of the Hilbert space) expressed (§A.3) as

$$\alpha \perp \beta, \quad \text{i.e. } \langle \alpha | \beta \rangle = 0.$$

There is also a completeness requirement, of relevance when the Hilbert space is of infinite dimension and, moreover, a condition of separability, which limits the infinite "size" of such a Hilbert space, but I shall not need to bother with these issues here.

In quantum mechanics, the complex-number scalars $a, b, c, \ldots$ for this vector space are the complex amplitudes that occur in the quantum-mechanical superposition law, this superposition principle itself providing the addition operation of the Hilbert vector space. The dimension of a Hilbert space can, indeed, be either finite or infinite. For the main purposes of concern here, it will be adequate to assume that the dimension is indeed some finite number $n$, which we might well allow to be extremely large, and I use the notation

$$\mathcal{H}^n$$

to denote this (essentially unique, for each $n$) Hilbert $n$-space. The notation $\mathcal{H}^\infty$ will be used for an infinite-dimensional Hilbert space. For simplicity, I restrict attention here mainly to the finite-dimensional case. Since the scalars are complex numbers, this dimension is a complex dimension (in the sense of §A.10), as a real (Euclidean) manifold $\mathcal{H}^n$ would be $2n$-dimensional. The norm of a state vector $\alpha$, as introduced in §2.5, is

$$\|\alpha\| = \langle \alpha | \alpha \rangle.$$

This is actually the squared length of the vector $\alpha$ in the ordinary real-number Euclidean sense, if we regard $\mathcal{H}^n$ as a $2n$-dimensional Euclidean space. The notion of *orthogonal basis* referred to above is then just that described in §A.4, in the finite-dimensional case, namely a set of $n$ non-zero state vectors $\boldsymbol{\varepsilon}_1, \boldsymbol{\varepsilon}_2, \boldsymbol{\varepsilon}_3, \ldots, \boldsymbol{\varepsilon}_n$, each of which is of unit norm,

$$\|\boldsymbol{\varepsilon}_1\| = \|\boldsymbol{\varepsilon}_2\| = \|\boldsymbol{\varepsilon}_3\| = \cdots = \|\boldsymbol{\varepsilon}_n\| = 1,$$

with mutual orthogonality,

$$\boldsymbol{\varepsilon}_j \perp \boldsymbol{\varepsilon}_k, \quad \text{whenever } j \neq k \ (j, k = 1, 2, 3, \ldots, n).$$

Every vector $\mathbf{z}$ in $\mathcal{H}^n$ (i.e. a quantum state vector) can then be expressed (uniquely) as a linear combination of basis elements,

$$\mathbf{z} = z_1 \boldsymbol{\varepsilon}_1 + z_2 \boldsymbol{\varepsilon}_2 + \cdots + z_n \boldsymbol{\varepsilon}_n,$$

the complex numbers $z_1, z_2, \ldots, z_n$ (amplitudes) being the *components* of $\mathbf{z}$ in the basis $\{\boldsymbol{\varepsilon}_1, \ldots, \boldsymbol{\varepsilon}_n\}$.

The vector space of possible states of a finite-dimensional quantum system is therefore some Hilbert space $\mathcal{H}^n$ of $n$ complex dimensions. Often, in quantum-mechanical discussions, infinite-dimensional Hilbert spaces are made use of, especially in quantum field theory (QFT; see §1.4, particularly). However, some sophisticated mathematical issues can arise, particularly when this infinity is "uncountable" (i.e. greater than Cantor's $\aleph_0$; see §A.2), in which case the Hilbert space fails to satisfy the axiom of separability (referred to briefly earlier) that is frequently imposed [Streater and Wightman 2000]. These matters play no significant role in what I wish to say here, so when I do discuss an infinite-dimensional Hilbert space, I shall mean the separable Hilbert space $\mathcal{H}^\infty$ for which there is a countably infinite orthonormal basis $\{\boldsymbol{\varepsilon}_1, \boldsymbol{\varepsilon}_2, \boldsymbol{\varepsilon}_3, \boldsymbol{\varepsilon}_4, \ldots\}$ (in fact many such bases). When the basis is infinite in this way, there will be issues of convergence to be respected (see §A.10), so that for an element $\mathbf{z} \ (= z_1 \boldsymbol{\varepsilon}_1 + z_2 \boldsymbol{\varepsilon}_2 + z_3 \boldsymbol{\varepsilon}_3 + \cdots)$ to have a finite norm $\|\mathbf{z}\|$ we require the series $|z_1|^2 + |z_2|^2 + |z_3|^2 + \cdots$ to converge to a finite value.

It should be recalled that a distinction was made in §2.5 between the physical state of a quantum system and its mathematical description, the latter using a state vector, say $\boldsymbol{\alpha}$. The vectors $\boldsymbol{\alpha}$ and $q\boldsymbol{\alpha}$, where $q$ is a non-zero complex number, are to represent the same physical quantum state. Thus, the various physical states are themselves represented by the various 1-dimensional subspaces of $\mathcal{H}$ (complex lines through the origin), or *rays*. Each such ray gives the complete family of complex multiples of $\boldsymbol{\alpha}$; in real-number terms, this ray is a copy of the Wessel plane (whose zero 0 is at the origin $\mathbf{0}$ of $\mathcal{H}$). We may think of the normalized state vectors as being of unit length in this Wessel plane, i.e. they represent points in the unit circle in this plane. All state vectors given by points on this unit circle represent the

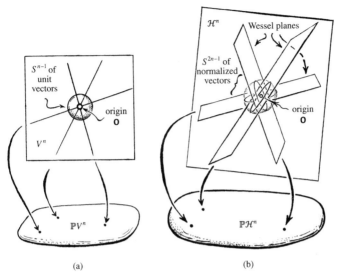

Figure 2-16: The *projective space* for an $n$-dimensional vector space $V^n$ is the $(n-1)$-dimensional compact space $\mathbb{P}V^n$ of rays (1-dimensional subspaces) of $V^n$, whence $V^n$ (without its origin) is a bundle over $\mathbb{P}V^n$. This is illustrated (a) in the real case and (b) in the complex case, where the rays are copies of the Wessel plane, with their unit circles shown. Case (b) is the situation of quantum mechanics, where $\mathcal{H}^n$ is the space of quantum state vectors, $\mathbb{P}\mathcal{H}^n$, the space of physically distinct quantum states, the normalized states forming a circle bundle over $\mathbb{P}\mathcal{H}^n$.

same physical state, so there is still the phase freedom, among unit state vectors, of multiplication by a complex number of unit modulus ($e^{i\theta}$ with $\theta$ real) without changing the physical state; see the end of §2.5. (It should be made clear, however, that this applies only to the entire state of the system. Multiplying different parts of a state by different phases may well change the overall physics of the state.)

The $(n-1)$-complex-dimensional space, each of whose individual points represents one of these rays, is referred to as the *projective* Hilbert space $\mathbb{P}\mathcal{H}^n$. See figure 2-16, which also illustrates the idea of a *real* projective space $\mathbb{P}V^n$, derived from a *real* vector $n$-space $V^n$ (figure 2-16(a)). In the complex case, the rays are copies of the Wessel plane (figure 2-16(b); see §A.10, figure A-34). Figure 2-16(b) also illustrates the subspaces of the *normalized* vectors (i.e. unit vectors) being a sphere $S^{n-1}$ in the real case and $S^{2n-1}$ in the complex case. Each physically distinct quantum state of a system is represented by some entire complex ray in the Hilbert space $\mathcal{H}^n$, and therefore by some individual point of the projective space

$\mathbb{P}\mathcal{H}^n$. The geometry of the space of physically distinct quantum possibilities of some finite physical system may be regarded as the complex projective geometry of some projective Hilbert space $\mathbb{P}\mathcal{H}^n$.

We are now in a position to see how the Born rule operates for a general – i.e. non-degenerate – measurement (see also §1.4). What quantum mechanics tells us is that for any such measurement there will be an orthonormal basis $(\boldsymbol{\varepsilon}_1, \boldsymbol{\varepsilon}_2, \boldsymbol{\varepsilon}_3, \dots)$ of possible outcomes, and the physical result of the measurement must always be one of these. Suppose that the quantum state before measurement is provided by the state vector $\boldsymbol{\psi}$, which for the moment we assume is normalized ($\|\boldsymbol{\psi}\| = 1$). We can express this in terms of the basis as

$$\boldsymbol{\psi} = \psi_1 \boldsymbol{\varepsilon}_1 + \psi_2 \boldsymbol{\varepsilon}_2 + \psi_3 \boldsymbol{\varepsilon}_3 + \cdots$$

and the complex numbers $\psi_1, \psi_2, \psi_3, \dots$ (the components of $\boldsymbol{\psi}$ in the basis) are the amplitudes that were referred to in §§1.4 and 1.5. The **R** procedure of quantum mechanics does not tell us which of the states $\boldsymbol{\varepsilon}_1, \boldsymbol{\varepsilon}_2, \boldsymbol{\varepsilon}_3, \dots$ the state $\boldsymbol{\psi}$ jumps to immediately following the measurement, but it does provide us with a probability $p_j$ for each possible outcome, this being given by the Born rule, which informs us that the probability of our measurement finding the state to be $\boldsymbol{\varepsilon}_j$ (so that the state has "jumped" to $\boldsymbol{\varepsilon}_j$) is the *square of the modulus* of the corresponding amplitude $\psi_j$, namely

$$p_j = |\psi_j|^2 = \psi_j \bar{\psi}_j.$$

It may be noted that our mathematical requirement that the state vector $\boldsymbol{\psi}$ and all the basis vectors $\boldsymbol{\varepsilon}_j$ are normalized has, as a consequence, the necessary property of probabilities that for all the different possibilities they must add up to unity. This very striking fact comes simply from the expression of the normalization condition

$$\|\boldsymbol{\psi}\| = 1$$

in terms of the orthonormal basis $\{\boldsymbol{\varepsilon}_1, \boldsymbol{\varepsilon}_2, \boldsymbol{\varepsilon}_3, \dots\}$, which by virtue of $\langle \boldsymbol{\varepsilon}_i | \boldsymbol{\varepsilon}_j \rangle = \delta_{ij}$ (see §A.10) gives

$$\|\boldsymbol{\psi}\| = \langle \boldsymbol{\psi} | \boldsymbol{\psi} \rangle = \langle \psi_1 \boldsymbol{\varepsilon}_1 + \psi_2 \boldsymbol{\varepsilon}_2 + \psi_3 \boldsymbol{\varepsilon}_3 + \cdots | \psi_1 \boldsymbol{\varepsilon}_1 + \psi_2 \boldsymbol{\varepsilon}_2 + \psi_3 \boldsymbol{\varepsilon}_3 + \cdots \rangle$$
$$= |\psi_1|^2 + |\psi_2|^2 + |\psi_3|^2 + \cdots$$
$$= p_1 + p_2 + p_3 + \cdots = 1.$$

This is one manifestation of the very remarkable synergy between the general mathematical framework of quantum mechanics and the consistency of the requirements of probabilistic quantum behaviour!

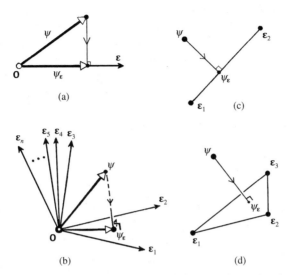

Figure 2-17: The Born rule in geometrical form (with non-normalized state vectors), where the probability of $\psi$ jumping to the orthogonally projected $\psi_\varepsilon$ is given by $\|\psi_\varepsilon\| \div \|\psi\|$, (a) in a non-degenerate measurement, (b) in a degenerate measurement which cannot distinguish $\varepsilon_1$ from $\varepsilon_2$ (projection postulate), (c) the essential geometry of the latter, depicted in the projective Hilbert space, and (d) the projective picture when the degeneracy is between the three states $\varepsilon_1$, $\varepsilon_2$, and $\varepsilon_3$.

We can, nevertheless, restate the Born rule in a way that does not require either $\psi$ or the basis vectors to be normalized by appealing to the Euclidean notion of *orthogonal projection*. Suppose that a measurement is performed on a state that, immediately prior to measurement, has state vector $\psi$, and the measurement ascertains that the state vector is (i.e. "jumps to") a multiple of $\varepsilon$ after being measured; then the probability $p$ of this outcome is the proportion whereby the norm $\|\psi\|$ of $\psi$ is reduced when we pass from $\psi$ to the orthogonal projection $\psi_\varepsilon$ of the vector $\psi$ along the (complex direction of) $\varepsilon$, i.e. it is the quantity

$$ p = \frac{\|\Psi_\varepsilon\|}{\|\psi\|}. $$

This projection is illustrated in figure 2-17(a). It should be noted that the vector $\psi_\varepsilon$ is the unique scalar multiple of $\varepsilon$ such that $(\psi - \psi_\varepsilon) \perp \varepsilon$. This is what *orthogonal projection* means in his context.

One advantage of this interpretation of the Born rule is that it directly extends to the more general situation of a *degenerate* measurement, for which a further rule known as the *projection postulate*, for such measurements, must also be incorporated. Some physicists have tried to argue that this postulate (even in its simple form of a quantum jump where there is no degeneracy in the measurement) is not a needed feature of standard quantum mechanics, since in a normal measurement the resulting state of the measured object is not likely to be an independent thing, having become entangled with the measuring device, as a consequence of its interaction with it. However, this postulate is indeed required, especially in those situations referred to as *null measurements* [see, for example, TRtR, §22.7], where the state necessarily jumps even when it does not disturb the measuring device.

A degenerate measurement is characterized by the fact that it is unable to distinguish between certain physically different possible outcomes. In this situation, rather than there being an essentially unique basis $(\varepsilon_1, \varepsilon_2, \varepsilon_3, \dots)$ of distinguishable outcomes, some of the $\varepsilon_j$ yield the same result of measurement. Suppose $\varepsilon_1$ and $\varepsilon_2$ are two such. Then the entire linear space of states spanned by $\varepsilon_1$ and $\varepsilon_2$ would also give the same result of measurement. This is a (complex) plane through the origin $\mathbf{0}$ in the Hilbert space $\mathcal{H}^n$ (spanned by $\varepsilon_1$ and $\varepsilon_2$), not just the single ray that we get for an individual physical quantum state (figure 2-17(b)). (There is a possible confusion of terminology here, since the term *complex plane* sometimes refers to what I am here calling the *Wessel plane*; see §A.10. The type of plane that is being referred to here has 2 complex dimensions, and therefore 4 real dimensions.) In terms of the projective Hilbert space $\mathbb{P}\mathcal{H}^n$ of physical quantum states, we now have a whole (complex) line of possible results of the measurement, not just a point – see figure 2-17(c). If such a measurement on $\psi$ yields a result that is now found to lie on this plane in $\mathcal{H}^n$ (i.e. line in $\mathbb{P}\mathcal{H}^n$), the particular state obtained by the measurement must be (proportional to) the projection $\psi$ into this plane (figure 2-17(b),(c)). A similar orthogonal projection applies in the case of three or more states that are involved in the degeneracy; see figure 2-17(d).

In some extreme situations of a degenerate measurement, this can apply to several different sets of states at once. Accordingly, instead of the measurement determining a basis of alternative possible outcomes, it would determine a family of linear subspaces of various dimensions, each of which is orthogonal to all of the others, these subspaces being what the measurement can distinguish between. Any state $\psi$ would be uniquely expressible as a sum of various projections determined by the measurement, these projections being the different alternatives to which $\psi$

might jump, upon measurement. The respective probabilities are again given by the proportion whereby the norm $\|\psi\|$ of $\psi$ is reduced when we pass from $\psi$ to each such projection.

Although, in the above description, I have been able to state all that we need about quantum measurement **R** while not ever referring to quantum operators, it is appropriate that we do finally make contact with that more conventional way of expressing things. The relevant operators are usually those referred to as *Hermitian* or *self-adjoint* (there being a slight difference between the two, of importance only in the infinite-dimensional case, and not relevant for us), whereby the operator **Q** satisfies

$$\langle \varphi | \mathbf{Q}\psi \rangle = \langle \mathbf{Q}\varphi | \psi \rangle$$

for any pair of states $\varphi$, $\psi$ in $\mathcal{H}^n$. (The reader familiar with the notion of a *Hermitian matrix* will recognize that this is just what is being asserted of **Q**.) The basis $\{\varepsilon_1, \varepsilon_2, \varepsilon_3, \dots\}$ then consists of what are called *eigenvectors* of **Q**, an eigenvector being an element $\mu$ of $\mathcal{H}^n$ for which

$$\mathbf{Q}\mu = \lambda\mu,$$

where the number $\lambda$ is called the *eigenvalue* corresponding to $\mu$. In quantum measurement, the eigenvalue $\lambda_j$ would be the *numerical value* revealed by the measurement that **Q** actually makes, when the state jumps to the eigenvector $\varepsilon_j$ upon measurement. (In fact $\lambda_j$ must be a *real* number for a Hermitian operator, corresponding to the fact that measurements in the usual sense of that word are indeed taken to reveal *real* numbers.) As a clarifying comment, it should be made clear that the eigenvalue $\lambda_j$ has nothing to do with the amplitude $\psi_j$. It is $\psi_j$'s squared modulus that gives the *probability* that the outcome of the experiment actually yields the numerical value $\lambda_j$.

This is all very formal, and the abstract complex geometry of a Hilbert space seems to be very far from the geometry of the ordinary space that we directly experience. Nonetheless, there is a distinct geometrical elegance in the general framework of quantum mechanics, and its relation to both **U** and **R**. Since the dimensions of the quantum Hilbert spaces tend to be rather high – not to mention the fact that it is a complex-number geometry rather than our familiar real-number geometry – direct geometric visualization is not usually easily achieved. However, in the next section we shall see that, with the particularly quantum-mechanical notion of *spin*, this geometry can be directly understood in relation to the geometry of ordinary 3-space, and this can be helpful in understanding what quantum mechanics is all about.

## 2.9. THE GEOMETRY OF QUANTUM SPIN

The most clear-cut relation between Hilbert-space geometry and that of ordinary 3-dimensional space indeed occurs with the spin states. This is particularly so, for a massive particle of spin $\frac{1}{2}$, such as an electron, proton, or neutron, or certain atomic nuclei or atoms. By studying these spin states we can get a better picture of how the measurement process in quantum mechanics actually operates.

A spin-$\frac{1}{2}$ particle always spins by a specific amount, namely $\frac{1}{2}\hbar$ (see §1.14), but the particle's spin *direction* behaves in a subtle, characteristically quantum-mechanical way. Thinking classically for the moment, we take this spin direction to be defined as the axis about which the particle spins, the axis being oriented outwards so that the spin is in a *right-handed* sense about this spin-axis (i.e. in an anticlockwise sense, as one looks in towards the particle along the outward spin axis). For any particular direction of spin, an alternative way for our particle to spin would *left*-handed about that same direction (by the same amount $\frac{1}{2}\hbar$), but it is conventional to describe that state, instead, as right-handed about the diametrically *opposite* direction.

The *quantum states of spin* for a particle of spin $\frac{1}{2}$ are exactly in accordance with these classical states – although subject to the strange rules that quantum mechanics demands. Thus, for any spatial direction, there will be a state of spin in which the particle spins in a right-handed sense about that direction, its spin having a magnitude of $\frac{1}{2}\hbar$. However, quantum mechanics tells us that all these possibilities can be expressed as linear superpositions of any two different such states, these spanning the space of all possible spin states. If we take these two as having opposite directions of spin about some particular direction, then they will be *orthogonal* states. Thus, we have a 2-complex-dimensional Hilbert space $\mathcal{H}^2$, and an *orthogonal basis* for states of spin $\frac{1}{2}$ would always be (right-hand spin about) such a pair of opposite directions. We shall be seeing shortly how any other direction of the particle's spin can indeed be expressed as a quantum linear superposition of these two oppositely spinning states.

In the literature, a commonly used basis takes these directions to be *up* and *down*, frequently written as

$$|\uparrow\rangle \quad \text{and} \quad |\downarrow\rangle,$$

respectively, where I have begun to adopt the Dirac *ket* notation for quantum state vectors, and where some descriptive symbol or letter(ing) appears between the symbols "$|\cdots\rangle$". (The full significance of this notation will not concern us here, though it may be mentioned that, in this form, the state vectors are referred to as *ket* vectors, and the *dual* vectors to these kets (see §A.4), called *bra* vectors, are written within "⟨··· |", so as to form a complete bracket "⟨··· | ···⟩" when an inner product is formed [Dirac 1947]. Any other possible spin state |↗⟩ of our particle must be linearly expressible in terms of these two basis states:

$$|\nearrow\rangle = w|\uparrow\rangle + z|\downarrow\rangle,$$

where we recall from §2.5 that it is only the ratio $z : w$ that serves to distinguish the physically different quantum states. This complex ratio is basically just the quotient

$$u = \frac{z}{w},$$

but we must also allow that $w$ might be zero, so as to accommodate the state |↓⟩ itself. This is achieved if we simply allow ourselves to write $u = \infty$ (formally) for this ratio when $w = 0$. Geometrically, including a point "∞" into the Wessel plane amounts to bending this plane down into a sphere (rather as the horizontal ground that we stand on bends round to be the sphere of the Earth), by closing it up with the point ∞. This gives us the simplest of the Riemann surfaces (see §§A.10 and 1.6), referred to as the *Riemann sphere* (although, in this kind of context such a sphere might sometimes be referred to as a *Bloch sphere* or *Poincaré sphere*).

A standard way of representing this geometry is to imagine a Wessel plane to be situated horizontally within a Euclidean 3-space, where the Riemann sphere is the unit sphere centred at the origin O, representing 0 of the Wessel plane. The unit circle in the Wessel plane is taken to be the *equator* of the Riemann sphere. Now we consider the point S at the *south pole* of the sphere and *project* the rest of the sphere from S to the Wessel plane. That is to say, the point Z on the Wessel plane corresponds to the point Z′ on the Riemann sphere if S, Z, and Z′ are all in a straight line (stereographic projection; see figure 2-18). In terms of standard Cartesian coordinates $(x, y, z)$ for the 3-space, the Wessel plane is $z = 0$ and the Riemann sphere is $x^2 + y^2 + z^2 = 1$. Then the complex number $u = x + iy$, representing the point Z in the Wessel plane with Cartesian coordinates $(x, y, 0)$, corresponds to Z′ on the Riemann sphere with Cartesian coordinates $(2\lambda x, 2\lambda y, \lambda(1 - x^2 - y^2))$, where $\lambda = (1 + x^2 + y^2)^{-1}$. The *north pole* N corresponds to the origin O of the Wessel plane, and this represents the complex number 0. All the points on the unit circle in the Wessel plane ($e^{i\theta}$, where $\theta$ is real; §A.10), including 1, i, −1, −i, correspond to the identical points around the Riemann sphere's equator. The south pole S of the Riemann sphere labels the additional point ∞ on the sphere, which maps to infinity on the Wessel plane.

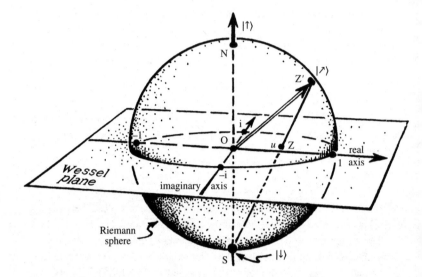

Figure 2-18: The Riemann sphere, related to (its equatorial) Wessel plane via stereographic projection, where Z on the plane projects to Z' on the sphere from its south pole S (its north pole being N and centre, O). This gives a geometrical realization of the spin direction $|\nearrow\rangle = |\uparrow\rangle + u|\downarrow\rangle$, for a spin-$\frac{1}{2}$ massive particle, where Z and Z' represent the complex number $u$, on the Wessel plane and Riemann sphere, respectively.

Now let us see how this mathematical representation of complex ratios relates to the spin states of our spin-$\frac{1}{2}$ particle, these being given by $|\nearrow\rangle = w|\uparrow\rangle + z|\downarrow\rangle$, with $u = z/w$. So long as $w \neq 0$, we can take $u$ as an ordinary complex number and, if we choose, we can scale things (abandoning any requirement that $|\nearrow\rangle$ be normalized) so that $w = 1$, whence $u = z$, and our state vector is now

$$|\nearrow\rangle = |\uparrow\rangle + z|\downarrow\rangle.$$

We can take $z$ to be represented as the point Z on our Wessel plane, which corresponds to the point Z' on the Riemann sphere. If we make suitable choices of phases for $|\uparrow\rangle$ and $|\downarrow\rangle$, then we find, satisfyingly, that the direction $\overrightarrow{OZ'}$ is the spin direction of $|\nearrow\rangle$. The state $|\downarrow\rangle$, corresponding to the south pole S of the Riemann sphere, corresponds to $z = \infty$, but for that we would need to normalize the state $|\nearrow\rangle$ differently, e.g. $|\nearrow\rangle = z^{-1}|\uparrow\rangle + |\downarrow\rangle$ (taking $\infty^{-1} = 0$).

The Riemann sphere, being simply the space of ratios $w : z$ of a pair of complex numbers $(w, z)$, not both zero, is actually the projective Hilbert space $\mathbb{P}\mathcal{H}^2$, describing the array of possible physically distinct quantum states that arise from superpositions of any two independent quantum states of any kind. But what is particularly striking for (massive) particles of spin $\frac{1}{2}$ is that this Riemann sphere corresponds precisely to the directions at a point in ordinary 3-dimensional physical space. (If the number of spatial directions had been different – as modern string theory seems to require; see §1.6 – then such a simple and elegant relation between spatial geometry and quantum complex superposition would not occur.) But even if we do not demand such a direct geometrical interpretation, the Riemann-sphere picture $\mathbb{P}\mathcal{H}^2$ is still useful. Any orthogonal basis for a 2-dimensional Hilbert space $\mathcal{H}^2$ is still represented by a pair of antipodal points A, B on an (abstract) Riemann sphere, and it turns out that, by use of a little simple geometry, we can always interpret the Born rule in the following geometrical way. Suppose C is the point on the sphere representing the initial (e.g. spin) state, and a measurement is made which decides between A and B, then we drop a perpendicular from C to the diameter AB to obtain the point D on this diameter, and we find that the Born rule can be interpreted in the following geometrical way (see figure 2-19):

$$\text{probability of C jumping to A} = \frac{\text{DB}}{\text{AB}},$$
$$\text{probability of C jumping to B} = \frac{\text{AD}}{\text{AB}}.$$

In other words, if we take our sphere to have diameter $= 1$ (rather than radius $= 1$), then these lengths DB and AD directly give us the respective probabilities for the state to jump to A or to B.

As this applies to any situation of measurement on a 2-state system, not just to massive particles of spin $\frac{1}{2}$, it is relevant in its simplest form to those considered in §2.3, where in the first experiment, shown in figure 2-4(a), a beam splitter puts a photon into a superposition of two possible routes, where either one of the photon detectors would be presented with a state which is an equal superposition of a photon and no photon. Let us (formally) compare this with a spin-$\frac{1}{2}$ particle initially taken to have spin state $|\downarrow\rangle$. This is to correspond, in figure 2-4(a), to the initial photon's momentum state as it emerges from the laser, moving towards the right (direction MA in figure 2-4(a)). After it encounters the beam splitter, the photon's momentum state is put into a superposition of this right-moving momentum state (still corresponding, formally, to $|\downarrow\rangle$), and which we can

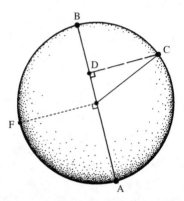

Figure 2-19: A measurement of a 2-state quantum system is set to distinguishing between the orthogonal pair of states A, B, these being respectively represented by antipodal points A, B on the Riemann sphere $\mathbb{P}\mathcal{H}^2$. The measuring device is presented with the state C, represented by C on $\mathbb{P}\mathcal{H}^2$. The Born rule gives the probability DB/AB for the device to find C jumping to A, and AD/AB for C jumping to B, where D is the orthogonal projection of C to the diameter AB.

correspond to the point A in figure 2-19, and an upwards-moving momentum state (direction MB in figure 2-4(a)), corresponding, now, to $|\uparrow\rangle$, and represented by the point B of figure 2-19. The photon's momentum is now an equal superposition of these two, and we can think of it as represented by the point F in figure 2-19, giving equal 50% probabilities for the two alternatives. In the case of the second experiment of §2.3 (the Mach–Zhender situation), shown in figure 2-4(b) with detectors at D an E, the mirrors and beam-splitter serve, effectively, to put the photon's momentum state back into its original form (corresponding to $|\downarrow\rangle$) and to the point A in figure 2-19) which the detector D, in figure 2-4(b), is geared to measure, providing a 100% detection probability and the one at E a 0% detection probability.

This example is very limited in the superpositions that it provides, but it is not hard to modify it so that the full spectrum of complex-weighted alternatives are exhibited. In many actual experiments, this is done in terms of photon polarization states instead of different momentum states. Photon polarization is also an example of quantum-mechanical spin, but here we find that the spin is either entirely right-handed about the direction of motion or entirely left-handed about this direction, corresponding to the two states of circular polarization considered in §2.6. Again we have just a 2-state system and a projective Hilbert space $\mathbb{P}\mathcal{H}^2$.[8] Accordingly, we can represent a general state as given by a point on a Riemann sphere, but the geometry is somewhat different.

To explore this, let us orient the sphere so that its north pole N is in the direction of the photon's motion, so the state $|\circlearrowright\rangle$ of right-handed spin will be represented by N. Correspondingly, the south pole S will represent the state $|\circlearrowleft\rangle$ of left-handed spin. A general state

$$|\leftrightsquigarrow\rangle = |\circlearrowright\rangle + w|\circlearrowleft\rangle$$

could be represented on the Riemann sphere by the point $Z'$, which corresponds to $z/w$, as in the case of a massive particle of spin $\frac{1}{2}$, above (figure 2-18). However, it is geometrically much more appropriate to represent the state by the point Q on the Riemann sphere, which corresponds to a square root of $z$, i.e. a complex number $q$ (basically the same $q$ as in §2.6) satisfying

$$q^2 = z/w.$$

The exponent "2" comes from the fact that the photon has spin 1, i.e. *twice the basic unit of spin*, which is the electron's $\frac{1}{2}\hbar$. Had we been concerned with a massless particle of spin $\frac{1}{2}n$, which is $n$ times this basic spin unit, then we would have been concerned with the values of $q$ that satisfy $q^n = z$. For a photon, $n = 2$; so $q = \pm\sqrt{z}$. To find the relation of Q to the ellipse of photon polarization (see §2.5), we first find the great circle that is the intersection with the Riemann sphere of the plane through O perpendicular to the line OQ (figure 2-20); then we project this vertically to an ellipse in the horizontal (Wessel) plane. This ellipse turns out to be the ellipse of polarization of the photon, inheriting the right-handed orientation about OQ of the great circle on the sphere. (The direction OQ is related to what is known as the *Stokes vector*, although more directly to the *Jones vector* (the technical explanation of what they each are is in chapter 3 of Hodgkinson and Wu [1998]; see also TRtR, §22.9, p. 559).) It may be noted that $q$ and $-q$ give the same ellipse and orientation.

It is of some interest to see how *massive states* of *higher spin* can also be represented in terms of the Riemann sphere by use of what is known as the *Majorana description* [Majorana 1932; TRtR, §22.10, p. 560]. For a massive particle (say an atom) of spin $\frac{1}{2}n$, where $n$ is a non-negative integer (so that the space of physically distinct possibilities is $\mathbb{P}\mathcal{H}^{n+1}$), and any physical spin state is

[8] In recent experiments, higher-dimensional Hilbert spaces have been experimentally constructed with individual photons, by making use of the orbital angular momentum degrees of freedom in the photon state [Fickler et al. 2012].

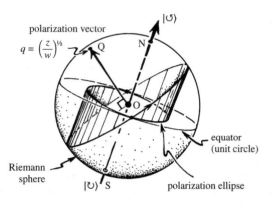

Figure 2-20: The general polarization state $w|\circlearrowright\rangle + z|\circlearrowleft\rangle$ of a photon, with direction of motion indicated towards the upper right, its two (normalized) states of circular polarization being $|\circlearrowright\rangle$ (right-handed) and $|\circlearrowleft\rangle$ (left-handed). Geometrically, we can represent this by the complex number $q$ on a Riemann sphere (marked Q), where $q^2 = z/w$ (allowing $q = \infty$ for the left-handed case, for Q at the south pole S, and $q = 0$ for the right-handed case, for Q at the north pole N, where ON is the photon's direction, O being the sphere's centre). The photon's polarization ellipse is the equatorial projection of the great circle perpendicular to the direction OQ.

given by an unordered set of $n$ points on the Riemann sphere (coincidences being allowed). We can consider each of these points to correspond to a contribution of spin $\frac{1}{2}$ in the direction of that point out from the centre (figure 2-21). I shall refer to each of these directions as a *Majorana direction*.

Such general states of spin, for spin $> \frac{1}{2}$, are not often considered by physicists, however, who tend to think of states of higher spin in terms of a commonly described type of measurement, which employs a *Stern–Gerlach* apparatus (figure 2-22). This makes use of a highly inhomogeneous magnetic field to measure the magnetic moment of a particle (which is normally aligned with the spin), deflecting a succession of such particles[9] by various amounts, depending upon how much of the spin (strictly the magnetic moment) is aligned with the magnetic field. For a particle of spin $\frac{1}{2}n$, there will be $n + 1$ different possibilities which, for the field oriented in the up/down direction, would be the Majorana states

$$|\uparrow\uparrow\uparrow \cdots \uparrow\rangle, \; |\downarrow\uparrow\uparrow \cdots \uparrow\rangle, \; |\downarrow\downarrow\uparrow \cdots \uparrow\rangle, \; \ldots, \; |\downarrow\downarrow\downarrow \cdots \downarrow\rangle,$$

---

[9] In fact, for technical reasons, this procedure does not work directly for electrons [Mott and Massey 1965], but is used successfully for many different kinds of atoms.

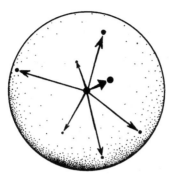

Figure 2-21: The Majorana description of the general spin state of a massive particle of spin $\frac{1}{2}n$ represents this state as an unordered set of $n$ points on the Riemann sphere, each of which can be thought of as the spin direction of a spin-$\frac{1}{2}$ component contributing to the entire spin-$n$ state.

$|\uparrow\uparrow\uparrow\rangle\ m = \frac{3}{2}$
$|\uparrow\uparrow\downarrow\rangle\ m = \frac{1}{2}$
$|\uparrow\downarrow\downarrow\rangle\ m = -\frac{1}{2}$
$|\downarrow\downarrow\downarrow\rangle\ m = -\frac{3}{2}$

Figure 2-22: A Stern–Gerlach apparatus uses a highly inhomogeneous magnetic field to measure the spin (or, more correctly, magnetic moment), in the chosen direction of the inhomogeneity, of an atom of spin $\frac{1}{2}n$. The different possible outcomes distinguish values for the component of spin in that direction, which are $-\frac{1}{2}n, -\frac{1}{2}n+1, -\frac{1}{2}n+2, \ldots, \frac{1}{2}n-2, \frac{1}{2}n-1, \frac{1}{2}n$ in units $\hbar$ of spin, these corresponding to the respective Majorana states $|\downarrow\downarrow\downarrow \cdots \downarrow\rangle, |\uparrow\downarrow\downarrow \cdots \downarrow\rangle, |\uparrow\uparrow\downarrow \cdots \downarrow\rangle, \ldots, |\uparrow\uparrow \cdots \uparrow\downarrow\downarrow\rangle, |\uparrow\uparrow\uparrow \cdots \uparrow\downarrow\rangle, |\uparrow\uparrow\uparrow \cdots \uparrow\uparrow\rangle$. We may envisage that the apparatus is rotatable about the direction of the beam, so that different selected spin directions, orthogonal to the beam direction, can be measured.

in which every Majorana direction is either up or down, but with various multiplicities. (In the standard terminology, each of these states is distinguished by its "$m$ value", which is one-half of the number of up arrows minus one half of the number of down arrows. This is basically the same "$m$ value" as arises in the harmonic analysis of the sphere, referred to in §A.11.) These particular $n + 1$ states are all orthogonal to one another.

A general $\frac{1}{2}n$-spin state, however, has no restriction on its Majorana directions. Yet the idea of a Stern–Gerlach type of measurement can be used to characterize where the Majorana directions actually are. Any Majorana direction ⬦ is determined by the fact that if a Stern–Gerlach-type measurement were to be made with its magnetic field pointing in the direction ⬦, it would find a *zero* probability of the state to be completely in the opposite direction $|\searrow\searrow\searrow\searrow \cdots \searrow\rangle$ [Zimba and Penrose 1993].

## 2.10. QUANTUM ENTANGLEMENT AND EPR EFFECTS

One remarkable family of implications of quantum mechanics, referred to as *Einstein–Podolsky–Rosen* (EPR) phenomena, provides us with what are among the most puzzling and large-scale confirmations of standard quantum theory. EPR phenomena arose out of Albert Einstein's attempts to show that the framework of quantum mechanics is basically flawed, or at least basically incomplete. He collaborated with two colleagues, Boris Podolsky and Nathan Rosen, to publish a now famous paper [Einstein et al. 1935]. What they pointed out, in effect, was that standard quantum mechanics has an implication that they, and many other people (even to this day), would have taken to be unacceptable. This implication was that a pair of particles, no matter how far apart the particles might be, still have to be considered as a single interconnected entity! A measurement performed on one of the particles appears to affect the other one *instantaneously*, putting this second particle into a particular quantum state that depends not only upon the result of the measurement made on the first particle, but – much more strikingly – upon the specific *choice* of measurement that is made on the first particle.

In order to appreciate, properly, the startling implications of this type of situation, it is particularly illuminating to consider the spin states of particles of spin $\frac{1}{2}$. The simplest example of an EPR phenomenon is an example of this type put forward by David Bohm in his 1951 book on quantum mechanics (that he apparently wrote to convince himself – unsuccessfully, as it turned out – of the full validity of the quantum formalism [Bohm 1951]. In Bohm's example, an initial particle of spin 0 splits into two other particles $P_L$ and $P_R$, each of spin $\frac{1}{2}$, travelling in opposite directions from the initial point O, finally reaching spin-measuring detectors at respective locations L (on the left) and R (on the right), widely separated from one another. We assume that each of the two detectors is able to rotate freely and independently, and that the choice of direction in which to measure the spin, for each detector, is not made until the particles are in free flight. See figure 2-23.

Figure 2-23: The EPR–Bohm spin-measurement situation. Two massive particles (atoms, say) $P_L$ and $P_R$ each of spin $\frac{1}{2}$ separate in opposite directions from an initially spin-0 state to a considerable distance, where their spins are separately measured by Stern–Gerlach-type devices at, respectively, L and R. The devices may be separately rotated so as to measure the spins independently in various directions.

Now, it turns out that if some particular direction were to be chosen that is the *same* for the two detectors, then the result of the spin measurement in that direction on the left-hand particle $P_L$ must be opposite to the corresponding spin measurement on the right-hand particle $P_R$. (This is just an instance of angular momentum conservation in that direction (see §1.14), since the initial state has zero angular momentum about any chosen direction.) Thus, if the chosen direction is up ↑, then finding the right-hand particle to be in the up state $|\uparrow\rangle$ at R would imply that a similar up/down measurement of the left-hand particle at L would necessarily result in the *down* state $|\downarrow\rangle$ at $L$; likewise, finding the down state $|\downarrow\rangle$ at R would necessarily imply finding the up state $|\uparrow\rangle$ at L. This would apply equally to any other direction, say $\nearrow$, so that a measurement in that direction of the spin of the L-particle would necessarily find **NO** – i.e. the state with the opposite direction $|\nearrow\rangle$ – if the measurement of the R-particle finds **YES**, i.e. the state $|\swarrow\rangle$; likewise, the L-particle's measurement finds **YES**, i.e. $|\swarrow\rangle$, if the R-particle's measurement finds **NO**, i.e. $|\nearrow\rangle$.

So far, there is nothing essentially non-local about this, even though the picture presented by the standard quantum formalism seems to be at variance with the normal expectations of local causality. We may ask in what way is it apparently at variance with locality? Well, suppose that we perform the R measurement just momentarily before the L measurement. If the R measurement finds $|\swarrow\rangle$, then instantaneously the L-particle's state must be $|\nearrow\rangle$; if the R measurement finds $|\nearrow\rangle$, then the L-particle's state instantaneously becomes $|\swarrow\rangle$. We could arrange that the time-interval between the two measurements is so small that even a light signal travelling from R to L, conveying the information of what the L-particle's state has been found to be, could not make it in time. The quantum information of what the L-particle's state has become violates the standard requirements of relativity theory (figure 2-23). So why is this behaviour "nothing essentially non-local"? It is not *essentially* non-local because we can readily cook up a classical "toy model" which exhibits this behaviour. We could envisage that each particle comes away from its creation point at O armed with instructions about how it is to respond to whatever spin measurement might be performed on it. All that is required, for consistency with the requirements of the previous paragraph, is that the instructions that the particles carry with them away from O must be pre-arranged to require each of the two particles to behave in exactly the *opposite* way, for each possible direction of measurement. We can achieve this if we imagine that the initial spin-0 particle contains a tiny sphere which splits, randomly, into two hemispheres at the moment the particle splits into two spin-$\frac{1}{2}$ particles, each of these carrying one of the hemispheres away in a translational motion (i.e. without either hemisphere being involved in any rotation). For each particle, the hemisphere represents the directions out from the centre which would evoke a **YES** response to a spin measurement in this direction. It is readily seen that this model always provides opposite results for any choice of direction for the measurement of the spin of both particles, this being exactly what is required, according to the quantum considerations of the previous paragraph.

This is an example of what the outstanding quantum physicist John Stewart Bell referred to as being analogous to *Bertlmann's socks* [Bell 1981, 2004]. Reinhold Bertlmann (now a professor of physics at the University of Vienna) had been a distinguished colleague of John Bell's at CERN, and Bell noted that he invariably wore socks of different colours. It was not always easy to catch sight of the colour of one of Dr Bertlmann's socks, but if you were fortunate enough to do so, and suppose that in this instance the colour is found to be green, then you were assured – *instantaneously* – that the colour of his other sock was some colour other than green. Do we conclude that there was some mysterious influence travelling from one foot to the other at superluminary speed as soon as the information of the colour of one of Dr Bertlmann's socks was received by an observer? Of course not. Everything is explained by the fact that Bertlmann had prearranged that the colours of his socks must always differ.

However, in the case of the pair of spin-$\frac{1}{2}$ particles in Bohm's example, the situation is radically altered if we allow the detectors at L and R to have spin-measurement directions which are allowed to vary *independently* of each other. In 1964, Bell established a startling and very fundamental result, which had the implication that no model of the Bertlmann-socks type can account for the joint probabilities that the quantum formalism provides (incorporating the standard Born rule), for independent spin measurements at L and R, for pairs of spin-$\frac{1}{2}$ particles coming from a common quantum source [Bell 1964]. In fact, what Bell demonstrated was that there are certain relations (inequalities) between the joint probabilities of the results of spin measurements in various directions at L and R – now called *Bell inequalities* – that are necessarily satisfied by any classical local model, but which are *violated* by the Born-rule joint probabilities of quantum mechanics. Various experiments were subsequently performed [Aspect et al. 1982; Rowe et al. 2001; Ma 2009] and agreement with the expectations of quantum mechanics is now convincingly confirmed, with violations of the Bell inequalities being thoroughly established. In fact, these experiments tend to use photon polarization states [Zeilinger 2010] rather than the states of spin for particles of spin $\frac{1}{2}$, but as we have seen in §2.9, the two situations are formally identical.

Many theoretical examples of Bohm-type EPR experiments, some of which are particularly simple, have been put forward, where discrepancies can be clearly seen between the expectations of quantum mechanics and those of classical local-realistic (i.e. Bertlmann-socks-type) models [Kochen and Specker 1967; Greenberger et al. 1989; Mermin 1990; Peres 1991; Stapp 1979; Conway and Kochen 2002; Zimba and Penrose 1993]. But rather than entering into any detail of various such examples, I shall present just one particularly striking EPR-type example, due to Lucien Hardy [1993], which is not quite the same as Bohm's situation but is somewhat similar to it. Hardy's example has the remarkable feature that all the probability values are simply 0 or 1 (i.e. "cannot happen" or "certain to happen") with the exception of one probability value, and all we need to know about that one is that it is non-zero (i.e. "sometimes happens"). As with Bohm's example, two particles of spin $\frac{1}{2}$ are emitted in opposite directions from a source at O to spin detectors at widely separated locations L and R. However, there is a difference now, in that the initial state at O is not spin 0, but a particular state of spin 1.

In the specific version of Hardy's example I am presenting here [see TRtR, §23.5], the two Majorana directions of this initial state are ← ("west") and ↗ (a little to the "north" of "northeast"). The exact slopes of these two directions are provided by figure 2-24: the ← direction is horizontal in the picture (with the negative orientation), and the ↗ direction has an upward slope of $\frac{4}{3}$ (with a positive orientation). The point about this particular initial state $|{\leftarrow}{\nearrow}\rangle$ is that whereas we find

$$|{\leftarrow}{\nearrow}\rangle \text{ is } not \text{ orthogonal to the pair } |{\downarrow}\rangle|{\downarrow}\rangle$$

(where ↓ is "south", and where → is "east" in what follows), we also find

$$|{\leftarrow}{\nearrow}\rangle \text{ } is \text{ orthogonal to each of the pairs } |{\downarrow}\rangle|{\leftarrow}\rangle, |{\leftarrow}\rangle|{\downarrow}\rangle, \text{ and } |{\rightarrow}\rangle|{\rightarrow}\rangle.$$

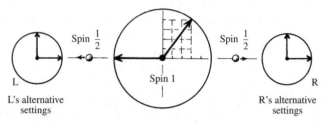

Figure 2-24:  The Hardy example: as in figure 2-23, but with the initial state a particular spin-1 state, with Majorana directions angled at $\tan^{-1}(-\frac{4}{3})$ to each other. Here, all the relevant probabilities are 0 or 1, except for one of them for which all that is needed is that it is non-zero (being actually $\frac{1}{12}$).

Here, what I mean by a pair $|\alpha\rangle|\beta\rangle$ is the state which is $|\alpha\rangle$ at L and $|\beta\rangle$ at R. By angular momentum conservation, the spin state of the combined pair of emitted particles remains the same state $|\leftarrow\nearrow\rangle$ up until one of the spin measurements is made, so the orthogonality relations with the initial state $|\leftarrow\nearrow\rangle$ apply also at the time of the later measurements. (As a comment to those who might worry that the spin-measuring devices depicted in figure 2-23 appear only to allow rotations about the axis defined by the flight directions of the particles, it may be pointed out that the relevant spatial directions needed for this example are all in one plane, so that this plane may be chosen to be orthogonal to the particle motions.)

The non-orthogonality statement in the first of the two displayed assertions above tells us (i) that if the spin-measuring detectors at L and R are both set to measure $\downarrow$, then *sometimes* (with a probability of $\frac{1}{12}$, in fact) it will indeed be found that both detectors obtain $\downarrow$ (i.e. **YES, YES**). The orthogonality statement in the second displayed assertion first tells us (ii) that if one detector is set to measure $\downarrow$ and the other to measure $\leftarrow$, then they cannot both find these results (i.e. at least one finds **NO**). Finally, it tells us (iii) that if both detectors are set to measure $\leftarrow$, then they cannot both find the opposite result $\rightarrow$, or, put another way, at least one of the detectors must find $\leftarrow$ (i.e. **YES**).

Let us see if we can make a classical local model (i.e. a Bertlmann-socks explanation) that can account for these requirements. We imagine that a succession of clockwork particles is emitted from O in the directions of the detectors at L and R, the particles being preprogrammed to give certain results upon encountering the detectors, giving respective results that depend upon how each detector turns out to be individually oriented – but the individual machine components governing each particle's behaviour are *not* allowed to signal to each other after the particles have become separated from O. In particular, our particles must be prepared for the possibility that both detectors have been oriented to measure ←, so if the machinery is to provide answers in agreement with (iii), it must arrange that one or the other of the particles will definitely yield the answer **YES** (i.e. ←) upon encountering a detector with the ← orientation. But, in that case, it might turn out that the *other* detector is found to have been oriented to measure ↓ instead, and (ii) demands that for the particle entering this ↓-measuring detector, the **NO** answer (i.e. ↑) must be obtained. Thus, for *every* emission of the pair of particles at O, one or the other of the particles must be prearranged to provide this ↑ answer to a ↓ measurement. But this falls foul of (i), namely that *sometimes* (on $\frac{1}{12}$ of the emissions, on average) if both detectors happen to be set to measure ↓, then the pair of **YES** answers (↓, ↓) *must* be obtained! Thus, there is no way to account for the expectations of quantum mechanics using any classical-type (i.e. Bertlmann-socks-type) local machinery.

The conclusion of all this is that, in many situations, separated quantum objects, no matter how far apart they may be, are still interconnected with one another and do not behave as independent objects. The quantum state of such a pair of separated objects is *entangled* (in Schrödinger's terminology), a notion that we have previously encountered in §2.7 (mention of it having been made already in §2.1). Quantum entanglements are not, in fact, unusual in quantum mechanics. Far from it: encounters between quantum particles (or previously unentangled systems) will almost invariably result in entangled states. And once they are entangled, they are exceedingly unlikely to become unentangled again merely through unitary evolution (**U**).

This dependence that an entangled pair of widely separated quantum objects have upon one another is a subtle thing, however. For it turns out that such entanglement necessarily falls short of providing an ability for new information to be transmitted from one object to the other. The facility to send actual information by superluminary means would violate the requirements of relativity. Quantum entanglement is something that has no analogue in classical physics. It lies in a strange quantum no-man's land between the two classical alternatives: communicability and complete independence.

Quantum entanglements are indeed subtle, as it requires some considerable sophistication to detect when such entanglements are actually present – as we have seen. Nevertheless, quantum-entangled systems, which would be close to ubiquitous consequences of quantum evolution, provide us with situations of *holistic* behaviour where, in a clear sense, the whole is greater than the sum of its parts. It is greater in a subtle and somewhat mysterious way, where in normal experience we do not notice the effects of quantum entanglements at all. It is indeed a puzzling issue as to why, in the universe we actually experience, such holistic features are present, yet hardly ever make themselves manifest. I shall return to this issue in §2.12, but in the meantime it will be instructive to examine how vast is the space of entangled states, in comparison with the sub-collection of states that are actually unentangled. This is again a matter of *functional freedom*, as discussed in §§A.2, A.8, 1.9, 1.10, and 2.2, but we shall find that there are some additional issues of foundational significance, concerning the interpretation of functional freedom in the quantum context.

## 2.11. QUANTUM FUNCTIONAL FREEDOM

As was indicated in §2.5, the quantum description of a single particle – referred to as the *wave function* of the particle – is something rather similar to a classical field, with a certain number of independent components per spatial point, and (like an electromagnetic field) propagating deterministically into the future by means of field equations. The wave function's field equation is, in fact, Schrödinger's equation (see §2.4). How much functional freedom is there in a wave function? In accordance with the ideas and notation of §A.2, a 1-particle wave function has a functional freedom of the form $\infty^{A\infty^3}$ for some positive integer $A$ (the dimensionality of ordinary space being 3).

The quantity $A$ is basically the number of independent components of the field (see §A.2), but there is the additional issue of wave functions being complex rather than real, which would lead to the expectation that $A$ would be twice the number of real components that one would get for a classical field. However, there is a further issue, discussed only briefly here, early in §2.6, that the wave function of a free particle should be described by a complex function of *positive frequency*, which in effect simply halves the freedom, reducing us back to the same $A$ that we would have had for the classical field. There is the additional issue of the overall multiplying factor that does not change the physical state, but this is completely insignificant in the context of functional freedom.

If we had two independent particles of different kinds, one with its state's functional freedom given by $\infty^{A\infty^3}$ and the other by $\infty^{B\infty^3}$, then the freedom in the *unentangled* quantum states of the pair would simply be the product of these two freedoms (since each state of one particle can be accompanied by any possible state of the other), namely

$$\infty^{A\infty^3} \times \infty^{B\infty^3} = \infty^{(A+B)\infty^3}.$$

But, as we saw in §2.5, in order to obtain *all* the different quantum states available to the pair of particles, including the entangled ones, we need to be able to provide a separate amplitude for each pair of locations (where the locations for the two particles vary independently), so our wave function is now a function of twice as many variables as the 3 that we had before (namely 6, now) – and, moreover, each pair of values from the respective possibilities provided by $A$ and $B$ counts separately (so we now get the product $AB$ for the total number of these possibilities, rather than the sum $A + B$). In terms of the notions introduced in §A.2, our wave function is a function on the *configuration space* (see §A.6, figure A-18) of the pair of particles, which (ignoring discrete parameters such as those describing the spin states) is the 6-dimensional product of ordinary 3-dimensional space with itself. (See §A.7, especially figure A-25, for the notion of *product space*.) Accordingly, the space on which our 2-particle wave function is defined is 6-dimensional, and we now find an enormously larger functional freedom

$$\infty^{AB\infty^6}.$$

In the case of three or four particles, etc., we would get respective functional freedoms

$$\infty^{ABC\infty^9}, \quad \infty^{ABCD\infty^{12}}, \quad \text{etc.}$$

In the case of $N$ identical particles the freedom in the functions would be somewhat restricted because of the issues of Bose–Einstein and Fermi–Dirac statistics referred to in §1.14, according to which the wave function must be, respectively, symmetric or antisymmetric. However, this does not reduce the functional freedom from what it would have been without this restriction, namely $\infty^{AN\infty^{3N}}$, because the restriction merely tells us that the wave function is determined by some subregion of the full product space (and of the same dimension), the values on the remaining region being determined from the symmetry or the antisymmetry requirement.

We see that the functional freedom that is involved in quantum entanglement utterly swamps the freedom in unentangled states. The reader may be justly concerned by the extraordinary fact that despite the overwhelming preponderance of entangled states among the results of standard quantum evolution, in our normal experiences we seem almost always to be able to ignore quantum entanglements completely! We must try to come to some understanding of this apparently enormous discrepancy, and other closely related matters.

In order that we can properly address issues of quantum functional freedom, such as the above seemingly blatant inconsistency between the quantum formalism and physical experience, we shall need to step back and try to understand the kind of "reality" that the quantum formalism is actually supposed to be providing us with. A good place to return to is the original situation, described in §2.2, that started the whole subject of quantum mechanics off, where – when we pay due attention to the principle of equipartition of energy – particles and radiation seem only to be able to coexist in a state of thermal equilibrium if physical fields and systems of particles are, in some sense, the same kind of entity, each having a similar kind of functional freedom. We recall the ultraviolet catastrophe, referred to in §2.2, that would result from a classical picture of (electromagnetic) field in equilibrium with a collection of (charged) classical particles. Because of the vast discrepancy between the functional freedom in the field (here $\infty^{4\infty^3}$) and that in the collection of particles treated classically (only $\infty^{6N}$ for $N$ structureless particles, which is far tinier), an approach to equilibrium would result in the complete draining of energy from the particles into the vast reservoir of functional freedom in the field – the *ultraviolet catastrophe*. This conundrum was resolved by Planck and Einstein through the postulate that the seemingly continuous electromagnetic field has to acquire a particle-like quantum character through the Planck formula $E = h\nu$, the energy $E$ being that of a mode of field oscillation of frequency $\nu$.

But, in view of what has been said above, we seem now to have to been driven to treating the particles themselves in a way that describes them as collectively having a description – namely the *wave function* for the entire system of particles – that has enormously more functional freedom than does a classical system of particles. This is especially noteworthy when we take into account the entanglements between the particles, where the functional freedom in $N$ particles is of the form $\infty^{\bullet\infty^{3N}}$ (where "$\bullet$" stands for some unspecified positive number), whereas for a classical field the freedom is (taking $N > 1$) the vastly smaller $\infty^{\bullet\infty^3}$. It would seem that the boot is now on the other foot, and that equipartition is now telling us that the degrees of freedom in the system of quantum particles would completely drain away the energy in the field. But this problem has to do with the fact that we have been inconsistent, as we have treated the field as a classical entity while at the same time relied on a quantum description for the particles. To resolve this, we have to examine how the degrees of freedom in a system should actually be counted, in quantum theory, from the proper physical point of view. Moreover, we must also briefly address the issue of how quantum theory actually treats physical fields, according to the procedures of quantum field theory (QFT; see §§1.3–1.5).

There are basically two ways of looking at QFT in this context. The procedure which lies at the background of many modern theoretical approaches to the subject is that of *path integrals*, based on an original idea by Dirac in 1933 [Dirac 1933] and developed by Feynman into a very powerful and effective technique for QFT [see Feynman et al. 2010]. (For a rapid outline of the main idea, see TRtR [§26.6].) However, this procedure, though powerful and useful, is very formal (and not actually mathematically consistent in a precise way). These formal procedures do, however, directly provide the Feynman-graph calculations (alluded to in §1.5) that underlie the standard QFT calculations that physicists use in order to obtain the quantum amplitudes that the theory predicts for basic particle-scattering processes. From the point of view of the issue that I am concerned with here, namely that of functional freedom, the expectation would be that the functional freedom in the quantum theory ought to be exactly the same as in the classical theory to which the path-integral quantization procedure is being applied. Indeed, the whole procedure is tuned to reproducing the classical theory as a first approximation, but with the appropriate quantum corrections (at order $\hbar$) to this classical theory, and such things would not affect the functional freedom at all.

The more directly physical way to address the implications of QFT is simply to think of the field as made up of an indefinite number of particles, namely the *field quanta* (photons in the case of the electromagnetic field). The total amplitude (i.e. the full wave function) is a *sum* – a quantum superposition – of different parts, each referring to a different number of particles (i.e. field quanta). The $N$-particle part would provide us with a *partial* wave function having a functional freedom of the form $\infty^{\bullet\infty^{3N}}$. We have to think of this number $N$ as not being definite, however, since field quanta are continually being created and destroyed through their interactions with sources, which would be electrically charged (or magnetic) particles, in the case of photons. This, indeed, is why our total wave function has to be a superposition of parts with differing values of $N$. Now, if we try to treat the functional freedom that is involved in each partial wave function in the same way that we would treat a classical system, and we try to apply energy equipartition as we did in §2.2, then we would have some severe difficulties. The functional freedom for larger numbers of field quanta would completely swamp the functional freedom for smaller numbers (since $\infty^{\bullet\infty^{3M}}$ is enormously larger than larger than $\infty^{\bullet\infty^{3N}}$ whenever $M > N$). If we were to treat this functional freedom in a wave function in the same way as we would for a classical system, then we would find that, for a system in equilibrium, energy equipartition informs us that all the energy would go into the contributions to the state in which there were more and more particles, completely draining it away from any contribution in which any fixed finite number of particles was involved, again leading to a catastrophic situation.

It is here that we must squarely face the issue of how the formalism of quantum mechanics relates to the physical world. We cannot take the functional freedom in wave functions to be on the same footing as the functional freedoms we find in classical physics, despite the fact that the (usually vastly entangled) wave function has a clear, albeit often subtle, influence on direct physical behaviour. Functional freedom still has a key role to play in quantum mechanics, but it must be combined with the critical idea introduced by Max Planck in 1900, with his famous formula

$$E = h\nu,$$

and with the further deep insights that Einstein, Bose, Heisenberg, Schrödinger, Dirac, and others later provided. Planck's formula tells us that the kind of "field" that actually occurs in nature has some sort of discreteness about it, which makes it behave like a system of particles, where the higher the mode of frequency of oscillation that the field might indulge in, the more strongly the energy in the field would manifest itself in this particle-like behaviour. Thus, quantum mechanics tells us that the kind of physical field that is actually encountered in nature's wave functions is not altogether like the classical field of §A.2 (and illustrated there, particularly, by the classical notion of a magnetic field). A quantum field begins to exhibit discrete, or particle-like, properties when it is probed at very high energies.

In the present context, the appropriate way to think of this is to regard quantum physics as providing us with some kind of "granular" structure to the phase space $\mathcal{P}$ of a system (see §A.6). It would not really be accurate to think of this as like replacing the continuum of space-time with something discrete, as with a discrete toy model universe, where the continuum $\mathbb{R}$ of real numbers might be replaced by a finite system $\mathbf{R}$ (as discussed in §A.2), consisting of an extremely large integer number $N$ of elements. Yet this kind of picture is not altogether inappropriate if we think of it being applied to the phase spaces that are relevant to quantum systems. As explained more fully in §A.6, the phase space $\mathcal{P}$ for a system of $M$ point-like classical particles, for which there are $M$ position coordinates and $M$ momentum coordinates, would have $2M$ dimensions. The unit of "volume" – a $2M$-dimensional hypervolume – would therefore involve $M$ distance measures which we could, for example, take as metres (m), and $M$ momentum measures, say grams (g) times metres per second (m s$^{-1}$). Accordingly, our hypervolume would be in terms of the $M$th power of the product of these, namely g$^M$ m$^{2M}$ s$^{-M}$, which would depend upon this particular choice of units. In quantum mechanics, however, we have a natural unit, namely *Planck's constant h*, and it turns out to be appropriate to use Dirac's "reduced" version "$\hbar = h/2\pi$", which has the tiny value

$$\hbar = 1.05457\cdots \times 10^{-31} \text{ g m}^2 \text{ s}^{-1}$$

in the particular units just referred to. The quantity $\hbar$ enables us to obtain a natural hypervolume measure for our $2M$-dimensional phase space $\mathcal{P}$, namely to use units of $\hbar^M$.

In §3.6 we shall come to the notion of *natural units* (or Planck units), which are chosen so that various fundamental constants of nature come out to have the value 1. It is not necessary to go all the way with this here, but if at least we arrange our choice of units of mass, length, and time so that

$$\hbar = 1$$

(which can be done in many different ways, in addition to the full choice of natural units given in §3.6), then we find that *any* phase-space hypervolume is just some *number*. We could indeed imagine that perhaps there is some natural "granularity" to $\mathcal{P}$, where each individual cell, or "grain", counts as just one unit, in terms of a choice of physical units for which $\hbar = 1$. Accordingly, phase-space volumes would always take just some integer value, basically simply "counting" the number of grains. The key point about this is that we can now directly compare phase-space hypervolumes of *different dimensionality* $2M$, basically just by counting these grains, no matter what value $M$ might have.

Why is this important for us? It is important because, for a quantum field in equilibrium with a system of particles interacting with the field and capable of altering the number of field quanta, we need to be able to compare phase-space hypervolumes of different dimensions. But, with classical phase spaces, the higher-dimensional hypervolumes totally overwhelm the lower-dimensional ones (e.g. the 3-volume of an ordinary smooth curve in Euclidean 3-space is always *zero*, no matter how long it is) and therefore those states with more dimensions of freedom would completely drain away all the energy from those with less, in accordance with the requirements of energy equipartition. The granularity that quantum mechanics provides solves this problem by reducing the volume measures simply to *counting*, so that whereas the higher-dimensional hypervolumes would tend to be very large in comparison with the lower-dimensional ones, they are not *infinitely* larger.

This applies directly to the situation that Max Planck was confronted with in 1900. Here we have a state which we now consider to be composed of coexisting components, where each component involves a different number of those field quanta that we now refer to as *photons*. For any particular frequency $\nu$, Planck's revolutionary principle (§2.2) implied that a photon of that frequency must have a corresponding particular energy, given by

$$E = h\nu = 2\pi\hbar\nu.$$

It was, in effect, by using such a counting procedure, that the previously little-known Indian physicist Satyendra Nath Bose, in a letter he sent to Einstein in June 1924, provided a direct derivation of the Planck radiation formula (without appealing to any electrodynamics), where, in addition to $E = h\nu$ and the fact that photon number would not be fixed (photon number not conserved), he just required that the photon have two distinct polarization states (see §§2.6 and 2.9) and, most importantly, would need to satisfy what we now call *Bose statistics* (or *Bose–Einstein statistics*; see §1.14), so that states differing from one another merely by interchanges of pairs of photons would not be counted as physically distinct. These last two features were revolutionary at the time, and Bose is deservedly remembered today in the name *boson* that is attached to any basic particle of integer spin (which is thereby subject to Bose statistics).

The other broad class of basic particles consists of those of half-odd-integral spin, namely the fermions (after the Italian nuclear physicist Enrico Fermi). Here the counting is slightly different, and follows *Fermi–Dirac statistics* instead, somewhat like the Bose–Einstein statistics, but where states with two (or more) such particles of the same kind and both in the same state are not counted sep-arately (Pauli exclusion principle). See §1.4 for a more detailed explanation of how bosons and fermions are treated in standard quantum mechanics (where the reader may ignore the extrapolation, described there, of standard theory into the speculative – but still highly fashionable – particle-physics scheme of supersymmetry).

With these provisos, the idea of functional freedom also holds good with quan-tum systems as well as classical ones, but one must be careful. The quantity "$\infty$" that features in our expressions is now not actually infinite, but may be thought of as something that would, in ordinary circumstances, be a very large number. It is not immediately evident how to address the functional-freedom issue in a general quantum context, especially since there may be many different components of the system in quantum superposition, involving different numbers of particles and which, classically, would have involved phase spaces of different dimension.

However, in the case of radiation in thermal equilibrium with charged particles, we may return to the considerations of Planck, Einstein, and Bose, which provide us with Planck's formula for the *intensity* of radiation (in equilibrium with matter), for each frequency $v$ (see §2.2):

$$\frac{8\pi h v^3}{c^3(e^{hv/kT} - 1)}.$$

In §3.4, we shall be seeing the huge relevance that this formula has to cosmology, as it is in extremely good agreement with the radiation spectrum of the *cosmic microwave background* (CMB).

In chapter 1, particularly in §§1.10 and 1.11, the issue of the role of the presence of extra space dimensions was raised, in relation to the plausibility of the considerable increase in spatial dimensionality that string theory demands, beyond the three that are directly observed. Proponents of such higher-dimensional theories sometimes argue that quantum considerations would prevent the excessive functional freedom from directly impinging upon normally observed physical processes, owing to the excessively high energies that would be needed for these degrees of freedom to come into play. In §§1.10 and 1.11, I argued that when we are considering degrees of freedom in the space-time geometry (i.e. gravitation), this contention is (at best) highly questionable. But I did not address the separate issue of the excessive functional freedom that would come about in the *non*-gravitational fields, such as electromagnetism – i.e. *matter* fields – that could be viewed as *inhabiting* those extra space dimensions. It is thus of some interest to consider whether the presence of such spatial supra-dimensionality might affect the cosmological use of the above formula.

With extra spatial dimensions – say $D$ spatial dimensions in total (where in conventional Schwarz–Green string theory, $D = 9$) – the *intensity* of radiation as a function of frequency $v$ would accordingly be given by by

$$\frac{Q h v^D}{c^D(e^{hv/kT} - 1)},$$

where $Q$ is a numerical constant (dependent on $D$), as compared with the 3-dimensional expression just given [see Cardoso and de Castro 2005]. In figure 2-25, I compare the $D = 9$ case with Planck's conventional case $D = 3$ we saw previously in figure 2-2. However, owing the enormous imbalance between the scales of size of the spatial geometry in the different directions, one would not expect such a formula to have any immediate cosmological relevance. Nevertheless, in the extremely early stages of the universe, in such models, at the general order of

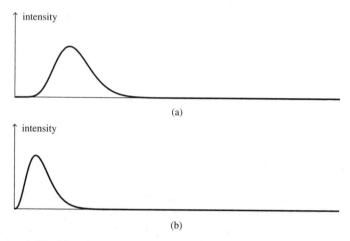

Figure 2-25: The shape of the Planck spectrum (a) for 9 spatial dimensions (const. $\times \, \nu^9 (e^{h\nu/kT} - 1)^{-1}$), as compared with (b) the normal 3-dimensional case (const. $\times \, \nu^3 (e^{h\nu/kT} - 1)^{-1}$).

the Planck time ($\sim 10^{-43}$ s) or a bit later, *all* spatial dimensions would be taken to have been curved at comparable scales, giving some parity between all 9 of these supposed spatial dimensions, so it might be considered that at such early times, this higher-dimensional version of the Planck formula could indeed have been the relevant one at such very early times.

As we shall be seeing in §§3.4 and 3.6, it turns out that there must have been another extraordinary imbalance, in those very early stages, namely that between the gravitational degrees of freedom and the degrees of freedom in all other fields. Whereas the gravitational degrees of freedom appear *not* to have been activated at all, those in the matter fields seem to have been maximally activated! At least that was the case at the time of *decoupling*, some 380000 years after the Big Bang. This extreme imbalance is evidenced directly from the nature of the CMB, as we shall be seeing in §§3.4 and 3.6. What we find is that whereas the degrees of freedom in matter and radiation were, in a highly *thermal* (i.e. maximally activated) state, those in the gravitational field – i.e. in the space-time geometry – seem to have remained almost entirely aloof from all that activity. It is hard to see how such an imbalance could have arisen only during the 280000 years following the Big Bang, as one would have expected that thermalization could only have *increased* during that period, as a direct consequence of the Second Law of Thermodynamics (see §3.3). Accordingly, we must conclude that this aloofness of the gravitational degrees of freedom must have dated back to those very early times (the general order of the Planck time, $\sim 10^{-43}$ s) the gravitational degrees of freedom only coming into play significantly at a much later stage (considerably later than decoupling time), arising from subsequent irregularities in the matter distribution.

It is, however, reasonable to raise a question as to whether the above higher-dimensional ($D = 9$) formula, even if considered to be applicable only very soon after the Big Bang, might actually be expected to retain some residue of its supposed early validity even until the time of decoupling (380000 years after the Big Bang), when the CMB radiation that we observe was actually produced (see §3.4). This actual CMB radiation does have a somewhat different intensity spectrum from that which would be anticipated from the above higher-dimensional expression (see figure 2-25), and agrees extremely closely with the $D = 3$ version (see §3.4), so we may take it that, as the universe expanded, the $D = 9$ version of the radiation spectrum would have changed over completely to this $D = 3$ version. What the curve represents is the frequency distribution that attains maximum entropy – i.e. maximum randomness in matter fields throughout their available degrees of freedom – *given* the space-time geometry that they find themselves within. If all the spatial dimensions had expanded at the same rate, then the spectrum with the $D = 9$ form would have been maintained and the entropy in the radiation would have been kept more-or-less constantly at that enormously higher value, which would be provided by the $D = 9$ version of the above formula.

However, what is now observed in the CMB is the $D = 3$ version, and from the point of view of the Second Law of Thermodynamics (§3.3), the presumed hugely excited degrees of freedom in the matter fields, arising from the 6 extra dimensions, must have gone somewhere, which presumably means into activating the tiny extra 6-dimensional spatial dimensions, either in the form of gravitational or matter degrees of freedom. In each case, I find it hard to see how to square this picture with the string-theoretic viewpoint that the extra 6-dimensions are currently at a stable minimum (see §§1.11 and 1.14). How is it that the highly thermalized matter degrees of freedom in the very early 9-spatial supra-dimensional universe could somehow have adjusted themselves so as to leave the extra 6 dimensions so apparently completely unexcited, as string theory appears to demand? One must also ask what gravitational dynamics could have produced such an enormous discrepancy in the different spatial dimensions, and question especially how it could have so cleanly separated the 6 curled-up unexcited dimensions from the 3 expanding ones.

I am not claiming that there is an obvious contradiction contained in these considerations, but it is certainly a very strange picture that demands some dynamical explanation. It is to be hoped that something more quantitative can be made out of all this. The origin of such a vast imbalance between the two classes of space-time dimensions is surely, in itself, a huge puzzle for the string-theory picture, and one may also thoroughly question how a proper thermalization of *gravitational* degrees of freedom did not occur at that stage, instead producing only the very curious picture of the two cleanly separated types of dimension that modern supra-dimensional string theory appears to demand.

## 2.12. QUANTUM REALITY

According to standard quantum mechanics, the information in the quantum state of a system – or the *wave function* $\psi$ – is what is needed for probabilistic predictions to be made for the results of experiments that might be performed upon that system. Yet, as we have seen in §2.11, the wave function involves a functional freedom that is far higher than that which manifests itself in reality, or at least in that aspect of reality that is revealed as the result of a quantum measurement. Are we to regard $\psi$ as actually representing physical reality? Or is it to be viewed as being merely a calculational tool for working out probabilities of the results of experiments that *might* be performed, the results of these being "real", but not the wave function itself?

As mentioned in §2.4, it was part of the Copenhagen interpretation of quantum mechanics to take this latter viewpoint, and, according to various other schools of thought also, $\psi$ is to be regarded as a calculational convenience with no ontological status other than to be part of the state of mind of the experimenter or theoretician, so that the actual results of observation can be probabilistically assessed. It seems that a good part of such a belief stems from the abhorrence felt by so many physicists that the state of the actual world could suddenly "jump" from time to time in the seemingly random way that is characteristic of the rules of quantum measurement (see §§2.4 and 2.8). We recall Schrödinger's despairing comments to this effect, in §2.8. As remarked above, the Copenhagen view is to take this jumping to be "all in the mind", as one's viewpoint about the world can indeed instantly shift as soon as new evidence about it (the actual result of an experiment) becomes revealed.

At this juncture, I ought to draw the reader's attention to an alternative viewpoint to that of the Copenhagen school, referred to as the *de Broglie–Bohm theory* [de Broglie 1956; Bohm 1952; Bohm and Hiley 1993]. I shall refer to it here by its commonly employed name *Bohmian mechanics*. This provides an interesting alternative ontology to that provided (or *not* really provided!) by the Copenhagen view, and it is fairly widely studied, though certainly not qualifying as a fashionable theory. It claims no alternative observational effects from that of conventional quantum mechanics, but provides a much more clear-cut picture of the "reality" of the world. In brief terms, the Bohmian picture presents *two* levels of ontology, the weaker of the two being provided by a universal *wave function* $\psi$ (referred to as the *carrier wave*). In addition to $\psi$, there is a definite location for all particles, specified by a particular point P in a *configuration space* $\mathcal{C}$ (as described in §A.6), which we can take to be $\mathbb{R}^{3n}$ if there are assumed to be to be $n$ (distinguishable scalar) particles in a flat background space-time. We take $\psi$ to be a complex-number-valued function on $\mathcal{C}$, which satisfies the Schrödinger equation. But the point P itself – i.e. the locations of all the particles – provides a *firmer* "reality" according to the Bohmian world. The particles have a well-defined dynamics that is determined by $\psi$ (so that $\psi$ has to be assigned some kind of reality, even if a "weaker" reality than that provided by P). There is no "back-reaction" on $\psi$ from the particle locations (as given by P). Specifically, in the two-slit experiment described in §1.4, each particle actually goes through one slit or the other, but it is $\psi$ that keeps track of the alternative route and which guides particles so that the correct diffraction pattern appears at the screen. Interesting though this proposal indeed is, from the philosophical perspective, it plays no necessary role for this book, in view of its expectations being identical to those of conventional quantum mechanics.

Even the conventional Copenhagen interpretation does not really evade the issue of having to take $\psi$ as an actual representation of *something* objectively "real", out there in the world. One argument for some such reality springs from a principle suggested by Einstein, which he put forward with his colleagues Podolsky and Rosen in their famous EPR paper referred to in §§2.7 and 2.10. Einstein argued for the presence of an "element of reality" in the quantum formalism whenever it implied some measurable consequence with *certainty*:

In a complete theory there is an element corresponding to each element of reality. A sufficient condition for the reality of a physical quantity is the possibility of predicting it with certainty, without disturbing the system. ... *If, without in any way disturbing a system, we can predict with certainty (i.e. with probability equal to unity) the value of a physical quantity, then there exists an element of physical reality corresponding to this physical quantity.*

However, it is part of the standard quantum formalism that, in principle, for any quantum state vector whatsoever, let us say $|\psi\rangle$, a measurement can be set up for which $|\psi\rangle$ is the only state vector that, up to proportionality, yields the result **YES** with certainty. Why is this? Mathematically, all we need to do is to find a measurement for which one of the orthogonal basis vectors $\varepsilon_1, \varepsilon_2, \varepsilon_3, \ldots$ referred to in §2.8, say $\varepsilon_1$, is actually the given state vector $|\psi\rangle$, and the measurement is set up to respond "**YES**" if it finds $\varepsilon_1$ and "**NO**" if it finds any of $\varepsilon_2, \varepsilon_3, \ldots$. This is an extreme example of a degenerate measurement; see towards the end of §2.8. (Those familiar with the Dirac notation for the operator notation of standard quantum mechanics, see §2.9 [Dirac 1947], will recognize this measurement as being achieved by the Hermitian operator $\mathbf{Q} = |\psi\rangle\langle\psi|$ for any given normalized $|\psi\rangle$, where **YES** corresponds to the eigenvalue 1 and **NO** to 0.) The wave function $\psi$ (up to some non-zero complex-number factor) would be uniquely fixed by the requirement that it gives the answer **YES** with certainty, with regard to this measurement, so by Einstein's above principle we conclude that there is, indeed, a general argument for there to be a clear element of reality to *any* wave function $\psi$, whatsoever!

In practice, however, it may well be completely out of the question to construct a measuring device of the required type, but the general framework of quantum mechanics asserts that such a measurement should be possible in principle. Moreover one would, of course, have to know ahead of time what the wave function $\psi$ actually was in order to know what measurement to make. But this could, in principle, have been worked out theoretically by the Schrödinger evolution of some previously measured state. Einstein's principle thus assigns an *element of reality* to any wave function that has been computed by Schrödinger's evolution equation, i.e. by $\mathbf{U}$, from some previously known state (it being "known" from the result of some previous measurement), where it is assumed that Schrödinger's equation (i.e. unitary evolution) is indeed true of the world, at least for the quantum systems under consideration.

Although for a good many possible $\psi$s the construction of such a measuring device would be well beyond the capabilities of current technology, there are numerous experimental situations in which it *is* perfectly practical. Accordingly, it will be instructive to examine a couple of simple situations for which this is indeed the case. The first is given by spin measurements of a spin-$\frac{1}{2}$ particle – or, let us say, atom of spin $\frac{1}{2}$ with a magnetic moment aligned with its spin. We can use a Stern–Gerlach apparatus (see §2.9 and figure 2-22) oriented in some direction "←" to measure the atom's spin in this direction, and if we obtain the answer **YES**, then we infer that its spin indeed has a state vector (proportional to) $|\leftarrow\rangle$. Suppose we then subject this state to a known magnetic field, and we calculate by means of the Schrödinger equation that, after one second, its state will have evolved to $|\nearrow\rangle$. Do we assign a "reality" to this resulting spin state? It certainly seems reasonable to do so, because a rotated Stern–Gerlach apparatus, geared now to measure spin in the direction $\nearrow$, at that moment will indeed obtain the result **YES** with certainty. This is, of course, a very simple situation, but it clearly generalizes to other situations of much greater complication.

Somewhat more puzzling, however, are situations where quantum entanglements are involved, and we might well consider various examples exhibiting EPR effects, such as those considered in §2.10. For definiteness, let us consider Hardy's example, where we may suppose that we have created an initially prepared spin-1 state with a Majorana description (§2.9) given by $|\leftarrow\nearrow\rangle$, as described explicitly in §2.10. Suppose that this state then decays into two spin-$\frac{1}{2}$ atoms, one moving to the left and the other to the right. We have seen that in this example there is no observationally consistent way to assign *independent* quantum states to each of these two atoms individually. Any such assignment necessarily gives incorrect answers to possible spin measurements that one might choose to apply to the left-hand and right-hand atoms. There is certainly a quantum state that applies to the atoms, but it is an *entangled* one that applies to the *pair* as a whole, and not to each of the two atoms individually. There would be a possible measurement that indeed confirms this entangled state – so the "$\psi$" of the above discussion would have to be this entangled 2-particle state. Such a measurement might consist of somehow reflecting the pair of atoms back together again and performing a measurement that confirms the original $|\leftarrow\nearrow\rangle$ state. This might well be difficult technically, but in principle possible. That would then assign an Einsteinian "element of reality" to the entangled state of the separated pair. However, this could not be achievable simply by measuring the spins independently by, say, a pair of separated Stern–Gerlach devices (see figure 2-26(a)), each measuring the spin of just one of the atoms. The quantum state must involve both, in a quantum-entangled way.

On the other hand, let us suppose that the left-hand atom is actually subjected to a Stern–Gerlach spin measurement, independently of the right-hand atom. This would automatically put the right-hand atom into a particular spin state. For example, suppose the left-hand atom's spin is measured in the $\leftarrow$ direction and a **YES** answer $|\leftarrow\rangle$ is obtained, then it turns out that the right-hand atom would be automatically put into the spin state $|\uparrow\rangle$, whereas if the left-hand measurement finds the **NO** answer instead, then the right-hand one automatically becomes $|\leftarrow\rangle$ (with the notation used in §2.10). These curious implications follow directly from the properties of the Hardy example given in §2.10.

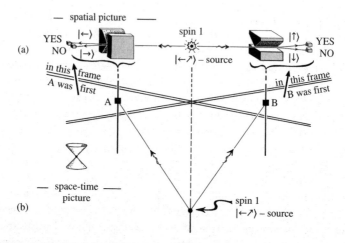

Figure 2-26: The non-local Hardy experiment of figure 2-24: (a) depicted spatially, but taken in conjunction with (b), shown in space-time terms, presenting a challenge to an objective space-time depiction of reality.

In each case we are guaranteed that *after* the left-hand measurement, the right-hand spin state now has a definite independent value assured. How might we go about confirming this? This asserted right-hand spin state could at least be provided with some confirmational support by an appropriate Stern–Gerlach measurement. However, such a measurement on the right-hand atom would not demonstrate that the right-hand spin state has a value, namely $|\uparrow\rangle$ if the left-hand measurement happened to register **YES** and $|\leftarrow\rangle$ if the left-hand measurement registered **NO**. The "reality" of this individual right-hand state seems now to be asserted, i.e. $|\uparrow\rangle$ if the left-hand $\leftarrow$ measurement found **YES** (and $|\leftarrow\rangle$ if it found **NO**). Suppose the left-hand $\leftarrow$ measurement indeed says **YES**. The appropriate **YES** answer to a right-hand $\uparrow$ measurement would not assure us that the right-hand spin state was actually $|\uparrow\rangle$ since the **YES** answer on the right might merely have come about by chance. What a right-hand $\leftarrow$ measurement giving a **YES** answer would tell us, for sure, would merely be that the measured state was *not* $|\downarrow\rangle$. Any other emerging spin state that was close to $|\downarrow\rangle$ would give rather a low probability of a **YES** answer to this $\uparrow$ measurement, with the probability increasing the closer to $|\uparrow\rangle$ the right-hand state happened to be. To obtain a convincing experimental case that the emerging right-hand state was indeed $|\uparrow\rangle$, one would have to repeat the entire experiment a very large number of times, in order for the statistics to build up. If every single time (when a **YES** response is found on the left) the right ↑ measurement does register **YES**, then a case for the actual "reality" of the emergent state on the right being |↑⟩ would seem to be strong (by Einstein's criterion), despite one's having to rely on such a statistical confirmation. After all, much of what we know in science about the reality of the world arises from confidence that is built up in such a statistical way.

This example illustrates another feature of quantum measurement. Our left-hand ← measurement served to "disentangle" the previously entangled state. Before the left-hand measurement was performed, the two atoms could *not* be treated as having individual quantum states, the "state" concept applying only to the pair as a whole. But a measurement on one component of the pair served to "cut the other one free", and allowed it to acquire a quantum state on its own. This is perhaps somewhat reassuring, as it gives some hint as to why quantum entanglements do not appear to permeate our whole existence, preventing us from considering anything as a separate entity on its own.

There is, however, another issue that many physicists find justifiably worrying. When a measurement is performed on one member A of a widely separated pair of entangled states, the question arises "when" does the other member B become disentangled from A and obtain an individual state of its own? A separate measurement might be performed on the other component B, and we might wonder whether it was the measurement on B that disentangled the pair instead of A. If the distance between the parts is great enough, we might envisage that the two measurements were *spacelike separated* (see §1.7) which (in special relativity) means "simultaneous" with respect to some choice of reference frame. However, in such circumstances there would be other reference frames in which the A measurement would be judged to have occurred first and other reference frames in which the B measurement occurred first (see figure 2-26(b)). To put this another way, the *information* of the result of either measurement would appear to have to travel faster than light in order to be in time to influence the result of the other measurement! We would have to think of the pair of measurements as acting on an essentially *non-local* entity, which is the entire entangled state of the pair of atoms.

This (frequent) non-locality is one of the most puzzling and intriguing aspects of entangled states. It is something with no analogue in classical physics. Classically, one might have a system with two separated parts A and B that had once been together, where A might be able to signal subsequent information about its later experiences to B, or B to A, or both, or they might behave completely independently of one another after separation. But quantum entanglement is something different. When A and B remain quantum entangled, they are *not* independent; yet they are unable to use this "dependence" upon one another, through this entanglement, to send actual information from one to the other. It is this incapability of sending actual information via entanglement that allows one to consider that the entanglement is transferred "instantaneously" without violating the tenets of relativity theory, which forbids superluminary transference of information. In fact, this entanglement transference is not really to be thought of as "instantaneous"; it is really "timeless", as it makes no difference whether this transference is considered to be from A to B or from B to A. It is just a restriction on the combined behaviour of A and B when subjected to independent measurement. (This "entanglement transference" is sometimes called *quantum information*. I have referred to it elsewhere as *quanglement* [TRtR, §23.10; Penrose 2002, pp. 319–31].) I shall need to return to this issue in the next section.

Before doing so, however, I wish to draw attention to another related argument for regarding the wave function as having a genuine ontological reality. This makes use of an ingenious idea due to Yakir Aharonov, and developed by Lev Vaidmen and others, which enables quantum systems to be explored in a different way from the conventional measurement procedures outlined in §2.8. Instead of thinking of a given quantum state as being subjected to a measurement which then puts it into a subsequent quantum state (the normal measurement process), the Aharonov procedure involves selecting systems with given, almost orthogonal, initial and final states. This enables what are called *weak measurements* to be considered, which do not disturb the system. Using such means one can probe features of quantum systems previously thought to be inaccessible. In particular, a picture of the actual spatial intensity in a stationary wave function can be mapped out. The details of this procedure are beyond the scope of this book, but it is worth a mention here, as it lends promise to the exploration of many other puzzling features of quantum reality [Aharonov et al. 1988; Ritchie et al. 1991].

## 2.13. OBJECTIVE QUANTUM STATE REDUCTION: A LIMIT TO THE QUANTUM FAITH?

Up to this point, although my descriptions have sometimes been from a slightly unconventional angle, I have not yet deviated in any significant way from the quantum faith with regard to the actual results of measurements. I have pointed out some of its most puzzling features, such as the fact that quantum particles must often be taken to be located at several different places simultaneously, owing to the ubiquitous quantum superposition principle, and that, also according to this principle, particles may appear as waves whereas waves appear to be composed of indefinite numbers of particles. Moreover, the vast majority of quantum states of systems containing more than one part would be expected to be *entangled*, so the parts cannot consistently be regarded as completely independent of one another.

All these puzzling aspects of the dogma of quantum mechanics I accept, at least in the regimes covered by current observation, as they have been amply confirmed in numerous precise experiments. Yet I have not refrained from pointing out that there appears to be a fundamental inconsistency between the two bedrock procedures of quantum theory, namely unitary (i.e. Schrödinger) evolution **U** and the state reduction **R** which takes place upon quantum measurement. To most quantum practitioners, this inconsistency is regarded as being something *apparent*, which is to be removed by the adoption of the right "interpretation" of the quantum formalism. In §§2.4 and 2.12, I have already referred to the *Copenhagen interpretation*, according to which the quantum state is not assigned an objective reality, but given merely a subjective calculational-aid status. However, I am very dissatisfied with this very subjective viewpoint, for various reasons where, particularly in §2.12, I have argued that the quantum state (up to proportionality) should actually be given a genuinely objective ontological status.

Another common viewpoint is that of environmental decoherence, whereby it is taken that the quantum state of a system should not be regarded as something isolated from its environment. It is argued that, under normal circumstances, the quantum state of a large quantum system – say, the quantum state of an actual detector of some kind – would rapidly become grossly entangled with a large part of its surrounding environment, including the molecules in the air, most of which would be effectively random in their motions, undetectable as to any refined detail, and basically irrelevant to its operation. Accordingly, the quantum state of that system (detector) would become "degraded", and its behaviour better treated as though it were simply a classical object.

For the precise description of such a situation, a mathematical construct referred to as a *density matrix* – an ingenious concept introduced by John von Neumann – is formulated, which allows the irrelevant environmental degrees of freedom to be eliminated from the description by being "summed over" [von Neumann 1932]. At that point, the density matrix then takes over the role as describing the "reality" of the situation. Then, by a clever piece of mathematical sleight of hand, this reality is enabled to be reinterpreted as a probability mixture of different alternatives from those under consideration before. The *observed* alternative that results from the measurement is taken to be one of these new alternatives, and it has been assigned a probability for its occurrence that has been correctly calculated according to the standard **R**-procedure of quantum mechanics, as described in §2.4.

A density matrix indeed represents a probability mixture of different quantum states, but it does so in many different ways at the same time. The sleight of hand referred to above involves what I have referred to as a *double ontology shift* [TRtR, end of §29.8, pp. 809–10]. Initially, the density matrix is interpreted as providing a probability mixture of different "real" alternative environment states. Then the ontology *shifts*, by assigning "reality" to the density matrix itself. This allows the freedom to pass to a different ontological interpretation (through a Hilbert-space basis rotation) where the same density matrix is now regarded, via a third ontological standpoint, as describing a probability mixture of the alternative outcomes of the measurement. The descriptions, normally given, tend to concentrate on the mathematics, with scant attention to the consistency of the *ontological* status of the various descriptions. In my own view, there *is* something of genuine significance in the environmental decoherence density-matrix picture, as there is indeed something remarkable in the way that the mathematics works out. But as regards what is really going on in the physical world, there is something profoundly missing. To get a proper solution to the measurement paradox, we need a change in the *physics*, not just some clever mathematics, brought in to cover the ontological cracks! As John Bell [2004] has put it:

> When they [the most sure-footed of quantum physicists] do admit some ambiguity in the usual formulations, they are likely to insist that ordinary quantum mechanics is just fine 'for all practical purposes'. I agree with them about that: ORDINARY QUANTUM MECHANICS (as far as I know) IS JUST FINE FOR ALL PRACTICAL PURPOSES.

Environmental decoherence just provides us with a provisional *FAPP* (Bell's acronym, denoting "for all practical purposes") picture; it may well be part of the answer – enough to get along with for the time being – but it is not the ultimate answer. For that, I believe we need something much deeper, which breaks away from our holding fast to the quantum *faith*!

If we try to maintain a consistent ontology, while still holding faithfully to **U** at all levels, then we are led, inevitably, to some kind of *many-worlds* interpretation, as was first made explicit by Hugh Everett III [Everett 1957].[10] Let us reconsider the situation of the (Schrödinger) cat described towards the end of §2.7 (and recall

---

[10] See the immediately subsequent note by J. A. Wheeler [Wheeler 1967; also DeWitt and Graham 1973; Deutsch 1998; Wallace 2012; Saunders et al. 2012].

figure 2-15), where we try to maintain a consistent **U** ontology throughout. There, we were to imagine a high-energy photon, emitted by a laser L, which is aimed at a beam splitter M. If the photon were to be transmitted *through* M to activate the detector at A, this would open the A-door, and the cat would pass through that door to obtain her food in a nearby room. On the other hand, if the photon were to be *reflected*, then the detector at B would open the B-door, and the cat would pass through that other door to obtain her food. However, M is a *beam splitter* and not simply a mirror, so the photon's state would emerge from M, according to its **U**-evolution, in a *superposition* of travelling the two routes MA and MB, resulting in a superposition of the A-door being open together with the B-door closed, and of the B-door being open together with the A-door closed. One might imagine that, according to this **U**-evolution, a human observer, sitting in the room with the cat's food, ought to perceive a *superposition* of the cat emerging through the A-door and of the cat emerging through the B-door. This, of course, would be an absurd situation never actually experienced and, moreover, it is not how **U** operates. Instead we are presented with the picture of the human observer being *also* put into a quantum superposition of two states of mind, one of which perceives the cat entering through the A-door and the other, perceiving the cat entering through the B-door.

These would be the two superposed "worlds" of the Everett-type interpretation, and it is argued (not very logically, in my view) that the observer's experiences "split" into two coexisting individual *un*-superposed experiences. My problem, here, is that I do not see why what we call "experiences" need to be un-superposed. Why should an observer not be able to experience a quantum superposition? It is not what we are used to, of course, but *why* is it not? It may be argued that we know so little about what actually constitutes a human "experience" that we are certainly entitled to speculate about such matters in one way or another. But we may definitely question why human experiences are to be allowed to un-superpose a given quantum state into two parallel world states, rather than maintaining just *one* superposed world state – which is what the **U**-description actually provides us with. We may recall the spin-$\frac{1}{2}$ states of §2.9. When we consider a spin state $|\nearrow\rangle$ as a superposition of $|\uparrow\rangle$ and $|\downarrow\rangle$, we do not think of there being two parallel worlds, one with $|\uparrow\rangle$ in it and the other with $|\downarrow\rangle$. We just have one world containing this state $|\nearrow\rangle$.

Moreover, there is the further issue of the probabilities. Why should a superposed human observer's experience "split" into two separate ones, with probabilities given by the Born rule? It doesn't really make too much sense to me what this might even mean! It is my own opinion that the extrapolation of **U**-evolution to such extreme situations as in the cat experiment is stretching our imaginations a great deal too far, and I would take the contrary position that situations of this nature simply constitute a *reductio ad absurdum* to the unlimited applicability of **U**. Well-tested as the implications of **U**-evolution indeed are, there has been no experiment that has yet reached anything approaching the level that would be required in such situations.

As already remarked in §2.7, the essential problem is the *linearity* of **U**. Such universal linearity is highly unusual in physical theories. It was noted in §2.6 that Maxwell's classical equations for the electromagnetic field *are* linear, but it should be pointed out that this linearity does not extend to the classical dynamical equations of an electromagnetic field together with charged particles or fluids that are interacting with it. The complete universality of the linearity demanded by the **U**-evolution of present-day quantum mechanics is utterly unprecedented. We may also recall, from §1.1 (and §A.11), that Newton's gravitational field also satisfies linear equations, but this linearity again does not extend to the motions of bodies under the action of Newtonian gravitational forces. Perhaps more pertinent to the current issue is the fact that in Einstein's more refined theory of gravitation – his general theory of relativity – the gravitational field *itself* is fundamentally *non*linear.

I shall here be arguing that there are indeed strong reasons for believing that the linearity of current quantum theory can be only *approximately* true of the world, so the *faith* that so many physicists appear to have in the universality of the overall framework of current quantum mechanics, including its linearity – and, therefore, its *unitarity* **U** – must be misplaced. It is often argued that no counterexamples to quantum theory have ever been observed, and that all experiments to date, over a huge variety of different phenomena and over a very considerable range of scales, have continued to give complete confirmation of quantum theory, and this includes the **U**-evolution of the quantum state. We may recall (§§2.1 and 2.4) that subtle quantum effects (entanglements) have been confirmed over a distance of 143 km [Xiao et al. 2012]. In fact, that 2012 experiment was confirming an implication of quantum mechanics more sophisticated than just the EPR effects (discussed in §2.10), namely what is referred to as *quantum teleportation* [see Zeilinger 2010; Bennett et al. 1993; Bouwmeester et al. 1997] but it also establishes that quantum entanglements *do* persist over such distances. Accordingly, the limits of quantum theory, whatever they may be, do not seem to be simply physical distance – and the distances involved in my Schrödinger-cat example would certainly be less than 143 km. No, I am asking for a limit to the accuracy of current quantum mechanics on a sort of scale of a quite different kind, namely where *displacements* *of mass* between components of a superposition are significant in a very particular sense.

My arguments for such a limitation come from a fundamental clash of principles between those of quantum mechanics (primarily quantum linear superposition) and those of Einstein's general theory of relativity. I shall outline one argument here [Penrose 1996] that dates from 1996, and is based on Einstein's principle of general covariance (see §§A.5 and 1.7). In §4.2 I indicate a more sophisticated and much more recent argument, based on Einstein's basic principle of equivalence (see §1.12).

The situation I am concerned with involves a quantum superposition of two states, each of which, if considered on its own, would be *stationary*, which means unchanging for all time. The idea is to show that, if we bring principles of general relativity to bear on the situation, we find that there is a specific limit on how stationary a superposition of the two can be. But in order to proceed with this argument, we must ask what, in quantum mechanics, the notion of *stationary* means and to address some related aspects of quantum mechanics. I have not yet gone very far into the formalism of quantum theory here, and we shall need a little more of its general ideas before we can proceed.

We recall from §2.5 that certain idealized quantum states can have very well-defined positions, namely the *position states* as given by the wave functions of the form $\psi(\mathbf{x}) = \delta(\mathbf{x} - \mathbf{q})$, where $\mathbf{q}$ is the position 3-vector of the spatial location Q at which the wave function is localized. Less localized would be states arising as superpositions of several such localized states, perhaps given by different 3-vectors $\mathbf{q}'$, $\mathbf{q}''$, etc. Such superpositions could even involve a *continuum* of alternative positions, and so could fill an entire 3-dimensional region of space. An extreme case, very different from a position state, would be a *momentum* state, considered in §2.6 and again in §2.9, represented by $y(\mathbf{x}) = e^{-i\mathbf{p}\cdot\mathbf{x}/\hbar}$ for some momentum 3-vector $\mathbf{p}$. These are again *idealized* states, like position states, not having finite norms; see end of §2.5. These are spread *uniformly* over the whole of space, though with a phase that rotates uniformly around the unit circle in the Wessel plane (§A.10), at a rate proportional to the particle's momentum, and directed along $\mathbf{p}$.

Momentum states are completely ill defined with regard to position and, conversely, any position state is completely ill defined as regards its momentum. Position and momentum are what are called *canonically conjugate* variables, and the better defined the state is with regard to one of these qualities, the worse defined it must be with regard to the other, in accordance with the Heisenberg *uncertainty principle*. This is usually stated in the form

$$\Delta\mathbf{x}\Delta\mathbf{p} \geqslant \tfrac{1}{2}\hbar,$$

where $\Delta\mathbf{x}$ and $\Delta\mathbf{p}$ are respective measures of *ill-definedness* of the position and momentum. This comes about because, in the algebraic formalism of quantum mechanics, canonically conjugate variables become *non-commutative* "operators" on the quantum state (see end of §2.8). For the operators $\mathbf{p}$ and $\mathbf{x}$ we find $\mathbf{xp} \neq \mathbf{px}$, where each of $\mathbf{x}$ and $\mathbf{p}$ behaves as *differentiation* with respect to the other (see §A.11). However, to go into this more fully would take us too far beyond the technical scope of this book; see Dirac [1930] or, for a modern and compact book on the basics of quantum-mechanical formalism, Davies and Betts [1994]. For a basic introduction, see TRtR, chapters 21 and 22.

In an appropriate sense (as follows from the requirements of special relativity), time $t$ and energy $E$ are also canonically conjugate, and we have the Heisenberg time–energy uncertainty principle, expressed as

$$\Delta t \Delta E \geqslant \tfrac{1}{2}\hbar.$$

The exact interpretation of this relation is sometimes disputed. However, there is one use of this relationship that is generally accepted, and this occurs with radioactive nuclei. With such an unstable nucleus, $\Delta t$ can be taken as a measure of its *lifetime*, and the above relation then tells us that there must be an energy uncertainty $\Delta E$, or equivalently a mass uncertainty of at least $c^{-2}\Delta E$ (by Einstein's $E = mc^2$).

Let us now return to our superposition of two stationary states. A stationary state, in quantum mechanics, is actually one whose *energy* is precisely defined, so by Heisenberg's time–energy uncertainty, the state must be spread out in time in a completely uniform way – which is, indeed, an expression of its stationarity (see figure 2-27). Moreover, as with a momentum state, its phase rotates uniformly around the unit circle in the Wessel plane, at a rate proportional to the state's energy value $E$, the time dependence being, in fact $e^{Et/i\hbar} = -\cos(Et/\hbar) - i\sin(Et/\hbar)$, so that the frequency of this phase rotation is $E/2\pi\hbar$.

I shall here consider a very basic situation involving a quantum superposition, namely a superposition of two states, each of which would be stationary if considered on its own. For simplicity, think of a rock in a superposition of two locations, given by states $|1\rangle$ and $|2\rangle$, resting on a horizontal plane surface, and where the two differ from one another only in that the rock has been moved from its location in $|1\rangle$ to its location in $|2\rangle$ by a horizontal translational displacement so that, in particular, the energy $E$ of each state is the same (figure 2-28). We consider a general superposition

$$|\psi\rangle = \alpha|1\rangle + \beta|2\rangle,$$

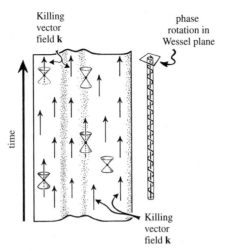

Figure 2-27: The classical and quantum notions of stationarity. In classical space-time terms, a stationary space-time possesses a timelike Killing vector field **k** along which the space-time geometry is unchanged by any (local) motion generated by **k**, and we take the time direction to be that of **k**. Quantum mechanically, the state has a precisely defined energy $E$, so that it varies in time merely by an overall phase $e^{Et/i\hbar}$, which rotates around the unit circle in the Wessel plane with frequency $E/2\pi\hbar$.

where $\alpha$ and $\beta$ are non-zero constant complex numbers. It follows that $|\psi\rangle$ is also stationary,[11] with this definite energy value $E$. When the energies of $|1\rangle$ and $|2\rangle$ are different, an interesting new situation arises, discussed in §4.2.

In general relativity, stationarity is expressed in a somewhat different (though related) way. We still think of a stationary state as one which is indeed completely uniform in time (though without any complex phase to rotate), but where the very notion of time is not uniquely defined. The general concept of temporal uniformity for a space-time $\mathcal{M}$ is normally expressed in terms of a *timelike Killing vector* **k**. A Killing vector is a vector field in the space-time (see §A.6, figure A-17), along which the space-time's metric structure remains completely unchanged, and **k** being timelike allows us to think of **k**'s direction as the direction of *time*, in a related reference frame; see §A.7, figure A-29. (Normally we would also impose

---

[11] To those familiar with the standard quantum formalism, this can be seen directly as follows. Taking $\mathbf{E} = (i\hbar)^{-1}\partial/\partial t$ to be the energy operator, we have $\mathbf{E}|1\rangle = E|1\rangle$ and $\mathbf{E}|2\rangle = E|2\rangle$, whence $\mathbf{E}|\psi\rangle = E|\psi\rangle$.

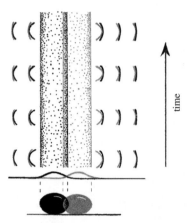

Figure 2-28: The gravitational field of a rock, in a superposition of two locations, horizontally displaced, indicated in black and grey. This provides a superposition of two space-times, with slightly different free fall accelerations, shown as black and grey space-time curves. The spatially integrated square of these acceleration differences provides the measure $E_G$ of the "error" involved in identifying the space-times.

an additional restriction on **k**, namely that it be *non-rotating*, i.e. *hypersurface orthogonal*, but this plays no particular role here.)

We have encountered the notion of a Killing vector before, in §§1.6 and 1.9, in relation to the original 5-dimensional Kaluza–Klein theory. In that theory, it was required that there be a continuous symmetry along the extra spatial dimension, and the Killing-vector field would "point" in the direction of that symmetry, so that the entire 5-dimensional space-time could be "slid over itself" in that direction without changing its metric geometry. The idea of the Killing vector **k** for a stationary 4-dimensional space-time $\mathcal{M}$ is the same, though now the 4-dimensional space-time can be slid over itself in the time-direction of **k**, preserving its space-time metric structure (see figure A-29 in §A.7).

This is very similar to the quantum-mechanical definition of stationarity (without the phase rotation), but now we must take this in the curved-space-time context of general relativity. In general relativity the Killing-vector field is *not* simply "given to us" as a motion along a presupposed time axis. On the other hand, it *is* a presumption of the standard formalism of quantum mechanics that we have a time evolution given to us (with respect to some preassigned time coordinate).

This is a specific ingredient of the Schrödinger equation. It is this distinction that will cause us a fundamental problem when we come to considering a quantum superposition in a general-relativistic context.

It should be made clear that in order to be able to consider issues of general relativity, we shall need to take the view that each *individual* state under consideration (here the states $|1\rangle$ and $|2\rangle$) can be appropriately treated as a *classical* object, subject to the classical laws of general relativity (at the relevant level of approximation). Indeed, if this is *not* the case, then we would already have a discrepancy with a faith in the universal applicability of the laws of quantum mechanics since classical behaviour is *observed* to hold, for macroscopic bodies, at an excellent level of approximation. Classical laws *do* work extraordinarily well for macroscopic objects, so *if* these cannot be accommodated within quantum procedures there would already have to be something wrong with the latter. This would have to also apply to the classical procedures of general relativity, and we have already taken note, in §1.1, of the extraordinary precision of Einstein's theory – for large gravitationally "clean" systems (e.g. the dynamics of binary neutron stars). Thus, if we are to accept that the **U**-procedures of quantum mechanics are inviolate, then we must also accept that it is legitimate to apply these procedures in a general-relativistic context, such as those under consideration here.

Now the stationarity of the individual states $|1\rangle$ and $|2\rangle$ would have to be described by Killing vectors $\mathbf{k}_1$ and $\mathbf{k}_2$ in *different* space-time manifolds $\mathcal{M}_1$ and $\mathcal{M}_2$ describing the gravitational fields of each. The two space-times must indeed be considered to be different, as the rocks are located differently with respect to the ambient geometry of the Earth. Accordingly, there is no unambiguous way to identify $\mathbf{k}_1$ with $\mathbf{k}_2$ (i.e. to think of $\mathbf{k}_1$ and $\mathbf{k}_2$ as being "the same") and whence assert the stationarity of the superposition. This is an aspect of Einstein's principle of *general covariance* (§§A.5 and 1.7), which forbids us from making a meaningful *pointwise identification* between two differently curved space-time geometries (e.g. by saying that a point in one space-time is the *same* as some point in the other space-time merely because they have the same space and time coordinates). Rather than attempting to resolve this issue at some deeper level, we simply estimate the *error* involved in trying to identify $\mathbf{k}_1$ with $\mathbf{k}_2$ in the Newtonian limit (given by $c \to \infty$). (The framework of this Newtonian limit is, technically, that given by Élie Cartan [1945] and Kurt Friedrichs [1927]; see also Ehlers [1991].)

How do we get a measure of this error? At each "identified" point we have two different accelerations of free fall $f_1$ and $f_2$ (taken with respect to the now common Killing vector $\mathbf{k}_1 = \mathbf{k}_2$ for the two space-times; see figure 2-28), these being the respective local Newtonian gravitational fields in the two space-times, and the square of this difference $|f_1 - f_2|^2$ is taken as the local measure of discrepancy (or *error*) in identifying the space-times. This local error measure is integrated (i.e. added up) over 3-dimensional space. The resulting *total* measure of error obtained in this way is a quantity $E_G$ that, for the present situation, can be shown by a relatively simple calculation to be the energy that it would cost to separate two instances of the rock, originally coincident and then moved to the locations specified by $|1\rangle$ and $|2\rangle$, where only the *gravitational* force between the two is taken into consideration. More generally, $E_G$ would be identified as the *gravitational self-energy* of the *difference* between the mass distributions in $|1\rangle$ and $|2\rangle$; for full details, see Penrose [1996] and also §4.2. Lajos Diósi [1984, 1987] came to a similar proposal several years earlier, but without the motivation from general relativity. (In §4.2, these issues are considered in greater detail, and a stronger argument is made for $E_G$ to represent an obstruction to the total stationarity of the superposition, based on Einstein's principle of equivalence.)

The error measure $E_G$ is taken to be a fundamental uncertainty in the *energy* of the superposition, so using Heisenberg's time–energy uncertainty principle, as for an unstable particle as described earlier, we conclude that the superposition $|\psi\rangle$ is *unstable* and would decay to either $|1\rangle$ or $|2\rangle$, in an average time $\tau$, of the general order of

$$\tau \approx \frac{\hbar}{E_G}.$$

Thus, we find that quantum superpositions do not last forever. If the mass displacement between a pair of states in superposition is very tiny – as is the case with every quantum experiment performed to date – then the superposition would be very long lived, according to these considerations, and no conflict with the principles of quantum mechanics would be noticed. But with a significant mass displacement between the states, such a superposition ought to decay spontaneously into one or the other, and this discrepancy with the basic quantum principles should be observable. As yet, no quantum experiment has reached the level at which this discrepancy could be observed, but such experiments have been in the process of development for a number of years now, and I shall briefly describe one of these in §4.2. It is hoped that results will be at hand well within the next decade, which could certainly represent an exciting development.

Even if the results of such experiments do point to discrepancies with the standard quantum faith, perhaps lending observational support to the above $\tau \approx \hbar/E_G$ criterion, this would be very far from the goal of an extended quantum theory in which **U** and **R** both arise as excellent approximations: **U** when mass displacements between superposed states are small, **R** when they are large. Yet this could indeed give some measure of a *limitation* (as yet untested) to current quantum faith. I would propose that *all* quantum state reductions arise as gravitational effects of the aforementioned type. In many standard situations of quantum measurement, the main mass displacements would occur in the *environment*, entangled with the measuring device, and in this way the conventional "environmental decoherence" viewpoint may acquire a consistent ontology. (This key feature of collapse models such as the one adopted here was pointed out by Ghirardi, Rimini, and Weber in their ground-breaking scheme of 1986 [Ghirardi et al. 1986, 1990].) But the ideas reach much farther than this and might well be experimentally tested in various currently active proposals [Marshall et al. 2003; Weaver et al. 2016; Eerkens et al. 2015; Pepper et al. 2012; see also Kaltenbaek et al. 2016; Li et al. 2011; Bedingham and Halliwell 2014], very likely within the next decade or so, or by use of other ideas not yet developed.

# 3

# Fantasy

---

## 3.1. THE BIG BANG AND FLRW COSMOLOGIES

Can *fantasy* have any genuine role to play in our basic physical understanding? Surely this is the very antithesis of what science is about, and should have no place in honest scientific discourse. However, it seems that this question cannot be dismissed as easily as might have been imagined, and there is much in the workings of nature that appears fantastical, according to the conclusions that rational scientific thought appear to have led us to when addressing sound observational findings. As we have seen, particularly in the previous chapter, the world actually does conspire to behave in a most fantastical way when examined at the tiny level at which quantum phenomena hold sway. A single material object can occupy several locations at the same time and like some vampire of fiction (able, at will, to transform between a bat and a man) can behave as a wave or as a particle seemingly as it chooses, its behaviour being governed by mysterious numbers involving the "imaginary" square root of $-1$.

Moreover, at the other end of the scale of size, there is again much that we find fantastical, perhaps even beyond the imaginations of writers of fiction. For example, entire galaxies are sometimes observed to be in collision, where we must believe that they had been dragged inexorably to one another, by the distortions of space and time that they each create. Indeed, such space-time distortional effects are sometimes even directly visible to us through the gross warping of the images of very distant galaxies. Moreover, the most extreme space-time distortions that we know of can lead to massive black holes in space, a pair of which being recently identified as having swallowed each other to make a bigger hole [Abbott et al. 2016]. Others have the the mass of many millions – even tens of thousands of millions – of suns, where such holes can easily swallow up entire solar systems. Such monsters are nevertheless of extremely tiny dimension, as compared with the galaxies themselves, at whose centres they tend to reside. Often such a black hole reveals its presence through the production of two closely collimated beams of energy and material particles, ejected outwards in opposite directions from the tiny central region of its host galaxy, with speeds which can approach 99.5% of that of light [Tombesi et al. 2012; Piner 2006]. In one example such a beam has been observed to be aimed at and impacting upon another galaxy, as though participating in a stupendous intergalactic war.

At an even greater scale, there are vast regions of an invisible something permeating the cosmos – taken to be some completely unknown substance appearing to constitute around 84.5% of the material content of the entire universe, and where there is yet something *else* that takes over at the farthest reaches of all, observed to be dragging everything apart at ever accelerating speeds. These two entities, rather despairingly given the somewhat uninformative respective names "dark matter" and "dark energy" are the major factors determining the overall structure of the known universe. Even more alarming than any of this is the fact that current cosmological evidence appears to be driving us to a firmly established understanding that the entire universe we know began with one gigantic explosion, where there had been nothing before – if the very idea of "before" makes any sense when applied to the origin of that time-space continuum that is taken to underlie all material reality. Surely this concept of a *Big Bang* is a truly fantastical idea!

Indeed it is; yet, there is much observational evidence to support the actuality of an incredibly dense and violently expanding very early stage of our universe, encompassing not merely the known universe's total material content, but also the entire space-time within which all physical substance now plays out its existence, and which appears to extend indefinitely in all directions. All that we know seems to have been created in this one explosion. What is the evidence for this? We must try to assess its credibility and to understand where it all leads us.

In this chapter we shall examine some of the current ideas concerning the origin of the universe itself, particularly in relation to the issue of how much fantasy may justifiably be introduced in order to explain the observational evidence. In recent years, numerous experiments have indeed supplied us with vast amounts of data of direct relevance to the very early universe, transforming what had previously been an assembly of largely untested speculation into an exact science. Most notable were the space satellites COBE (Cosmic Background Explorer), launched in 1989, WMAP (Wilkinson Microwave Anisotropy Probe), launched in 2001, and the Planck space observatory, launched in 2009, which explored the cosmic microwave background (see §3.4) in ever increasing detail. Yet, deep questions remain, and some profoundly puzzling issues have served to drive some cosmological theorists in directions that may reasonably be described as particularly fantastical.

Some degree of fantasy is undoubtedly justified, but have today's theorists strayed too far in this direction? In §4.3 I shall present my own rather unconventional answer to many of these puzzles, having its own concoction of outrageous-seeming ideas, and I briefly outline my case for taking these seriously. But in this book I am more concerned with the currently conventional pictures of the earliest stages of our remarkable universe, and to examine the plausibility of some of the directions into which some modern cosmologists have found themselves driven.

To begin with, we have Einstein's magnificent general theory of relativity, which, as we now know, describes the structure of our curved space-time and the motions of celestial bodies with extraordinary precision (see §§1.1 and 1.7). Following Einstein's early attempts to apply this theory to the structure of the universe as a whole, the Russian mathematician Alexander Friedmann, in 1922 and 1924, first found the appropriate solutions to Einstein's field equations, in the context of a completely spatially uniform (homogeneous and isotropic) distribution of expanding material – this material being taken to be approximated by a *pressureless fluid*, referred to as *dust*, representing the smoothed-out mass–energy distribution of the galaxies [Rindler 2001; Wald 1984; Hartle 2003; Weinberg 1972]. This description indeed appears, observationally, to give a reasonably good overall approximation to the smoothed-out distribution of matter in the actual universe, and it provides the energy tensor $\mathbf{T}$ that Friedmann required as the gravitational source term in Einstein's equations $\mathbf{G} = 8\pi\gamma\mathbf{T} + \Lambda\mathbf{g}$ (see §1.1). Friedmann's models have a characteristic feature that the origin of the expansion is a *singularity*, now referred to as the *Big Bang*, where the curvature of the space-time had started out as *infinite*, the mass–energy density of the material source $\mathbf{T}$ diverging to infinity as we proceed backwards in time to this space-time singularity. (Curiously, the now universally used terminology "Big Bang" was originally intended as a derogatory term, introduced by Fred Hoyle (himself a fervent supporter of the rival *steady-state* theory; see §3.2) in his BBC radio talks, delivered in 1950. These talks are also referred to in a different context, in §3.10, and were subsequently compiled into a book [Hoyle 1950].)

Just for the moment, I shall provisionally take Einstein's tiny cosmological constant $\Lambda$ – the seeming source of the *accelerated* expansion of the universe, referred to above (and see §1.1) – to be *zero*. There are then just three different cases to consider, these depending upon the nature of the spatial geometry whose curvature $K$ can be positive ($K > 0$), zero ($K = 0$), or negative ($K < 0$). In standard cosmology books, it is conventional to *normalize* the value of $K$ to one of the three values $1, 0, -1$. Here I shall find it easier to phrase things as though $K$ is a real number describing the actual degree of spatial curvature. We can think of $K$ as specifying the value of this spatial curvature at some canonically chosen value of the time parameter $t$. For example, we could choose that canonical $t$ value to be the time of *decoupling* (see §3.4) when the cosmic microwave background radiation was created, but the specific choice will not be important for us here. The essential point is that the *sign* of $K$ does not change with time, so whether $K$ is positive, negative, or zero is a feature of the model as a whole, irrespective of any choice of "canonical time".

It should, however, be mentioned that the value of $K$ alone does not quite characterize the spatial geometry. There are also non-standard "folded-up" versions of all these models for which the spatial geometry might be something rather complicated, where in some examples the spatial geometry can be *finite* even when $K = 0$ or $K < 0$. Interest has sometimes been expressed in such models (see Levin [2012], Luminet et al. [2003], and originally Schwarzschild [1900]). However, such models will not have importance for us here, this issue having no significant effect on most of the arguments presented. Ignoring such topological complications, we are presented with just three types of uniform geometry which are very elegantly illustrated, in the 2-dimensional case, by the Dutch artist M. C. Escher; see figure 3-1 (and compare figure 1-38 in §1.15). The 3-dimensional situation is similar.

The case $K = 0$ is easiest to understand, because the spatial sections are simply ordinary Euclidean 3-spaces, though in order to express the expanding nature of the model we must think of these 3-dimensional Euclidean sections as being related to one another, in an expanding way; see figure 3-2 for $K = 0$. (This expansion can be understood in terms of the diverging timelike lines that represent the world-lines of the idealized galaxies described by the model, these being the *time-lines* that we shall be coming to shortly.) The 3-spaces which are the spatial sections for $K > 0$ are only slightly harder to appreciate, being *3-spheres* ($S^3$), each being the 3-dimensional analogue of an ordinary spherical surface ($S^2$), and the expansion of the universe is expressed in the increase with time of the sphere's radius; see figure 3-2 for $K > 0$. In the negatively curved case $K < 0$, the spatial sections have *hyperbolic* (or *Lobachevskian*) *geometry*, this geometry being neatly described using the (Beltrami–Poincaré) *conformal representation*, described in the 2-dimensional case as the space interior to a circle $\mathcal{S}$, in a Euclidean plane, where the "straight lines" of the geometry are described as arcs of circles which meet the boundary $\mathcal{S}$ at right angles (see figure 3-2 for $K < 0$ and figure 1-38(b) in §1.15 [see, for example, TRtR, §§2.4–2.6; Needham 1997]). The picture of 3-dimensional hyperbolic geometry is similar, where the circle $\mathcal{S}$ is replaced by a sphere (an ordinary 2-sphere), bounding a portion (a 3-ball) of Euclidean 3-space.

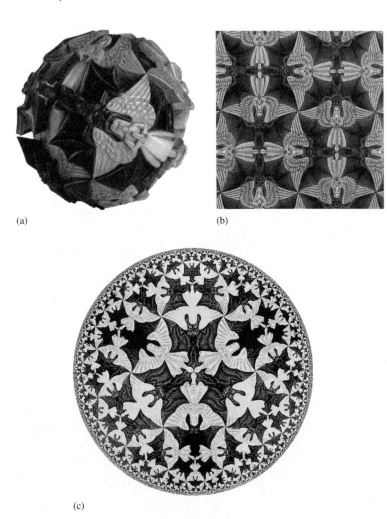

(a)

(b)

(c)

Figure 3-1: Escher's way of illustrating, in the 2-dimensional case, the three types of uniform geometry (a) positive curvature ($K > 0$), (b) the flat Euclidean case ($K = 0$), (c) negative curvature ($K < 0$) using Beltrami's conformal representation of hyperbolic geometry, also shown in figure 1-38.

The terminology "conformal", applied to these models, comes from the fact that the measure of *angle* that the hyperbolic geometry assigns to two smooth curves at a point of intersection is the same as would be assigned in the background

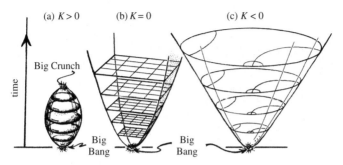

Figure 3-2: The Friedmann dust-filled cosmological models, illustrated for zero cosmological constant $\Lambda$: (a) $K > 0$, (b) $K = 0$, (c) $K < 0$.

Euclidean geometry (so, for example, the angles at the tip of the fishes' fins in figure 1-38(a) or of the devils' wings, in figure 3-1(c) are correctly represented, no matter how close they are to the bounding circle). Another (rough) way of stating this is that the *shapes* (though usually not the sizes) of very small regions are accurately depicted in these representations (see also figure A-39 in §A.10).

As remarked earlier, there is now some impressive evidence that in our universe $\Lambda$ actually has a small *positive* value, so we must consider the corresponding Friedmann models with $\Lambda > 0$. In fact, tiny though $\Lambda$ actually is, its observed value is easily large enough (assuming that it is indeed constant, as Einstein's equations demand) to overcome the collapse to a "big crunch" depicted in figure 3-2(a). Instead, in *all three* possible situations for the values of $K$ that are allowed by current observation, the universe is expected ultimately to indulge in an *accelerated* expansion. With such a constant positive $\Lambda$, the expansion of the universe will continue to accelerate indefinitely, giving an *exponential* ultimate expansion (see figure A-1 in §A.1). In accordance with this, our current expectations for the overall history of the universe are depicted in figure 3-3, where the vagueness that is depicted at the back of the picture is intended to allow for all three possibilities for the spatial curvature $K$.

The *remote futures* of all these models, for $\Lambda > 0$, even when they are perturbed by irregularities, are very similar, and appear to be well described by the particular space-time model that has become known as *de Sitter space*, whose Einstein tensor **G** takes the simple form $\Lambda$**g**. This model was found by Willem de Sitter (and independently, by Tullio Levi-Città) in 1917 [see de Sitter 1917a,b; Levi-Città 1917; Schrödinger 1956; TRtR, §28.4, pp. 747–50]. It is indeed now commonly considered to provide a good approximation to the remote future to our *actual*

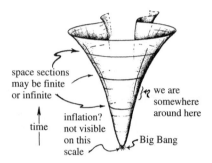

space sections
may be finite
or infinite

↑
time
|

inflation?
not visible
on this
scale

we are
somewhere
around here

Big Bang

Figure 3-3: Space-time depiction of our universe's expected evolution, as modified to incorporate the observed (sufficiently large) $\Lambda > 0$. The depicted uncertainty in the behaviour at the back of the picture is to reflect an uncertainty in the overall spatial geometry, which has no significant evolutionary role.

universe, where the remote-future energy tensor is expected to be completely dominated by $\Lambda$, giving $\mathbf{G} \approx \Lambda\mathbf{g}$ in the future limit.

This, of course, assumes that Einstein's equations ($\mathbf{G} = 8\pi\gamma\mathbf{T} + \Lambda\mathbf{g}$) continue to hold indefinitely, so that our presently ascertained value of $\Lambda$ remains a constant. In §3.9, we shall see that, according to the exotic ideas of inflationary cosmology, the de Sitter model is also taken to describe the universe during a much earlier time immediately following the Big Bang, though with an enormously larger value of $\Lambda$. These issues will be of considerable importance for us later (especially in §§3.7–3.9 and 4.3), but will not particularly concern us for the moment.

De Sitter space is a highly symmetrical space-time, which can be described as a (pseudo-)sphere in Minkowski 5-space; see figure 3-4(a). Explicitly, it arises as the locus $t^2 - w^2 - x^2 - y^2 - z^2 = -3/\Lambda$, obtaining its local metric structure from that of the ambient Minkowski 5-space with coordinates $(t, w, x, y, z)$. (For those familiar with the standard way of writing metrics using differentials, this Minkowski 5-metric takes the form $\mathrm{d}s^2 = \mathrm{d}t^2 - \mathrm{d}w^2 - \mathrm{d}x^2 - \mathrm{d}y^2 - \mathrm{d}z^2$.) The de Sitter space is fully as symmetrical as Minkowski 4-space, each having a 10-parameter symmetry group. We may also recall the hypothetical anti–de Sitter space that was considered in §1.15 (see footnote 9 on p. 113). It is very closely related to de Sitter space, and it also has a symmetry group of this size.

De Sitter space is an empty model, its energy tensor $\mathbf{T}$ being zero, so it has no (idealized) galaxies to define time-lines, whose orthogonal 3-space sections could have been used to determine specific 3-geometries of "simultaneous time". In fact, rather remarkably, it turns out that in de Sitter space we can choose such

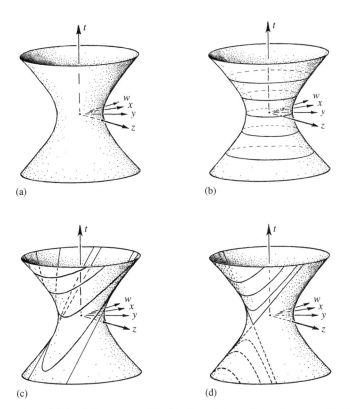

Figure 3-4: (a) De Sitter space, (b) with $K > 0$ time-slicing ($t =$ const.), with (c) $K = 0$ time-slicing ($t - w =$ const.) as in steady-state cosmology, and (d) with $K < 0$ time-slicing ($-w =$ const.).

3-dimensional spatial (simultaneous-time) sections in three essentially different ways, so that de Sitter space can be interpreted as an expanding spatially uniform universe with each of the three alternative types of spatial curvature, depending upon the way that it is sliced with such 3-surfaces, taken as being of constant cosmic time: $K > 0$ (given by $t =$ const.), $K = 0$ (given by $t - w =$ const.), and $K < 0$ (given by $-w =$ const.); see figure 3-4(b)–(d). This was elegantly shown by Erwin Schrödinger in his 1956 book, *Expanding Universes*. The old steady-state model that we shall come to in §3.2 is described by de Sitter space, according to the $K = 0$ slicing shown in figure 3-4(c) (and conformally represented in figure 3-26(b) in §3.5). Most versions of inflationary cosmology (that we shall be coming to in §3.9) also use this $K = 0$ slicing, as this allows the inflation to continue in a uniformly exponential way for an indefinite time.

In fact, with regard to our actual universe on an extremely large scale, present observations do not point unambiguously to which of these spatial geometries might provide the most appropriate picture. But whatever the ultimate answer, it does now appear that the case $K = 0$ is very close to being correct (somewhat remarkably, in view of strong-seeming evidence for $K < 0$ towards the end of the twentieth century). In one sense, this is the least satisfactory observational situation, since if all that can be said is that $K$ is very close to zero, we still cannot be sure that more refined observations (or a more convincing theory) may not later point to one of the other spatial geometries (spherical or hyperbolic, that is) being more appropriate for our universe. If, for example, good evidence for $K > 0$ were finally to emerge, this would have genuine philosophical significance, since it would have the implication that the universe is not spatially infinite. As things stand, however, it is normally simply asserted that the observations tell us that $K = 0$. This may well be a very good close approximation, but in any case we do not know how close to actual spatial homogeneity and isotropy the overall universe might be, particularly in view of certain counter indications in the CMB observations [e.g. Starkman et al. 2012; Gurzadyan and Penrose 2013, 2016].

To complete the picture of the entire space-time, according to Friedmann's models and their generalizations, we need to know how the "size" of the spatial geometry would evolve with time, right from the start. In the standard cosmological models, like Friedmann's – or the generalizations known as *Friedmann–Lemaître–Robertson–Walker* (FLRW) models, where all in this general class have spatial sections that are homogeneous and isotropic, the whole space-time sharing the symmetry of these sections – there is a well-defined notion of a *cosmic time t* to describe the evolution of the universe model. This cosmic time is the time measure, starting with $t = 0$ at the Big Bang, that would be measured by an ideal clock following the world-lines of the idealized galaxies; see figure 3-5 (and figure 1-17 in §1.7). I shall refer to these world-lines as the *time-lines* of the FLRW model (sometimes referred to as the world-lines of the *fundamental observers*, in cosmology texts). The time-lines are the geodesic curves orthogonal to the spatial sections, those being the 3-surfaces of constant $t$.

The case of de Sitter space is somewhat anomalous, in this respect, because, as mentioned earlier, it is empty in the sense that $\mathbf{T} = \mathbf{0}$, in Einstein's $\mathbf{G} = 8\pi\mathbf{T} + \Lambda\mathbf{g}$, so there are no matter world-lines to provide time-lines or, consequently, to define spatial geometries, and consequently we have a choice, locally, as to whether we regard the model as describing a $K > 0$, $K = 0$, or $K < 0$ universe. Nevertheless,

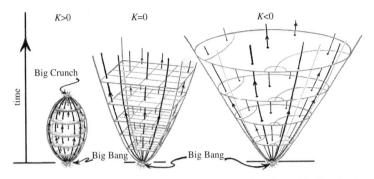

Figure 3-5: The Friedmann models of figure 3-2 with time-lines (idealized galaxy world-lines) drawn in.

globally the three situations are different, as we can see from figure 3-4(b)–(d) that in each case a different portion of the entire de Sitter space is covered by the slicing. In the discussions below, I shall assume that **T** is non-zero, providing a positive energy density of matter, so that the time-lines are well defined, and so are the spacelike 3-surfaces of constant time for each $t$ value, as shown in figure 3-2.

In the positive spatial curvature case $K > 0$, for a standard dust-filled Friedmann universe, we can use the radius $R$ of the 3-sphere spatial sections to characterize the "size", and examine this as a function of $t$. When $\Lambda = 0$, we find a function $R(t)$ that describes a *cycloid* in the $(R, t)$-plane (taking the speed of light $c = 1$), this being the curve having the simple geometrical description of being traced out by a point on the circumference of a circular hoop (of fixed diameter equal to the maximum value $R_{\max}$ attained by $R(t)$) that is rolling along the $t$-axis (see figure 3-6(b)). We note that (after a time given by $\pi R_{\max}$) the value of $R$ reaches the value 0, again, as it had at the Big Bang, so the entire universe model (with $0 < t < \pi R_{\max}$) collapses down to a second singular state, often referred to as the *Big Crunch*.

In the remaining cases $K < 0$ and $K = 0$ (and $\Lambda = 0$), the universe model expands indefinitely, and there is no Big Crunch. For $K < 0$ there is an appropriate notion of "radius", analogous to $R$, but for $K = 0$ we can just pick an arbitrary pair of idealized galaxy world-lines and take $R$ to be their spatial separation. In the case $K = 0$, the expansion rate slows down to zero asymptotically, but it reaches a limiting positive value in the case $K < 0$. With present observations leading to the belief that $\Lambda$ is actually positive and is large enough ultimately to dominate

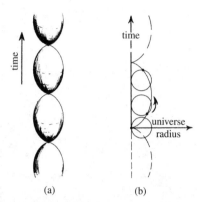

(a)                              (b)

Figure 3-6:  (a) Oscillating Friedmann ($K > 0$, $\Lambda = 0$) model; (b) its radius as a function of time is a cycloid.

the expansion rate, the value of $K$ becomes unimportant to the dynamics, and the universe finally indulges in the accelerated expansion indicated in figure 3-3.

In the early days of relativistic cosmology the positive-$K$ model (with $\Lambda = 0$) was frequently referred to as the *oscillating model* (figure 3-6(a)), since the cycloidal curve continues indefinitely if we allow the rolling of the hoop to continue beyond one cycle (broken curve in figure 3-6(b)). One might envisage that the continually repeating cycles of the cycloid might possibly represent successive cycles of the actual universe where, through some kind of bounce, each crunch that the universe experiences is somehow converted into an expansion. A similar possibility occurs with the cases $K \leqslant 0$, where we could envisage a previous collapsing phase of space-time, identical to the time-reverse of the expanding phase, whose Big Crunch coincides with the Big Bang of what we would take to be the current expanding phase of the universe. Again we would have to imagine some kind of bounce which is somehow able to convert the collapse into an expansion.

For this to be physically plausible, however, some believable mathematical scheme would need to be presented, which is consistent with current physical understanding and procedures, and which is able to accommodate such a bounce. For example, we might envisage altering the equations of state that Friedmann adopted in order to describe the overall matter distribution of his "smoothed-out galaxies". Friedmann used the approximation referred to as *dust*, in which there is taken to be no interaction (other than gravity) between the constituent "particles" (i.e. the "galaxies"), whose world-lines are the time-lines. A change

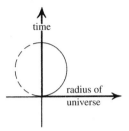

Figure 3-7: In Tolman's radiation-filled ($K > 0$, $\Lambda = 0$) model, the radius as a function of time can be depicted as a semicircle.

in the equations of state can considerably alter the behaviour of $R(t)$ near $t = 0$. In fact, what would appear to be a better approximation than Friedmann's dust, near the Big Bang, is the equation of state later used by the American mathematical physicist and cosmologist Richard Chace Tolman [1934]. In Tolman's (FLRW) models, the equation of state that he adopted was that of *pure radiation*. This may be expected to provide a good approximation to the state of matter in the very early universe, when temperatures get so high that the energy per particle greatly exceeds the energy $E = mc^2$ for the mass $m$ of the most massive of the particles likely to be present immediately following the Big Bang. In Tolman's scheme, in the $K > 0$ case, the shape of the $R(t)$-curve is not one arch of a cycloid, but (with the appropriate scaling of $R$ and $t$) a *semicircle* (figure 3-7). In the case of dust, we might have regarded the transition from Crunch to Bang as being justified by an appeal to *analytic continuation* (see §A.10), since we can indeed evolve from one arch of the cycloid curve to the next by such mathematical means. But in the case of the semicircle of Tolman's pure radiation, the analytic continuation procedure would simply complete the semicircle to a circle, which makes no sense at all, if we are looking to this procedure as a means to a bounce, allowing continuation to negative values of $t$.

If a bounce is to come about merely through a change in the equation of state, something much more radical than Tolman's radiation would be needed. One serious point to consider is that if the bounce is to take place through some non-singular transition, in which there is a smooth space-time throughout and where the spatial symmetry of the model is preserved, then the converging time-lines of the collapsing phase must become converted to diverging ones in the subsequent expanding phase, along the neck connecting one phase with the next. If this neck is to be smooth (non-singular), then the turnaround of this extreme convergence of the time-lines to an extreme divergence can only be achieved by the presence of enormous curvature at the neck, of a strongly repulsive kind which violently contradicts the standard energy-positivity conditions that would be satisfied by ordinary classical matter (see §§1.11, 3.2, and 3.7; Hawking and Penrose [1970]).

Because of this, we cannot expect any reasonable classical equations of state to provide us with a bounce within the context of FLRW models, and the issue must be raised as to whether the equations of quantum mechanics will enable us to fare better. We must bear in mind that, near the classical FLRW singularity, the space-time curvatures get indefinitely large. If we describe such a curvature in terms of a radius of the curvature, this radius (as the reciprocal of a curvature measure) would get correspondingly small. So long as we continue to use the notions of classical geometry, space-time curvature radii would become indefinitely small near the classical singularity, eventually becoming smaller even than the Planck scale of $\sim 10^{-33}$ cm (see §§1.1 and 1.5). Considerations of quantum gravity lead most theorists to anticipate that at this scale there would have to be drastic departures from the normal smooth-manifold picture of space-time (although, in §4.3, I shall be making a very different argument concerning this issue). Whether or not this is so, it is not unreasonable to expect that the procedures of general relativity will necessarily have to be modified in order to fit in with those of quantum mechanics in the vicinity of such a violently curved space-time geometry. That is to say, some appropriate *quantum gravity* theory appears to be needed to cope with those situations where the classical procedures of Einstein lead to singularity (but contrast with §4.3).

It has indeed been commonly argued that there is a precedent for this kind of thing. As was pointed out in §2.1, there had been a serious problem at the turn of the twentieth century with the classical picture of the atom, since theory had seemed to predict that atoms should collapse catastrophically to a singular state, with electrons spiralling into the nuclei (with a burst of radiation), and it required the introduction of quantum mechanics to resolve the issue. May we not expect something similar for the catastrophic collapse of an entire universe, where the procedures of quantum mechanics might lead us to a resolution of the issue? A trouble here is that, even now, there is no generally accepted quantum-gravity proposal. More serious is that most of the proposals that have been put forward do not actually resolve the singularity problem, singularities remaining even in the quantized theory. There are some notable exceptions, in which a non-singular quantum bounce is claimed [Bojowald 2007; Ashtekar et al. 2006] but I shall need to return to this matter in §§3.9 and 3.11 (and also §4.3), where I argue that this

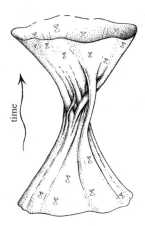

Figure 3-8: A hypothetical bouncing universe, where extreme irregularity is imagined to allow a singularity-free transition from collapse to expansion.

kind of proposal does not really provide much hope for resolving the singularity issue for our actual universe.

A completely different possibility for avoiding the singularity arises from the expectation that small deviations from exact symmetry, present in a collapsing universe phase, would be likely to become greatly magnified as the Big Crunch is approached, so that in the vicinity of the fully collapsed state the structure of the space-time is not at all well approximated by an FLRW model. The hope had frequently been expressed, therefore, that the singularity manifested in the FLRW models might be spurious, and that in a more general asymmetrical situation such classical space-time singularities would simply not occur, leading to an expectation that a general collapsing universe could, through some complicated intermediate space-time geometry (see figure 3-8), emerge in an irregular expanding state. Even Einstein himself tried to argue this way, so that singularities might be avoided, leading to a bounce from an irregular collapse [Einstein 1931; Einstein and Rosen 1935], or by orbital motions of bodies somehow preventing the final singular collapse [Einstein 1939].

It would be argued that following such a near-singular (but supposedly nonsingular) collapse, this emerging state, in turn, through a consequent ironing out of irregularities, might be expected soon to approximate, closely, an expanding FLRW model as indicated in figure 3-8. Indeed, some detailed analysis carried out in 1963 by two Russian theoretical physicists, Evgeny Mikhailovich Lifshitz and

Isaak Markovich Khalatnikov [Lifshitz and Khalatnikov 1963] appeared to show that singularities would *not* arise in general situations, supporting the validity of a non-singular bounce of the kind just described. The claim, therefore, was that in general relativity the space-time singularities arising in gravitational collapse – as exhibited in known exact solutions such as collapsing Friedmann or other FLRW models – arise only because these known solutions possess unrealistic special features, such as exact symmetries, so such singularities would not persist when asymmetrical generic perturbations are introduced. However, this turned out *not* to be correct, as we shall be seeing in the next section.

## 3.2. BLACK HOLES AND LOCAL IRREGULARITIES

In 1964, I had begun thinking seriously about the closely related problem of a more local gravitational collapse of a star, or of a collection of stars, to what we now refer to as a *black hole*. The notion of a black hole had been behind the scenes ever since the very remarkable 19-year-old Indian astrophysicist Subrahmanyan Chandrasekhar, in 1930, had first demonstrated [see Wali 2010; Chandrasekhar 1931] that there is a limit – of about 1.4 times the mass of the sun – to the mass that a white-dwarf star can attain, without its collapsing catastrophically under its own gravitational pull. A white dwarf is an exceedingly compact star. One of the first to be discovered was the puzzling companion of the brightest star in the sky, Sirius. This tiny companion, Sirius B, has a mass roughly equal to that of the Sun, yet a diameter no greater than that of the Earth, which is some $10^6$ times smaller than the Sun by volume. Such a white dwarf would have basically used up its resources of nuclear fuel and now held apart merely by what is called *electron degeneracy pressure*. This pressure arises from the Pauli exclusion principle (see §1.14), as applied to electrons, this having the effect of constraining the electrons from getting too crowded together. Chandrasekhar showed that if his mass limit is exceeded, there is a fundamental obstruction to the efficacy of this process, as the electrons begin to move at speeds approaching that of light, so that when the star cools down sufficiently, further collapse cannot be stopped by this means.

An even more condensed state can occur, where collapse can sometimes be halted through what is called *neutron degeneracy pressure*, in which electrons get squashed into protons to form neutrons, and the Pauli exclusion principle now acts on these neutrons [Landau 1932]. In fact, many such *neutron stars* have now been observed, in which the star's density reaches an incredible value, comparable with (or sometimes even exceeding) that of an atomic nucleus itself, where a mass of a little more than that of the Sun becomes concentrated within a sphere of a radius as small as about 10 km, so over $10^{14}$ times smaller than the Sun by volume. Neutron stars often carry enormous magnetic fields and can spin rapidly, the effects of such rotating magnetic fields on local charged material causing electromagnetic signals that can be detected at Earth, even to over $10^5$ light years away, as the blip-blip-blip of a pulsar. But again there is a limit – the *Landau limit*, like that of Chandrasekhar's – to the mass that a neutron star can attain. There is still some uncertainty as to the exact value of this limit, but it is unlikely to be much above two solar masses. As it turns out, the most massive neutron star yet discovered (at the time of writing), a pulsar in close orbit with a white dwarf (of a $2\frac{1}{2}$-hour period), constituting the system J0348+0432, appears to have a mass of, indeed, twice that of the Sun.

With regard to local physical processes, as theory is presently understood, there is no further way to halt the collapse of a more massive version of such a highly compressed body. Yet, many far more massive stars – and concentrated collections of stars – are observed, and fundamental questions arise as to what can be their ultimate fate when gravitational collapse eventually takes over with such entities, such as when a very large star's nuclear fuel runs out. In Chandrasekhar's modest words, in his ground-breaking 1934 paper on the subject, he commented:

> The life-history of a star of small mass must be essentially different from the life-history of a star of large mass. For a star of small mass the natural white-dwarf stage is an initial step towards complete extinction. A star of large mass > critical mass m cannot pass into the white-dwarf stage, and one is left speculating on other possibilities.

On the other hand, many others remained sceptical. Most particularly the distinguished British astrophysicist (Sir) Arthur Eddington [1935] commented:

> The star has to go on radiating and radiating and contracting and contracting and contracting until, I suppose, it gets down to a few km radius, when gravity becomes strong enough to hold in the radiation, and the star can at last find peace. ... I think there should be a law of nature to prevent a star from behaving in this absurd way!

This issue had become of particular concern in the early 1960s, and had been stressed especially by the distinguished American physicist John Archibald Wheeler, notably because of the then recent discovery by the Dutch astronomer Maarten Schmidt, in 1963, of the first *quasar* (as such objects would later be named) 3C 273. From its evident great distance from us (as ascertained by redshift measurements), the intrinsic brightness of this object was judged to be extraordinary, over $4 \times 10^{12}$ times the brightness of the Sun, so that its output is about one hundred times the total output of our own entire Milky Way galaxy! Coupled to its relatively tiny size – something comparable with our Solar System, which could be deduced from its distinct and rapid variations in output over periods of a few days – this extraordinary mass–energy output led astronomers to conclude that the central object responsible for these emissions would have to involve a huge, yet extremely compact mass, this even being compressed to something nearly as small as its own *Schwarzschild radius*. This critical radius, for a spherically symmetric body of mass $m$, has the value

$$\frac{2\gamma m}{c^2},$$

where $\gamma$ is Newton's gravitational constant and $c$ is the speed of light.

Some words of explanation are appropriate here, with regard to this radius. It concerns the well-known *Schwarzschild solution* of Einstein's equations ($\mathbf{G} = \mathbf{0}$; see §1.1) for the vacuum gravitational field surrounding a static spherically symmetrical massive body (an idealized star). This solution was found by the German physicist and astronomer Karl Schwarzschild very shortly after Einstein had fully formulated his general theory in late 1915 (and very shortly before Schwarzschild himself tragically died from a rare disease contracted at the Russian Front in World War I). If we imagine that the collapsing body were to shrink symmetrically inwards, and that Schwarzschild's solution continues inwards uniquely in accordance with this, as the equations indeed demand, then the coordinate expression for the metric encounters a singularity at the Schwarzschild radius, and it was thought by most physicists (including Einstein himself) that the actual space-time geometry would necessarily become singular in its structure, at this place.

However, it was later realized that the Schwarzschild radius is *not* a space-time singularity, but that this is the radius at which collapsing (spherically symmetric) matter would, upon reaching it, enter what we now call a black hole. Any spherical object compressed to within its Schwarzschild radius would collapse irretrievably and rapidly fade from view. It was argued that the energy emissions seen in 3C 273 would have to arise from the violent processes associated with such a gravitational collapse, in regions just external to the Schwarzschild radius. Stars or other kinds of material would become grossly distorted and heated in the violent processes involved prior to their actually becoming swallowed by the hole.

Gravitational collapse to a black hole, under an assumption of exact spherical symmetry, has some considerable similarity to the situation which occurs with the

Friedmann models, there being a known exact solution of Einstein's equations – this time, one due to Oppenheimer and Snyder in 1939 – which gives a full geometrical space-time picture of such a collapse in the spherically symmetrical case. The energy tensor **T** for the collapsing matter is again that of Friedmann's dust. Indeed the "matter" part of their solution is exactly a portion of a Friedmann dust model – like part of a collapsing universe. In the Oppenheimer–Snyder solution, there is a spherically symmetrical matter (dust) distribution which collapses right down to and through the Schwarzschild radius, leading to a space-time singularity at the centre, where the density of the collapsing matter – and also the curvature of space-time – becomes infinite.

The Schwarzschild radius itself turns out to be a singularity only in the static-type coordinates that Schwarzschild used, though for a long time it had been mistakenly thought to be a genuine physical singularity. Curiously, it appears that the first person to realize that the Schwarzschild radius is only a coordinate singularity, and that it is possible to extend the solution smoothly through this region to an actual singularity at the centre may have been the mathematician Paul Painlevé in 1921, who had briefly been the prime minister of France in 1917 and was to be again in 1925 [Painlevé 1921]. However, his theoretical conclusion appears to have been largely unappreciated by the relativity community and, indeed, there was much confusion at the time about how Einstein's theory was to be interpreted. Subsequently, in 1932, Abbé Georges Lemaître showed explicitly that freely falling matter could pass within this radius without encountering a singularity [Lemaître 1933]. A simpler description of this geometry was provided much later by David Finkelstein in 1958 [Finkelstein 1958], using a form of the Schwarzschild metric that had, curiously, been found much earlier by Eddington himself in 1924 for a different purpose [Eddington 1924], and unrelated to gravitational collapse!

The Schwarzschild radius surface is now referred to as an (absolute) *event horizon*. For reasons that will become clearer shortly, material can fall inwards through this radius, but once inside it cannot escape. A question that arises is whether the presence of deviations from exact spherical symmetry and/or the use of equations of state more general than the pressureless dust employed here by Oppenheimer and Snyder might be able to enable the collapse to avoid arriving at a singular state, so that we could envisage a very complicated – though actually *non*-singular – intermediate configuration, via which the collapse might "bounce", to become converted into an irregular expansion of material, formed out of the matter which initially fell in.

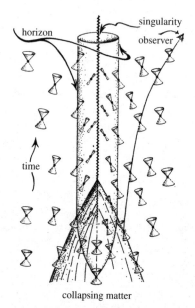

Figure 3-9:  Standard space-time picture of gravitational collapse to a black hole.
An observer outside the horizon cannot see events inside the horizon.

The space-time description of the Oppenheimer–Snyder model is illustrated
in figure 3-9 (one spatial dimension being suppressed). The key aspects of the
geometry are provided by the null cones (see figure 1-18(b)), which restrict all
propagation of information to be within these cones (see §1.7). This picture is
based on the description given by Finkelstein referred to above [Finkelstein 1958].
We take note of the fact that the presence of the highly concentrated collapsing
material results in the cones being substantially distorted inwards, by increasing
amounts as they become more central, so that at a certain radius the outward
edges of the future cones become vertical in the picture, so that signals within
this radius are incapable of escaping to the outside world. This, indeed, is the
Schwarzschild radius for the collapse. We can ascertain from the figure that the
collapsing material can pass inwards through this radius, but that, after it does,
it loses all ability to communicate with the outside. At the centre, we find the
space-time singularity, where space-time curvatures *do* diverge to infinity, and
the collapsing material reaches infinite density. All (timelike) world-lines, after
they have crossed to within the Schwarzschild radius, whether they lie within the

collapsing material or are following it in afterwards, end up at the singularity. There is no escape!

There is an issue of interest here, concerning the comparison with Newtonian theory. It has frequently been pointed out that the very same radius also has significance in Newtonian gravity, as had been noted, way back in 1783, by the British scientist Rev. John Michell [1783]. He had arrived at precisely Schwarzschild's radius value using Newtonian theory, based on the view that light emitted from within that surface at the speed of light would fall back and be unable to escape. This was remarkably prescient on Michell's part, but the conclusion can be severely questioned since, in Newtonian theory, the speed of light is not constant, and it would be argued that for a Newtonian body of that size the light speed would be much larger, as would be the case for light falling in on the body from a great distance. The notion of a black hole really only arises from the particular features of general relativity, and it does not occur in Newtonian theory; see Penrose [1975a].

The question now arises, just as with the collapsing FLRW cosmologies, as to whether deviations from spherical symmetry might lead to a dramatically different picture. Indeed, we might anticipate that when the collapsing material does not quite have the exact spherical symmetry assumed here, these deviations could well increase as the central regions are approached, so that the infinite densities and infinite space-time curvatures are consequently avoided (see figure 3-8 in §3.1). On this view, the singularities would arise only because there is an exact focusing of the incoming matter towards the central point. Accordingly, without this focal precision in the collapse, although the densities might still get very large, they would be anticipated not to become infinite, and after some violent and complicated swirlings and splashings, the material might be expected to re-emerge in some form, according to this picture, without an actual singularity having been reached. At least, that was the idea.

In the autumn of 1964, I began to worry about this issue in a serious way, and I wondered whether I could possibly answer the question by employing some mathematical ideas that I had developed earlier in the context of the steady-state model of the universe (this model having been put forward in the 1950s by Hermann Bondi, Thomas Gold, and Fred Hoyle; see Sciama [1959, 1969]), in which the universe has no beginning, its expansion being continual and never ending, and where the thinning out of the matter, due to the expansion, is compensated by new material (mainly in the form of hydrogen) being continually created at a very low rate throughout the universe. I had wanted to see whether an apparent contradiction between the steady-state model and standard general relativity (with an energy-positivity requirement on the matter of the general kind referred to in §3.1 above) might be averted by the presence of deviations from the complete symmetry that is employed in the usual steady-state picture. By the use of a geometrical/topological argument, I had convinced myself that such deviations from symmetry could *not* remove this contradiction. I never published this argument, but I had also used similar ideas in a different context, applying them (in a largely, but not completely, rigorous fashion) to the asymptotic structure of a gravitationally radiating system [Penrose 1965b, Appendix]. These methods are very different from those that had been normally applied in general relativity, those usually involving finding explicit special solutions or performing massive numerical computations.

With regard to gravitational collapse, the aim was to show, likewise, that in any situation where the collapse is sufficiently severe, the presence of deviations from symmetry (and the adoption of more general equations of state than simply Friedmann's dust or Tolman's radiation, etc.) would not substantially change the conventional Oppenheimer–Snyder picture, and that the presence of some kind of singularity obstructing any completely smooth evolution would be inevitable. It needs to be borne in mind that there are many other situations in which a body could contract inwards in a relatively mild way due to gravitation, where perhaps the presence of other forces might lead to a stable configuration or else to some kind of bounce. Thus, we need an appropriate criterion to characterize the kind of irretrievable collapse exhibited in the Oppenheimer–Snyder type of situation, as shown in figure 3-9. It is, of course, essential for what is required that this criterion cannot be one that relies upon any assumption of symmetry.

After much thought, I came to realize that no entirely *local* characterization could achieve what is needed, nor could anything of the nature of a total or average measure of (say) space-time curvature be of any real use. Eventually, I hit upon the notion of a *trapped surface*, whose presence in the space-time is a good signal that irretrievable collapse has indeed taken place. (The interested reader is referred to Penrose [1989, p. 420] for a description of the curious circumstance under which this idea was indeed hit upon.) Technically, a trapped surface is a closed spacelike 2-surface all of whose *null normal* directions – a concept illustrated in figure 3-10(a) (see also §1.7) – converge in future directions. The term *normal* means "at right angles" in ordinary Euclidean geometry (see figure 1-18 in §1.7) and we see in figure 3-10 that the (future-directed) null normals give the directions of light rays (i.e. null geodesics) that come away from the 2-surface at right angles to it, as viewed in any instantaneous spacelike 3-surface containing the given 2-surface.

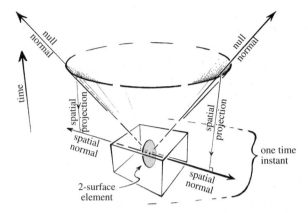

Figure 3-10: The null normal directions to a spacelike 2-surface are the two light-ray directions that come away from the 2-surface at right angles to it, as viewed in any (instantaneous) spacelike 3-surface containing the given 2-surface.

To get an idea of what this means in spatial terms, think of a smooth 2-dimensional curved surface $\mathcal{S}$ in ordinary Euclidean 3-space. Imagine a flash of light that occurs simultaneously over $\mathcal{S}$ and examine how the wavefront of the emitted light propagates away from $\mathcal{S}$, either to one side of $\mathcal{S}$ or to the other (figure 3-11(a)). At a place where $\mathcal{S}$ is curved, the wavefront on the concave side will have an area which immediately begins to shrink, whereas on the convex side it begins to expand. What happens with a trapped surface $\mathcal{S}$, however, is that on *both* sides of $\mathcal{S}$ the wavefronts begin to shrink! See figure 3-11(b). At first sight, this may appear to be an unrealizable local condition for an ordinary spacelike 2-surface, but in fact this is not so in space-time. Even in flat space-time (Minkowski space; see figure 1-23 in §1.7) 2-surfaces which are *locally* trapped can easily be constructed. The simplest example is to take $\mathcal{S}$ as the intersection of two past light cones, with spacelike-separated vertices P and Q; see figure 3-11(c). Here the null normals to $\mathcal{S}$ are all converging into the future, towards either P or Q (and the reason that this defies our immediate intuitions about 2-surfaces in Euclidean 3-space is that this $\mathcal{S}$ cannot be contained within a single flat Euclidean 3-space, or "time-slice"). This particular $\mathcal{S}$ is not a trapped surface, however, because it is not a *closed* (i.e. compact; see §A.3) surface. (In some accounts, the term *closed trapped surface* is used for what I refer to simply as a trapped surface [see, for example, Hawking and Ellis 1973].) Thus, the condition on a space-time that it contain a trapped surface is indeed not a local one. The Oppenheimer–Snyder

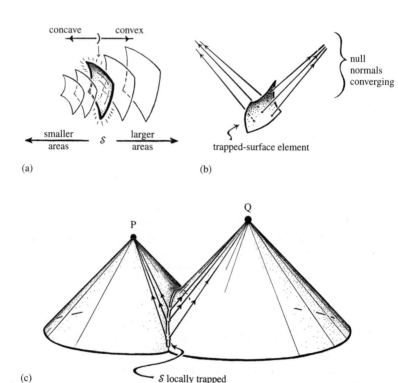

Figure 3-11: The trapped-surface condition. (a) In ordinary Euclidean 3-space, a flash of light occurring simultaneously over a curved 2-surface $\mathcal{S}$ would reduce in area if coming from the concave side and increase in area if coming from the convex side. (b) For any local patch $\mathcal{S}$ of a trapped surface, on the other hand, the convergence of the light rays occurs on *both sides*. (c) This "locally trapped" behaviour is not anomalous in space-time, for a non-compact $\mathcal{S}$, since it occurs already in Minkowski space in the intersection of two past light cones.

space-time contains actual (i.e. closed) trapped surfaces in the region *within* the Schwarzschild radius, after the collapse has taken place. And by the very nature of the trapped-surface condition, any reasonably small perturbation of initial data leading to such a collapse must also contain trapped surfaces, irrespective of any symmetry considerations. (Confusingly, this is, technically, what is called an *open* condition, meaning that sufficiently small changes do not violate the condition.)

The theorem [Penrose 1965a] I was able to establish in late 1964 showed, in essence, that when a trapped surface appears in a space-time it leads to a space-time singularity. A little more precisely, what is shown is that if a space-time (subject to certain physically reasonable restrictions that I shall come to in a moment) contains a trapped surface, then it cannot be extended indefinitely into the future. This inextendability is what signals the presence of a singularity. No demonstration of infinite curvatures or infinite densities is achieved by such a theorem, but it is hard to see, in general circumstances, what other type of obstruction there could be that prevents the evolution of the space-time into the future. There are other theoretical possibilities, but they do not occur in generic circumstances (i.e. they occur only with restricted functional freedom; see §§A.2 and A.8).

The theorem also depends on the assumption that Einstein's equations hold (with or without a cosmological constant $\Lambda$) with the energy source tensor **T** satisfying what is referred to as the *null energy condition* (asserting that for any null vector **n** the quantity obtained by contracting **n** twice into **T** is never negative).[1] This is a very weak requirement on the gravitational sources, and it holds true for any physically sensible classical material. The other assumption that I needed to make was that the space-time is taken to arise as an ordinary time evolution from some spatially unbounded initial state – i.e. technically, from a non-compact (i.e. "open" – see §A.5) initial spacelike 3-surface. This basically established that for local situations of gravitational collapse, once a trapped surface has occurred, singularities cannot be avoided, for physically reasonable classical material, quite irrespective of any assumptions of symmetry.

Of course, one can still question whether trapped surfaces are ever likely to arise in plausible astrophysical situations, particularly because one might take the view that for bodies that are even more concentrated than neutron stars, our understanding of the relevant particle physics at such enormous densities might well not suffice to provide a reliable picture of what would actually be going on. However, this is not the real issue, since other situations of gravitational collapse are expected to occur in which trapped surfaces arise with perfectly familiar densities. Basically, this is because of the way that general relativity behaves under an overall change of scale. If we have any space-time model whose metric is given by the tensor field **g** (see §§1.1 and 1.7), satisfying Einstein's equations with energy source tensor **T** (and cosmological constant $\Lambda$) and we replace **g** by $k$**g**, where $k$ is some constant positive number, then we find that the Einstein equations are again satisfied with energy tensor **T** (and with cosmological constant

---

[1] In index notation, this condition is $T_{ab}n^a n^b \geqslant 0$, whenever $n^a n_a = 0$. In some of my writings [see, for example, Penrose 1969a, p. 264], I have referred to this as the *weak energy condition*, which has confused some people, since Hawking and Ellis [1973] use *that* term in a different (stronger) sense.

$k^{-1}\Lambda$, but we can ignore this tiny contribution). The way that matter density $\rho$ is encoded in **T** implies[2] that $\rho$ must, accordingly, be replaced by $k^{-1}\rho$. Thus, if we have any collapse model in which a trapped surface occurs when the density reaches some particular value $\rho$, then we can obtain another model in which there is still a trapped surface but for which the density has as small a value as we desire, simply by scaling up the metric appropriately. If we have a collapse model in which trapped surfaces arise only when densities reach a certain extraordinarily high value (e.g. something far larger than the nuclear densities of neutron stars), then there will be another highly scaled-up model, in which distances are much greater – say, the scales of central galactic regions rather than neutron stars – for which the densities are no higher than those normally experienced here on Earth. This is indeed expected to be the case in the neighbourhood of the 4 million solar mass black hole that is believed to be located at the centre of our own Milky Way Galaxy. Certainly in the case of the quasar 3C 273, the average densities in the neighbourhood of the event horizon would be likely to be far smaller, and there should be no obstruction to the formation of trapped surfaces under these conditions.

For other points of view, concerning the formation of trapped surfaces, and for very mathematical treatments, see Schoen and Yau [1993] and Christodoulou [2009]. In Penrose [1969a, see figure 3 particularly], I give a simple intuitive argument that the effectively equivalent condition of a reconverging light cone could easily come about at comparatively low densities in a gravitational collapse. This alternative condition characterizing irreversible gravitational collapse (leading to a singularity) is one of those discussed mathematically in Hawking and Penrose [1970].

Since violent collapse processes of this kind can happen at a reasonably local level even in an expanding universe, it would be expected that processes of this general kind would certainly also be involved on a much larger scale, such as within a collapsing universe, when there are significant irregularities in the mass distribution of collapsing material. Accordingly, the above considerations would indeed apply also in the situation of the global collapse of an entire universe, singularities being a generic feature of gravitational collapse in classical general relativity. Indeed, early in 1965, the then young graduate student Stephen Hawking [1965] noticed that a standard FLRW model in its collapsing phase would also possess trapped surfaces but now enormous ones – on the scale of our entire

---

[2] In terms of the index notation, we have $\rho = T_{ab}t^a t^b$, where $t^a$ defines the observer's time direction, normalized according to $t^a t^b g_{ab} = 1$. Thus, $t^a$ scales with the factor $k^{-1/2}$, whence $\rho$ scales with the factor $k^{-1}$.

observable part of the universe – so we must again conclude that singularities are inevitable for a spatially open collapsing universe. (We need "open" because my 1965 theorem assumed a non-compact initial surface.) In fact, he phrased his argument in the reverse time direction, so that it would apply to the early stages of an expanding open universe – i.e. to a generally perturbed Big Bang – rather than the late stages of a collapsing one, but the message is basically the same: the introduction of irregularities into the standard symmetrical open cosmologies does not remove their singularities any more than it does in local collapse models [Hawking 1965]. In a succession of later papers [Hawking 1966a,b, 1967], Hawking was able to develop the techniques further, primarily so that the resulting theorems could apply globally to spatially closed universe models (in which case, the trapped-surface condition is not needed). Then in 1970 we combined forces to produce a very general theorem which included as special cases practically all the singularity results that we had before [Hawking and Penrose 1970].

How did the conclusions of Lifshitz and Khalatnikov, mentioned at the end of §3.1, square with all this? There would have been a serious inconsistency, but after hearing about the first singularity theorem referred to above (from work made known to them at the 1965 London international general relativity conference GR4), they, following an important input from Vladimir Belinskiĭ, were able (with Belinskiĭ) to correct an error in their earlier work and found that there were more general solutions than those they had found before. Their new conclusion was that singularities would, after all, arise in general collapse situations, in accordance with the conclusions that I (and subsequently Hawking) had come to. The detailed analysis that Belinskiĭ, Lifshitz, and Khalatnikov carried out led them to put forward a very complicated picture of what the general singularity might be like [Belinskiĭ et al. 1970, 1972]. This is now often referred to as the *BKL conjecture*. I shall refer to it here as the BKLM proposal, in view of the early influence from the work of the distinguished American general relativity theorist Charles W. Misner, who independently put forward a cosmological model exhibiting a singularity with these same complicated features a little before the Russians did [Misner 1969].

## 3.3. THE SECOND LAW OF THERMODYNAMICS

The upshot of §3.2 is, in essence, that we cannot resolve the space-time singularity issue within the framework of the equations of classical general relativity theory. For we have seen that the singularities are *not* merely special features of certain particular known exact symmetrical solutions of these equations; they will also occur in completely general situations of gravitational collapse. However, there still remains the possibility, raised towards the end of §3.1, that we might anticipate more success by appealing to the procedures of quantum mechanics. These procedures refer, basically, to some form of Schrödinger equation (see §§2.4, 2.7, and 2.12), where the relevant classical physical processes – here those arising from Einstein's notion of curved space-time, according to general relativity – would have to be quantized appropriately, according to some scheme of quantum gravity.

One key issue is that Schrödinger's equation shares with the equations of standard classical physics, including general relativity, the property of being *time-reversal symmetric*, and this should hold in any form of quantum gravity that follows standard procedures. Thus, whatever solution of the quantum equations we might have, we ought always to be able to construct another in which the parameter $t$ that is to represent the "time" is replaced by $-t$, and this should always give us another solution of the equations. It should be noted, however, that in the case of Schrödinger's equation (as opposed to standard classical equations), we must accompany this replacement by an interchanging of the imaginary units i and $-$i. That is, in time reversal, we must pass to the complex conjugates of all complex quantities that are involved. (If we are asking that this time measure $t$ refers to "time since the Big Bang", and demand that $t$ remain positive, then time symmetry would refer to the replacement of $t$ by $C - t$, where $C$ is some large positive constant.) In any case, time-reversal symmetry would be a clear expectation of any remotely conventional quantization procedures applied to gravitational theory.

Why is this time symmetry in the equations important and puzzling to us in our discussion of the space-time singularity problem? The central issue is that of the second law of thermodynamics – abbreviated here as simply "the 2nd Law". We shall find that this fundamental law is deeply and intimately connected with the nature of the singularities in space-time structure, and that it causes us to question whether standard quantum-mechanical procedures hold much hope for the full resolution of the singularity problem.

To obtain an intuitive understanding of the 2nd Law, imagine some familiar activity which would appear to be completely irreversible in time, such as the spilling of a cup of water, the water being then absorbed into the carpet below. We may think of this entirely in Newtonian terms, where the water molecules individually act according to the standard Newtonian dynamics, where particles are accelerated in accordance with forces acting between them and with the gravitational field of the Earth. At the level of individual particles, all the actions of the particles follow laws that are completely reversible in time. Yet if we try to imagine the actual situation of the spilling water reversed in time, we get the absurd-seeming picture of water molecules spontaneously removing themselves in a highly organized way from the carpet and flinging themselves upwards in an extraordinarily precise manner, so as to be collected all at once into the cup. This process is still completely consistent with Newton's laws (e.g. the energy required to raise the water molecules to the cup comes from their heat energy of random motions while in the carpet). But such a situation is never experienced in practice.

The way that physicists describe this macroscopic time asymmetry, despite the time symmetry inherent in all relevant submicroscopic actions, is through the notion of *entropy*, which, roughly speaking, is a measure of the *manifest disorder* of a system. The 2nd Law basically asserts that in all macroscopic physical processes, the entropy of a system increases with time (or, at least, does not decrease, apart from possible tiny fluctuations from this general trend). Thus, the 2nd Law appears to be merely asserting the familiar and seemingly rather depressing fact that, when left to themselves, things simply become more and more manifestly disordered as time progresses!

We shall be finding, shortly, that this interpretation somewhat overstates the negative aspects of the 2nd Law, and a closer examination of the matter leads us to a much more interesting and positive picture. Let us first try to be a little more accurate about the notion of the entropy of the state of a system. I should clarify this notion of *state* here, especially as it has rather little to do with the concept of a quantum state we encountered in §§2.4 and 2.5. What I am referring to here is what I shall call the *macroscopic state* of a (classical) physical system. In defining the macroscopic state of a particular system, we are not concerned with the fine details of where individual particles might be or how particular particles might be moving. We are concerned, instead, with averaged quantities, like the temperature distribution within a gas or fluid, and its density and general flow of motion. We are concerned with the general composition of the material in different locations, such as the various concentrations and motions of, say, nitrogen ($N_2$) or oxygen ($O_2$) molecules, or of $CO_2$ or $H_2O$, etc., or whatever might be the constituents of the system under consideration, but not the details of the individual locations or motions of these molecules. A knowledge of the values of all such macroscopic parameters would define the macroscopic state of the system. Admittedly this is a little vague, but what one finds in practice is that refinements in the choices of these macroscopic parameters (such as those arising from an improved technology of measurement) seem to make little difference to the resulting entropy value.

A point that often confuses people should be made clear here. In colloquial terms, we could say that the low entropy states, being "less random", are therefore "more organized", and therefore the 2nd Law is telling us that the organization in the system is continually becoming reduced. However, from another point of view, we could say that the organization in the high-entropy state that the system ends up in is just as organized as was the initial low-entropy state. The reason for this claim is that (with deterministic dynamical equations) the organization is never lost, because the final high-entropy state contains vast numbers of detailed correlations in the particle motions, these being of such a nature that if we were to reverse every motion exactly, then the entire system would work its way back until it reaches the "organized" initial state of low entropy. This is just a feature of dynamical determinism, and it tells us that simply referring to "organization" by itself gains us nothing in our understanding of entropy and the 2nd Law. The key point is that low entropy corresponds to *manifest* or *macroscopic* order, and that subtle correlations between the locations or motions of the submicroscopic ingredients (particles or atoms) are *not* things which contribute to the entropy of the system. This, indeed, is a central issue in the definition of entropy, and without such a term as *manifest* or *macroscopic* in the above descriptions of the entropy notion, we would not have been able to make any headway towards an understanding of entropy and of the physical content of the 2nd Law.

What, then is this entropy measure? Roughly speaking, what we do is to count all the different possible submicroscopic states that could go to make up the given macroscopic state, and the number $N$ of these provides a measure of the entropy of the macroscopic state. The larger $N$ turns out to be, the larger will be the entropy. However, it is not really reasonable to take something proportional to $N$ as our measure of entropy, essentially because we want some kind of quantity which behaves additively when two systems, independent of one another, are considered together. Thus, if $\Sigma_1$ and $\Sigma_2$ are two such independent systems, we would want the entropy $S_{12}$ of the two systems taken together to be equal to the sum $S_1 + S_2$ of the respective entropies $S_1$ and $S_2$:

$$S_{12} = S_1 + S_2.$$

However, the number of submicroscopic states $N_{12}$ that go to make up $\Sigma_1$ and $\Sigma_2$ together would be the product $N_1 N_2$ of the number $N_1$ that go to make up $\Sigma_1$ and the number $N_2$ that go to make up $\Sigma_2$ (since each of the $N_1$ ways of making up $\Sigma_1$ may be accompanied by any one of the $N_2$ ways of making up $\Sigma_2$). To convert the product $N_1 N_2$ to the sum $S_1 + S_2$, all we need to do is to use a logarithm in the definition of the entropy (§A.1):

$$S = k \log N,$$

where we choose some convenient constant $k$.

This is, indeed, essentially the famous definition of entropy given by the great Austrian physicist Ludwig Boltzmann in 1872, but there is one further point that needs to be made clear in this definition. In classical physics, the number $N$ is normally going to be *infinite*! In accordance with this issue, we really have to think of this "counting" in a rather different (and more continuous) way. In order to express this procedure succinctly, it is best that we return to the notion of *phase space*, which was introduced in its essentials in §2.11 (and is explained more fully in §A.6). We recall that the phase space $\mathcal{P}$, of some physical system, is a conceptual space, normally of a very large number of dimensions, each of whose points represents a complete description of the *submicroscopic state* of the (say, classical) physical system being considered, this state encompassing all the *motions* (as given by their momenta) as well as all the *positions* of all particles that constitute the system. As time evolves, the point P, in $\mathcal{P}$, which represents the submicroscopic state of the system, will describe a *curve* $\mathcal{C}$ in $\mathcal{P}$, whose location within $\mathcal{P}$ will be fixed by the dynamical equations, once the location within $\mathcal{P}$ of any particular (initial) point $P_0$, on $\mathcal{C}$, has been chosen. Any such point $P_0$ will fix which actual curve $\mathcal{C}$ gives us the time evolution of our particular system (see figure A-22 of §A.7) described by P (where $P_0$ describes the *initial* submicroscopic state of the system). This is the nature of the determinism that is central to classical physics.

Now, in order to define the entropy, we need to collect together – into a single region called a *coarse-graining region* – all those points in $\mathcal{P}$ which are considered to have the same values for their macroscopic parameters. In this way, the whole of $\mathcal{P}$ will be divided into such coarse-graining regions; see figure 3-12. (We may have to think of these regions as having fairly "fuzzy" boundaries, as there will always be slightly problematic issues of defining precisely where the boundaries of these coarse-graining regions actually are.) The points of $\mathcal{P}$ that lie in the neighbourhoods of such boundaries are normally considered to be an insignificant fraction of the total, and so can be ignored. (See §1.4 of Penrose [2010], especially figure 1.12.) Thus, the phase space $\mathcal{P}$ will be divided up into these regions, and we can think of the *volume V* of such a region as providing a measure of the number of different ways that different submicroscopic states can go to make up the particular macroscopic state defined by its coarse-graining region.

Fortunately, there is a natural $2n$-dimensional volume measure on phase space $\mathscr{P}$ determined by classical mechanics (see §A.6), for a system with $n$ degrees of freedom. Each position coordinate $x$ is accompanied by its corresponding momentum coordinate $p$, and the symplectic structure of $\mathscr{P}$ provides an area measure for each such coordinate pair, as indicated in figure A-21. When all coordinates are taken together we get the $2n$-dimensional Liouville measure referred to in §A.6. For a quantum system this $2n$-volume is a numerical multiple of $\hbar^n$ (see §§2.2 and 2.11). With a very large number of degrees of freedom in the system under consideration, this will be a very large-dimensional volume. However, the natural quantum-mechanical measure of volume allows us to compare volumes of different-dimensional phase spaces in a natural way (see §2.11). We are now in a position to provide Boltzmann's remarkable definition of the entropy $S$ of a macroscopic state as

$$S = k \log V,$$

where $V$ is the volume of the coarse-graining region defined, in $\mathscr{P}$, by the values of the macroscopic parameters that specify the state. The number $k$ is a fundamental constant, of tiny value $1.28 \times 10^{-23}$ J K$^{-1}$ (joules per kelvin), referred to as *Boltzmann's constant* (already encountered in §§2.2 and 2.11).

In order to see how this helps in our understanding of the 2nd Law, it is important to appreciate how stupendously different in size the various coarse-graining regions are likely to be, at least in the kind of situation that is normally encountered in practice. The logarithm in Boltzmann's formula, together with the smallness of $k$ in commonplace terms, tends to disguise the vastness of these volume differences (cf. §A.1), so that it is easy to overlook the fact that tiny entropy differences actually correspond to absolutely enormous differences in coarse-graining volumes. Let us think of a point P moving along a curve $\mathcal{C}$ in phase space $\mathscr{P}$, where P represents the (submicroscopic) state of some system that we are concerned with, and $\mathcal{C}$ describes its time evolution according to the dynamical equations. Let us suppose that P moves from one coarse-graining region $\mathcal{V}_1$ to a neighbouring one $\mathcal{V}_2$, with respective volumes $V_1$ and $V_2$ (see figure 3-12). As noted above, even with only slight differences in the entropies assigned to $\mathcal{V}_1$ and $\mathcal{V}_2$ the likelihood would be that one of their respective volumes, $V_1$, $V_2$, would be stupendously larger than the other. If $\mathcal{V}_1$ were the enormously larger of the regions, then there would be only a very minute proportion of its points for which the curves $\mathcal{C}$ leading from them go immediately into $\mathcal{V}_2$ (shown as $\mathcal{V}_2'$ in figure 3-12). The *largest* volume neighbouring $V_1$ would normally dwarf even the *total* of its other neighbours, and would be entered by almost every

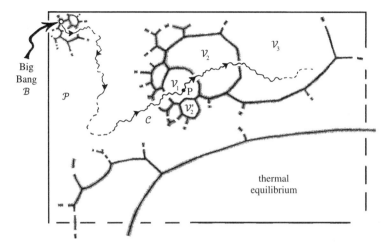

Figure 3-12: Phase space $\mathcal{P}$, the large-dimensional manifold whose points represents the entire (classical) state of a system (all positions and momenta; see figure A-20), is here indicated divided up into course-graining regions (with fuzzy boundaries), each of which groups together all states with the same macroscopic parameters (to some given degree of precision). The Boltzmann entropy assigned to P, in a course-graining region $\mathcal{V}$ of volume $V$, is $k \log V$. The second law of thermodynamics is understood as the tendency of the volumes to increase enormously along P's evolution curve $\mathcal{C}$ (see figure A-22), these vast volume differences being merely hinted at by the moderate size differences in the picture. Ultimately, the 2nd Law arises because $\mathcal{C}$ is constrained to have its origin in the extremely tiny region $\mathcal{B}$, representing the Big Bang.

evolution from $V_1$. Although the curve representing the time evolution of the (submicroscopic) state is guided through $\mathcal{P}$ by deterministic classical equations, these equations show little concern for the coarse graining, so we do not go far wrong if we treat this evolution as being effectively random, in relation to the coarse-graining regions. Accordingly, if $\mathcal{V}_1$ is indeed vastly larger than $\mathcal{V}_2$, we regard it as extremely unlikely that the future evolution of P, in $\mathcal{V}_1$, would be to find its way next into $\mathcal{V}_2$. On the other hand, if $\mathcal{V}_2$ were enormously larger than $\mathcal{V}_1$ (the case illustrated in figure 3-12), then it could be extremely probable that a $\mathcal{C}$-curve which originates $\mathcal{V}_1$ might next find itself in $\mathcal{V}_2$, and once lost in $\mathcal{V}_2$ it would be vastly more likely to find its way next into an even larger coarse-graining volume $\mathcal{V}_3$ than it would to return to a minutely tinier region like $\mathcal{V}_1$. Since the (vastly) larger volume corresponds to an (albeit usually only slightly) larger entropy, we see, in general rough terms why the expectation is for the entropy to increase with time in a relentless way. Indeed, this is just what the 2nd Law is telling us to expect.

However, this explanation is giving us only about half the story, and essentially only the easy half at that. It tells us, roughly, why it is – given that our system starts in a macroscopic state of comparatively low entropy – that the vast majority of submicroscopic states underlying that given macroscopic state will experience a steady increase in entropy (with perhaps the occasional dip, in a small fluctuation) as time increases. This entropy increase is what the 2nd Law tells us, and the rough argument just given provides us with some kind of justification for this entropy increase. Yet, on reflection, we may perceive something rather paradoxical about this deduction, as we appear to have deduced a time-asymmetrical conclusion concerning systems subject to completely time-symmetric dynamical laws. In fact, we have *not*. The time asymmetry arises simply from the fact that we have been asking a time-asymmetrical question of our system, namely that we have asked for the system's probable *future* behaviour, *given* its present macroscopic state, and in relation to this question we have arrived at a conclusion consistent with the time-asymmetrical 2nd Law.

But let us see what happens if we try to ask the time-reverse of this question. Suppose we have a macroscopic state of comparatively low entropy – say, our cup full of water, held high, but a little unstably, above the carpet. Let us now ask not for the most probable future behaviour of the water but for the probable way in which this state might have come about through its *past* activities. Consider two neighbouring coarse-graining regions $\mathcal{V}_1$ and $\mathcal{V}_2$, of phase space $\mathcal{P}$, just as before, but where we now consider that the given submicroscopic state is represented by the point P is now in $\mathcal{V}_2$ in figure 3-12. If $\mathcal{V}_2$ were enormously larger than $\mathcal{V}_1$, then there would be only an extremely tiny proportion of points within $\mathcal{V}_2$ giving locations for P for which the $\mathcal{C}$-curve could have entered from $\mathcal{V}_1$, whereas if $\mathcal{V}_1$ were the vastly larger of the two, then there would be hugely many possibilities for $\mathcal{C}$ to have entered $\mathcal{V}_2$ from $\mathcal{V}_1$. Thus, using the same kind of reasoning that had been successfully applied in the forward time direction, we appear to find that it would be far more probable for our point to have found its way into $\mathcal{V}_2$ via a coarse-graining region of vastly larger volume than via one of a vastly tinier volume, i.e. from one of a somewhat higher entropy than from one of somewhat smaller entropy. Repeating this argument farther and farther back in time, we conclude that by far the majority of routes to points within $\mathcal{V}_2$ will be by $\mathcal{C}$-curves that have higher entropy, and with higher and higher entropy still (with perhaps the occasional fluctuation), the farther back in time we proceed.

This, of course, is in direct contradiction with the 2nd Law, as we now appear to have deduced that, as we go back in time from the present situation, we are likely to find larger and larger entropies the farther back we go. In other words, given any situation of fairly low entropy, we should be overwhelmingly likely to find that the very *reverse* of the 2nd Law should have held in times prior to the present! This is clearly a nonsensical conclusion, if we are asking for behaviour that accords with experience, since all evidence points to there being nothing special about the present time, with regard to the 2nd Law, this law having been just as true of our universe in situations prior to the present as in situations in the future. Indeed, more than that, since all of our direct observational evidence on the matter obviously comes from the past, and it is what we see of physical behaviour in past time directions that gives us our belief in the 2nd Law. This observed behaviour seems to be in direct contradiction with what we have just theoretically deduced!

Take the example of our cup of water. We now ask for the most likely way that the water found its way into the cup, held high and rather unstably above the carpet below. The theoretical argument just given presents, as the kind of picture that would "most probably" have preceded this situation, a sequence of events in which the entropy was *reducing* as time moves forward (i.e. increasing backward in time), such as the water starting as dispersed throughout a patch in the carpet, then spontaneously collecting itself together, with the random motions of the fluid organizing themselves in such a way that they become projected coherently upwards in the direction of the cup, so that all the water settles simultaneously within the cup. This, of course, is very far from what would have actually happened in practice. What would have actually happened would have been a sequence of events with entropy *increasing* with time, fully consistent with the 2nd Law, such as the water being poured into the cup from a jug held by some person or, if we prefer to avoid reference to direct human intervention, by a tap turned on and off by some piece of automatic machinery.

So, what has gone wrong with our argument? Nothing is wrong if we are looking for the most probable sequence of events achieving the desired macroscopic state out of a completely random fluctuation. But this is not how things actually come about in the world that we know. The 2nd Law tells us to expect that the remote future will be very disordered, in a macroscopic way, and this represents no constraint that would invalidate our arguments about probable sequences of events into the future. But so long as the 2nd Law did indeed hold true at all times since the universe's inception, then the remote past must have been completely different, having been constrained to be extremely highly organized macroscopically.

If we incorporate into our probability assessments this single additional constraint on the initial macroscopic state of the universe – namely that it was one of *exceedingly tiny entropy* – then we have to reject the above reasoning about probable behaviour as we proceed into the past, since it falls foul of this constraint, and instead we can now accept a picture in which the 2nd Law does indeed hold true at all times.

The key to the 2nd Law, therefore, is the existence of an extraordinarily macroscopically organized initial state of the universe. But what was that state? As we have seen in §3.1, current theory – supported by the compelling observational evidence that we shall be coming to shortly in §3.4 – tells us that this was the gigantic all-encompassing explosion known as the *Big Bang*! How can it be that such an unimaginably violent explosion actually represented an exceptionally low-entropy incredibly macroscopically organized state? Let us come to this in the next section, where we shall seem to find an extraordinary paradox lurking within this singular event.

## 3.4. THE BIG BANG PARADOX

Let us first ask an observational question. What direct evidence tells us that there was indeed an immensely compressed and enormously hot state encompassing the entire observable universe, in a way that would have been consistent with the Big Bang picture of §3.1? Most convincing is the remarkable *cosmic microwave background* (CMB), sometimes referred to as the *flash of the Big Bang*. The CMB consists of electromagnetic radiation – i.e. light, but of a wavelength much too long to be visible to the eye – that comes in towards us from all directions, extremely uniformly (but basically incoherently). This is thermal radiation of a temperature of $\sim$2.725 K, which means merely about 2.7 degrees (on the centigrade or Celsius scale) above the absolute zero of temperature. In fact, the "flash" that is seen is considered to be of the enormously hot ($\sim$3000 K) universe at a time, some 379000 years later than the Big Bang itself, referred to as *decoupling*, when the universe first became fully transparent to electromagnetic radiation. (Though certainly not quite *at* the Big Bang, this event was at only around 1/40000 of what is considered to be the entire time span of the universe, right up to the present day.) The expansion of the universe since the time of decoupling has stretched out the wavelength of the light by a factor that corresponds to the amount by which the universe has expanded – a factor of about 1100 – so that its energy density

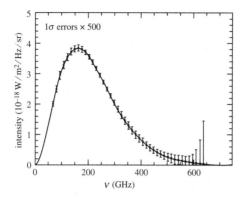

Figure 3-13: The extremely close fit of the CMB to a *thermal* (Planckian) spectrum (continuous line) as seen by COBE: the error bars in the CMB observations are exaggerated by a factor of 500.

has been correspondingly vastly reduced, whence the temperature we now see is only the 2.725 K that the CMB exhibits.

The fact that this radiation is essentially incoherent, or thermal, is impressively confirmed by the nature of its frequency spectrum, as shown in figure 3-13. The graph plots, vertically up the page, the intensity of the radiation for each particular frequency of the radiation, where the frequency increases off to the right. The continuous curve is the *Planck black-body curve* we encountered in §2.2 (figure 2-2) for a temperature of 2.725 K. The little marks along the curve are the actual observations, with error bars indicated. In fact, the error bars are exaggerated by a factor of 500, so the height of what would be the actual error bar could not be perceived by the eye, even at the far right, where the uncertainty is at its greatest. This agreement between observation and a theoretical curve is quite remarkable, undoubtedly the best fit to a thermal spectrum seen to occur naturally in the external world.[3]

But what is this agreement telling us? It is telling us that what we are looking at appears to be a state that is extraordinarily close to thermal equilibrium (and this is what the above term *incoherent* refers to). But what would it actually mean to say that this very early universe state was in thermal equilibrium? The reader

---

[3] It is often asserted that the CMB provides the best agreement between an observed phenomenon and the Planck spectrum. However, this is misleading, because COBE merely compares the CMB spectrum with an artificially produced thermal one, so the actual CMB spectrum is only established to be as Planckian as that artificial one.

is referred back to figure 3-12 in §3.3. This largest coarse-graining region, so labelled, would normally be far larger than any other coarse-graining region – and in usual circumstances is so large, compared with the others, that it would greatly exceed the total volume of them all! Thermal equilibrium represents the macroscopic state that one envisages a system finally settling into, sometimes referred to as the *heat death of the universe* – though here, puzzlingly, it seems to refer to a *heat birth* for the universe. There is a complicating factor that the early universe was rapidly expanding, so we are not looking at a state that is actually in equilibrium. Nevertheless, the expansion can here be regarded as being essentially *adiabatic* – a point fully appreciated by Tolman, way back in 1934 [Tolman 1934] – which tells us that the entropy is not changing during the expansion. (In a situation such as in the case here, where there is adiabatic expansion preserving thermal equilibrium, this would be described in phase space by a family of coarse-graining regions of equal volume, each merely being labelled by a different size of universe. It is indeed appropriate to think of this early state as being one of essentially *maximum entropy*, despite the expansion!)

We seem to have been presented with an extraordinary paradox. The argument given in §3.3 tells us that the 2nd Law requires – and is basically explained by – the Big Bang being a macroscopic state of extraordinarily *small* entropy. Yet the CMB evidence seems to be telling us that the macroscopic state of the Big Bang had an enormous entropy, even equal to the maximum among all possibilities. So what is it that has gone so seriously wrong?

One form of explanation of this paradox, often suggested, is to take the view that since the very early universe would have been exceedingly "small", there must be some kind of "ceiling" on possible entropies, and that the state of thermal equilibrium that apparently held in these early stages had simply the largest entropy that was available at that time. However, this is *not* the correct answer. Such a picture might indeed be appropriate for a completely different kind of situation in which the size of the universe is determined by some external constraint, such as with a gas in a cylinder closed by some airtight piston, where the degree of compression imposed by the piston is governed by some external mechanism, there being an external energy source (or sink). But this is not the situation in the case of the universe as a whole, where its geometry and energy, including its overall dimension, is entirely "internally" governed, through Einstein's dynamical equations of general relativity (including the equations of state for the matter; see §§3.1 and 3.2). In such circumstances (the equations being fully deterministic and invariant under reversal of the direction of time – see §3.3), there can be no change in the overall volume of phase space as time evolves. Indeed, the phase

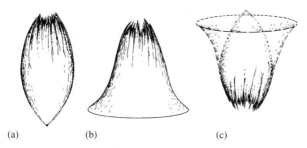

Figure 3-14: (a) A generically perturbed Friedmann model with $K > 0$, $\Lambda = 0$ (in contrast with figure 3-6(a)), acting in accordance with the 2nd Law would be expected to collapse via the congealing of numerous black holes, resulting in an extremely messy singularity, very unlike that of FLRW. (b) Similar behaviour, expected for a generically perturbed collapsing model. (c) The time-reverse of these situations, as expected for a generic big bang.

space $\mathcal{P}$, after all, is not *itself* supposed to "evolve"! All evolution is simply described by the location of the $\mathcal{C}$-curve within $\mathcal{P}$, in this case representing the entire evolution of the universe (see §3.3).

The issue is perhaps made clearer if we contemplate the late stages of a universe model in *collapse*, when it approaches its big crunch. Recall the Friedmann model for $K > 0$, $\Lambda = 0$, as illustrated in figure 3-2(a) in §3.1. We now take this model to be perturbed by irregular matter distributions, some of which have ultimately undergone individual collapses to black holes. Then we must consider that some of these black holes will eventually merge with one another, and that the collapse to a final singularity will be something exceedingly complicated, with little resemblance to the highly symmetrical big crunch of the exact spherically symmetrical Friedmann model depicted in figure 3-6(a). The collapse situation would, instead, be qualitatively much more like the great mess sketched in figure 3-14(a), in which a final singularity might well be something in accord with the BKLM proposal referred to at the end of §3.2. The final collapsed state would be one of enormous entropy, despite the fact that the universe has again reached a very tiny scale. Although this particular (spatially closed) recollapsing Friedmann model is *not* now considered to be a very plausible candidate for modelling our own universe, the same considerations would apply to any other of the Friedmann models, with or without cosmological constant. The collapsing version of each such model, similarly perturbed by irregular matter distributions, must be again expected to result in an all-consuming mess of the black-hole type of singularity (figure 3-14(b)). Time-reversing each of these states, we find a possible initial singularity (a possible big bang) with a correspondingly *enormous* entropy, in contradiction with the ceiling proposal being suggested here (figure 3-14(c)).

At this point I should mention some alternative possibilities that are sometimes suggested. Some theorists have proposed that the 2nd Law might somehow have to *reverse* in such collapse models, the universe's total entropy becoming smaller and smaller (following the state of maximal expansion) as its big crunch is approached. However, such a picture is particularly difficult to maintain in the presence of black holes which, once formed, will by themselves define a direction of entropy *increase* (owing to the temporal asymmetry of the null-cone arrangement at the horizon, shown in figure 3-9), at least until the hugely remote times of their disappearance due to Hawking evaporation; see §§3.7 and 4.3. In any case, this kind of possibility does not nullify the argument presented in the text. As another issue of relevance, some readers might concern themselves that in such complicated collapse models, the black-hole singularities might well be initiated at very different times, so that their time-reverses could not be thought of as constituting a big bang that goes off "all at once". However, it is a feature of the (generally believed, though yet unproven) *strong cosmic censorship hypothesis* [Penrose 1998a, TRtR, §28.8] that in the general case, such a singularity would be *spacelike* (§1.7), so can indeed actually be regarded as a simultaneous event. Moreover, quite apart from the issue of the general truth of strong cosmic censorship, there are many solutions that are known that satisfy this condition, and all such possibilities, in expanding form, would represent relatively high-entropy alternatives. This, in itself, would very much weaken the significance of this worry.

Thus we find no evidence of a low ceiling to the universe's entropy which would necessarily have to be present on account of the universe's small spatial dimension. In a general way, the accumulation of matter into black holes, and the congealing of these black-hole singularities into one final singular mess represents a process that is very much consistent with the 2nd Law, and an enormous entropy increase would be expected to accompany this final process. The geometrically "tiny" final state of the universe can indeed have an enormous entropy, far larger than that of the earlier stages of such a collapsing universe model, the spatial tininess representing, in itself, no ceiling on the entropy that one might have attempted to use, in time-reversed form, as a reason for the Big Bang being of extremely small entropy. In fact, this picture (figure 3-14(a),(b)) of a generic collapsing universe provides the key to resolving the paradox of how the Big Bang can actually have been of an extraordinarily low entropy – compared with what it might have been –

despite its appearance of having been in a *thermal* (i.e. maximum-entropy) state. The answer lies in the fact that there can be an enormous entropy gain once we allow significant deviations from spatial uniformity, the greatest gain resulting from those irregularities leading to black holes. A spatially uniform Big Bang, therefore, can indeed be of hugely low entropy, relatively speaking, despite the thermal nature of its contents.

One of the most impressive pieces of evidence for the Big Bang actually being of a spatially rather uniform nature, closely consistent with the geometry of an FLRW model (and inconsistent with the much more general messy type of singularity sketched in figure 3-14(c)), comes again from the CMB, but this time from its angular uniformity rather than from its closely thermal nature. This uniformity shows up in the fact that the temperature of the CMB is very nearly exactly equal in all directions in the sky, with deviations from uniformity occurring only at a proportional level of around $10^{-5}$ (once we have corrected for a small Doppler effect resulting from our own motion through the ambient material). In addition, there is a fairly general regularity in the distribution of galaxies and other material, so the spread of baryons (see §1.3), over very large scales, has a good deal of uniformity, though there are some noteworthy irregularities, such as vast so-called voids, where the visible matter density is enormously lower than the overall average. In general terms, one may say that the regularity appears to be the greater the earlier in the universe's history we look back to, and the CMB provides evidence for the earliest matter distribution that we can observe directly.

This picture is consistent with a viewpoint that the very early universe was indeed extremely uniform, but with very slight irregularities in the density. Over the passage of time (and helped by "frictional" processes of various kinds, which tend to slow down relative motions), these density irregularities became gravitationally enhanced, consistently with a picture in which clumping of material gradually increased with time to produce stars, these being gathered in galaxies, with massive black holes in galactic centres, this clumping being ultimately driven by relentless gravitational influences. This indeed would have represented a vast entropy increase, illustrating that, when gravity is brought into the picture, the primordial fireball that is evidenced by the CMB must actually have been very far from a maximum-entropy state. The thermal nature of this fireball that is presented by the Planckian spectrum shown in figure 3-13 tells us only that if we consider the universe (at decoupling time) to be just a system consisting of matter and radiation in interaction, then it could be regarded as being essentially in thermal equilibrium. But when gravitational influences are brought into consideration, the picture changes dramatically.

gas in a box

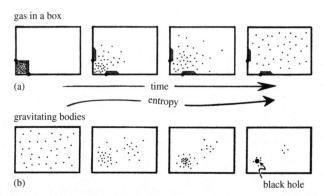

(a)                                time
                                  entropy

gravitating bodies

(b)                                                    black hole

Figure 3-15:  (a) For molecules of gas in a box, spatial uniformity is obtained with maximum entropy; (b) for gravitating stars in a galactic-scale "box", large entropy is achieved by clumped states, ultimately with a black hole.

If we imagine a gas in a sealed box, for example, then it is natural to think of its maximum entropy as being achieved by a macroscopic state in which the gas is spread uniformly throughout the box (figure 3-15(a)). In this respect, it would resemble the fireball that produced the CMB, spread uniformly over the sky. But if we replace the molecules in our gas by a vast system of gravitating bodies such as individual stars, then we get a very different picture (figure 3-15(b)). Gravitational effects would cause the distribution of stars to become irregular and clumpy. Eventually, an enormous increase in entropy would be achieved when many of the stars collapse or congeal into black holes. Although this might take a long time (though helped along by the frictional presence of gas among the stars), we see that, with gravity ultimately having a dominating presence, there is much gain in the entropy obtained by moving *away* from the uniform distribution.

We see the effects of this even at the level of everyday experience. One may ask how the 2nd Law operates in relation to the sustaining of life on Earth. It is often said that we survive on this planet by getting energy from the Sun. But this is not a properly accurate statement of the situation when we think of the Earth as a whole, because basically all the energy that the Earth receives in the daytime is returned before long into space again, into the dark sky of the night. (Of course there will be slight corrections to an exact balance, coming from global warming and radioactive heating inside the Earth, etc.) Otherwise the Earth would simply get hotter and hotter and become uninhabitable in a few days! However, the photons that we get directly from the Sun are of relatively high frequency (being roughly in the

Figure 3-16: Life on Earth is maintained by the great temperature imbalance in our sky. Incoming low-entropy energy from the Sun, in relatively fewer higher-frequency (∼yellow) incoming photons, is converted by the green plants to far more numerous lower-frequency (infrared) outgoing photons, removing an equal energy from the Earth in high-entropy form. By this means, plants, and thence other terrestrial life, can build up and maintain their structure.

yellow part of the spectrum), whereas those returned to space are the much lower-frequency infrared photons. By Planck's $E = h\nu$ (see §2.2), the incoming ones are *individually* of much higher energy than those returning to space, so there must be many fewer coming in to the Earth than going out for the balance to be achieved (see figure 3-16). Fewer photons coming in mean fewer degrees of freedom for the incoming energy and more for the outgoing energy, and therefore (by Boltzmann's $S = k \log V$) the photons coming in have much lower entropy than those going out. The green plants take advantage of this and use the low-entropy incoming energy to build up their substance, while emitting high-entropy energy. We take advantage of the low-entropy energy in the plants, to keep our own entropy down, as we eat plants, or as we eat animals that eat plants. By this means, life on Earth can survive and flourish. (These points were apparently first clearly made by Erwin Schrödinger in his ground-breaking 1967 book, *What Is Life* [Schrödinger 2012].)

The crucial fact, in this low-entropy balance, is that the Sun is a hot spot in an otherwise dark sky. But how did that come about? Many complicated processes, such as those involving thermonuclear reactions, etc., are involved in the overall picture, but the key point is that the Sun is there at all – and this came about because the material of the Sun (as with other stars) evolved through the process of gravitational clumping, from the relatively uniform initial distribution of gas and dark matter.

The mysterious substance known as *dark matter* needs to be mentioned here, as it apparently constitutes some 85% of the material (non-$\Lambda$) content of the universe, but it is only detectable through its gravitational effects, its exact constitution being unknown. For our current considerations it merely affects the total mass value, somewhat, that will come into some of the numerical quantities involved (see §§3.6, 3.7, and 3.9; but for a possibly more significant theoretical role for dark matter, see §4.3). Irrespective of the dark matter issue, we can see how crucial to our current existence was the low-entropy nature of this initially uniform distribution of matter. Our existence, as we know it, depends on the low-entropy gravitational reservoir inherent in the initially uniform matter distribution.

This leads us to consider a remarkable – indeed *fantastical* – thing about the Big Bang. It is not merely the mystery of its very occurrence, but that it was an event of extraordinarily low entropy. Moreover, the remarkable thing is not merely that but the fact that the entropy was low in a very particular way, and apparently *only* in that way, namely that the *gravitational* degrees of freedom were, for some reason, *completely suppressed*. This is in stark contrast with the matter degrees of freedom and those of (electromagnetic) radiation, as they appear to have been maximally excited in the form of a thermal, maximal entropy state. To my mind, this is perhaps the most profound mystery of cosmology and, for some reason, it is a still largely unappreciated mystery!

We shall need to be more specific about how special this Big Bang state was, and how much entropy can be gained by this process of gravitational clumping. Accordingly, we shall need to come to terms with the enormous measure of entropy that is actually involved in black holes (figure 3-15(b)). We shall be coming to this in §3.6. But in the meantime, it will be necessary to address another issue, namely that arising from the quite likely possibility that the universe is actually spatially *infinite* (as would be the case with the FLRW models with $K \leqslant 0$ – see §3.1), or at least that most of it might lie beyond the scope of direct observation. Accordingly, we shall need to confront the issue of *cosmological horizons*, which we shall come to in the next section.

## 3.5. HORIZONS, COMOVING VOLUMES, AND CONFORMAL DIAGRAMS

Before we can enter into a more precise measure of the degree to which our Big Bang was special, among the totality of all possible space-time geometries and matter distributions, we need to confront the distinct possibility that many models with infinite spatial geometry would have an *infinite total entropy*, and this provides us with a complication. Yet the gist of the argument presented above is not really seriously affected if we consider not the total entropy of the universe, but something like the entropy per comoving volume. The idea of a *comoving region*, in an FLRW model, is that we consider a spatial region evolving with time, whose boundary follows the *time-lines* of the model (the world-lines of the idealized galaxies; see figure 3-5 in §3.1). Of course, when black holes are under consideration – which, as we shall be seeing in the next section, provide the crucial input to the entropy issue – and significant deviations from an exact FLRW form take place, it may not be so clear what the notion of a "comoving volume" should actually mean. However, when we consider things on a large-enough scale, this uncertainty becomes relatively unimportant.

It will be useful, in what follows later, to consider what does happen in the exact FLRW models on a very large scale. In all those FLRW models I have been discussing in this chapter, there is a notion of what is referred to as a *particle horizon*, a notion first clearly defined by Wolfgang Rindler in 1956 [Rindler 1956]. To obtain its usual definition, we consider some point P in the space-time and examine its past light cone $\mathcal{K}$. A good many of the time-lines (see §3.1) will intersect $\mathcal{K}$, and it may be considered that the portion $\mathcal{G}(P)$ of the space-time swept out by these time-lines constitutes P's family of *observable galaxies*. But some time-lines may be too distant from P to intersect $\mathcal{K}$, providing a boundary $\mathcal{H}(P)$ to $\mathcal{G}(P)$ which is a timelike hypersurface ruled by time-lines. This 3-surface $\mathcal{H}(P)$ is the *particle horizon* of P (figure 3-17).

For any particular cosmic time $t$, the slice of $\mathcal{G}(P)$ through the universe model given by that constant value of $t$ will have a finite volume, and the maximum entropy value for that region would be finite. If we consider the *entire* time-line $l_P$ through P, then the region $\mathcal{G}(P)$ of observable galaxies would be likely to get larger the farther into the future along $l_P$ we take P to be. For the standard FLRW models with $\Lambda > 0$, there will be a limiting "largest" region $\mathcal{G}(l_P)$ which is spatially finite for any fixed value of the cosmic time $t$, and we may consider that the maximum entropy attainable within $\mathcal{G}(l_P)$ is all that we need consider for the above arguments. For some purposes, it is helpful to also consider another type of cosmological horizon (also first clearly defined by Rindler [1956]), namely the *event horizon* of an infinitely future-extended world-line such as $l_P$, which is the (future) boundary of the set of points lying to the past of $l_P$. This ties in with the event horizon occurring in the normal picture of collapse to a black hole, where now $l_P$ would be replaced by the world-line of a remote external (eternal) observer who does not fall into the black hole (see figure 3-9 in §3.2).

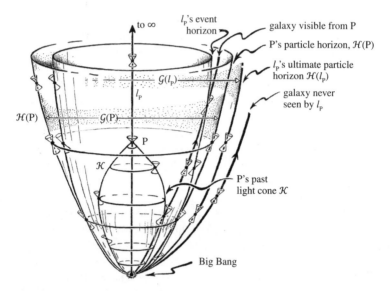

Figure 3-17: Space-time picture of an FLRW model, indicating various types of horizon and galaxy world-lines.

The issues raised by these horizons are often quite confusing in pictures such as those of figures 3-2 and 3-3 in §3.1 – see figure 3-17 – and compare also figure 3-9 in §3.2. We get a much clearer understanding of these notions if we display them by use of *conformal diagrams* [Penrose 1963, 1964a, 1965b, 1967b, TRtR, §27.12; Carter 1966]. A particularly useful feature of these diagrams is that they frequently allow us to represent *infinity* as a finite boundary to the space-time. We have already seen this aspect of conformal pictures in figures 1-38(a) and 1-40 in §1.15. Another feature of such diagrams is to make more transparent the causal aspects (e.g. particle horizons) of the big-bang *singularities* of FLRW models.

These pictures employ a conformal rescaling of the metric tensor **g** (see §§1.1, 1.7, and 1.8) of the physical space-time $\mathcal{M}$ to provide a new metric $\hat{\mathbf{g}}$ for a conformally related space-time $\hat{\mathcal{M}}$, according to

$$\hat{\mathbf{g}} = \Omega^2 \mathbf{g},$$

where $\Omega$ is a (generally positive) smoothly varying scalar quantity on the space-time, so the null cones (and local time direction) are unchanged when **g** is replaced by $\hat{\mathbf{g}}$. Under quite general circumstances $\hat{\mathcal{M}}$, with its smooth metric $\hat{\mathbf{g}}$, indeed acquires a smooth boundary, at which $\Omega = 0$, representing *infinity* for the original space-time $\mathcal{M}$. The value $\Omega = 0$ represents an infinite "squashing down" of $\mathbf{g}$ in the infinite regions of $\mathcal{M}$ to provide a finite boundary region $\mathscr{I}$ for $\hat{\mathcal{M}}$. Of course this procedure will only provide us with a smooth boundary for $\hat{\mathcal{M}}$ under suitable circumstances of fall-off for the metric (and perhaps the topology) of $\mathcal{M}$, but it is rather remarkable how well this procedure does work for space-times $\mathcal{M}$ of particular physical interest.

Complementing this procedure for the representation of space-time infinity is a related one according to which, under suitable circumstances, we may infinitely "expand out" a *singularity* in the metric of $\mathcal{M}$ so as to obtain a boundary region $\mathscr{B}$ for $\hat{\mathcal{M}}$ representing this singularity. With a cosmological model $\mathcal{M}$, we may well obtain a smoothly adjoined boundary region $\mathscr{B}$, for $\hat{\mathcal{M}}$, to represent the Big Bang. We may also be fortunate enough that the inverse scale factor $\Omega^{-1}$ smoothly approaches zero as $\mathscr{B}$ is reached, and this is indeed the case for the big bang of the most important FLRW cosmologies considered here. (Note: the capitalized form "Big Bang" is reserved for that specific singular event that appears to have initiated the universe we know; "big bang", on the other hand, refers to initial singularities of cosmological models generally; see also §4.3.) For the Tolman radiation-filled models, $\Omega^{-1}$ has a simple zero at the boundary, but for the Friedmann dust-filled models, it has a double zero. Having $\mathcal{M}$'s infinite future and its singular origin both represented as smooth boundary regions attached to the conformally related $\hat{\mathcal{M}}$, we can get a good picture of the types of horizon referred to above.

A common convention adopted with such conformal space-time pictures is to have the null cones pointing upwards and (normally) with their surfaces sloping roughly, and as far as is possible, at 45° to the vertical. This is illustrated in figures 3-18 and 3-19, which, like figure 1-43 in §1.15, are examples of a *schematic* conformal diagram, which are *qualitative* pictures, where one tries to arrange things so that the slopes of the null cones are all more-or-less at 45° to the vertical. (We can also imagine a schematic conformal diagram representing a fully perturbed model universe, containing many black holes.) When there is a positive cosmological constant, $\mathscr{I}$ turns out to be *spacelike* [Penrose 1965b; Penrose and Rindler 1986], and this implies that the region contained within the event horizon of any world-line is spatially finite for any given (cosmic) time. To be consistent with current observations, but without incorporating a (conventionally believed) *inflationary* phase to the very early universe (see §3.9), the point P in figure 3-18 would be around three-quarters of the way up the line $l_{\mathrm{P}}$. This would be the case if we take the future evolution of the universe to be consistent with Einstein's equations with the observed value for $\Lambda$ (assumed constant) and use the observed matter content [Tod 2012; Nelson

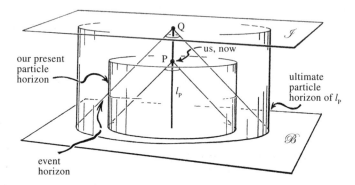

Figure 3-18: This schematic conformal diagram indicates the universe's entire history according to current theory, although drawn without the inflationary phase commonly believed to have taken place almost immediately following the Big Bang (see §3.9). Without inflation, our current time location P would be around three-quarters of the way up the diagram (roughly as shown); with inflation, the overall picture would be qualitatively similar, but P would be almost at the very top of the picture, just below where Q is located.

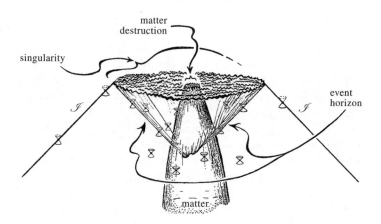

Figure 3-19: Schematic conformal picture of a collapse to a black hole, as in §3.9, but spherical symmetry need not be assumed. Note that the (irregular) singularity is drawn spacelike, in accordance with strong cosmic censorship.

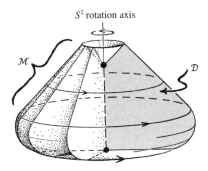

Figure 3-20: Strict conformal diagrams describe space-times with spherical symmetry. A planar region $\mathcal{D}$ is $S^2$-rotated to provide the required 4-dimensional space-time $\mathcal{M}$. Each point of $\mathcal{D}$ represents ("sweeps out") a sphere $S^2$ in $\mathcal{M}$, except that any point on an *axis*, a vertical broken-line boundary of $\mathcal{D}$ or else a black spot, represents a single point of $\mathcal{M}$.

and Wilson-Ewing 2011]. If, in addition, we incorporate an inflationary phase, the picture would be qualitatively similar to that of figure 3-18 but the point P would be almost at the top of $l_P$, only just below its terminal point Q. The two past cones in the picture would then almost coincide. (See also §4.3.)

Figure 3-19 is a schematic conformal diagram representing a (not necessarily spherically symmetrical) gravitational collapse. Some null cones are shown, and future infinity $\mathscr{I}$ is seen to be null, the picture describing an asymptotically flat space-time, with $\Lambda = 0$. For $\Lambda > 0$, the picture would be basically similar, but with a spacelike $\mathscr{I}$, as in figure 3-18.

When considering space-times possessing spherical symmetry (such as with FLRW models of figures 3-2 and 3-3 of §3.1, or the Oppenheimer–Snyder collapse to a black hole of figure 3-9 in §3.2), we may obtain more precision and compactness with a *strict* conformal diagram (as basically formulated by Brandon Carter in his 1966 PhD thesis [Carter 1966]). This is a flat plane figure, $\mathcal{D}$ depicting a region bounded by lines (representing infinite regions, singularities, or symmetry axes), where each internal point of $\mathcal{D}$ is to be thought of as representing an ordinary (spacelike) 2-sphere ($S^2$), so we may think of the entire space-time $\hat{\mathcal{M}}$ as being swept out by "rotating" the region about its symmetry axis (or sometimes *axes*). See figure 3-20. The *null* directions within $\mathcal{D}$ are to be all depicted as being at 45° to the vertical; see figure 3-21. In this way, a very good conformal picture of the desired space-time $\mathcal{M}$ can be obtained, extended to $\hat{\mathcal{M}}$ by the attachment of its conformal boundary.

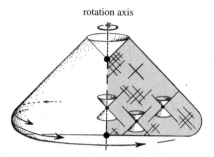

Figure 3-21:  In a strict conformal diagram, null directions within $\mathcal{D}$ are aligned at 45° to the vertical, these being the intersections with $\mathcal{D}$ of the null cones in $\mathcal{M}$.

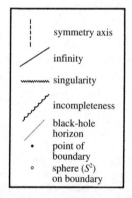

Figure 3-22:  Standard conventions for strict conformal diagrams.

In our visualizations, it is useful to think in terms of a *3-dimensional* $\mathcal{M}$, constructed by rotating $\mathcal{D}$ in a circular motion ($S^1$) about a vertical (i.e. timelike) rotation axis. However, we must bear in mind that to obtain the full 4-dimensional space-time, we must imagine this rotation being actually carried out in accordance with a *2-dimensional* spherical ($S^2$) action. Occasionally – when we consider models which have spatial sections that are *3-spheres* ($S^3$) – we need to consider cases when there are *two* rotation axes, and this is somewhat harder to visualize! There are various conventions that are useful to adopt concerning strict conformal diagrams, these being depicted in figure 3-22.

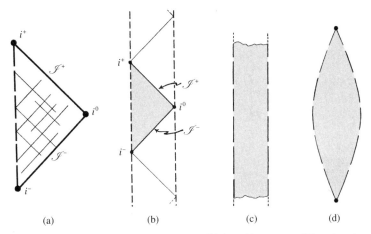

Figure 3-23: Strict conformal diagrams for Minkowski space and its extensions: (a) Minkowski space; (b) Minkowski space as part of a vertical sequence of Minkowski space diagrams that go to make up the Einstein universe $\mathcal{E}$; (c) the Einstein universe $\mathcal{E}$ with topology $\mathbb{R}^1 \times S^3$; (d) $\mathcal{E}$ again, with black spots to show its points at future and past infinities.

In figure 3-23(a), Minkowski 4-space with its conformal boundary (see figure 1-40 in §1.15) is represented as a strict conformal diagram. In figure 3-23(b), this is seen as a portion of Einstein's ($S^3 \times \mathbb{R}$) static model universe (see figure 1-43 in §1.15), in accordance with figure 1-43(b). Einstein's model (as described in figure 1-42) is itself represented as a strict conformal diagram in figure 3-23(c) (where we note the above-mentioned use of *two* rotation axes to generate the $S^3$), or else as in figure 3-23(d) if we wish to include its (conformally singular) boundary points at past and future infinity.

Many other universe models can also be so represented by strict conformal diagrams. In figure 3-24, I have drawn such diagrams for the three $\Lambda = 0$ Friedmann models (sketched previously in figure 3-2 in §3.1) and, in figure 3-25, we find conformal diagrams of those with adequately large $\Lambda > 0$ (collectively indicated in figure 3-3 in §3.1). The strict conformal diagram for de Sitter 4-space (recall figure 3-4(a)) is shown in figure 3-26(a), where figure 3-26(b) depicts that portion which is the old steady-state model of Bondi, Gold, and Hoyle (see §3.2). Parts (c) and (d) of figure 3-26 are strict conformal diagrams for the wrapped and unwrapped anti–de Sitter spaces $\mathcal{A}^4$ and $\Upsilon \mathcal{A}^4$ (compare §1.15). In figure 3-27

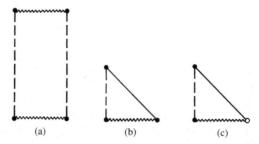

Figure 3-24: Strict conformal diagrams of the $\Lambda = 0$ Friedmann dust-filled models of figure 3-2: (a) $K > 0$, (b) $K = 0$, (c) $K < 0$, where the spatial $S^2$-conformal infinity of hyperbolic geometry (in accordance with Beltrami's representation in figures 3-1(c) and 1-38(b)) is captured by the open spot at the right.

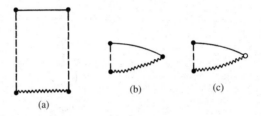

Figure 3-25: Strict conformal diagrams of the $\Lambda > 0$ Friedmann models, illustrating a spacelike future infinity $\mathscr{I}$: (a) $K > 0$, with large enough $\Lambda$, so that there is an eventual exponential expansion; (b) $K = 0$; (c) $K < 0$.

we see a strict conformal diagram of the earlier picture figure 3-17, and the roles of the various horizons are now brought out much more clearly than before.

The Schwarzschild solution (see §3.2), in its original form, terminating at its Schwarzschild radius, is shown in the strict conformal diagram figure 3-28(a). In figure 3-28(b), we find the extension through its event horizon, in accordance with the metric form known as the *Eddington–Finkelstein form*, referred to in §3.2, describing the space-time of a black hole. Figure 3-28(c) shows the *maximally extended* Synge–Kruskal form of the Schwarzschild solution, found originally by John Lighton Synge in 1950 [Synge 1950] and by some others about a decade later [see particularly Kruskal 1960; Szekeres 1960]. Figure 3-29(a) depicts the Oppenheimer–Snyder collapse to a black hole of figure 3-9 in a strict conformal diagram, and in figure 3-29(b),(c) we see how to construct this space-time by piecing together portions of figure 3-24(b) (time-reversed) and figure 3-28(b).

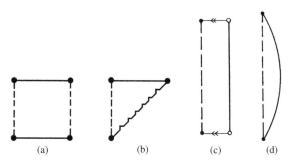

Figure 3-26: Strict conformal diagrams of (a) entire de Sitter space, (b) portion of de Sitter space describing the steady-state model (see figure 3-4(c)), (c) anti–de Sitter space $\mathcal{A}^4$, where the top edge is identified with the bottom edge, yielding a cylinder, (d) unwrapped anti–de Sitter space $\mathcal{Y}\mathcal{A}^4$ (see §1.15). The diagrams for $\mathcal{A}^5$ and $\mathcal{Y}\mathcal{A}^5$ are the same as (c) and (d) here, but where the rotation is through $S^3$, rather than $S^2$.

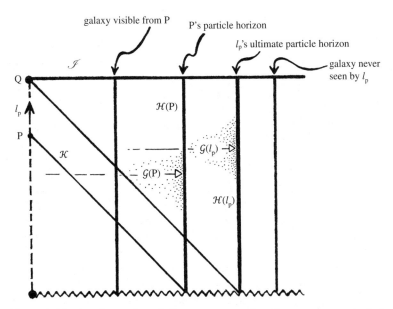

Figure 3-27: A strict conformal diagram representing the same cosmological features as shown in figure 3-17, but with greater clarity.

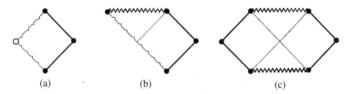

Figure 3-28: Strict conformal diagrams for the Schwarzschild metric and its extensions: (a) original Schwarzschild space-time; (b) Eddington–Finkelstein extension through upper horizon; (c) Synge–Kruskal–Szekeres maximal extension.

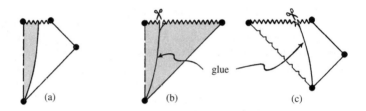

Figure 3-29: (a) Strict conformal diagram of Oppenheimer–Snyder collapse to a black hole, obtained by gluing a portion (b) of the time-reverse of Friedmann picture (figure 3-24(b)) to a portion (c) of Eddington–Finkelstein picture (figure 3-28(b)). Matter-filled regions (dust) are shaded.

I shall need to return to the issue of main concern for us in the previous section §3.4, namely the apparent paradox that whereas we know from the 2nd Law that the Big Bang *must* have been a state of extraordinarily low entropy, our direct evidence of that event, as signalled by the impressively *thermal* nature of the CMB, was of a state of apparently *maximum* entropy. As we have seen in §3.4, the key to the resolution of the paradox lies in potential for spatial irregularity, and in the nature of the collapse to a singularity that would be expected in a spatially irregular collapsing universe, as sketched in figure 3-14(a),(b), where the converging of many black-hole singularities would result in an overriding singularity of extreme complication. It is hard to illustrate such a messy crunch in a schematic conformal diagram, particularly if strong cosmic censorship (see §3.4) is expected to remain true (as seems rather probable). Such a collapse might well be in some kind of accordance with the BKLM behaviour referred to at the end of §3.2, and in figure 3-30 I have attempted to provide some hint of how the wild singular behaviour might begin to build up in such a

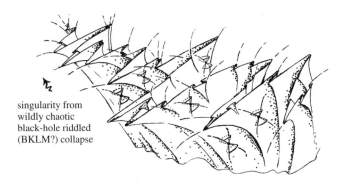

singularity from
wildly chaotic
black-hole riddled
(BKLM?) collapse

Figure 3-30: The picture is an attempt to provide some hint of the wild space-time behaviour arising as a generic BKLM-type singularity is approached from the past.

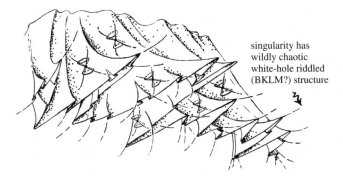

singularity has
wildly chaotic
white-hole riddled
(BKLM?) structure

Figure 3-31: Time-reverse of figure 3-30: to provide some hint of the wild space-time emerging from generic BKLM-type initial singularity.

BKLM collapse. The *time reverse* figure 3-31 of this enormously high-entropy situation (see also figure 3-14(c)) would have a singularity of a very different structure from that of an FLRW model and which is certainly not of a character that would allow a realistic description in terms of a *strict* conformal diagram, though there seems to be nothing against the use of a schematic conformal diagram to represent this extremely complicated kind of generic situation (figure 3-31). Our actual Big Bang does seem to be of such a nature that it can be pretty well modelled by an FLRW singularity (and its possible conformal extendibility into the past will be a key issue for §4.3), this fact representing an enormous constraint. It is this constraint that restricts its entropy to something extraordinarily tiny, compared with the vast entropy values that a space-time singularity of general type would allow. In the next section, we shall see how stupendously restrictive, indeed, is a singularity that is closely of an FLRW type.

## 3.6. THE PHENOMENAL PRECISION IN THE BIG BANG

In order to get some idea of the enormousness of possible entropy gain, once we incorporate gravity and allow ourselves to depart from FLRW uniformity, we must turn our attention back to black holes. They seem to represent some kind of maximum of gravitational entropy, so we must ask what entropy is actually to be assigned to them. There is, in fact, a wonderful formula for the entropy $S_{bh}$ of a black hole, first roughly obtained by Jacob Bekenstein [1972, 1973], using some general and particularly persuasive physical arguments, and then refined by Stephen Hawking [1974, 1975, 1976a] (to obtain the exact number "4" in the resulting formula) by means of a classic discussion involving quantum field theory in a curved space-time background describing collapse to a black hole. This formula is

$$S_{bh} = \frac{Akc^3}{4\gamma\hbar},$$

where $A$ is the area of the black hole's event horizon (or, rather, of a spatial cross-section of it; see figure 3-9 in §3.2). The constants $k$, $\gamma$, and $\hbar$ are, respectively, those of Boltzmann, Newton, and Planck (in Dirac's form), and $c$ is the speed of light. It should be pointed out that, for a non-rotating black hole of mass $m$, we find

$$A = \frac{16\pi\gamma^2}{c^4}m^2,$$

so that

$$S_{bh} = \frac{4\pi m^2 k\gamma}{\hbar c}.$$

A black hole can also rotate, and if its angular momentum has the value $am$, then we find [see Kerr 1963; Boyer and Lindquist 1967; Carter 1970]

$$A = \frac{8\pi\gamma^2}{c^4}m(m + \sqrt{m^2 - a^2}), \quad \text{so } S_{bh} = \frac{2\pi k\gamma}{\hbar c}m(m + \sqrt{m^2 - a^2}).$$

In what follows, it will be convenient to adopt what are called *natural units* (often referred to as *Planck units* or *absolute units*) of length, time, mass, and temperature, for which we arrange the definitions of each so that

$$c = \gamma = \hbar = k = 1,$$

these natural units being related to our more familiar practical units, according (approximately) to

$$\text{metre} = 6.3 \times 10^{34},$$
$$\text{second} = 1.9 \times 10^{43},$$
$$\text{gram} = 4.7 \times 10^{4},$$
$$\text{kelvin} = 7.1 \times 10^{-33},$$
$$\text{cosmological constant} = 5.6 \times 10^{-122},$$

so that all measures are now simply pure numbers. Then the above formulae (for a non-rotating black hole) are simply expressed as

$$S_{\text{bh}} = \tfrac{1}{4}A = 4\pi m^2, \quad A = 16\pi m^2.$$

This entropy value turns out to be enormous, for the black holes that we expect to arise through astrophysical processes (and, for this reason, the actual choice of units turns out to matter rather little, though it is best to be explicit). This hugeness of the entropy is perhaps not surprising when one considers how "irreversible" processes involving black holes actually are. It used to be pointed out how large the entropy in the CMB is, namely around $10^8$ or $10^9$ for each baryon (see §1.3) in the universe, and this would be far greater than the entropy of normal astrophysical processes. But this entropy value begins to pale into insignificance when we compare it with the entropy that must be attributed to the black holes, most particularly those huge ones residing in galactic centres. An ordinary stellar-mass black hole might be expected to have an entropy per baryon of around $10^{20}$ or so. But our own Milky Way galaxy has a black hole of around four million solar masses, which gives an entropy per baryon of some $10^{26}$ or more. It is hardly likely that most of the *mass* in the universe is currently in the form of black holes, but we can see how the black holes actually come to dominate the *entropy* if we consider a model for which the observable universe is populated by galaxies, like our Milky Way, each made up of $10^{11}$ ordinary stars and with a $10^6$ solar mass central black hole (which probably underrates the present-day average black-hole contribution). Now, we obtain an overall entropy per baryon ratio of around $10^{21}$, which completely dwarfs the $10^8$ or $10^9$ value assigned to the CMB.

From the above expressions we see that the entropy per baryon is likely to get much larger for big black holes, scaling basically in proportion the mass of the hole. Thus, for a given mass of material, the largest entropy that we can achieve this way would be for that entire mass to be concentrated in one black hole. If we take that black-hole mass to be the total baryonic mass within our currently observable universe, which is usually taken to be that within our present particle horizon, then this gives a figure of around $10^{80}$ baryons, which would give us a total entropy of $\sim 10^{123}$, rather than the far tinier value of around $10^{89}$ or so that the fireball evidenced by the CMB appears to have had.

In these considerations, I have so far ignored the fact that baryonic matter appears to represent only around 15% of the material content of the universe, the remaining 85% being what is called *dark matter*. (I am not including *dark energy* here – in other words $\Lambda$ – in these considerations, since I am taking $\Lambda$ to be the cosmological constant, which does not constitute an actual "substance" which could contribute to a gravitational collapse. The issue of an "entropy" associated with $\Lambda$ will be considered in §3.7.) We could well imagine that our hypothetical black hole, encompassing the entire content of the observable universe, ought also to include the mass provided by this dark matter. This would push up our maximum entropy figure to something more like $10^{124}$ or $10^{125}$. For the purposes of the present discussion, however, I shall adopt the more conservative figure of $10^{123}$, partly because it appears to be completely unknown what the constitution of the dark matter actually is. A further reason for being slightly cautious about the larger figure is that there might be some genuine geometrical issue about actually constructing an appropriate expanding universe model in which the entire matter content could be reasonably considered to lie within a single hole. It might make more physical sense to allow for several somewhat smaller black holes, spaced over the entire observable universe. For such purposes, something like a factor of 10 or so to play with here gives the picture much greater plausibility.

One further point should be clarified here. The term *observable universe* usually refers to the material intercepted by the past light cone of our current space-time location P, as shown in figures 3-17 and 3-27. Provided that we are considering the standard classical cosmological models, this is fairly unambiguous, although there is a minor issue of whether we include events occurring prior to the decoupling 3-surface, while lying within our past light cone. This makes little difference unless we involve the commonly assumed *inflationary* phase to the early universe that we shall come to in §3.9. That would vastly increase the distance out of the particle horizon and the amount of matter that it would encompass. It

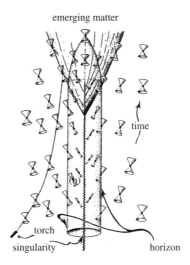

emerging matter

time

torch

singularity    horizon

Figure 3-32: Space-time picture of a hypothetical "white hole": the time-reverse of figure 3-9. Before its explosion into matter, externally emitted light (e.g. from the depicted torch), cannot enter through the horizon.

appears to be normal practice not to include such an early inflationary period in the definition of the term *particle horizon*. I shall follow that practice here.

We should bear in mind that when considering the way in which the Big Bang was special, what we had been looking at in §3.5 was the *time-reverse* of a collapsing model. Our picture of that hypothetical collapsing model would be of a final singularity that had come about through many smaller black holes having been first produced, these subsequently congealing into larger ones. It seems very plausible to suppose that this would allow us to get reasonably close, in the final reckoning, to a single black hole encompassing the lot, even if we end up a little short of this. It should be noted that when we perform the time-reverse of this thoroughly messy collapse, what we arrive at for the maximum-entropy Big Bang is not an explosion containing (say) a big black hole but the *time-reverse* of such a hole, frequently referred to as a *white hole*. To get a picture of the space-time describing a white hole, we consider figure 3-9 turned upside down; see figure 3-32. Its strict conformal diagram, shown in figure 3-33, is therefore figure 3-29(a) upside-down. It is not at all mathematically unreasonable that such a configuration could be part of a far more general big bang than that described by an FLRW model, and the initial entropy here could be a stupendous $\sim 10^{123}$,

Figure 3-33:  Strict conformal diagram of a spherically symmetric white hole.

rather than the comparatively tiny $\sim 10^{89}$ that we seem to see in the primordial fireball, as manifested in the CMB.

In the preceding paragraph, I have phrased the argument in terms of black holes rather than white holes, but from the point of view of calculating entropy values the results are the same. The Boltzmann definition of entropy, as we recall from §3.3, simply depends upon the volumes of coarse-graining regions in phase space. The nature of the phase space itself is insensitive to the direction of time (as reversing the time direction merely replaces the momenta with their negatives) and the macroscopic criteria for defining the coarse-graining regions would not be dependent on the time direction. Of course white holes are not to be expected in the universe we actually inhabit, as they grossly disobey the 2nd Law. However, they are perfectly legitimate in the above considerations for calculating the degree of "specialness" of the Big Bang, since it is those states that *do* violate the 2nd Law that must enter our considerations.

Thus, we find that an entropy value of at least around $10^{123}$ would have been a possibility for a space-time singularity constituting the initial state of the universe, where we require merely consistency with the (time-symmetric) equations of general relativity, with ordinary matter sources, and with a total baryon number of some $10^{80}$ within the observable universe (with its accompanying dark matter). Accordingly, we must allow for a phase space $\mathscr{P}$ with a total volume of at least

$$V = e^{10^{123}}$$

(because the $k = 1$ Boltzmann formula $S = \log V$ of §3.3 needs to be able to accommodate $S = 10^{123}$; see §A.1). In fact, as shown in §A.1, it makes very little difference (certainly to the accuracy of the number "124" in this expression) if the "e" is replaced by "10", so I shall refer to $\mathscr{P}$ as having a total volume of at least a size

$$10^{10^{123}}.$$

What we actually find in our universe is a primordial fireball, at the time of decoupling, whose entropy had a value no bigger than about $10^{90}$ (taking $10^{80}$ baryons with an entropy per baryon of $10^9$, and throwing in a good measure of dark matter also), so it would occupy a coarse-graining region $\mathcal{D}$ of the much tinier volume

$$10^{10^{90}}.$$

How much tinier is this volume $\mathcal{D}$, as a fraction of the entire phase space $\mathcal{P}$? The answer is clearly just

$$10^{10^{90}} \div 10^{10^{123}},$$

which, as shown in §A.1, is virtually indistinguishable from

$$\frac{1}{10^{10^{123}}}, \quad \text{i.e. } 10^{-10^{123}},$$

so that we do not even notice the actual volume of $\mathcal{D}$, so enormously large is the total volume $10^{10^{123}}$ that we need to take for $\mathcal{P}$. This gives us some idea of the utterly extraordinary *precision* that was involved in the creation of the universe, as we presently understand it. In fact, processes involving a considerable increase of entropy may well have been involved at times between the actual initial singularity – which we may represent as a particularly tiny coarse-graining region $\mathcal{B}$ within the phase space $\mathcal{P}$ – and decoupling. (See figure 3-34, where the magnitudes given in the caption include the dark matter contributions.) Thus, we must expect that there was even greater precision involved in the creation of the universe, which is measured now by the size of $\mathcal{B}$, within the phase space $\mathcal{P}$. This is still $10^{-10^{123}}$, the even tinier nature of the region $\mathcal{B}$, being again masked by the very hugeness of the figure which describes the size $10^{10^{123}}$ of $\mathcal{P}$ itself.

## 3.7. COSMOLOGICAL ENTROPY?

There is one further issue that I should address in the context of contributions to the entropy of some ("dark") substance, namely how one should rate the contribution of what is commonly referred to as *dark energy* – i.e. $\Lambda$ (in my interpretation of this term). Many physicists take the view that the presence of $\Lambda$ has the effect

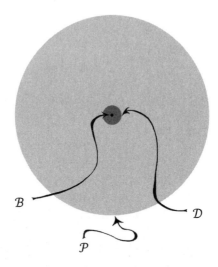

Figure 3-34: Our entire observable universe's phase space $\mathcal{P}$ has a volume in Planck units (or any other normally used units) of around $10^{10^{124}}$. The region $\mathcal{D}$ of states representing decoupling has the absurdly smaller volume of around $10^{10^{90}}$, and the volume of the region $\mathcal{B}$ of states representing the Big Bang, would be a good deal even smaller, each being a mere $\sim 10^{-10^{124}}$ of the total. Such a diagram cannot remotely do this disparity justice!

of providing an enormous entropy in the remote future of our continually expanding universe, that somehow "kicks in" at some very late (but unspecified) stage in the universe's history. The rationale for this viewpoint arises mainly from a common, and well-authenticated, belief [Gibbons and Hawking 1977] that the cosmological event horizons that occur in these models should be treated in the same way as black-hole horizons, and since we are looking at horizons of stupendous dimension – of an area vastly exceeding that of the largest black hole yet observed (that being apparently of around $4 \times 10^{10}$ solar masses) by a factor of roughly $10^{24}$ – we would arrive at a stupendously huge "entropy" $S_{cosm}$ whose value is approximately

$$S_{cosm} \approx 6.7 \times 10^{122}.$$

This figure is calculated directly from the presently estimated observational value of $\Lambda$

$$\Lambda = 5.6 \times 10^{-122},$$

and it uses the Bekenstein–Hawking entropy formula (assuming that we accept its use under these circumstances) applied to the area $A_{\text{cosm}}$ of the cosmological event horizon, that area being given exactly by

$$A_{\text{cosm}} = \frac{12\pi}{\Lambda}.$$

It should be noted that if this horizon argument is to be trusted – in the same way that we trust the Bekenstein–Hawking argument for black-hole entropy – then this should represent a *total* entropy value, not merely some contribution that comes from the "dark energy". Yet, we find that, in fact, the horizon area provides this quantity "$S_{\text{cosm}}$" purely from the value of $\Lambda$, and quite independently of matter distributions or other detailed departures from the exact de Sitter geometry of figure 3-4 [see Penrose 2010, §B5]. Nevertheless, even though $S_{\text{cosm}}$ ($\sim 6 \times 10^{122}$) appears to fall a little short of the total entropy value (when we include the dark matter contribution) of $\sim 10^{124}$ that I have been contemplating above, it is likely to be enormously larger than the maximum total entropy of perhaps about $10^{110}$ that might well actually be ultimately achieved in black holes if we only include the baryonic matter in our present observable universe, or the $10^{112}$ or so that the dark matter could perhaps provide.

Yet, we must ask what this ultimate "total entropy" value of $S_{\text{cosm}}$ ($\sim 6 \times 10^{122}$) actually refers to. Since it just depends on $\Lambda$, and has nothing to do with the details of the material content of the universe, we might well take the view that $S_{\text{cosm}}$ should be an entropy assigned to the *entire* universe. But if the universe is spatially infinite (as is a common opinion among cosmologists), then this single "entropy" value would have to be spread over this infinite spatial volume and so would contribute only a vanishingly small amount to the *finite* comoving region under consideration here. On this interpretation of the cosmological entropy, "$6 \times 10^{122}$" would count as a *zero* entropy *density*, and should therefore be ignored completely in considerations of entropy balance in our dynamical universe.

We might, on the other hand, try to consider that this entropy value refers only to the comoving volume based on the matter content of our observable universe, i.e. the comoving volume $\mathcal{G}(\text{P})$ within our particle horizon $\mathcal{H}(\text{P})$ (see §3.5; figures 3-17 and 3-27), P being our present space-time location. But this has no plausible rationale, particularly since "the present time", which determines the point P on our world-line $l_{\text{P}}$, has no particular significance in this context. It would seem more appropriate to take the comoving volume $\mathcal{G}(l_{\text{P}})$ considered in §3.5, where our world-line $l_{\text{P}}$ has to be extended indefinitely into the future. The measure of matter within this region is not sensitive to the "time" at which we presently observe the universe. It is the totality of the matter that will *ever* come into our observable universe. In the conformal picture (figure 3-18 in §3.5), the fully extended $l_P$ meets future conformal infinity $\mathscr{I}$ (where we recall that $\mathscr{I}$ is a *spacelike* hypersurface when $\Lambda > 0$) at some point Q, and we are now concerned with the total quantity of material intersected by the past light cone $\mathcal{C}_Q$ of Q. This past light cone is, indeed, our *cosmological event horizon*, and it has a more "absolute" character than the matter just within our present particle horizon. As time progresses, our particle horizon spreads out, and the matter within the volume $\mathcal{G}(l_P)$ represents the ultimate limit to which it spreads.

In fact (assuming the time evolution provided by Einstein's equations, with the observed positive value of $\Lambda$, which we take to be a constant), we find a total value of material intercepted by $\mathcal{C}_Q$ that is nearly $2\frac{1}{2}$ times that within our present particle horizon [Tod 2012; Nelson and Wilson-Ewing 2011]. The kind of value for the maximum possible entropy that this material could attain, taking that value to be what would be reached if all this material were to lie within a single black hole, is thus somewhat more than five times the maximum that we obtained before when we simply used the material within our present particle horizon. This larger figure gives the value $\sim 10^{124}$, instead of the $\sim 10^{123}$ that we had above. When we include the contribution from the dark matter, we get $\sim 10^{125}$. This value is several hundred times larger than $S_{\text{cosm}}$, so choosing a universe model with the same overall matter density as ours, but with big enough black holes, we appear to be able to violate the ultimate value that $S_{\text{cosm}}$ is alleged to provide us with, in gross contradiction with the 2nd Law! (There is an issue arising from the eventual Hawking evaporation of these black holes, but this does not invalidate the reasoning involved here; see Penrose [2010, §3.5].)

Given that there are some uncertainties in these figures, it is still perhaps just about plausible to argue that the value $6 \times 10^{122}$ that we obtain for the cosmological entropy is really the "true" maximum that is available for the amount of material that lies within $\mathcal{C}_Q$. But there are stronger reasons than those just given for doubting that $S_{\text{cosm}}$ is to be taken seriously as a true ultimate entropy for this portion of the universe – or, indeed, to be regarded as any physically relevant "entropy" at all. Let us return to the basic argument that suggested the analogy between the cosmological horizon $\mathcal{C}_Q$ and a black-hole event horizon $\mathcal{E}$. When we try to use this analogy to address the question of which portion of the universe this cosmological entropy might actually refer to, we find a curious contradiction. We have already seen, by the arguments above, that this portion cannot be the *entire* universe, and it had seemed plausible to suppose that the portion being referred to is simply the region interior to the cosmological horizon. However, when we compare this cosmological situation with that of a black hole, we find

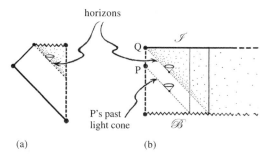

horizons

Q

𝒥

P

P's past
light cone

ℬ

(a)                    (b)

Figure 3-35: Strict conformal diagrams illustrating the region (shown dotted) corresponding to the entropy of (a) a black hole and (b) a positive-$\Lambda$ cosmology. For a spatially infinite universe, the *density* of "cosmological entropy" would have to be zero.

that this interpretation is not at all logical. In the case of a collapse to a black hole, the Bekenstein–Hawking entropy is normally thought of as being the entropy *of* the black hole, which is a perfectly reasonable interpretation. But when we make the comparison with the cosmological situation, the black hole's event horizon $\mathcal{E}$ being compared with the cosmological horizon $\mathcal{C}_Q$ as shown in the strict conformal diagrams of figure 3-35, we see that what the space-time region *within* the black hole horizon $\mathcal{E}$ corresponds to is the region of the universe *outside* the cosmological horizon $\mathcal{C}_Q$. These are the regions lying on the "future" sides of the respective horizons, i.e. to the sides into which the future null cones point. As we have seen above, for a spatially infinite universe, this would give us a zero entropy *density* throughout the entire external universe! This again seems to make little sense, if we are trying to interpret $S_{\text{cosm}}$ as providing a dominant physical contribution to the physical entropy of the universe. (Other suggestions might be made as to "where" the entropy in $S_{\text{cosm}}$ is to reside, such as the space-time region lying to the causal future of $\mathcal{C}_Q$, but this also seems to make very little sense, since $S_{\text{cosm}}$ is completely independent of any material or black holes that might enter that region.)

A point about this argument that might perhaps concern some readers is that, in §3.6, the concept of a white hole was introduced, and there the null cones point outwards, into the future, away from the central region, in a way that is similar to the case of a cosmological horizon. It was also pointed out that the Bekenstein–Hawking entropy ought to be equally valid for a white hole as for a black hole, owing to the fact that the Boltzmann entropy definition is insensitive to the direction of time. One could, therefore, try to argue that the *white*-hole analogy with the cosmological horizon might justify the interpretation of $S_{\text{cosm}}$ as an actual physical entropy. However, white holes are unphysical objects, in the universe we know, as they violently disobey the 2nd Law, being introduced in §3.6 for hypothetical reasons only. For comparisons directly relating to operation of the temporal increase of entropy in the 2nd Law, as was the relevant one in the paragraph preceding the one above, the comparison must be between the cosmological horizon and *black*-hole horizons, not those of white holes. Accordingly, the "entropy" that $S_{\text{cosm}}$ is supposed to measure would have to refer to the region external to the cosmological horizon, not internal to it, which, as remarked earlier, would give a vanishing spatial entropy density, for a spatially infinite universe, as argued above.

There is, however, a further point that should be raised, in connection with the use of Boltzmann's formula $S = k \log V$ (see §3.3) in the context of black holes. It has to be confessed that the entropy $S_{\text{bh}}$ for a black hole (see §3.6) has not yet, in my view, been fully and persuasively equated to a Boltzmann-type entropy, where the relevant phase space volume $V$ is clearly identified. There are various approaches aimed at achieving this [see, for example, Strominger and Vafa 1996; Ashtekar et al. 1998], but I am not really happy with any of them. (See also §1.15, in relation to the motivating ideas for the holographic principle, which I am also unhappy with.) The reasons for taking $S_{\text{bh}}$ very seriously as an *actual* measure of a black hole's entropy are different from any that have, so far, come from a direct use of Boltzmann's formula. Nevertheless, these reasons [Bekenstein 1972, 1973; Hawking 1974, 1975; Unruh and Wald 1982] are, in my view, very powerful and are demanded by an overall consistency of the 2nd Law in a quantum context. Although they did not directly use the Boltzmann formula, this does not imply any inconsistency with it, and indicate merely an inherent difficulty in coming to grips with currently unresolved quantum phase-space notions in the context of general relativity (compare also §§1.15, 2.11, and 4.3).

It should be clear to the reader that, in my own view, this assignment of an entropy contribution $(12\pi/\Lambda)$ to the universe from $\Lambda$ in this way is physically extremely dubious. But this is not merely for the reasons presented above. If one were to regard $S_{\text{cosm}}$ as playing any role in the dynamics of the 2nd Law, which somehow manifests itself only at very late stages of our universe's de Sitter-like exponential expansion, then we would need some theory of "when" this entropy would "kick in". The de Sitter space-time has a very high degree of symmetry (a 10-parameter group, as big as that of Minkowski 4-space; see §3.1 and, for example, Schrödinger [1956] and TRtR [§§18.2 and 28.4]) and does not in itself allow for such a time to be naturally specified. Even if one takes the "entropy" given by $S_{\text{cosm}}$ seriously as having some kind of meaning (such as from vacuum fluctuations), it does not appear to have any dynamical role to play in interrelating with other forms of entropy. $S_{\text{cosm}}$ is just a *constant* quantity, whatever meaning one might choose to give it, and it makes no difference to the operation of the 2nd Law *whether or not* we choose to refer to it as an entropy of some kind.

In the case of an ordinary black hole, on the other hand, there was the original Bekenstein argument [Bekenstein 1972, 1973], which involved the performing of *thought experiments* in which lukewarm material might be lowered slowly into a black hole so that one might envisage converting the energy in its heat into useful work. It turns out that if one had not assigned an entropy to a black hole, in rough agreement with the formula for $S_{\text{bh}}$ above, then it would have been possible, in principle, to violate the 2nd Law in this way. Thus, the Bekenstein–Hawking entropy is shown to be an essential ingredient of the overall consistency of the 2nd Law in the context of black holes. This entropy has a clear interrelation with other forms of entropy and is essential for the overall consistency of thermodynamics in the context of black holes. This has to do with the dynamics of black-hole horizons, and the fact that they can be enlarged by processes that might otherwise appear to reduce the entropy, such as lowering lukewarm material into the hole and extracting its entire mass/energy content as "useful" energy, thereby violating the 2nd Law.

The situation with cosmological horizons is completely different. Their actual locations are very observer-dependent, quite unlike the *absolute* event horizons of stationary black holes in asymptotically flat space (see §3.2). Yet the *area A* of a cosmological horizon is just a *fixed* number, being simply determined by the value of the cosmological constant $\Lambda$ by the above expression $12\pi/\Lambda$ and having nothing to do with any dynamical processes going on in the universe, such as how much mass–energy passes through the horizon or how that mass might be distributed, all of which can certainly affect the *local* geometry of the horizon. This is quite unlike the situation with a black hole, whose horizon area inevitably increases when material passes through it. No dynamical processes have any affect on $S_{\text{cosm}}$, which steadfastly remains at the value $12\pi/\Lambda$, no matter what.

This, of course, depends upon $\Lambda$ actually being a constant, and not some mysterious unknown dynamical "dark energy field". Such a "$\Lambda$-field" would have to possess an energy tensor $(8\pi)^{-1}\Lambda\mathbf{g}$, so Einstein's equations $\mathbf{G} = 8\pi\gamma\mathbf{T} + \Lambda\mathbf{g}$ can be written in the form

$$\mathbf{G} = 8\pi\gamma\left(\mathbf{T} + \frac{\Lambda}{8\pi\gamma}\mathbf{g}\right),$$

as though without the cosmological term of Einstein's modified theory of 1917, but with the $(8\pi)^{-1} \Lambda \mathbf{g}$ being considered to be simply the $\Lambda$-field contribution, added to the energy tensor $\mathbf{T}$ of all the remaining matter, giving the total energy tensor (in parentheses on the right). However, this additional term is completely unlike that provided by ordinary matter. Most notably, it is gravitationally repulsive although having a positive mass/energy density. Moreover, permitting $\Lambda$ to vary carries with it many technical difficulties, one of the most noteworthy being a serious danger of violating the null energy condition referred to in §3.2. If we try to think of dark energy as some kind of substance, or collection of substances that do not interact with other fields, then we find that equations of differential geometry (the contracted Bianchi identities, in fact) tell us that $\Lambda$ must be a constant, but if we allow the total energy tensor to deviate from this form, then this null energy condition is very likely to be violated, since it is only marginally satisfied when the energy tensor has a $\lambda \mathbf{g}$ form.

Closely related to the entropy issue is that of the so-called cosmological temperature $T_{\text{cosm}}$. In the case of a black hole, the fact that there should be a black-hole temperature associated with the Bekenstein–Hawking black-hole entropy (and conversely) follows from very basic thermodynamic principles [Bardeen et al. 1973]. Indeed, in his initial articles establishing his precise formula for black-hole entropy, Hawking [1974, 1975] also obtained a formula for a black hole's *temperature*, which, in the non-rotating (spherically symmetric) case, gives the value

$$T_{\text{bh}} = \frac{1}{8\pi m}$$

in natural (Planck) units. For a black hole of the kind of size that could arise in normal astrophysical processes (for which the mass $m$ would not be smaller than a solar mass) this temperature is extremely low, and would be highest for the least massive of black holes, and then only at a general level not all that much greater than the lowest temperature ever produced artificially on Earth.

The cosmological temperature $T_{\text{cosm}}$ is motivated by the analogy with a black hole whose horizon size would be that of the cosmological horizon $\mathcal{C}_Q$, and we obtain

$$T_{\text{cosm}} = \frac{1}{2\pi} \sqrt{\frac{\Lambda}{3}}$$

in natural units, which in kelvins would be

$$T_{\text{cosm}} \approx 3 \times 10^{-30} \text{ K}.$$

This, indeed, is an absurdly low temperature, far lower even than the Hawking temperature of any black hole that one could seriously conceive of as arising in the universe we know. But is $T_{cosm}$ really an actual temperature in the normal physical sense of that word? There seems to be a common opinion, among those cosmologists who have seriously considered this issue, that it ought surely to be considered so.

There are various arguments purporting to support this interpretation, some better motivated than simply appealing to a black-hole analogy, but all are seriously questionable in my view. Perhaps the most mathematically attractive of these (used also for a stationary black hole) depends upon *complexifying* the space-time 4-manifold $\mathcal{M}$, enlarging it to a *complex* 4-manifold $\mathbb{C}\mathcal{M}$ [Gibbons and Perry 1978]. The notion of *complexification* is one that applies to real manifolds that are defined by smooth-enough equations (technically, *analytic* equations), and the complexification procedure simply involves replacing all the real-number coordinates by complex numbers (§§A.5 and A.9), while keeping the equations completely unchanged, so that we obtain a complex 4-manifold (which would be 8-real-dimensional; see §A.10). All the standard stationary black-hole solutions of the Einstein equations, with or without the cosmological constant $\Lambda$, admit such complexifications, and what one obtains is a space with a complex periodicity of just such a scale of size that yields, through subtle thermodynamic principles [Bloch 1932], a temperature that rather remarkably agrees, precisely, with the value that Hawking had earlier obtained for a black hole (rotating or otherwise). This makes it extremely tempting to conclude that this temperature value should be assigned to a cosmological horizon in the same way, and when the argument is applied to empty de Sitter space, for cosmological constant $\Lambda$, one indeed obtains $T_{cosm} = (2\pi)^{-1}(\Lambda/3)^{1/2}$, agreeing with the value given above.

A puzzle arises, however, for those solutions of Einstein's equations (with $\Lambda$-term) where there is both a cosmological event horizon and a black-hole event horizon, for then the procedure yields both complex periodicities at the same time, providing the inconsistent interpretation of two different temperatures existing simultaneously. It is not exactly a mathematical inconsistency, as the complexi-fication procedure can be carried out (somewhat inelegantly) in different places in different ways. Accordingly, one might perhaps argue that one temperature is relevant close to a black hole, whereas the other becomes relevant far from any black holes. However, the case for a convincing physical conclusion is decidedly weakened.

The original (more directly physical) argument for interpreting $T_{cosm}$ as an actual physical temperature came from considerations of quantum field theory in a curved background space-time, applied to de Sitter space-time [Davies 1975; Gibbons and Hawking 1976]. However, it turned out that this depends rather critically on which particular coordinate system is used for the QFT background [Shankaranarayanan 2003; see also Bojowald 2011]. This kind of ambiguity can be understood in relation to a phenomenon known as the *Unruh effect* (or the *Fulling–Davies–Unruh effect*), predicted by Stephen Fulling, Paul Davies, and particularly William Unruh [Fulling 1973; Davies 1975; Unruh 1976] in the mid 1970s. According to this effect, an accelerating observer experiences a *temperature*, owing to considerations of quantum field theory. This temperature would be extremely tiny for ordinary accelerations, and is given by the formula,

$$T_{\text{accn}} = \frac{\hbar a}{2\pi k c}$$

for an acceleration of magnitude $a$ or, in natural units, by simply

$$T_{\text{accn}} = \frac{a}{2\pi}.$$

In the case of a black hole, an observer suspended above the hole by a rope attached to a distant fixed object would feel this (absurdly tiny) temperature of the Hawking radiation, which attains the value $T_{\text{bh}} = (8\pi m)^{-1}$ at the horizon. Here the value of $a$ at the horizon is calculated as the "Newtonian" acceleration $m(2m)^{-2} = (4m)^{-1}$, at the radial distance $2m$ of the horizon. (At the horizon, the acceleration actually felt by the observer would strictly be *infinite*, but the calculation takes into account a time-dilation factor, which is also infinite at the horizon, which motivates the finite resulting "Newtonian" value used here.)

On the other hand, an observer falling directly into the hole would feel a *zero* Unruh temperature, since freely falling observers feel no acceleration (by the Galilei–Einstein principle of equivalence, which indeed asserts that an observer in free fall under gravity would not feel any force of acceleration; see §4.2). Thus, whereas we can interpret the Hawking black-hole temperature as being an example of the Unruh effect, we see that this temperature can be cancelled by free fall. When we apply the same idea in the cosmological context, and try to interpret the cosmological "temperature" $T_{\text{cosm}}$ in the same way, we must again conclude that a freely falling observer would *not* "feel" this temperature. Indeed, this applies to any *comoving observer* in the standard cosmological models – and in particular in de Sitter space – so we conclude that a comoving observer should experience no acceleration, and therefore no Unruh temperature. Accordingly, from this perspective the "temperature" $T_{\text{cosm}}$, low as it is, would *not* be actually felt at all by a comoving observer!

This gives an additional reason to be suspicious of the associated "entropy" $S_{cosm}$, as having any dynamical role to play, with regard to the 2nd Law, and I would myself regard both $T_{cosm}$ and $S_{cosm}$ with considerable suspicion. This is not to say that I take $T_{cosm}$ as having no physical role at all. I imagine that it might well represent some kind of critical lowest temperature, and this could perhaps have some role to play in relation to the ideas described in §4.3.

## 3.8. VACUUM ENERGY

In the previous chapters I have been taking what modern cosmologists tend to refer to as *dark energy* (not to be confused, of course, with the completely different dark matter) as being simply Einstein's 1917 *cosmological constant* $\Lambda$, which is a perfectly reasonable stance to take, being consistent with all current observations. Einstein originally introduced this term into his equations $\mathbf{G} = 8\pi\gamma\mathbf{T} + \Lambda\mathbf{g}$ (see §1.1) for what admittedly turned out to be an inappropriate reason. He proposed this modification of his equations in order to achieve a model of a *static* spatially closed 3-sphere universe ($\mathcal{E}$ in §1.15), whereas the fact that the universe is actually *expanding* was, around a decade later, convincingly demonstrated by Edwin Hubble. Einstein then regarded the introduction of $\Lambda$ as his greatest blunder, perhaps because it led him to miss the opportunity to *predict* this expansion! Indeed, according to the cosmologist George Gamow [1979], Einstein once remarked to him that "the introduction of the cosmological term was the biggest blunder he ever made in his life". However, from our present understandings, it is extremely ironic that Einstein regarded the introduction of $\Lambda$ as a blunder, in view of the fundamental role that $\Lambda$ now appears to be playing in modern cosmology, as is witnessed by the award of the 2011 Nobel Prize in physics to Saul Perlmutter, Brian P. Schmidt, and Adam G. Riess [Perlmutter et al. 1998, 1999; Riess et al. 1998] "for the discovery of the accelerating expansion of the Universe through observations of distant supernovae", this accelerated expansion being most directly explained by Einstein's $\Lambda$.

Nonetheless, the possibility that the cosmic acceleration might actually have some other cause should not be dismissed. A common view among physicists, whether or not they view $\Lambda$ as being a constant (and, in fact, perhaps being nevertheless interpretable as Einstein's cosmological constant), is that the presence of $\Lambda$, or rather the tensor $\Lambda\mathbf{g}$, where $\mathbf{g}$ is the metric tensor, in Einstein's $\mathbf{G} = 8\pi\gamma\mathbf{T} + \Lambda\mathbf{g}$, arises from what is called *vacuum energy*, which should permeate all of empty space. The reason that physicists expect the vacuum to have a non-zero (positive) energy, and therefore a mass (in accordance with Einstein's $E = mc^2$) comes from some very basic considerations of quantum mechanics and quantum field theory (QFT; see §§1.3–1.5).

It is common practice, in QFT, to resolve a field into vibrational *modes* (see §A.11), each of which would have its definite energy. Among these various vibrational modes (each oscillating with its corresponding specific frequency, in accordance with Planck's $E = h\nu$) would be one that has the *minimum* energy value, but it turns out that this energy value is *not zero* – and is referred to as *zero-point energy*. Even in a vacuum, therefore, the *potential* presence of any field leads to its manifesting itself in at least a tiny minimum amount of energy. For the different vibrational possibilities, there would be different such energy minima, and the total of all these, for all different fields, would provide what is called the *vacuum energy*, i.e. an energy in the vacuum itself.

In a non-gravitational context, the normal view would be that this vacuum energy background can be safely ignored, because it just provides a universal constant quantity that can simply be subtracted out from the total of all contributing energies, the energy *differences* from this background value being all that plays a role in (non-gravitational) physical processes. But when gravity comes into the picture, things are very different, because this energy ought to have a mass ($E = mc^2$) and mass is the source of gravity. This might be all very well, at a local level, if the background value were tiny enough. Although this background gravitational field might strongly influence one's considerations of cosmology, it ought not to play any significant role in local physics, owing to the extreme weakness of gravitational influences. However, when all the various zero-point energies are added together one tends to find the disturbing answer, *infinity*, because this summation, for all the different vibrational modes, is a *divergent series*, like those discussed in §A.10. How are we to deal with this apparently catastrophic situation?

Often, a case can be made for a divergent series such as $1 - 4 + 16 - 64 + 256 - \cdots$ to have a finite "sum" (in this case $\frac{1}{5}$, as shown in §A.10), an answer that cannot be obtained by simply adding up the terms, but which can be mathematically justified in various ways, most importantly by appealing to the notion of *analytic continuation*, briefly referred to in §A.10. Similar arguments can be put forward to justify the even more remarkable assertion $1 + 2 + 3 + 4 + 5 + 6 + \cdots = -\frac{1}{12}$. Using such means (and other procedures related to these), physicists involved with QFT are frequently able to assign *finite* answers to such divergent series, so that they can often provide finite values to calculations that would otherwise seem to yield the unhelpful conclusion "$\infty$". Curiously, this second summation (of all the natural numbers) even plays a role in determining the 26-dimensionality of space-time demanded by the original bosonic string theory (see §1.6). It is the space-time *signature* that is relevant here, namely the *difference* between the number of space and time dimensions, namely $24 = 25 - 1$, and this "24" relates to the "12" in the divergent sum.

Such procedures are also used in attempts to provide a finite answer to the vacuum energy problem. It should be mentioned that the physical reality of vacuum energy is often claimed as an experimentally observed phenomenon, as it makes itself directly evident in a famous physical phenomenon referred to as the *Casimir effect*. This effect shows up as a force between two parallel conducting metallic plane plates, which are not electrically charged. When the plates are held extremely close to one another, but not actually touching, then an attractive force occurs between them, and this is found to agree well with what was originally calculated [in 1948] by the Dutch physicist Hendrik Casimir, on the basis of the vacuum energy effects referred to above. Experiments which confirm this effect – and also generalizations due to the Russian physicist Evgeny Lifshitz and his students – have been successfully performed many times [Lamoreaux 1997]. (This is the same E. M. Lifshitz referred to in §§3.1 and 3.2, in connection with the singularities of general relativity.)

None of this requires an understanding of the actual value of the vacuum energy, however, as the effect would come about as mere *differences* from the background energy, in accordance with the comments above. Moreover, the distinguished American mathematical physicist Robert L. Jaffe [2005] has pointed out that the Casimir force can be obtained (albeit in a somewhat more complicated way) by standard QFT techniques, without any reference to vacuum energy at all. Accordingly, quite apart from the divergence issue that arises when one attempts to calculate the actual value of the vacuum energy, the experimentally established existence of the Casimir effect does *not* actually establish the physical reality of vacuum energy. This goes against a commonly expressed view that the physical reality of vacuum energy has already been established.

Nevertheless, the very significant possibility of vacuum energy having a gravitating effect (i.e. acting as the source of a gravitational field) must be taken seriously. If gravitating vacuum energy does indeed exist, then it certainly cannot have an infinite density. If it is to be obtained by summing all the modes of oscillation in the kind of way described earlier in this section, then there needs to be some way of "regularizing" the infinite value that the summing of modes seems inevitably to lead us into, one such procedure indeed being that of analytic continuation referred to above. Analytic continuation is certainly a well-defined and powerful mathematical technique whereby seeming infinities can be made finite. However, I do believe that in a physical context, such as the one of relevance here, some words of definite caution are in order.

Let us recall this procedure, as briefly remarked upon in §A.10. Analytic continuation has to do with those functions of a complex-number variable $z$ which are holomorphic, the term *holomorphic* meaning *smooth* in the sense of complex numbers (see §A.10). It is a remarkable theorem (§A.10) that any function $f$ that is holomorphic in some neighbourhood of the origin 0 of the Wessel plane can be expressed by means of a *power series*

$$f(z) = a_0 + a_1 z + a_2 z^2 + a_3 z^3 + a_4 z^4 + \cdots,$$

where $a_0, a_1, a_2, \ldots$ are complex constants. If such a series converges, for some non-zero value of $z$, then it will also converge for any other value of $z$ that lies closer to the origin 0 in the Wessel plane. There is some fixed circle in this plane, centred at 0, whose radius $\rho$ ($>0$) is called the *radius of convergence* of the series – so that it converges if $|z| < \rho$ and it diverges if $|z| > \rho$. If the series diverges for all non-zero $z$, we say $\rho = 0$, but we also allow this radius to be infinite ($\rho = \infty$), in which case the function is defined by the series over the *entire* Wessel plane – being referred to as an *entire function* – and there is nothing more to be said about it with regard to analytic continuation.

But if $\rho$ is some finite positive number, then the possibility of extending the function $f$ by analytic continuation presents itself. An example of this arises with the series

$$1 - x^2 + x^4 - x^6 + x^8 - \cdots$$

considered in §A.10, and as example B in §A.11, which – when complexified by allowing the real variable $x$ to be replaced by the complex variable $z$ – sums to the explicit function $f(z) = 1/(1 + z^2)$ within the convergence radius $\rho = 1$. However, the series diverges for $|z| > 1$ even though the answer to the sum of this series, namely $f(z) = 1/(1 + z^2)$, is perfectly well defined all over the Wessel plane except at the two *singular* points $z = \pm i$, when $1 + z^2$ vanishes (giving $f = \infty$). See figure A-38 in §A.10. Putting $z = 2$ in this justifies the answer $1 - 4 + 16 - 64 + 256 - \cdots = \frac{1}{5}$.

In more general situations, we may not have an explicit expression for the sum of the series, yet it may still be the case that we can extend the function $f$ to regions outside the circle of convergence, while retaining $f$'s holomorphic nature. One way of doing this (usually not very practical) is to recognize that since $f$ is holomorphic everywhere within its circle of convergence, we can choose any *other*

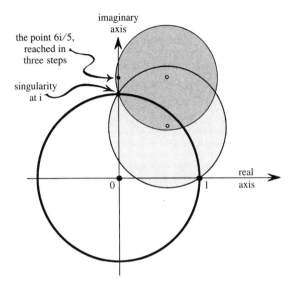

the point 6i/5,
reached in
three steps

singularity
at i

imaginary
axis

real
axis

0                    1

Figure 3-36: Analytic continuation illustrated. The circle of convergence for the power series for $f(z) = 1/(1 + z^2)$ is the unit circle, the series diverging outside it and therefore at the point $z = 6i/5$. Moving the centre first to $z = 3(1 + i)/5$, and then to $3(1 + 2i)/5$ (marked with tiny circles in the picture), the radii of the circles of convergence being determined by the singularity at $z = i$, we can extend the reach of the power series to the point $z = 6i/5$.

point $Q$ within the circle of convergence (a complex number $Q$ with $|Q| < \rho$), so that $f$ must still be holomorphic there, and use a power series expansion for $f$ about $Q$, i.e. an expression of the form

$$f(z) = a_0 + a_1(z - Q) + a_2(z - Q)^2 + a_3(z - Q)^3 + \cdots$$

to represent $f$. We can think of this as a standard power series expansion about the origin $w = 0$ in the Wessel plane for the complex number $w = z - Q$, this new origin point being the same as the point $z = Q$ in the $z$-plane, so that the circle of convergence is now centred at the point Q in $z$'s Wessel plane. This may extend the definition of the function to a larger region, and we can then repeat the procedure, and so on for larger and larger regions. This is illustrated in figure 3-36 for the particular function $f(z) = 1/(1 + z^2)$, where at the third stage, the procedure has allowed us to extend the function to the other side of the singularity at $z = i$ (explicitly to $z = 6i/5$), the successive centres ($Q$-values) being 0, $3(1 + i)/5$, and $3(1 + 2i)/5$.

For functions containing what are called *branch singularities*, we find that the analytic continuation procedure leads us to ambiguous answers, depending upon which path is chosen when we go around such a branch singularity in order to extend the function. Elementary examples of such branching occur with fractional powers, such as $(1 - z)^{-1/2}$ (which would, in this case, provide a different sign depending on which way we go around the branch singularity at $z = 1$) or $\log(1+z)$, which has a branch at $z = -1$, leading to additive ambiguities of integer multiples of $2\pi i$, depending upon how many times we wind around the branch singularity at $z = -1$. Apart from ambiguities resulting from such branching (which is a non-trivial matter), analytic continuation is, in an appropriate sense, always *unique*.

This uniqueness is a somewhat subtle matter, however, and is best understood in terms of the Riemann surfaces referred to in §A.10. Basically, the procedure is to "unravel" all the branching in a way that replaces the Wessel plane by some multiple-layered version of it, which is then reinterpreted as some Riemann surface, on which the multiple extensions of the function $f$ become single valued. The analytically continued $f$ then becomes completely unique on this Riemann surface (see [Miranda 1995]; for a brief introduction to these ideas, see TRtR, §§8.1–8.3). Nevertheless, we need to accustom ourselves to very odd-looking ambiguous "sums" sometimes arising from perfectly ordinary-looking (not always divergent) series. For example, with the function $(1 - z)^{-1/2}$ considered above, the value $z = 2$ gives us the curiously ambiguous divergent sum

$$1 + \frac{1}{1} + \frac{1 \times 3}{1 \times 2} + \frac{1 \times 3 \times 5}{1 \times 2 \times 3} + \frac{1 \times 3 \times 5 \times 7}{1 \times 2 \times 3 \times 4} + \frac{1 \times 3 \times 5 \times 7 \times 9}{1 \times 2 \times 3 \times 4 \times 5} + \cdots = \pm i,$$

and with $\log(1 + z)$ the value $z = 1$ gives the *convergent* but, by the above procedure, far *more* ambiguous

$$1 - \tfrac{1}{2} + \tfrac{1}{3} - \tfrac{1}{4} + \tfrac{1}{5} - \tfrac{1}{6} + \cdots = \log 2 + 2n\pi i,$$

where $n$ can be any integer whatsoever. Taking answers like this seriously in a genuine physical problem might be considered to be dangerously close to *fantasy*, and a clear-cut theoretical motivation would be required for such a thing to convey physical credibility.

Subtleties of this kind do indicate that we must be particularly cautious about using analytic continuation to obtain physical answers to questions that appear to require the appropriate summation of divergent series. Nevertheless, the issues just referred to are normally well appreciated by theoretical physicists concerned with such matters (especially those physicists involved with string theory and related topics). But these are not the issues that I wish to stress here. My question has to do more with one's faith in the particular coefficients $a_0, a_1, a_2, a_3, \ldots$ in some series $a_0 + a_1 z + a_2 z^2 + a_3 z^3 + \cdots$. If these numbers are ones that arise from a fundamental theory which demands that the values are exactly determined by the theory, then the procedures outlined above might turn out to have physical relevance in appropriate situations. However, if they arise from calculations which involve approximations, uncertainties, or detailed dependence on external circumstances, then we must be much more cautious about results that depend upon analytic continuation procedures (or, as far as I can see, upon any other method of the summation of badly divergent infinite series).

As an illustration, let us return to the series $1 - z^2 + z^4 - z^6 + z^8 \cdots (= 1 + 0z - z^2 + 0z^3 + z^4 + 0z^5 - \cdots)$, recalling that it has a radius of convergence $\rho = 1$. If we imagine that we perturb the coefficients $(1, 0, -1, 0, 1, 0, -1, 0, 1, \ldots)$ of this series randomly, but only very slightly, and in such a way that the series remains convergent within the unit circle (which can be assured by keeping the perturbed coefficients bounded by some number), then we are almost certain to obtain a series for a holomorphic function that has a *natural boundary* at the unit circle [Littlewood and Offord 1948; Eremenkno and Ostrovskii 2007] (see figure 3-37 for a particular kind of function of this nature). That is to say, the unit circle has the property that the perturbed function becomes *so* singular there that no analytic continuation is possible anywhere along the circle. Thus, in a clear sense, among holomorphic functions defined on the (open) unit disc (i.e. $|z| < 1$), those that can be holomorphically extended *anywhere at all* across this circle form a vanishingly small proportion. This tells us that, in a clear sense, we would be very lucky to be able to apply the procedure of analytic continuation in order to sum divergent series, when they are perturbed in general physical situations.

This is not to say that such procedures for regularizing infinities are necessarily meaningless in general physical situations. As a possibility for vacuum energy, it might be that there is a "background" defined by a divergent series that is meaningfully summable to a finite answer using analytic continuation procedures, to which can be added a highly *convergent* part which can be summed separately in the normal way. For example, if we add to our above series for $1/(1 + z^2)$, defined by coefficients $(1, 0, -1, 0, 1, -1, 0, 1, \ldots)$, another "small" part, defined by coefficients $(\varepsilon_0, \varepsilon_1, \varepsilon_2, \varepsilon_3, \varepsilon_4, \ldots)$ for which the series $\varepsilon_0 + \varepsilon_1 z + \varepsilon_2 z^2 + \varepsilon_3 z^3 + \varepsilon_4 z^4 + \cdots$ sums to an *entire* function ($\rho = \infty$), then the analytic extension to outside the unit circle can work just as it did without the perturbation, and the

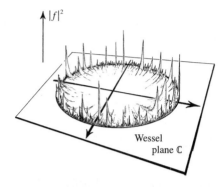

Figure 3-37: The picture schematically illustrates a type of obstruction to holomorphic continuation: a *natural boundary* for a holomorphic function $f$. At *any* point on the unit circle, continuation of this function $f$ to outside the circle is impossible, even though $f$ is holomorphic everywhere within the circle ($|f|$ plotted).

analytic continuation argument can proceed as before. These comments are made here, however, to warn the reader that there are many pitfalls and subtle issues that may have to be faced when applying such procedures for summing divergent expressions. It may be, in particular situations, that they turn out to be appropriate, but one must be extremely cautious about drawing physical conclusions when they are applied.

Let us now return to the specific issue of vacuum energy, and try to see whether it is plausible to regard Einstein's $\Lambda$ as actually being a measure of the energy of empty space-time. Particularly in view of the multitude of physical fields that might be considered to have to contribute (potentially) to this vacuum energy, performing such a calculation explicitly seems an almost hopeless task. Nevertheless, one can apparently say a few things with relative confidence, without going into the details of it all. The first point is that, from considerations of *local Lorentz invariance* – which is basically the assertion that there should not be any "preferred" space-time directions – it is normally strongly argued that the vacuum energy tensor $\mathbf{T}_{vac}$ ought to be proportional to the metric tensor $\mathbf{g}$:

$$\mathbf{T}_{vac} = \lambda \mathbf{g}$$

for some $\lambda$. The hope would be that an argument could be found showing that $\mathbf{T}_{vac}$ actually comes out to give the contribution to the Einstein equations that corresponds to the *observed* value of the cosmological constant, in other words:

$$\lambda = \frac{\Lambda}{8\pi}$$

in natural units, so that this gives the appropriate contribution to the right-hand side of Einstein's equations; for then $\mathbf{G} = 8\pi\mathbf{T} + \Lambda\mathbf{g}$ (in natural units) can now be written

$$\mathbf{G} = 8\pi(\mathbf{T} + \mathbf{T}_{\text{vac}}).$$

However, about all that one gets from considerations of QFT is that either $\lambda = \infty$, this being the most "honest", albeit useless, answer – if one eschews the use of mathematical tricks like those considered above – or $\lambda = 0$, which was the most favoured viewpoint before it became clear from observations that something of the nature of a positive cosmological constant must actually be present, or else something of the order of *unity* in natural units (with perhaps a few simple powers of $\pi$ thrown in, for good measure). Clearly, this last answer would have been found to be very satisfactory had it not been for the fact that the observed data provides a value of something like

$$\Lambda \approx 6 \times 10^{-122}$$

in natural units, surely a discrepancy of totally fantastical proportions!

To me, this sheds serious doubt on the interpretation of $\Lambda$ as being actually a measure of vacuum energy. But most physicists seem strangely reluctant to relinquish such an interpretation. Of course, if $\Lambda$ is not vacuum energy, one needs some other theoretical reason for the (positive) value of Einstein's $\Lambda$-term (especially in view of the fact, noted in §1.15, that string theorists might appear to claim such a theoretical preference for a *negative* $\Lambda$). In any case, $\Lambda$ would be a funny kind of energy, as it is gravitationally *repulsive* despite supplying, in effect, a *positive* energy value. This is a result of the very curious form of energy tensor, namely $\lambda\mathbf{g}$, which is quite unlike the energy tensor of any other known (or seriously considered) physical field, as remarked upon earlier, in §3.7.

Any reader who is puzzled how a positive $\Lambda$ can act as positive mass with regard to spatial curvature, while providing the *repulsive* force that gives rise to the universe's accelerated expansion is referred to a point made towards the end of §3.8. The "$\Lambda\mathbf{g}$" is not really a physically sensible energy tensor, despite the term "dark energy". Specifically, it is the three negative pressure components of $\Lambda\mathbf{g}$ which provide the repulsion, the one positive density term contributing to the spatial curvature.

Moreover, this peculiar form of energy tensor also underlies the curious "paradox" involved in the frequent claim that "over 68% of the matter content of the universe is in the form of an unknown *dark energy*". For, unlike all other forms of mass–energy, $\Lambda$ is gravitationally *repulsive*, rather than attractive, so it behaves completely oppositely from ordinary matter in this respect. Moreover, being constant in time (if indeed it *is* Einstein's cosmological *constant* $\Lambda$), this "68%" proportion will increase steadily as the universe expands, the average *densities* of all "ordinary" forms of matter (including dark matter) reducing relentlessly with time to comparative *total insignificance*!

This last feature raises another issue, that cosmologists have expressed disquiet about: that it is only around now in the universe's history (where "now" is to be interpreted very broadly indeed, in ordinary terms – namely some time within the range $10^9$–$10^{12}$ years after the Big Bang) when the "energy density" provided by $\Lambda$ is comparable with that of ordinary matter (including dark matter, whatever that is). In much earlier stages of the universe's history (say $<10^9$ years) the contribution of $\Lambda$ would have been insignificant, whereas in much later stages (say $>10^{12}$ years) $\Lambda$ will dominate completely over everything else. Is this a remarkable coincidence – something very odd that is in need of an explanation of some kind? Many cosmologists seem to believe so, and some would prefer that "$\Lambda$" be really some kind of evolving field, often referred to as *quintessence*. I shall return to these issues in the next two sections, but my own view would be that it is completely misleading to think of dark energy as any kind of material substance, or even as vacuum energy. There may well be a mystery to be resolved underlying $\Lambda$'s actual value (see, for example, §3.10), but it should be borne in mind that Einstein's $\Lambda$-term is basically the only modification to his original equations ($\mathbf{G} = 8\pi\mathbf{T}$) that can be made without drastically altering some of the underlying features of his magnificent theory. I see no reason why nature should not have taken advantage of this glaring possibility!

## 3.9. INFLATIONARY COSMOLOGY

Let us next consider some of the reasons why most cosmologists have felt a strong need to support the introduction of the fantastical-seeming proposal of *inflationary cosmology*. What is this proposal? It is an extraordinary scheme initially put forward in around 1980, independently by a Russian, Alexei Starobinsky (although in a rather different context), and an American, Alan Guth. According to their ideas, our actual universe, almost immediately following its Big Bang origin, in an extremely tiny period between about $10^{-36}$ and $10^{-32}$ seconds just after *that* momentous event, became subject to an *exponential expansion* – called *inflation* – that resembled the effect of the presence of a huge cosmological constant $\Lambda_{\text{infl}}$ which vastly exceeded the presently observed value of $\Lambda$ by an enormous factor, which would be something of the very rough order of $10^{100}$:

$$\Lambda_{\text{infl}} \approx 10^{100} \Lambda$$

(though there are many different versions of inflation which give somewhat different values). It should be noted that this $\Lambda_{\text{infl}}$ is still extremely tiny (by a factor of $\sim 10^{-21}$) as compared with the value $\sim 10^{121} \Lambda$, expected from vacuum-energy considerations. See Guth [1997] for a popular account; for more technical accounts, see also Blau and Guth [1987], Liddle and Lyth [2000], and Muckhanov [2005].

Before considering the reasons that most cosmologists today do accept this astonishing notion (inflation being included in all serious accounts of modern cosmology, both technical and popular), I should warn the reader that the term "fantasy" in the title of this chapter and, indeed, of this book, is aimed most particularly at this scheme. We shall be finding, especially in §3.11, that there are many *other* ideas currently under discussion by cosmologists that may well be considered a good deal *more* fantastical than inflationary cosmology. But what is particularly noteworthy about inflation is its almost universal acceptance by the cosmological community!

It should be said, however, that inflation is not just one universally agreed-upon proposal. There are numerous different schemes that go under the general banner of inflation and experiments are often aimed at distinguishing one such scheme from another. In particular, the issue of the $B$-modes in polarization of light in the CMB that was very publicly reported by the BICEP2 team, in late March 2014 [Ade et al. 2014], being claimed as strong evidence for one broad class of inflation models – even referred to as their smoking gun – was to a large extent aimed at distinguishing between different versions of inflation. The presence of such $B$-modes was argued to be a signal of the presence of primordial gravitational waves, these being predicted by some versions of inflation (but see also the end of §4.3 for an alternative possibility for the generation of $B$-modes). At the time of writing, the interpretation of these signals remains highly controversial, there being other possible interpretations of the observed signals. Nevertheless, few cosmologists today appear to doubt that the fantastical overall idea of inflation has some kind of actual truth, concerning the very early stages of the universe's expansion.

Yet, as I have explained earlier in the preface and §3.1, I do *not* mean to imply that inflation's fantastical nature ought to bar it from serious consideration. It was indeed put forward in an attempt to explain certain very remarkable – even "fantastical" – *observed* aspects of our actual universe. Moreover, inflation's current general acceptance comes from its explanatory power in accounting for other seemingly independent and previously unexplained remarkable features of the universe. Accordingly, it is important to take note that if inflation is not actually true of our universe – and I shall shortly be presenting my arguments for believing that it may well *not* be – then something else must be true, probably involving comparably exotic ideas of seeming fantasy!

Inflation has been put forward in numerous different versions, and I have neither the knowledge nor the fortitude to describe, in this section, any but the original one and the general version currently popular (but see also §3.11, where I shall briefly mention some of the wilder versions of cosmic inflation). In the original scheme as first put forward, the source of cosmic inflation was an initial state of "false vacuum" which, by a phase transition – analogous to the phenomenon of boiling, where a liquid becomes a gas, as overall conditions change – the universe *tunnelled*, quantum mechanically, into a different vacuum state. These different vacua would be characterized by different values of an effective $\Lambda$-term in Einstein's equations.

I have not previously discussed the (well-established) phenomenon of *quantum tunnelling* in this book. In its normal employment, it occurs with a quantum system that has two energy minima **A** and **B** separated by an *energy barrier*, where a system initially in the higher-energy state **A**, may spontaneously tunnel into **B** without there being energy available to surmount the barrier. I shall not concern the reader with the details of this procedure here, but I believe that its very questionable relevance in the current context should certainly be pointed out.

In §1.16, I brought to the reader's attention the issue of the choice of a vacuum state, which is a necessary ingredient in the specification of a QFT. One can have two different proposals for a QFT which would be identical (i.e. having identical specifications for their algebra of creation and annihilation operators, etc.) were it not for the fact that there is a different vacuum specified in each case. This was the crucial issue, in §1.16, with regard to the unwanted multitude of different string (or M-) theories that constituted the so-called landscape. The problem, pointed out in §1.16, was that each of the QFTs in this vast landscape would constitute an entirely separate universe, where no physical transition from a state in one such universe to a state in another could take place. The process that is involved in quantum tunnelling is a perfectly acceptable quantum-mechanical action, however, so that it is *not* normally thought of as something that would allow a transition from a state in one "universe" to a state in another, where the two differ from each other in their choice of vacuum.

Nevertheless, the proponents of the idea that cosmic inflation can result from the tunnelling from a state of false vacuum, with one choice of vacuum state, with vacuum energy given by $\Lambda_{\text{infl}}$, to another vacuum, with vacuum energy given by the currently observed cosmological constant $\Lambda$, is still a popular one among certain sections of the cosmological community [Coleman 1977; Coleman and De Luccia 1980]. From my own particular perspective, I should also remind the reader of the difficulties, pointed out in §3.8, that I have even with the very notion that the cosmological constant should be regarded as a manifestation of vacuum energy. Of course, within the context of fantastical ideas that might be needed for a proper understanding of the origin of our universe, with its many very remarkable features that seem to be aspects of its extremely early stages (e.g. §3.4), ideas of this kind, which break free of the strictures of normal physics, should certainly not be ruled out of consideration. However, I do believe that careful note should be taken when ideas usually considered to be outside the scope of established procedures seem to need to be employed.

Although returning to the matter in §3.11, I shall not say much in this section about this original notion that inflation arose from a tunnelling from one vacuum to another, most particularly because this seems not to be the version now most commonly adhered to, owing to theoretical difficulties with the "graceful exit" of the inflationary phase (at around $10^{-32}$ s) of the original scheme, where the universe's inflation needed to fade away everywhere all at once, and to undergo what is called *reheating*. In order to circumvent these difficulties, a subsequent scheme referred to as *slow-roll inflation* was independently introduced, in 1982, by Andrei Linde [1982] and by Andreas Albrecht and Paul Steinhardt [1982], and most of my comments will be specifically concerned with this type of idea. In slow-roll inflation, there is a scalar field $\varphi$ referred to as the *inflaton field* (although in some early publications $\varphi$ was referred to as a *Higgs field*, an inappropriate terminology now abandoned) this being taken to be responsible for the inflationary expansion in the extremely early universe.

The term *slow roll* refers to a feature of the *graph* of the potential function $V(\varphi)$ for the $\varphi$-field *energy* (see figure 3-38), the state of the universe being represented as a point rolling down the slope of the curve. In different versions of slow-roll inflation (there being many), this potential function is specifically designed, rather than deduced from more primitive principles, so that rolling down it should provide intended properties required for inflation to work. There

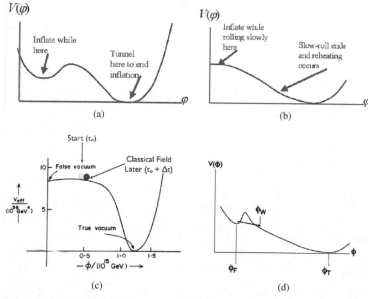

Figure 3-38: Some of the numerous proposals for the inflaton field $\varphi$'s potential function, suggested so as to provide all its desired properties. The considerable variety in the suggested $\varphi$-curve shapes is indicative of the lack of an underlying theory for the (inflaton) $\varphi$-field.

are no motivations from accepted particle physics or, as far as I can see, from any other parts of physics for the shape of this $V(\varphi)$ curve. The slow-roll part of the curve is incorporated so as to allow the universe to inflate for as long as is needed, after which the curve approaches a minimum, enabling the inflation to tail off in a reasonably uniform way, as the energy potential $V(\varphi)$ settles into its stable minimum, ending inflation. Indeed, different authors [see, for example, Liddle and Leach 2003; Antusch and Nolde 2014; Martin et al. 2013; Byrnes et al. 2008] suggest many different shapes for $V(\varphi)$, the arbitrariness of this procedure perhaps revealing a weakness in the inflationary proposal.

Nevertheless, such ideas would clearly not have been introduced without some strongly felt motivations. Let us, therefore, start with the conundrums which the inflationary scheme initially arose to address, in around 1980. One of these was an uncomfortable consequence of various popular grand unified theories (GUTs) of particle physics. Several GUTs predict the existence of *magnetic monopoles* [Wen and Witten 1985; Langacker and Pi 1980], i.e. individual *separate* north

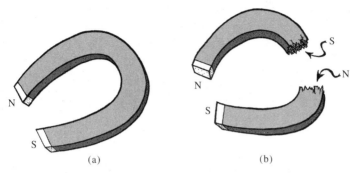

(a)                              (b)

Figure 3-39: Individual magnetic monopoles – isolated north or south poles – do not occur in conventional physics. An ordinary magnet has a north pole at one end and a south pole at the other, but if we break it in two, new poles appear at the break, so that the overall magnetic "charge" of each part remains balanced at zero.

or south magnetic poles. In conventional physics (and in observation also, so far at least) magnetic poles never occur as individual entities, but arise as part of a *dipole*, the poles always being paired with their opposites, north with south, as with ordinary magnets. If we break a magnet, separating the poles, then a pair of new poles appears at the break, a new south pole on the part with the original north pole and a new north pole on the other part. See figure 3-39. In effect, what we call magnetic poles arise as artefacts, resulting from circulating internal electric currents. Individual particles can also behave as magnets (dipoles), but have never been observed as monopoles, i.e. *individual* north or south poles.

However, the theoretical existence of such monopoles was, and still is, strongly argued for by some theorists, and most notably by some prominent in string theory, as is evidenced by the remark made by Joseph Polchinski in 2003 [see Polchinski 2004]:

the existence of magnetic monopoles seems like one of the safest bets that one can make about physics not yet seen.

In the very early universe, according to such GUTs, there would be a great proliferation of these predicted magnetic monopoles, but no such entities have ever been actually observed, nor is there indirect observational evidence that they have ever been present in the universe. To avoid a disastrous discrepancy with observation, the inflationary idea was suggested, whereby the exponential early expansion would thin out any initial preponderance of such magnetic monopoles to a level so tiny as to escape actual conflict with observation.

Of course, this motivation alone would not carry a great deal of weight with many theorists (such as myself) because the resolution of this discrepancy could well simply be that none of the GUTs in question is actually true of our universe – no matter how appealing such a theory may have seemed to its promoters. Nevertheless, we do conclude from this that without inflation there would be an additional serious issue to be faced by many current ideas about the nature of fundamental physics, most notably various versions of string theory. However, the magnetic monopole problem does not particularly feature in current claims for the necessity of inflation, these coming primarily from elsewhere. Let us examine these claims next.

The other arguments that were early put forward as key motivations for the introduction of cosmic inflation closely relate to points I raised in §3.6, namely those concerning the great *uniformity* of the matter distribution in the early universe. There is, however, a key difference between my own line of argument and those of the inflationists. My emphasis has been very much from the point of view of the puzzle presented by the Second Law of Thermodynamics (see §§3.3, 3.4, and 3.6) and the curiously lopsided way in which the extraordinary lowness of the initial entropy is delineated by the CMB, where just *gravitational* degrees of freedom appear to have been uniquely singled out among all those possible degrees of freedom that might have been suppressed (see end of §3.4). On the other hand, the proponents of inflation concentrate on just some particular aspects of this grand puzzle, albeit with significant connections with those of §§3.4 and 3.6, but put forward from a completely different perspective.

As I shall be describing later, there are indeed other observational reasons for the inflationary idea to be genuinely taken seriously, demanding at least some very exotic truth for their explanation, but these did not feature in the original motivations for inflation. In the early days, just three particular puzzling observational cosmological facts tended to be singled out. These were referred to as the *horizon problem*, the *smoothness problem*, and the *flatness problem*. It has been a common belief – a success frequently asserted by inflationists – that all three are actually solved by inflation. But are they?

Let us start with the *horizon problem*. The particular matter of concern here is the fact (already noted in §3.4) that the CMB, coming to us from all directions in the sky, has a temperature that is everywhere almost exactly the same – with differences of only a few parts in $10^5$, once correction is made for a Doppler effect due to the Earth's proper motion with respect to this radiation. A possible explanation

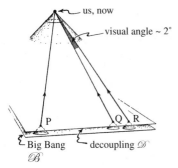

Figure 3-40: In the standard cosmological picture without inflation, the Big Bang 3-surface $\mathscr{B}$ in a conformal diagram would precede the decoupling 3-surface $\mathscr{D}$ by only a small amount (as indicated). Accordingly, two events Q and R on $\mathscr{D}$, whose visual separation from our space-time vantage point exceeds about 2°, can never have been in causal contact, since then their past light cones would not intersect, as we follow them into the past, before $\mathscr{B}$ intervenes.

for the uniformity, particularly in view of the extremely closely *thermal* nature of this radiation (figure 3-13 in §3.4), might be that this entire fireball of a universe had been the result of some great early thermalization process, bringing the entire universe, at least as far out as we can observe, into an expanding thermalized (i.e. maximum-entropy) state.

However, one difficulty with this picture was argued to be the fact that, according to the expansion rate of standard Friedman/Tolman cosmological models (§3.1), events on the *decoupling* 3-surface $\mathscr{D}$ (where the radiation was, in effect, produced, see §3.4), which are sufficiently separated, would lie so far outside each other's particle horizons that they would be causally independent. This would be so, even for points P and Q on $\mathscr{D}$ whose separation in the sky from our own vantage point is only about 2°, as illustrated in the schematic conformal diagram of figure 3-40. Such points P and Q would not have any kind of causal connection, according to this cosmological picture, because their past light cones are completely separated, right back to the Big Bang (the 3-surface $\mathscr{B}$). Accordingly, there would not have been any occasion for thermalization as a procedure for equalizing temperatures at P and Q.

This matter had puzzled cosmologists, but after Guth (and Starobinsky) put forward their extraordinary idea of *cosmic inflation*, a possible line of argument appeared to be at hand for resolving this conundrum. The introduction of an early inflationary phase has the remarkable effect of vastly increasing the separation

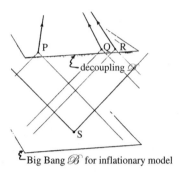

Figure 3-41: When inflation is included in the models (figure 3-40), the Big Bang 3-surface $\mathscr{B}$ is displaced much further down in the conformal picture (indeed, normally a good deal further down than indicated here), which has the effect that the past light cones of P and Q will always intersect before reaching $\mathscr{B}$, no matter how far apart they may be from our vantage point (e.g. P and R in the picture).

between the 3-surfaces $\mathscr{B}$ and $\mathscr{D}$, in the conformal diagram, so that any pair of points P and Q on the decoupling surface $\mathscr{D}$ visible to us in the CMB sky (even if in opposite directions from our vantage point) would have past light cones with a very significant overlap as they are traced back in time until they encounter the Big Bang 3-surface $\mathscr{B}$ (figure 3-41). This extended region between $\mathscr{B}$ and $\mathscr{D}$ – in fact, far greater than actually depicted in figure 3-41 for an inflationary expansion of $10^{26}$ – is a portion of de Sitter space-time (see §3.1). In figure 3-42 I provide a scissors-and-paste construction of the inflationary model, which I hope makes this conformal pushing back of the Big Bang intuitively clear. Thus, with inflation, there would have been ample time for complete thermalization. This resolution of the horizon problem would provide an easily sufficient period of communication for the entire part of the primordial fireball within our particle horizon to be brought into (an expanding) thermal equilibrium, all CMB temperatures thereby becoming almost exactly equal.

Before raising what I believe to be fundamental objections to this argument, it will be helpful, first, to consider the second of the problems claimed to have been solved by inflation, namely the *smoothness problem*. The problem has to do with the more-or-less uniform matter distribution and space-time structure that we seem to see throughout the universe (the presence of voids, etc., being regarded as relatively minor departures from this uniformity). Here it is argued that the $10^{26}$-fold (or so) exponential expansion would itself serve to iron out any significant irregularities that might have been present in the initial (supposedly

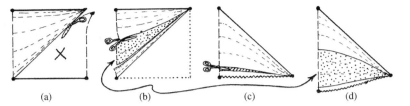

Figure 3-42:  How to construct a model of inflationary cosmology in terms of strict conformal diagrams. (a) Take a steady-state portion (figure 3-26(b)) of de Sitter space (figure 3-26(a)) and (b) cut a very long time stretch (shown dotted) from this portion; (c) remove a very short and early time stretch from the (let's say $\Lambda = 0$) $K = 0$ Friedmann model (figure 3-26(b)) and (d) glue the steady-state time stretch into it.

very irregular) state of the universe. The idea would be that whatever non-uniform features might have been present initially, these would have become stretched by an enormous linear factor (say $\times 10^{26}$), and so a considerable degree of smoothness would be the result, consistently with what is observed.

Both of these arguments represent attempts to explain the uniformity of the universe by invoking a phase of enormous expansion at a very early stage of its existence. I shall argue, however, that both are fundamentally fallacious [see also Penrose 1990; TRtR, chapter 28]. The most overreaching reason for this has to do with the fact that, as we have seen in §3.5, the extraordinary lowness in the entropy that we find in our universe – a lowness that is essential for the very existence of our 2nd Law – is actually expressed in this very uniformity. The whole idea underlying the inflationary argument seems to be that we can begin our universe with an essentially randomly irregular (and therefore maximal-entropy) starting point and, from that, arrive at the extraordinarily uniform, and therefore *gravitationally low-entropy*, situation that we find in the CMB fireball and in the considerable uniformity that we still find in the present universe (see end of §3.4). The key issue is the 2nd Law and where it came from. It cannot plausibly have come about simply from an ordinary physical evolution, defined by time-reversible dynamical equations, where we start from comparatively *random* – i.e. *high*-entropy – starting point.

The important fact to consider is that all the dynamical processes underlying inflation are actually time symmetric. I have not yet touched upon the type of equations that are used for slow-roll inflation. There are various ingredients required, the most important being the scalar inflaton field $\varphi$, which satisfies equations specifically concocted to make inflation work. There are procedures that come into the discussion, such as phase changes (like the "boiling" referred to earlier) which have the appearance of being time asymmetric, but these are macroscopic entropy-increasing procedures which depend upon time-symmetric submicroscopic processes, the time asymmetries providing manifestations of the 2nd Law, not explanations for it. The way that the inflationary argument is phrased is to make it look intuitively plausible that an overall entropy-*lowering* process could be dynamically achievable – but the 2nd Law tells us that it is not!

Let us start with the second inflationary argument above, aimed at showing that a smoothed-out universe would inevitably result from the inflationary process. Suppose that it is indeed true that, with a generic starting point, the inflationary processes will almost invariably lead to a smoothed-out expanding universe after inflation has finished. This concept fundamentally conflicts with the 2nd Law. There will be many states that, if simply conjured up at that *later* time, would *not* be smooth (and if that were not the case, there would have been no need for the inflation to have got rid of them). Let us reverse the direction of time from such a macroscopic state – but with generically perturbed submicroscopic ingredients – and let the reversed-time dynamical evolution (with equations still allowing for the possibility of inflation, with $\varphi$-field, etc.) take over. This must lead us somewhere, but now with entropy increasing in the *collapsing* direction. Where it leads us would be generally some very complicated high-entropy black-hole congealed state, which is not at all like an FLRW model, but much more like the situations depicted in figure 3-14(a),(b) in §3.3. Reversing time again to get an initial picture like figure 3-14(c), we find that $\varphi$ would have been powerless to smooth this (far more probable) initial situation. In fact, inflationary calculations are almost invariably done in an FLRW background, which completely begs the question since it is the *non*-FLRW initial states that form the stupendous majority, as we have seen in §§3.4 and 3.6, and there is no reason whatever to believe that these will actually inflate.

What about the first argument, that the near isotropy of the CMB temperature is explained by inflation bringing the points on the decoupling sky into causal contact with each other? Again, the problem is that we have to explain how a low-entropy situation has come about starting from what is supposed to be a "general" (and therefore *high*-entropy) initial state. For this purpose, bringing these points into causal contact does us no good at all. Perhaps this causal contact indeed allows a thermalization to take place, but what good does that do us? What we need to explain is why, and in what way, the entropy was so extraordinarily low. A thermalization process *raises* the entropy (making previously unequal temperatures equal, this being just one manifestation of the 2nd Law). So by invoking thermalization at this stage, we have actually pushed the entropy down even *lower* in the past, and this makes the problem of how the very special initial state of the universe came about even worse!

In fact, none of this aspect of inflation really touches the problem at hand. The observed isotropy in the CMB temperatures is, in my view of things, really a *secondary* effect in the context of a far larger issue, namely the extraordinary lowness of the initial entropy as exhibited in the uniformity of the Big Bang singularity. Just in itself, the CMB temperature isotropy does not make a significant total contribution to this issue of the entropy lowness of the universe as a whole (as we have seen in §3.5). It is merely a reflection of a much more important isotropy, namely that of the detailed *spatial geometry* of the earliest universe (i.e. the initial singularity). It is the fact that the gravitational degrees of freedom were not excited in the earliest universe – as is signalled by the complete absence of white-hole singularities forming part of the Big Bang – that is the basic issue we must face. For some deep reason, a reason totally untouched by the inflationary proposal, the initial singularity was indeed exceedingly uniform, and *this* uniformity was responsible for a very uniform, and uniformly evolving, primordial fireball that we see in the CMB, and could well be what is responsible in the temperature uniformity over the CMB sky, thermalization having nothing to do with it.

To the non-believer in inflation – such as myself – it is actually fortunate that, in standard non-inflationary cosmologies, there is no possibility of a thermalization having taken place over the whole CMB sky. That would only serve to obliterate information about the actual nature of the initial state $\mathscr{B}$, so an assurance that this total thermalization had no time to occur tells us that the CMB sky is directly revealing some of the geometry in $\mathscr{B}$. In §4.3, I shall give my own viewpoint on the significance of these issues.

The third role claimed for inflation is a resolution of what is referred to as the *flatness problem*. Here I have to concede that inflation appears to have chalked up a genuinely predictive success on the observational side, whatever may be the theoretical merits (or demerits) of the actual inflationary argument. When the flatness argument was initially being put forward (in the 1980s), there had seemed to be clear observational evidence that the matter content of the universe, including dark matter, could not be more than about one third of the total needed to provide a spatially flat universe ($K = 0$), so it had seemed that the evidence pointed to a spatially *negative*-curvature universe ($K < 0$), whereas a clear prediction of inflation had seemed to point to overall spatial flatness as one of its necessary consequences. Yet, several committed inflationists had confidently predicted that when observations were refined, more material would have to be discovered, giving agreement with $K = 0$. What changed the observational situation, in 1998, was evidence for a positive $\Lambda$ (see §§1.1, 3.1, 3.7, and 3.8) which provided an effective additional matter density that turned out to be of just about the right size to lead to the theoretical conclusion $K = 0$. This could certainly be regarded as some sort of observational confirmation of one of the key predictions that many inflation theorists had been forcefully making for some while.

The flatness argument from inflation was basically similar to that used for addressing the smoothness problem. Here, the reasoning was that even if there had been a substantial spatial curvature present in the universe prior to the inflationary phase, the enormous stretching that the inflation would have imparted (by a linear factor of $10^{26}$ or so, depending on the version of inflation employed) would have resulted in a geometry in which the spatial curvature is indistinguishable from $K = 0$. Nevertheless, I am again distinctly unhappy with this actual argument, basically for the same reason as in the case of the smoothness argument. If "now" we had happened to find a universe whose structure were different from this, e.g. enormously irregular or roughly smooth but with $K \neq 0$, we could evolve this backwards in time (according to time-symmetric equations allowing for a "$\varphi$" with the potentiality to give rise to inflation) and see what initial singularity this provides us with. The forward evolution of that initial singularity would, in consequence, not give us a roughly smooth $K = 0$ universe.

There is another related fine-tuning argument that is sometimes presented as providing a compelling reason for believing in inflation. It concerns the ratio $\rho/\rho_c$ of local matter density $\rho$ to that critical value $\rho_c$ which would give a spatially flat universe. The argument is that in the very early universe, $\rho/\rho_c$ must have been exceptionally close to 1 (perhaps to up to 100 decimal places), for, if not, then the universe now would not have the current value of $\rho/\rho_c$ that is observed to be still pretty close to 1 (about up to 3 decimal places). Accordingly, we have a problem to explain the origin of this exceptional closeness of $\rho/\rho_c$ to 1 at an early stage of the universe's expansion. The inflation theorist's argument for this early extreme closeness of $\rho/\rho_c$ to 1 is that it was the result of an even earlier inflationary phase, which would have ironed out any deviations of $\rho/\rho_c$ from 1, if they were present in the Big Bang itself, yielding the exceptional closeness that would be needed just after the subsequent termination of the inflationary phase. However, there is a genuine question as to whether inflation would actually necessarily achieve this required ironing out, for the same reason as before (see §3.6).

I can see that there is, nevertheless, an issue here to be faced, and if one rejects inflation one must suggest an alternative theoretical argument. (I shall be indicating my alternative perspective in §4.3.) A further problem is the turning off of the inflationary phase (the graceful exit problem referred to earlier) which would have to take place with a simultaneity of quite extraordinary precision, in order to obtain the very precisely required $\rho$ value in a spatially uniform way, just at the "moment" when inflation turns off. There would also appear to be severe difficulties with the requirements of relativity in satisfying such a simultaneity requirement.

In any case, the issue of "fixing" the value of $\rho$ is only a very small part of the problem, because there is here an implicit assumption that there is just a single number "$\rho$" that needs to have a special value. This is only a very tiny part of the whole problem, this being the *spatial uniformity* of this density, which is an issue of the assumed closeness of the early universe to an FLRW model. As pointed out in §3.6, it is this spatial uniformity, and its relation to the extraordinary lowness in the gravitational contribution to the entropy, that is the real issue, and as argued earlier here, inflation does not really address this at all.

Yet, whatever flaws there might indeed be in these underlying motivations behind inflation, there are at least two further observational facts that have provided the theory with some remarkable support. One of these is the presence of correlations that are actually observed in tiny deviations from uniformity in the CMB which extend over wide angles of the sky, being strongly indicative of there actually being causal influences connecting widely separated points on the CMB sky (such as the P and Q of figure 3-40). This important fact indeed is inconsistent with a standard Friedmann/Tolman Big Bang cosmology, yet fully consistent with inflation (figure 3-41). If inflation is wrong, these correlations need explanation from some other scheme, apparently involving pre–Big Bang activity! Schemes of this nature will be discussed in §§3.11 and 4.3.

The other important observational support for inflation comes from the nature of the tiny deviations from temperature uniformity over the whole CMB sky (usually referred to as *temperature fluctuations*). Observations show that they are very close to being *scale invariant* (i.e. have the same degree of variation at different scales). Evidence for this had been noticed independently by Edward R. Harrison and Yakov Borisovich Zel'dovich [Zel'dovich 1972; Harrison 1970] many years before the inflationary ideas came about, but subsequent observations of the CMB [Liddle and Lyth 2000; Lyth and Liddle 2009; Mukhanov 2005] have much extended the range over which scale invariance is seen. The exponential (whence self-similar) character of inflationary expansion provides a general explanation

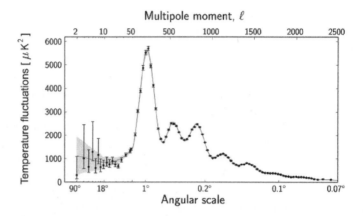

Figure 3-43:  The power spectrum of the CMB as measured by the Planck satellite. The vertical axis measures the temperature fluctuation and the horizontal axis (as marked at the top of the figure), the overall spherical harmonic parameter $\ell$ (the same as the $k$ of §A.11).

for this, where in the inflationary scheme the initial seeds of the irregularities are taken to be early quantum fluctuations in the $\varphi$-field, somehow becoming classical as the expansion progresses. (This is one of the weakest ingredients of the theoretical argument, since no logical argument within the standard framework of quantum mechanics for this quantum–classical transition is provided [see Perez et al. 2006].) Not only does inflation claim an explanation for this close scale invariance, but also for a slight deviation from it determined by what is referred to as the *spectral parameter*. These fluctuations provide the key initial input for the calculation of what is referred to as the *power spectrum* of the CMB (arising from a CMB harmonic analysis over the celestial sphere; see §A.11). Figure 3-43 shows the very remarkable fit (at least for large $\ell$ values – the $k$ values of §A.11) between the observed CMB data (obtained from the Planck space observatory, launched in 2009) and theoretical calculations. It should be borne in mind, however, that the numerical input from inflation in these calculations is very tiny (basically just two numbers), the detailed shape of the curve coming otherwise from standard cosmology, particle physics, and fluid mechanics relevant to the physical activity in the time period between the turn-off of inflation and decoupling. This is a long period (around 380000 years) of non-inflationary cosmology, as represented by the region between the 3-surfaces $\mathcal{B}$ and $\mathcal{D}$ as shown in figure 3-40, but now where $\mathcal{B}$ represents the moment of turn-off of inflation rather than the Big Bang itself. This is a period where the physics is well understood, and the input from inflation is quite minimal [see Peebles 1980; Börner 1988].

Set against these admittedly impressive successes are some puzzling anomalies with regard to inflation, though also somewhat puzzling irrespective of inflation. One of these is the fact that the correlations in CMB temperatures between widely separated points do not appear to extend beyond a separation angle (from our vantage point) of about $60°$, despite the inflationary argument that there should be no such angular limit on such correlations. Moreover, there are some irregularities in the large-scale mass distribution, such as the vast voids already referred to in §3.5 and, on the largest scales of all, asymmetries and inhomogeneities [Starkman et al. 2012; Gurzadyan and Penrose 2013] that seem to defy the conventional inflationary picture, where the initial source of the density fluctuations is taken to have a *random* quantum origin. Such matters are in serious need of explanation, and they do not seem to fit in well with the ideas of conventional inflation. These questions will be returned to in §4.3.

A point worth remarking upon here concerns the very distinctive way that this analysis has been carried out, namely in terms of a *harmonic analysis* (see §A.11) of the whole CMB sky, where interest has centred almost entirely on the power spectrum (i.e. that contribution to the entire CMB intensity that comes from all modes for each individual $\ell$ value). Whereas this procedure has undoubtedly led to some remarkable success, as evidenced by the extraordinarily close fit between the theoretical and observational, as shown in figure 3-43 (for $\ell$ values larger than around $\ell = 30$), it should be pointed out that this type of analysis has certain limitations. These may perhaps have skewed our interests in certain particular directions at the expense of others.

It should be pointed out, first, that, by merely concentrating upon the powers spectrum, we are ignoring a greater and greater proportion of the available information, the larger that $\ell$ gets. Let us look at this a little more carefully. In §A.11, the quantities $Y_{\ell m}(\theta, \phi)$, called *spherical harmonics*, are referred to, which are the different modes into which the pattern of temperatures seen in the CMB sky is to be decomposed. If we fix $\ell$ (a non-negative integer $\ell = 0, 1, 2, 3, \dots$), then the integer $m$ is allowed any of the $2\ell + 1$ alternative values $-\ell, -\ell + 1, -\ell + 2, -\ell + 3, \dots, \ell - 2, \ell - 1, \ell$. For each such pair $(\ell, m)$, the spherical harmonic $Y_{\ell m}(\theta, \phi)$ is some specific function on the sphere – which we are here taking to be the *celestial sphere* (with spherical polar coordinates $\theta, \phi$; see §A.11). For a given maximum value $L$ of $\ell$, the total number of different $m$s altogether will be $L^2$, which is far more than the number of $\ell$s, which is only $L$. The power spectrum exhibited in figure 3-43, obtained by the Planck satellite, takes $\ell$ values up to the maximum $L = 2500$, so it provides us with some 6250000 different numbers to characterize the temperature distribution over the CMB sky. But if we used all the information in the CMB sky up to this accuracy, we would have $L^2 = 6250000$ numbers. Thus we see that this power spectrum takes note of merely $1/L = 1/2500$ of the total information available!

In any case, despite the admitted success achieved in comparing theory with observed data, by means of the power spectrum, there are certainly other ways that the CMB might be fruitfully analysed. The decomposition of the CMB sky into the modes, as provided by the spherical harmonics, is the sort of analysis that one might apply, for example, to the elastic vibrational modes of a balloon. This could be argued to be some sort of analogy with what one might imagine the Big Bang to be like, but there are other kinds of analogy that could also be appropriate. Let us consider the sky of our own Earth. For that, a decomposition into spherical harmonics would hardly have been helpful! It is difficult to imagine how the subject of astronomy could have arisen at all if the night sky had been analysed using just the power spectrum. It would have been hard enough to have detected the Moon as a localized object, and certainly not the important nature of its periodic changes in apparent shape – its *phases* – that are so obvious to us as we simply look at it, let alone stars or galaxies by such means. The strong reliance on harmonic analysis, in the case of the CMB is, I believe, largely a feature of preconceived ideas about the Big Bang itself, and there are alternatives, one of which will be discussed in §4.3.

## 3.10. THE ANTHROPIC PRINCIPLE

It appears that at least some inflationists [see, for example, Guth 2007] have now come to appreciate that inflation cannot, simply of itself, explain the gravitation-ally extremely low-entropy smoothed-out state that we see in the early universe, and that this uniformity of the universe needs something more than just the dynam-ical possibility for inflation to take place. Even if inflation correctly forms part of the evolutionary history of the universe, it needs something more, such as a condition that provides an initial singularity of a closely FLRW type. If we try to hold to what appeared to be a central part of the inflationists' original philosophy – namely that the starting point for our universe should be essentially *random*, i.e. not fine tuned in some essential low-entropy way – then we need either a serious violation of the 2nd Law or else some other kind of selection criterion for the possible early state of the universe. One possibility for such a criterion, frequently mooted, is the *anthropic principle* [Dicke 1961; Carter 1983; Barrow and Tipler 1986; Rees 2000], referred to briefly at the end of §1.15.

What the anthropic principle depends upon is the idea that whatever is the nature of the universe, or universe portion that we see about us, being subject to whatever dynamical laws that seem to govern its actions, this must be something strongly favourable to our very existence. For, surely, if it were not favourable to us, we would not be here but somewhere else, either spatially (e.g. on some other planet), temporally (maybe at some radically different time), or perhaps even in some quite different universe. Of course, the "we" in this consideration need not actually be human beings or even any kind of creature that humanity has ever encountered, but some sort of sentient being capable of perceiving and reasoning. Usually the term *intelligent life* is used to express what is required.

Thus, as is not uncommonly argued, for the universe we actually perceive, it is necessary that the initial conditions were of the very special kind that allowed intelligent life to come about. A fully random initial state like that sketched in figure 3-14(c) of §3.3 could well be reasoned to be totally antagonistic to the development of intelligent life. For a start, it does not lead to the low-entropy highly organized situations that appear to be absolutely essential to anything remotely like intelligent information-processing life that the anthropic principle appears to demand. We might, therefore, take the view that the anthropic argument does indeed demand some strong restriction on the Big Bang's geometry, if this is to be part of a universe that could be inhabited by, and therefore perceived by, intelligent life forms.

But is such an anthropic requirement likely to be sufficient for narrowing down the possibilities for $\mathscr{B}$'s geometry (i.e. for the geometry of our Big Bang) so that, perhaps, an inflationary process might do the rest? Indeed, a case for inflation playing a role of this kind is not infrequently put forward [Linde 2004]. Accordingly, one is to imagine that the initial 3-surface $\mathscr{B}$ is (was!) actually a complicated mess, as in figure 3-14(c) of §3.3, but $\mathscr{B}$, being infinite in extent, would contain, simply by chance, odd places where it is smooth enough that inflation could take over. The argument is that these particular places would expand out exponentially, in an inflationary manner, to provide eventually habitable portions of the overall universe. Despite the inherent difficulties about providing anything approaching rigorous arguments in relation to this, I believe that a pretty strong case can, nevertheless, be made against such possibilities.

In order to make any kind of serious argument about this, I shall need to assume that a strong version of "cosmic censorship" holds (see §3.4), which implies, in effect, that $\mathscr{B}$ can be regarded as a *spacelike* (3-)surface (see figure 1-21 in §1.7), so that the different parts of $\mathscr{B}$ would be causally independent of one another. The surface $\mathscr{B}$ would not be expected to be necessarily very smooth, however. Nevertheless, it turns out that there would be a future-half "light cone" originating at each point of $\mathscr{B}$ – according to a precise definition of what is to be meant by $\mathscr{B}$'s "points" [see Penrose 1998a]. (The "points" of the singular boundary $\mathscr{B}$ are precisely defined in terms of the causal structure of the non-singular part of the space-time, being specified as *terminal indecomposable future-sets* (TIFs) in the space-time. See also Geroch et al. [1972].)

The arguments of §3.6 would suggest that, irrespective of the effects of inflation, the fraction of the total "volume" (in an appropriate sense) of $\mathscr{B}$ that could give rise to a universe expanse suitably resembling what we seem to find ourselves in – up to our particle horizon – would not be more than something like $10^{-10^{124}}$, since we need a region $\mathfrak{R}$ of $\mathscr{B}$ of that improbability, so that its entropy could be low enough to fit in with the discussion at the end of §3.6. (Just for definiteness, I am here including a dark matter contribution, see §3.6, but the argument that follows is insensitive to this issue.) This calculation simply depends upon the Bekenstein–Hawking black-hole entropy formula together with some estimate of the total mass that is involved, and is not sensitive to the effects of inflation, except that any entropy-increasing process involved in inflation would simply increase the rarity of regions of $\mathscr{B}$ that would inflate appropriately, i.e. *decrease* the number $10^{-10^{124}}$. If $\mathscr{B}$ is *infinite* in extent then, despite this improbability, such an exceptionally smooth, hugely low entropy region $\mathfrak{R}$ must exist somewhere within $\mathscr{B}$. Then, according to this inflationary proposal, that part $\mathfrak{R}$ would inflate to an entire universe, of the nature of our own (figure 3-44(a)), and intelligent life could arise in such an exponentially inflated region, and only in such an inflated region. In accordance with this picture, the entropy issue would be solved – or so it is claimed.

But can it really be solved in this kind of way? A very striking thing about the low-entropy nature of our actual universe is that it is not just a local thing, operative only in our own neighbourhood, but the basic structures – planets, stars, galaxies, galactic clusters – seem to proliferate in a roughly similar form (as far as we are able to ascertain) throughout the entire observable universe. Most particularly, the 2nd Law operates in the same way everywhere we look, in our vastly spatially extensive universe, just in the same way as in our neighbourhood. We find matter initially distributed in a fairly uniform way, often collecting into stars, galaxies, and black holes. We see great temperature variation (between that of hot stars and of empty space) arising by virtue of the eventual effects of gravitational clumping. It is this that provides the low-entropy stellar energy

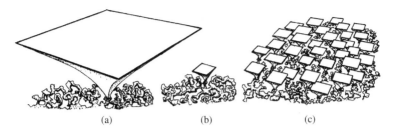

Figure 3-44: An inflationist's picture of how, very very rarely, there is a smooth enough region for inflation to take place, creating a universe like our own, with its 2nd Law, favourable to the production of intelligent life. (b) It is absurdly cheaper, in improbability terms, to inflate a considerably smaller region, but fewer intelligent beings would be created. (c) To get the number of such beings up to that obtained with the larger region, it would be vastly cheaper to do it with many smaller regions instead.

source essential to the production of life, whence (presumably) intelligent life arises here and there (see the latter part of §3.4).

Yet, intelligent life on our own Earth needs only a very tiny proportion of this volume of gravitational low-entropy. It is hard to see that our own lives depend upon such similar conditions holding in the Andromeda galaxy, for example, though perhaps some mild restrictions on it might be needed to prevent it from emitting anything dangerous to our existence. More to the point is that we see no limit to the broad similarity of the distant universe with the kind of conditions that we are familiar with in our local universe region, and this appears to hold true no matter how far out we look. If we are indeed simply requiring suitable conditions for the evolution of intelligent life just here, then the figure of $\sim 10^{-10^{124}}$ that we appear to find for the improbability of the universe conditions that we actually seem to find ourselves in is ridiculously smaller than the much more modest figure needed just for ourselves. We don't need favourable conditions in the Andromeda galaxy for our own existence; still less do we need such conditions within the distant reaches of the Coma cluster, nor within any of the other vastly remote parts of the visible universe. It is the relatively low entropy in those very distant regions that contributes primarily to the smallness of the figure $10^{-10^{124}}$, its absurd tininess being unimaginably smaller than anything that might be required for the intelligent life found here on Earth.

To illustrate this point, let us imagine that we don't see such a large volume of universe resembling our own general neighbourhood, but only up to some distance, say one tenth of the way out, perhaps because of a closer particle horizon or because, farther out, the universe resembles nothing like the gravitationally low-entropy state we are familiar with. That would cut down the mass content in our $10^{10^{124}}$ calculation by the factor $10^3$, thereby reducing the maximum black-hole entropy by $(10^3)^2 = 10^6$. This reduces $10^{124}$ to merely $10^{118}$, so we now get the absurdly smaller improbability (absurdly larger probability) $10^{-10^{118}}$ for locating such a region, say $\mathfrak{Q}$, within $\mathscr{B}$ that will inflate to a universe of the kind just described (see figure 3-44(b)).

One might try to argue that this more limited universe region, inflating from $\mathfrak{Q}$, will not contain so many intelligent beings, so our enhanced probability region is not doing as good a job of creating them as the larger universe that we actually see. But this argument carries no weight, because although we are now getting only $\frac{1}{1000}$ of the number of such beings in the smaller habitable region, we can get the required number up to whatever it is in our actual universe by simply considering 1000 of the more restricted inflating universes (figure 3-44(c)), at an improbability of

$$10^{-10^{118}} \times 10^{-10^{118}} \times 10^{-10^{118}} \times \cdots \times 10^{-10^{118}}$$

multiplied 1000 times, i.e. $(10^{-10^{118}})^{10^3} = 10^{-10^{121}}$, which is nowhere near the improbability given by the figure of $10^{-10^{124}}$ we appear to need for our actual universe. Thus, it is much "cheaper" in terms of improbabilities to make lots of small habitable universe-regions (i.e. 1000 $\mathfrak{Q}$-like subregions of $\mathscr{B}$) than it is to make just one big one (from $\mathfrak{R}$). The anthropic argument helps us here not at all!

Some inflationists might argue that the picture I have been presenting above does not give an appropriate view of the way that a limited region of inflation is supposed to act, and something more like an inflationary *bubble* is the relevant picture. In intuitive terms, one might have been thinking of the "boundary" of the inflating bubble to be, roughly speaking, a comoving 2-surface, as illustrated in figure 3-45(a), where the exponential increase in scale taking place during inflation would be represented simply as an exponentially increasing conformal factor $\Omega$, where $\Omega$ relates the metric of the diagram to that of the inflationary universe portion that the picture represents. But proponents of this kind of bubble inflation, where only a portion of the total universe is supposed to be involved, are not always clear about how the boundary between the inflating part and the non-inflating part is to be treated.

Often, the verbal descriptions suggest that the boundary of the inflationary region (such as a new "false vacuum") would spread outwards with the speed of light, engulfing the surrounding space-time, as it goes. This would appear to

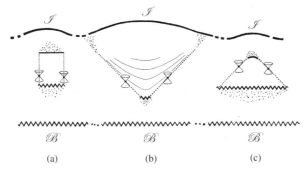

Figure 3-45: Schematic conformal diagrams for various ideas of inflationary bubbles: (a) the bubble boundary follows the time-lines; (b) the bubble boundary
expands out with the speed of light; (c) the bubble boundary moves inwards with
the speed of light, though its volume may nevertheless increase with time. The
broken line at the top represents an uncertainty as to how the future infinities of
these models are to relate to that of the ambient background.

demand a picture that resembles figure 3-45(b), but then one might well expect
that random influences from regions of $\mathscr{B}$ outside $\mathfrak{R}$ would drastically spoil
the purity of the inflationary picture. Moreover, this seems to do us no good at
all from the point of view of the preceding arguments, because the time-lines
in the inflating region now all emerge from what is effectively a single point
of $\mathscr{B}$, and the 3-surfaces of constant (inflationary) time are now represented as
hyperbolic 3-surfaces, depicted as the fainter lines in the diagram, and these 3-sur-
faces would have *infinite* volume – giving a picture, within the inflating region, of
a spatially *infinite* (hyperbolic) universe. The inflationary region now describes
an infinite universe that is supposedly of the general character of our own, so the
improbability figure of $10^{-10^{124}}$ is now $10^{-10^{\infty}}$, which is hardly an improvement!
In any case, even if this region is taken not to be actually spatially infinite, but
with a boundary of some sort, there remain severe unanswered problems about
how to treat the discontinuous unexplained physics of this boundary.

The remaining possibility seems to be of the kind given in figure 3-45(c).
Here, the entire inflationary evolution, and the universe history that follows it –
perhaps ultimately resulting in the kind of non-inflationary de Sitter-like exponential expansion that is currently observed in our universe – is depicted in the small
truncated pyramid-like region. Despite its rather improbable-looking appearance,
this picture is, in some respects, the most logical, because if the inflationary phase
is regarded as a *false* vacuum it might be expected to decay into the other. The

argument here would be that its small apparent "size" would be compensated by the hugeness of the conformal factor $\Omega$, which converts its geometry into a huge, perhaps twice exponentially expanded metric region (as is argued to be the case for our current observed universe)! This kind of picture certainly has severe difficulties and, like the others above, does not really address the objections raised about the absurdity of the above improbability figure $10^{-10^{124}}$.

Matters such as the improbability factors, that I have been considering, crude as they may be, do directly relate to conventional notions of entropy, as described by Boltzmann and by Bekenstein and Hawking [Unruh and Wald 1982], but such considerations seem rarely to be entered into in a serious way by inflation theorists. I remain thoroughly unconvinced that inflation helps in any way to resolve the key conundrum introduced in §3.6, whereby the initial state is given to us as having extremely low entropy, this restriction being in gravitational degrees of freedom only, thereby presenting us with a closely uniform FLRW type of universe. The anthropic argument is of no help to inflation in addressing this conundrum.

In fact, the anthropic argument is far less competent to address the conundrum of the 2nd Law even than I have indicated above. It is true that life as we know it has come about in harmony with the 2nd Law as we know it, but an anthropic argument based on the existence of life contributes almost nothing to an argument for the *existence* of a 2nd Law. Why does it not?

It is indeed the case that life on this planet has arisen through the relentless action of the evolutionary process of natural selection, building up more and more elaborate structures requiring low entropy for their continued existence and development. Moreover, all this depends crucially on the low-entropy reservoir provided by the hot sun in a background dark sky, and that demanded an initial (gravitationally) extremely low-entropy state (§3.4). It is important to appreciate that all this is in accordance with the 2nd Law. The total entropy is indeed going up all the time despite the wonderfully organizational effects – exotic plants, exquisitely constructed animals, etc. – that natural selection presents us with. So we might be tempted to come to an *anthropic* conclusion that the presence of life somehow explains the fact of the 2nd Law, giving an anthropic necessity to this law from our very existence.

In our familiar world, this is the kind of way that life has indeed arisen. We have become used to that. But in terms of low-entropy requirements, is this the "cheapest" (i.e. "most probable") way to produce the world that we see around us? Most certainly it is *not*! One can make a very rough estimate of the probability that life, as it now exists on Earth, with all its detailed molecular and atomic locations and motions, came about simply by chance encounters from particles coming in from space in, let us say, six days! For this to happen spontaneously would involve an improbability of, perhaps, the rough general order of $10^{-10^{60}}$, which, in terms of probabilities, would be a far "cheaper" way of producing intelligent beings than the way in which it was actually done! Something of this sort is really obvious simply from the nature of the 2nd Law itself. The lower-entropy earlier states of the universe that initially gave rise to humanity in its earliest stages (being of lower entropy simply by virtue of the 2nd Law) must have been far more improbable (in this sense) than is the situation now. This is just the 2nd Law in action. So it must be "cheaper" (in terms of improbabilities) for the state to have come about as it is now simply by chance, than for it to have arisen from an earlier much lower entropy state – if *that* had come about merely by chance! This argument continues, right back to the Big Bang. If we are looking for a chance-based anthropic argument, like the one being considered (involving subregions $\mathfrak{R}$ and $\mathfrak{Q}$ of $\mathscr{B}$), then the later it is that we consider the creation to have taken place, the "cheaper" it will be! Clearly, there must have been a different reason for the absurdly low-entropy initial state (the Big Bang) than one based on pure chance. The absurdly lopsided nature of that initial state (apparently with gravitational degrees of freedom, only, being completely suppressed) must have come about from a totally different, and much deeper, reason. The anthropic argument adds nothing to our understanding of these issues, any more than does inflation. (This type of conundrum is sometimes referred to as that of "Boltzmann brains". We shall return to this issue in §3.11.)

On the other hand (as may be argued to be the case also for inflation), the anthropic argument does appear to have some other significant explanatory roles to play, with regard to certain deep features of basic physics. The earliest example that I became aware of was presented in a lecture, given by the well-known astrophysicist and cosmologist Fred Hoyle (referred to in §3.2, in relation to the steady-state cosmological model). I attended this lecture in Cambridge when I was a young Research Fellow at St John's College. The lecture was given under the title "Religion as a Science", if my memory serves correctly, and I think it likely that it was one given in the University Church in the autumn of 1957. In that talk, Hoyle was concerned with the subtle issue of whether the physical laws might be fine-tuned in a way favourable to the existence of life.

It had been only a few years earlier, in 1953, that Hoyle had made a very remarkable prediction that there had to be a hitherto apparently unnoticed energy level of carbon (at about 7.68 MeV) in order that carbon (and thence many other elements heavier than carbon) could be built up in stars – red giant stars, which on exploding as supernovae would disperse the carbon into space. With some difficulty, Hoyle had persuaded the nuclear physicist William Fowler (of Caltech, in California) to look to see if this energy level existed. When Fowler was finally persuaded to look, he immediately found that Hoyle's prediction was true! The current observational value of this energy level (around 7.65 MeV) is a little lower than Hoyle's prediction, but well within the needed range. (Strangely, Hoyle did not share in the 1983 Nobel Prize, won by Fowler and Chandrasekhar.) Curiously, although now clearly an observational fact, the *theoretical* existence of this energy level seems to remain somewhat problematic as present understandings of theoretical nuclear physics go [Jenkins and Kirsebom 2013]. Hoyle, in his 1957 lecture, remarked upon the fact that if this energy level of carbon and another one of oxygen (that had been previously observed) had not been tuned so precisely with one another, then oxygen and carbon would not have come in remotely the proportions that are necessary for life to come about.

Hoyle's extraordinary and successful prediction of this carbon energy level is often cited as being a prediction, based on the *anthropic principle* – indeed the only clearly predictive success of this principle to date [Barrow and Tipler 1986; Rees 2000]. However, others have argued [Kraagh 2010] that Hoyle's prediction was not initially motivated by anthropic ideas. To my thinking, this is a bit of a moot point. It is clear that Hoyle had very good reasons for his prediction, since carbon has actually been found on the Earth in very significant proportions (clearly!), and had to have come about somehow. There is no need to appeal to the fact that these proportions also happen to be favourable for the evolution of life on Earth – and hence intelligent life. It might even have been regarded as a distraction from the force of his argument to concentrate on its biological implications. The carbon *is* here in copious proportions and, according to the physical understandings that had been reached at that time, it would have been very hard to see how it could have come about by means other than it having been produced in (red-giant) stars. Nevertheless, the importance of this issue, in relation to the existence of life, must surely have featured strongly in Hoyle's drive to get to the bottom of the *origin* of the very significant amount of carbon on the Earth.

Indeed, it is clear to me that Hoyle did also have an interest in "anthropic reasoning" at that time. In 1950, when I was a mathematics undergraduate at University College London, I listened to an inspirational series of radio talks given by Hoyle, on "The Nature of the Universe". I distinctly recall that at one point he raised the issue, in relation to the favourable conditions for life on this Earth, that some people regard it as "providential" that all the conditions on this planet are ideal, in so many ways, for the evolution of life – to which his response was that if this had not been the case, "then we wouldn't be here; we would be

somewhere else." Hoyle's curiously "anthropic" phrasing,[4] struck me particularly
– although we should bear in mind that the specific terminology *anthropic* was
introduced only much later by Brandon Carter [1983], who formalized the idea
of an anthropic principle much more clearly.

In fact, the version of the anthropic argument referred to in Hoyle's radio
talk would be what Carter refers to as the *weak* anthropic principle, namely the
(almost tautological) issue, mentioned at the beginning of this section, asserting
the need for a favourable location within our given space-time universe, spatially
or temporally. Carter's *strong* anthropic principle, on the other hand, is the issue
of whether the laws of nature, or numerical constants (such as the proton/electron
mass ratio) with which those laws have to operate, might somehow be "fine tuned"
in order for intelligent life to come about. It would be this strong version of the
anthropic principle that Hoyle's remarkable prediction of the 7.68 MeV energy
level of carbon could be considered to exemplify.

Another important example of anthropic reasoning, this time eventually turning
out to be of the weak version, although addressing deep theoretical issues of basic
physics, arose from the *Dirac large numbers hypothesis* [Dirac 1937, 1938]. Paul
Dirac had been examining some of the pure numbers that come into physics,
that is, dimensionless numbers that do not depend on the units being used. Some
of them are reasonable-sized numbers that we might imagine being explained
according to some mathematical formula (say, involving combinations of $\pi$, $\sqrt{2}$,
etc.). The reciprocal fine structure constant

$$\frac{\hbar c}{e^2} = 137.0359990 \cdots$$

($-e$ being the electron's charge), or the ratio of the proton's mass $m_p$ to the
electron's mass $m_e$, namely

$$\frac{m_p}{m_e} = 1836.152672 \cdots ,$$

are such possibilities, though in neither case is any actual mathematical formula
known.

However, Dirac argued that other pure numbers in basic physics are so large
(or small) that it seems implausible that a formula might exist for them. One such
would be the ratio of electric to gravitational attraction between an electron and a
proton, as in a hydrogen atom. This very large ratio (independent of their distance

---

[4] This use of "providential" is distinctly the form of words that I recall Hoyle using in his actual radio
talk, though I was unable to find it in the subsequent write-up of these talks [Hoyle 1950].

apart, as both are inverse square forces), is approximately

$$2.26874 \times 10^{39} = 2268740000000000000000000000000000000000,$$

where, of course, we do not expect most of these digits to be actually zeros! Dirac pointed out that using a *natural time unit* defined (say) by the proton mass $m_p$ or else the electron mass $m_e$, namely the respective quantities $T_{prot}$ or else $T_{elect}$, defined by

$$T_{prot} = \frac{\hbar}{m_p c^2} = 7.01 \times 10^{-25} \text{ seconds},$$

$$T_{elect} = \frac{\hbar}{m_e c^2} = 1.29 \times 10^{-21} \text{ seconds},$$

we find the age of the universe ($1.38 \times 10^{10}$ years $= 4.35 \times 10^{17}$ seconds, approximately) to be roughly

$$6.21 \times 10^{41} \text{ in proton units},$$

$$3.37 \times 10^{38} \text{ in electron units}.$$

These huge pure numbers (depending, somewhat, upon which particle we choose to use as our natural clock) are remarkably close to that for the ratio of the electric to gravitational force.

Dirac took the view that there must be a deep reason for the similarity of these very large numbers – and also some others we shall be coming to shortly. He therefore argued, in accordance with his *large numbers hypothesis*, that there must be a (yet unknown) physical reason that these numbers are remarkably closely related, differing in proportion from one another only by relatively tiny factors (such as the ~1836 proton/electron mass ratio), or through simple powers of this very large number. Such powers are exemplified by the values of $m_p$ or $m_e$ in *Planck* (i.e. absolute) units:

$$m_p = 7.685 \times 10^{-20},$$

$$m_e = 4.185 \times 10^{-23},$$

which are roughly like the reciprocal of the square root of the numbers we have just been considering. We can think of all these numbers being reasonably small multiples of simple powers of a large number $N$ of the general order of

$$N \approx 10^{20}$$

and then we find that ordinary particles (electron, proton neutron, pi-meson, etc.) are all $\sim N^{-1}$ in Planck units. The ratio of the electric to gravitational force for ordinary particles is $\sim N^2$. The age of the universe in time units of ordinary particles is $\sim N^2$, so the age of the universe in Planck units is $\sim N^3$. The total mass in the universe within our current (or ultimate) particle horizon is also $\sim N^3$, in Planck units, the number of actual massive particles within this region being $\sim N^4$. Moreover, the rough value of the cosmological constant $\Lambda$ is $\sim N^{-6}$, in Planck units.

Whereas most of the numbers, such as electric to gravitational force ratio or masses of particles in Planck units, seem to be constants (very closely, at least) inbuilt, as ingredients of the dynamical laws of the universe, the actual age of the universe, starting with the Big Bang, *cannot* be a constant, since it must obviously be increasing with time! Accordingly, so Dirac reasoned, the number $N$ cannot be a constant, so nor can any of these other large (or correspondingly small) numbers be constant; they must vary, at a rate determined by the power of $N$ relevant in each case. In this way, Dirac hoped that a fundamental physical/mathematical explanation for the "unreasonably" large number $N$ might not be needed after all, since $N^3$ is simply the *date*!

This is certainly an elegant and ingenious proposal, and when Dirac put it forward, it was consistent with observations. Basically, Dirac's proposal required that the strength of gravity would be slowly weakening with time, whence Planck units, which depend on the gravitational constant $\gamma$ being taken to be unity, would themselves also have to change with time. However, unfortunately for this proposal, subsequent more precise measurements [Teller 1948; Hellings et al. 1983; Wesson 1980; Bisnovatyi-Kogan 2006] demonstrated that $\gamma$ is *not* varying – certainly not at the rate the proposal required. Moreover, this appeared to leave us with the conundrum of the extraordinary fluke of our current date in Planck units corresponding closely to $N^3$, a number determined by physical laws apparently unchanging with time.

It is here that the (weak) anthropic principle comes to our rescue. It had been pointed out by Robert Dicke in 1957, and subsequently in more detail by Brandon Carter in 1983 [Dicke 1961; Carter 1983], that if one considers all the main physical processes that determine the lifetime of an ordinary main-sequence star, like our Sun, where these will indeed involve the strengths of the electric to gravitational forces on electrons and protons, that one may calculate what would be the general order of lifetime of such a star. This turns out to be $N^2$, or thereabouts, so that intelligent beings dependent on such a star, needing a steady and reliable output of radiation from it, upon being able to look out at the universe and to form some reliable estimate of the age of that universe, are likely to find the remarkable coincidence that the age that they find is indeed around $N^2$, in time units defined by normal particles, and $N^3$ in absolute units.

This is a classic use of the *weak* anthropic principle, providing a resolution of a problem that had seemed deeply puzzling. Yet there are worryingly few such examples (and, indeed, I know of no others). Of course, the argument does suppose that these intelligent beings are of the general kind we are familiar with, dependent for their evolution upon a stable and appropriate planetary solar system centred on a suitable main sequence star. Moreover, in the universe we know, with perhaps various Hoyle-type coincidences needed in order to allow the adequate production of chemical elements, the building up of these elements depending on seemingly fortuitous arrangements of energy levels, we may question whether life would have been possible at all if these apparently fortuitous detailed features of our physical laws had been just a little different – or perhaps even completely different. All this is the territory of the *strong* anthropic principle, which I consider next.

The strong principle is sometimes proposed in a quasi-religious form, as though the physical laws had been providentially fine-tuned in the construction of the universe, in order that (intelligent) life could come about. A slightly different way of formulating what is basically the same argument is to envisage that there might exist vast numbers of parallel universes, where each might have a different collection of values for its physical constants, or even a different set of (presumably mathematical) laws controlling behaviour. The idea of the strong anthropic principle can then be phrased in terms that all these different universes might somehow coexist, most being dead in the sense that no conscious (intelligent) beings exist within them. Only in those universes where such beings can come about would the needed coincidences for their very existence be discovered and marvelled at by the beings themselves.

It is, to my mind, disturbing how frequently theoretical physicists eventually come to rely on such arguments in order to compensate for a lack of predictive power that their various theories turn out to have. We have already witnessed this in the landscape of §1.16. Whereas originally the hopes for string theory, and its descendants, were that some kind of uniqueness would be arrived at, whereby the theory would supply mathematical explanations for the various measured numbers of experimental physics, the string theorists were driven to find refuge in the strong anthropic argument in an attempt to narrow down an absolutely vast number of alternatives. In my own view, this is a very sad and unhelpful place for a theory to find itself.

Moreover, we know extraordinarily little about what the requirements of (intelligent) life actually are. These are often phrased in terms of the needs of human-like creatures, such as Earth-like planets, liquid water, oxygen, carbon-based structures, etc., or even just the basic requirements of ordinary chemistry. We must bear in mind that from our human perspective, we may have a very limited and biased view of what may be possible. We see intelligent life around us and tend to forget how little we know of life's actual requirements, or initial conditions for it to come about. Occasionally, science fiction can be helpful in reminding us how little we understand about what might be essential to the development of intelligence, two remarkable examples being Fred Hoyle's *The Black Cloud* and Robert Forward's *Dragon's Egg* (and its sequel *Starquake*) [Hoyle 1957; Forward 1980, 1985]. Both are fascinating reads and full of original, scientifically based ideas. Hoyle suggests a fully developed individual intelligence within a galactic cloud. Forward develops, in remarkable detail, how a life form might evolve on the surface of a neutron star, operating at an enormously greater speed than we do. Yet, these are humanly imagined intelligent life forms, still very much within the scope of structures that we have so far come across in the universe.

As a final comment, it cannot really be said that the conditions for intelligent life *are* all that favourable in the universe we actually inhabit. There is some intelligence on planet Earth, but we have no direct evidence that it is other than extremely rare elsewhere in our universe. We might well ask to what extent the actual universe is really very favourable for conscious existence!

## 3.11. SOME MORE FANTASTICAL COSMOLOGIES

I should again remind the reader that the term *fantastical* is not intended to be understood in a necessarily derogatory sense. As I have stressed earlier, particularly in §§3.1 and 3.5, our actual universe is, in various ways, itself something quite fantastical, seemingly in need of fantastical ideas for its comprehension. Much of this is already directly revealed in the CMB, providing not just our most direct evidence for the Big Bang's very existence, but also revealing certain curious aspects of the Big Bang's particular nature. We find this nature as having been an extraordinary combination of two opposites: almost complete randomness (as revealed by the CMB's thermal spectrum) coupled with an extraordinary order, with an improbability at least as extreme as $10^{-10^{123}}$ (revealed by the CMB's uniformity over the sky). The main trouble with the models that people have suggested so far is not so much that they are crazy (although most are, indeed, crazy to some degree), but that they are not nearly sufficiently crazy to explain both these extreme observational facts at the same time – and, indeed, most theorists do not even seem to recognize the magnitude or even the outlandish nature of these particular facts that we are presented with in the very early universe, although some do thoroughly address other puzzling matters revealed in the CMB.

In recent years, I have myself tried my hand at providing my own crazy-looking cosmological model, in a direct attempt to accommodate specifically these observed aspects of the Big Bang. My intention in these chapters so far has, however, been generally to refrain from imposing specific schemes of my own upon the reader. Nevertheless, I am allowing myself the luxury of providing a very brief description of this "crazy" scheme in the penultimate section §4.3. My treatment of the proposals that I am relating in the current section will be even briefer, as I think that it would be both inappropriate and too difficult for me to attempt to address them in any detail. This is largely because of the variety and scope – and often sheer implausibility – of the multitude of schemes under consideration.

One broad class of extraordinary proposals does need some attention here, as they are very commonly scientifically discussed, and appear to be taken so seriously as to be regarded by a good proportion of the general public as scientifically *accepted* ideas! The theories I refer to here are those which regard our universe as being just one of a vast number of parallel universes. There are basically two, or perhaps three, different strands of reasoning that are taken as driving us into this form of belief.

One of these comes from a key issue discussed in chapter 2, and brought to a head in §2.13, namely the interpretation of the formalism of quantum mechanics that leads us to the so-called Everett interpretation or many-worlds interpretation of quantum mechanics, which is where we are, in a sense, logically driven if we hold to the view that unitary evolution **U** applies accurately to the universe as a whole without any physically real action of state reduction **R**. Accordingly, as with the Schrödinger cat described in §2.13, the two alternatives of the cat walking through the A-door or the cat walking through the B-door are considered *both* to take place, but in parallel worlds. Since such bifurcations are considered to be taking place all the time, according to this view, we are led to an utterly stupendous multitude of such worlds all coexisting simultaneously. As I described at the end of §2.13, I do not myself take such a picture seriously as providing a sensible viewpoint with regard to physical reality, though I do appreciate why many of those who hold unshakably to a total faith in the physical truth of the quantum formalism are led to holding to such a standpoint.

However, this is not the parallel-universe picture that I wish to describe here. The alternative motivational line of reasoning of relevance here is a different one (although some may take the view that, in an appropriate sense, the two pictures are the same, or at least related in some way to each other). This reasoning was described towards the end of §3.10 as one interpretation of the *strong anthropic principle*, where it is considered that there might indeed be parallel universes actually existing though in no way able to communicate with ours, in which the pure numerical constants of nature (or even laws of nature) might differ from one to another, and that those in the universe that we directly perceive would have to be particularly favourable for life. The argument is that the seemingly "providentially" favourable values of the pure-number constants of nature can be understood if we imagine that universe histories not too different from our own *do* actually exist "in parallel" with each other, but with different collections of values for these pure numbers. Only those universes for which the pure numbers take favourable values would be inhabited by conscious intelligent beings, and since we *are* such beings – so the argument goes – these collections of numbers must necessarily be found to be favourable.

Closely related to this second view, but perhaps with a more direct physical motivation, is a standpoint that has sprung from the ideas of inflationary cosmology. We may recall from the discussion given at the beginning of §3.9 that the original point of view, with regard to inflation, was that very soon after the universe started, at around $10^{-36}$ seconds following the Big Bang itself, there was an initial state of the universe (a "false vacuum") where the cosmological constant had, in effect, a value very different (by a factor of, perhaps, very roughly $\sim 10^{100}$) from that which $\Lambda$ has now, and that the universe then "tunnelled" into the vacuum that we have now at the end of the inflationary period (at $10^{-32}$ s) – but take note of my cautionary notes about such tunnelling, made early in §3.9. We recall that the "Dirac considerations" of §3.10 suggest that the large pure numbers should all behave like simple powers of a particular large number $N$ (where for "age of the universe" we now substitute "average lifetime of main-sequence star") and in particular we find, for the cosmological constant, that $\Lambda \approx N^{-6}$ in Planck units. In accordance with this viewpoint, we have for the *inflationary* cosmological constant $\Lambda_{\text{infl}} \approx 10^{100}\Lambda$, which suggests that the inflationary version $N_{\text{infl}}$ of $N$ ought to be given very roughly by

$$N_{\text{infl}} \approx 2000,$$

since then $N_{\text{infl}}^{-6} \approx (2 \times 10^3)^{-6} \approx 10^{-20} = 10^{100} \times 10^{-120} \approx 10^{100}\Lambda \approx \Lambda_{\text{infl}}$, as required, and so the effective pure large numbers in the inflationary phase should

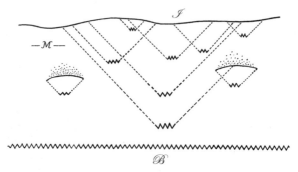

Figure 3-46: Schematic conformal picture of eternal inflation: it is argued that, although exceptionally infrequent, occasional events occur, setting off local inflationary bubbles (here taken mostly in accordance with figure 3-45(b)).

be modified accordingly, in relation to what they are now. This would be the expectation in accordance with the Dirac large-numbers hypothesis, as modified by the Dicke–Carter anthropic argument for the age of the universe. Presumably, the value $2 \times 10^3$ for $N_{\text{infl}}$ would not be favourable for the development of intelligent life during the inflationary phase – but it's worth a thought!

There are various extensions of the original inflationary ideas, the most influential going under names like *eternal inflation* [Guth 2007; Hartle et al. 2011], *chaotic inflation* [Linde 1983], or *eternal chaotic inflation* [Linde 1986]. (An explanatory paper on the above terms is given by Vilenkin [2004].) The general idea of these proposals seems to be that inflation can be set off at various places throughout space-time, such occurrences resulting in (very rare) spatial regions that quickly dominate over everything else in that neighbourhood, owing to the rapid spatial exponential expansion. Such regions are often referred to as *bubbles* (see also §3.10) and it is supposed that our own perceived universe region is such a bubble. In some versions of this kind of proposal, it is envisaged that there is actually no beginning to this activity, and it is normally taken that there is to be no end. The rationale for this type of activity seems to come from the inflationary expectation that, whereas the tunnelling from one vacuum to another is taken to be something with a very tiny probability, such events must necessarily occur from time to time in an indefinitely exponentially expanding infinite universe (modelled by de Sitter space; see §3.1). Conformal diagrams for this kind of activity are sometimes drawn resembling those of figure 3-46 (being based, essentially, on figure 3-45(b) of §3.10). It is sometimes envisaged that such expanding bubbles might intersect, with observational consequences not very clearly delineated, and which are hard to make geometrical sense of, but for which observational evidence is sometimes claimed [Feeney et al. 2011a,b].

These cosmological schemes are often thought of as having certain characteristics of the parallel-universe proposals because in the various different bubbles there would be expected to be different values of the cosmological constant $\Lambda$ (and with a negative $\Lambda$ in some bubbles). In accordance with the (Dicke–Carter-modified) Dirac large-numbers hypothesis considered earlier, one might well expect that there would be changes in other pure-number values, so some bubbles might be favourable to life and others not. The type of anthropic discussions considered in §3.10 would again have relevance. However, the reader ought, by now, to have clearly gathered my own lack of sympathy with schemes that depend for their viability on anthropic arguments of this kind!

One of the admitted difficulties with these bubble-universe inflationary pictures is an issue that goes under the heading of "the problem of Boltzmann brains". (The reason that Boltzmann's name is attached to this idea seems to be that in a short article [Boltzmann 1895] he considered the possibility that our 2nd Law might have arisen from an extremely unlikely random fluctuation. However, Boltzmann did not put the idea forward as something that he believed, or even originated, but attributed it to his "old assistant Dr Schuetz". In fact, this basic consideration is one that I already mentioned in §3.10 as a demonstration that the anthropic argument is of no use whatever for explaining the existence of our actual 2nd Law. But a similar type of argument is often introduced as a serious problem encountered with proposals like eternal inflation.)

The difficulty is expressed as follows. Let us suppose, as inflationary schemes apparently require, that there was an exceptionally improbable space-time region $\mathfrak{R}$ – possibly part of the Big Bang 3-surface $\mathscr{B}$, but alternatively just located at somewhere within the *depths* of the space-time, as is envisaged in the eternal-inflation picture – where $\mathfrak{R}$ was the "seed" that triggered the inflationary phase from which our perceived universe was supposed to have come about. The absurdity of the anthropic type of explanation for our necessarily being in such a bubble is made manifest when we consider how ridiculously cheaper (in the sense of improbabilities; see §3.9) it would be simply to produce, by mere random collisions of particles, the entire solar system with all its life ready-made, or even just a few conscious *brains* – referred to as *Boltzmann brains*. So the problem is: why did we not come about *this* way, rather than from an absurdly less probable Big Bang, after $1.4 \times 10^{10}$ tedious years of unnecessary evolution? It seems to me that this conundrum simply points to the futility of seeking explanations of this anthropic type for the low-entropy requirements of our actual universe – and it clearly indicates to me the incorrectness of the bubble-universe idea. As I already argued in §3.10, it also demonstrates the impotence of the anthropic argument as an explanation of the universe we see about us, with its 2nd Law operating in the way that we see it operates all about us. We need a completely different explanation for why our Big Bang had the very remarkable form that it appears in fact to have had (see §4.3). If the ideas of eternal inflation or chaotic inflation do actually require such an anthropic argument for their viability, then I would say that these ideas are simply *not viable*.

To end this chapter, I mention two other cosmological proposals that are not so wild as the ones just considered, but which are, in their somewhat different ways, intriguingly fantastical. They both depend, at least in their original formulations, on certain ideas from higher-dimensional string theory so, in view of arguments I expressed in chapter 1, it might well be felt that I would not be in total sympathy with such proposals. Nevertheless, as I have expressed many times in this chapter, I believe it is essential (inflation or no inflation) that some theory is needed which provides us with an extremely special initial geometry (expressed here as structure at the 3-surface $\mathscr{B}^-$). It is not really at all unreasonable that theorists should turn to string-theoretic ideas for inspiration with regard to some kind of geometry that breaks free of the strictures of classical general relativity, especially in relation to the relevant physics at $\mathscr{B}^-$. Moreover, I do believe that there are ideas of considerable importance within both of the proposals I shall be describing, even though I am not altogether happy with either. They are both pre–Big Bang schemes, but differing from each other in various respects. One is a scheme introduced by Gabrielle Veneziano, and developed in detail by him and Gasperini [Veneziano 1991, 1998; Gasperini and Veneziano 1993, 2003; Buonanno et al. 1998a,b], and the other is the ekpyrotic/cyclic cosmology of Steinhardt, Turok, and collaborators [Khoury et al. 2001, 2002b; Steinhardt and Turok 2002, 2007].

One may ask why it should be considered helpful to extend our universe model to before the Big Bang, particularly in view of *singularity theorems* (see §3.2) that tell us that if Einstein's classical equations are to be maintained (with reasonable physical assumptions, such as standard local energy positivity assumptions on the matter content), then a singularity-free continuation backwards through the Big Bang is not possible. Moreover, there is no generally accepted proposal whereby an appeal to quantum gravity allows us to achieve such a continuation in general circumstances, although there are some interesting ideas for this; see Ashtekar et al. [2006] and Bojowald [2007] for continuing proposals within the theory of loop quantum gravity. Yet, if one chooses *not* to adopt the standard inflationary picture

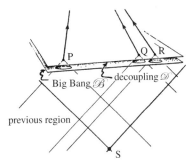

Figure 3-47: In the absence of inflation, correlations between CMB sources can occur outside the horizon limit of classical cosmology if there were a pre–Big Bang region. In this schematic conformal diagram, the pre–Big Bang event S could act to correlate Q and R, and even P, at a considerable angular distance from them.

(which I regard as a rational standpoint, in view of issues raised in §§3.9 and 3.10), there is a definite need to consider seriously a possibility that our Big Bang 3-surface $\mathscr{B}$ might well have been *preceded* by some sort of "earlier" space-time region.

Why is this? As mentioned already in §3.9, and illustrated in figure 3-40, in the standard (Friedmann/Tolman) cosmologies, correlations in the CMB, beyond just $2°$ or so in the sky, should not be observed. Yet there is now strong observational evidence for such correlations, even up to around $60°$. Standard inflation deals with this matter by enormously extending the "conformal distance" between the 3-surfaces $\mathscr{B}$ and $\mathscr{D}$ (of decoupling); see figure 3-41 in §3.9. However, if there is sufficient actual space-time *prior* to $\mathscr{B}$, then such correlations can certainly arise through activity in this pre–Big Bang region, as shown in figure 3-47. Thus, if inflation is to be rejected, there is a clear motivation from observation to consider that there might well have been something actually going on before the Big Bang!

In the Gasperini–Veneziano proposal, the ingenious idea is put forward that inflation *itself* was something that occurred before the Big Bang – which is an excellent example of moving the goal-posts! These authors have their own reasons for making this temporal displacement of inflation, depending on some string-motivated considerations involving a string-theoretic degree of freedom referred to as the *dilaton field*. This is closely related to the "$\Omega$" that occurs in the conformal rescalings of the metric ($\hat{\mathbf{g}} = \Omega^2 \mathbf{g}$) that were considered in §3.5 – these rescalings being referred to in this work as passing from one conformal frame to another. In higher-dimensional string theory, there is also the issue that there are both "internal" dimensions (i.e. the tiny curled-up unseen ones) and also the ordinary "external" ones, and these can behave differently under rescalings. But irrespective of these particular reasons for considering conformal rescalings, it is certainly an interesting possibility (of essential relevance also to the scheme I shall describe in §4.3). In the Veneziano scheme, for example, there might appear to be a geometrical oddity of wanting to have a dilaton-driven inflation setting in during the *collapsing* pre–Big Bang phase, but how one reads this is a matter of which choice of conformal frame is being used. An inflationary *contraction* in one conformal frame can look like an *expansion* in another. The scheme makes a serious attempt to address the issue of the highly improbable structure of the Big Bang (the initial 3-surface $\mathscr{B}$), and there are arguments for deriving the observed near-scale invariance of the temperature fluctuations in the CMB, obviating the need for conventional inflation.

The ekpyrotic[5] proposal of Paul Steinhardt, Neil Turok, and their colleagues borrows from string theory the introduction of a 5th space dimension, which connects two copies of the 4-dimensional space-time, these being referred to as *branes* (presumably something of the nature of D-branes, or *brane-worlds*, as discussed in §1.15, although not described as such in the papers, the terminology *M-theory branes* and *orbifold branes* being used). The idea is that just before the bounce that occurs in this scheme, which converts a Big Crunch to Big Bang, the distance between these two branes is reducing rapidly, becoming zero at the moment of the bounce, immediately increasing again after that event. The structure of the 5-geometry is taken to remain non-singular, with coherent equations, despite the singular nature of the projected 4-space-time. Even though inflation in the ordinary sense is absent, arguments are presented that can indeed supply the close to scale invariance exhibited by the temperature fluctuations in the CMB [Khoury et al. 2002a].

One may reasonably ask how the problem raised in §3.9 – whereby the messy crunch of a gravitational-entropy-increasing collapse (see figure 3-14(a),(b) in §3.4) is somehow to be converted into a gravitationally low-entropy Big Bang – is to be evaded (figure 3-48). The idea here is that prior to the ultimate collapse to its Big Crunch, the pre-bounce phase was involved in a de Sitter-like exponential expansion phase (our observed $\Lambda$-driven expansion), which, according to this scheme, would last for some $10^{12}$ years, during which this expanding phase would

---

[5] From the ancient Greek *ekpyrosis*, "a Stoic belief in the periodic destruction of the cosmos by a great conflagration every Great Year. The cosmos is then recreated only to be destroyed again at the end of the new cycle".

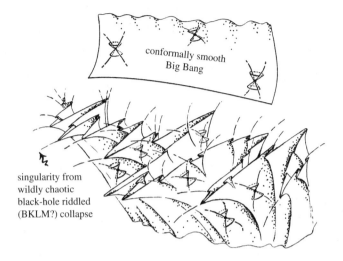

conformally smooth
Big Bang

singularity from
wildly chaotic
black-hole riddled
(BKLM?) collapse

Figure 3-48: A key problematic issue arising in pre–Big Bang theories is how a collapsing universe phase "bounces" to become an expanding one remotely resembling our own. If the initial state of the expanding phase is very low in gravitational entropy (i.e. with closely uniform spatial geometry), as appears to be the case for our own universe, then how can this come about if the collapsing phase has the expected chaotic behaviour (perhaps BKLM) of very high gravitational entropy?

thoroughly dilute the density of black holes and of any other remaining high-entropy debris. (It should be noted, however, that this expansion time would not be nearly long enough for black holes to have disappeared by Hawking evaporation, which would need a much longer period of around $10^{100}$ years; see §4.3.) What is being "diluted" here is the entropy *density* with respect to the comoving volume; the total entropy per comoving volume cannot reduce, without violating the 2nd Law. This applies also in the subsequent collapsing phase, which is taken to occur after around $10^{12}$ years, so the total entropy per comoving volume still cannot reduce. So how does this actually satisfy the 2nd Law in an appropriate way? To understand how this is intended to work, it is best that we move on to the *cyclic* version of this theory.

So far, I have just been describing the original ekpyrotic scheme, which (as with the Veneziano proposal) treats just a *single* bounce from a contraction phase to an expanding one. But Steinhard and Turok extended their model to a continuing *succession* of cycles, where each starts with its big bang, this initially evolving closely according to a conventional FLRW $\Lambda$-cosmology (without an early inflationary stage), but after some $10^{12}$ years of (mostly exponential) expansion turns round to become a contracting model that ends in a big crunch, which then undergoes an *ekpyrotic bounce* to a new big bang, thereby starting the whole process over again. This gives us an unending sequence of cycles, infinite in both directions. All the non-standard FLRW behaviour (i.e. that not in accordance simply with Einstein's $\Lambda$-equations) is controlled by the 5th dimension, bounded by the branes, as considered above in the case of a single ekpyrotic bounce. The brane separation reduces to zero at each bounce, controlling that activity in a non-singular way.

We need to address the question: how does this cyclic model avoid contradiction with the 2nd Law? As far as I can see, there are two aspects to this. One of these is addressed by the fact that the comoving volumes under consideration above, though they can actually be continued *through* the bounce by following the comoving time-lines through each bounce, do not need to correspond in *size* from one cycle to the next – and, indeed, they are taken *not* to do so. Let us consider some particular time-slice $\mathcal{S}_1$, given by a time $t = t_0$ in one particular cycle, and then choose the exactly corresponding time-slice $\mathcal{S}_2$ in the next cycle, again given by $t = t_0$ (measuring the time from the big bang of each cycle). We can follow a chosen comoving region $Q_1$ in the earlier time-slice and follow the time-lines right through the bounce until we reach the time-slice $\mathcal{S}_2$. We now find that, following the time-lines faithfully, the region $Q_2$ that we arrive at within $\mathcal{S}_2$ is enormously larger, so that the entropy value, though very much increased from what it was in $Q_1$ is now spread over the far greater volume $Q_2$, so that the entropy density is able to be just as it was before at $\mathcal{S}_1$, but consistently with the 2nd Law.

It is reasonable to ask, however, whether the absolutely enormous increase in entropy that is to be anticipated throughout the entire history of our universe cycle can be accommodated through a volume increase of this kind. This issue is related to the second issue referred to above; for by far the greatest entropy content of our universe, even today, let alone in the remote future, is in supermassive black holes in galactic centres. In the time period envisaged for the total lifetime of our universe cycle – namely $10^{12}$ years or so (at least for the expanding phase) – these black holes should still be around and would represent, by an absolutely vast factor, the major contribution to the universe's entropy. Whereas they will get very dispersed through the exponential expansion, they would be brought together again in the final collapse, and would appear to have to form a significant part of the final crunch. It is not at all clear to me why they could be ignored in the proposed ekpyrotic crunch-to-bang transition!

It should be mentioned that there are other proposals that represent serious attempts to describe the special nature of the Big Bang, the most noteworthy, in my opinion, being the *no-boundary scheme* of Hartle and Hawking [1983], which, despite its ingenious originality, I do not actually regard as *adequately* fantastical. None of these proposals, to my knowledge, explains the *fantastic* discrepancy between (a) the wild high-entropy geometry of black-hole singularities and (b) the extraordinarily special geometry of the Big Bang. Something more, perhaps with an even greater element of fantasy, is needed!

To sum up, these schemes are genuinely fantastical, and are aimed at resolving serious issues raised by the curious nature of the Big Bang. They tend to depend upon areas of physics that are fashionable for reasons other than those of cosmology (string theory, extra dimensions, etc.). They contain interesting and thought-provoking ideas, with some serious motivations behind them. Nevertheless they remain somewhat artificially implausible, in my view, at least in their present form, and they still do not adequately address the fundamental issues raised in §3.4, with regard to the 2nd Law's basic role in relation to the *singularly peculiar* nature of the Big Bang.

# 4

# A New Physics for the Universe?

## 4.1. TWISTOR THEORY: AN ALTERNATIVE TO STRINGS?

Following the first of my Princeton lectures (on fashion), as I recall, I was approached by a prospective graduate student in theoretical physics who seemed evidently troubled about what line of research he should pursue, and he wanted some advice. It seems that I had rather disturbed his enthusiasm, which was aimed at moving into the exciting world of pushing forward the boundaries of fundamental scientific knowledge. Like many others, he had found the ideas of string theory to be enticing; but he had been somewhat discouraged by the negative assessment, in my talk, of the direction in which the subject was appearing to be relentlessly moving. At the time, I had been unable to provide him with any significantly positive or constructive advice. I was reluctant to suggest my own area of twistor theory as an appropriate alternative, not only because there appeared to be no one he could work with on the subject, but also because it was a difficult area for a student with aspirations to make some real progress, particularly for someone with just a physics and not a mathematics background. As twistor theory had evolved, it required a kind of mathematical sophistication dependent upon notions not normally part of a physics student's training. Moreover, this theory had foundered, for some 30 years, on a seemingly insurmountable difficulty that we referred to as the *googly problem*, which I shall be describing towards the end of this section.

This encounter was a day or so before a scheduled lunch appointment with Princeton's star mathematical physicist, Edward Witten, and I was nervous that Witten might have been displeased that I had been expressing my doubts about the directions that string theory had taken. As it turned out, Witten surprised me by describing some recent work of his in which he had actually been able to combine some ideas of string theory with those of twistor theory, in order to make what seemed to be some remarkable progress in handling the intricate mathematics of strong interactions. I had been taken aback particularly because

Witten's formalism was specifically aimed at handling processes taking place within *4-dimensional* space-time. As should be clear to readers of chapter 1, my negative reactions to modern string theory stem almost entirely from the apparent need for excessive space(-time) dimensionality. I have my own difficulties with supersymmetry also (which was still present in Witten's twistor–string scheme), but these are less entrenched with me, and in any case Witten's new ideas seemed to be far less dependent on supersymmetry than the degree to which mainstream string theory had come to depend upon the notion of higher space-time dimensionality.

What Witten showed me interested me greatly, since it not only applied to what I regarded as the correct dimensionality of space-time, but was directly applicable to basic *known* particle-physical processes. These are the scatterings of gluons from each other, such processes being fundamental to strong interactions (see §1.3). Gluons are the carriers of the strong forces, in the same way that photons are the carriers of electromagnetic forces. Photons, however, do not directly interact with each other, since they interact only with charged (or magnetic) particles, not other photons. This is the basis of the *linearity* of Maxwell's electromagnetic theory (see §§2.7 and 2.13). But strong interactions are profoundly *non*linear (satisfying the Yang–Mills equations; see §1.8), the interactions of gluons with each other being basic to the nature of strong interactions. Witten's new ideas [Witten 2004], which had roots in some previous work of others [see, for example, Nair 1988; Parke and Taylor 1986; Penrose 1967], showed how the then standard procedures for calculating these gluon-scattering processes, which used conventional Feynman-diagram methods (see §1.5), could be vastly simplified and, as it turned out, sometimes reduced from what appeared to be a book's worth of computer calculations to a few lines.

Since that time, many people have picked up on these developments, initially largely because of Witten's very considerable esteem within the mathematical physics community, and twistor theory has since acquired a new life within a very active movement, where more and more powerful techniques have been found for the calculation of particle scattering amplitudes, in the high-energy limit where the masses (i.e. *rest* masses) of particles become relatively unimportant and the particles can be treated, effectively, as though they have no mass. Not all of these techniques involve twistor theory, and there are numerous different schools of thought about this, but the general conclusion has been that the new methods for doing these calculations are now enormously more efficient than standard Feynman-graph techniques. Despite the importance of string-theoretic notions in the initial ideas behind these developments, these seem to have faded somewhat, in favour of other newer developments, though some elements of string-related concepts (now in standard 4-dimensional space-time) retain their importance.

It should be mentioned, however, that many of these calculations are performed for a particular class of theories that have very special, enormously simplifying, and not altogether physically realistic properties, most particularly $n = 4$ super-symmetric Yang–Mills theory (see §1.14). A viewpoint commonly expressed is that these models are analogous to very simple idealized situations in classical mechanics, that one has to master first, like the simple harmonic oscillator for ordinary quantum physics, and that our understanding of more complicated realistic systems is something that can come later, after the simpler systems are properly understood. For myself, while I can certainly appreciate the value of studying simpler models, for which genuine progress can be made and from which insights may indeed be gained, I feel that the analogy with the harmonic oscillator is very misleading. Simple harmonic oscillators are close to ubiquitous in the small vibrations of non-dispersive classical systems, whereas $n = 4$ super-symmetric Yang–Mills fields do not appear to play any corresponding role in the actual physical quantum fields of nature.

It will be appropriate to give a relatively brief introduction to the basics of twistor theory here, just to give the main ideas and to touch upon some of the details, but I shall not be able to discuss these scattering-theory developments, nor go into the twistor theory in a great deal of depth. For more detail, see Penrose [1967a], Huggett and Tod [1985], Ward and Wells [1989], Penrose and Rindler [1986], Penrose and MacCallum [1972], and TRtR [chapter 33]. The central idea is that space-time itself is to be regarded as a *secondary* notion, constructed from something more primitive, with quantum aspects to it, referred to as *twistor space*. As an underlying guiding principle, the formalism of the theory unites basic notions of quantum mechanics with (conventional 4-dimensional) relativistic space-time physics, bringing these subjects together through the magical properties of complex numbers (§§A.9 and A.10).

In quantum mechanics we have the superposition principle, whereby different states are combined via the use of complex numbers, namely the amplitudes, which are fundamental to the theory (see §§1.4 and 2.7). In §2.9, we saw, in the notion of quantum-mechanical spin (especially for spin-$\frac{1}{2}$), how these complex numbers are intimately tied in with the geometry of 3-dimensional space, where the *Riemann sphere* (figure A-43 in §A.10 and figure 2-18 in §2.9) of different possible ratios of a pair of complex amplitudes can be identified with the different directions in ordinary 3-space – these being the possible spin-axis directions of a spin-$\frac{1}{2}$ particle. In relativistic physics, there is an apparently quite separate

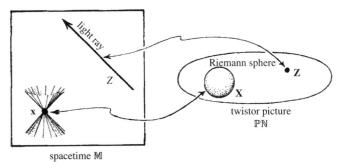

Figure 4-1: The basic twistor correspondence. Each point $\mathbf{Z}$ of the twistor space $\mathbb{PN}$ corresponds to a light ray $Z$ (null straight line) in Minkowski space $\mathbb{M}$ (possibly at infinity). Each point $\mathbf{x}$ of $\mathbb{M}$ corresponds to a Riemann sphere $\mathbf{X}$ in $\mathbb{PN}$.

role for the Riemann sphere, which is again specific to the 3-dimensionality of space (but taken along with the 1-dimensionality of time). In this case, it is the *celestial sphere* of different directions along an observer's past light cone that, as it turns out, can be naturally identified with a Riemann sphere [Penrose 1959].[1] In a certain sense, twistor theory brings together the quantum-mechanical role of complex numbers with the relativistic one, via these two physical roles for the Riemann sphere. Thus, we may begin to see how the magic of complex numbers might indeed provide some link in unifying the quantum world of the small with the relativistic principles of the space-time physics of the large.

How might this work? As an initial picture of twistor theory, we consider a space $\mathbb{PN}$ (where the full reason for this particular notation will emerge shortly, the "$\mathbb{P}$" standing for *projective*, in the same sense as with Hilbert spaces in §2.8). Each individual point of $\mathbb{PN}$ is to represent, *physically*, an entire *light ray* – which in space-time terms is a *null* straight line: the entire history of a freely moving massless (i.e. light-like) particle, such as a photon (figure 4-1). This light ray would be the picture presented by conventional space-time physics, with physical processes being regarded as taking place within the Minkowski space $\mathbb{M}$ of special relativity (see §1.7; notation as in §1.11), but in the twistor picture, this entire

---

[1] There is a certain subtlety, which might worry some readers, that the quantum-mechanical Riemann sphere has the more restricted symmetry group, $SU(2)$, than has the relativistic group $SL(2, C)$. However, the latter group is fully involved in the relation to quantum spin in the spin-raising and lowering roles of twistor operators, as the *fourth physical approximation* of Penrose [1980]; see also Penrose and Rindler [1986, §6.4].

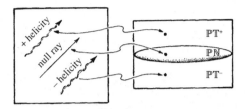

Figure 4-2:  Projective twistor space $\mathbb{PT}$ is composed of three parts: $\mathbb{PT}^+$, representing right-spinning massless particles, $\mathbb{PT}^-$, representing left-spinning ones, and $\mathbb{PN}$, representing non-spinning ones.

light ray is geometrically presented as just a single *point* of $\mathbb{PN}$. Conversely, to represent a *space-time point* (i.e. an *event*) **x** in $\mathbb{M}$, in terms of structures within this twistor space $\mathbb{PN}$, we simply consider the family of all the light rays in $\mathbb{M}$ that pass through **x**, and see what kind of structure this family is to have within $\mathbb{PN}$. From what has been said above, the locus in $\mathbb{PN}$ that represents a space-time point **x** is simply a *Riemann sphere* (essentially the celestial sphere of **x**) – the simplest Riemann surface. Since Riemann surfaces are just complex curves (§A.10), it would appear to make some sense for $\mathbb{PN}$ actually to be a *complex manifold*, these Riemann spheres arising as complex-1-dimensional submanifolds. However, this cannot quite work because $\mathbb{PN}$ is *odd*-dimensional (5-dimensional) and it needs to be *even* real-dimensional in order to have a chance to be representable as a complex manifold (see §A.10). We need another dimension! But then we find, very remarkably, that when we include the energy and the helicity (i.e. spin) of a massless particle, $\mathbb{PN}$ does extend, in a physically natural way, to a real 6-manifold $\mathbb{PT}$ which has a natural structure as a complex 3-manifold, in fact, a complex projective 3-space ($\mathbb{CP}^3$) called *projective twistor space*. See figure 4-2.

How does this work in detail? In order to understand the twistor formalism, it is best that we consider the complex 4-dimensional *vector space* $\mathbb{T}$ (see §A.3), sometimes referred to as *non-projective twistor space*, or simply just *twistor space*, of which the above space $\mathbb{PT}$ is the projective version. The relation between $\mathbb{T}$ and $\mathbb{PT}$ is exactly the same as the relation between a Hilbert space $\mathcal{H}^n$ and its projective version $\mathbb{P}\mathcal{H}^n$ that we saw in §2.8 (illustrated in figure 2-16(b) in §2.8); that is to say, all non-zero complex multiples $\lambda\mathbf{Z}$ of a given non-zero twistor **Z** (element of $\mathbb{T}$) give us the *same* projective twistor (element of $\mathbb{PT}$). In fact, the twistor space $\mathbb{T}$ is very similar to a 4-dimensional Hilbert space as regards its algebraic structure, although its physical interpretation is completely different from that of Hilbert spaces in quantum mechanics. Generally speaking, it is the *projective* twistor space $\mathbb{PT}$ that is useful to us if we are thinking of geometrical matters, whereas the space $\mathbb{T}$ is appropriate if we are concerned with the algebra of twistors.

As with a Hilbert space, the elements of $\mathbb{T}$ are subject to notions of *inner product*, *norm*, and *orthogonality*, but rather than adopting a notation like $\langle \cdots \rangle$ that was used in §2.8, we shall find it more convenient to denote the inner product of a twistor $\mathbf{Y}$ with a twistor $\mathbf{Z}$ as

$$\bar{\mathbf{Y}} \cdot \mathbf{Z},$$

where the complex conjugate twistor $\bar{\mathbf{Y}}$, of $\mathbf{Y}$, is an element of the *dual* twistor space $\mathbb{T}^*$, so that the norm $\|\mathbf{Z}\|$ of a twistor is

$$\|\mathbf{Z}\| = \bar{\mathbf{Z}} \cdot \mathbf{Z},$$

and orthogonality between twistors $\mathbf{Y}$ and $\mathbf{Z}$ is $\bar{\mathbf{Y}} \cdot \mathbf{Z} = 0$. However, twistor space $\mathbb{T}$ is *algebraically* not quite a Hilbert space (in addition to serving a different kind of purpose from Hilbert spaces in quantum mechanics). Specifically, the norm $\|\mathbf{Z}\|$ is *not* positive definite (as it would be for a proper Hilbert space),[2] which is to say that for a non-zero twistor $\mathbf{Z}$ we can have all three alternatives:

$\|\mathbf{Z}\| > 0$  for a *positive* or *right-handed* twistor $\mathbf{Z}$, belonging to the space $\mathbb{T}^+$,

$\|\mathbf{Z}\| < 0$  for a *negative* or *left-handed* twistor $\mathbf{Z}$, belonging to the space $\mathbb{T}^-$,

$\|\mathbf{Z}\| = 0$  for a *null* twistor, belonging to the space $\mathbb{N}$.

The entire twistor space $\mathbb{T}$ is the disjoint union of the three parts $\mathbb{T}^+$, $\mathbb{T}^-$, and $\mathbb{N}$, as is its projective version $\mathbb{PT}$ the disjoint union of the three parts $\mathbb{PT}^+$, $\mathbb{PT}^-$, and $\mathbb{PN}$ (see figure 4-3).

It is the *null* twistors that provide the direct link with light rays in space-time, the projective version $\mathbb{PN}$, of $\mathbb{N}$, indeed representing the space of light rays in Minkowski space $\mathbb{M}$ (but including some special "idealized" light rays at *infinity* $\mathscr{I}$, when $\mathbb{M}$ is extended to the *compactified* Minkowski space $\mathbb{M}^{\#}$, referred to in §1.15; see figure 1-41). In the case of null twistors, we have a very direct geometrical interpretation of the orthogonality relation $\bar{\mathbf{Y}} \cdot \mathbf{Z} = 0$ (or

---

[2] In §2.8 we had the notion of a (finite-dimensional) Hilbert space. This is a complex vector space with a Hermitian structure of *positive definite* signature $(+ + + \cdots +)$. Here we require a $(+ + - -)$ signature instead, which means that in terms of conventional complex coordinates, the (squared) norm of a vector $\mathbf{z} = (z_1, z_2, z_3, z_4)$ would be $\|\mathbf{z}\| = z_1\bar{z}_1 + z_2\bar{z}_2 - z_3\bar{z}_3 - z_4\bar{z}_4$. In standard twistor notation, however, it turns out more convenient to use (completely equivalent) twistor coordinates $\mathbf{Z} = (Z^0, Z^1, Z^2, Z^3)$ (not to be read as powers of a single quantity $Z$) for which we take $\|\mathbf{Z}\| = Z^0\overline{Z^2} + Z^1\overline{Z^3} + Z^2\overline{Z^0} + Z^3\overline{Z^1}$.

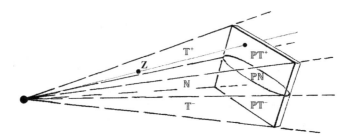

Figure 4-3: The complex lines through the origin of non-projective twistor space $\mathbb{T}$ correspond to the points of projective twistor space $\mathbb{PT}$.

equivalently $\bar{\mathbf{Z}} \cdot \mathbf{Y} = 0$). This orthogonality condition simply states that the light rays represented by $\mathbf{Y}$ and $\mathbf{Z}$ *intersect* (possibly at infinity).

As is the case with elements of an ordinary Hilbert space, each element $\mathbf{Z}$ of $\mathbb{T}$ has a kind of *phase*, changed by multiplication by $e^{i\theta}$ ($\theta$ real). Although this phase does have a kind of geometrical meaning, I shall ignore this here and consider the physical interpretation of a twistor $\mathbf{Z}$ *up to* such phase multiplication. What we find is that $\mathbf{Z}$ represents the *momentum* and *angular momentum* structure of a free *massless particle*, according to the normal prescriptions of classical special relativity (where this includes certain limiting situations for which the 4-momentum vanishes and the massless particle is at infinity). Thus, we get a genuinely appropriate physical structure for our freely moving massless particle, which is more than just a light ray, as this interpretation now includes the *non-null* twistors as well as the null ones. We find that a twistor indeed correctly defines an energy–momentum and an angular momentum for a massless particle, incorporating its *spin* about its direction of motion. However, this actually gives a *non-localized* description of a massless particle when it possesses a non-zero intrinsic spin, so that in this case its light-ray world-line is only *approximately* defined.

It should be emphasized that this non-locality is not an artefact, arising from the unconventional nature of the twistor descriptions; it is a (frequently unrecognized) aspect of the *conventional* description of a massless particle with spin, if represented in terms of its momentum and angular momentum (the latter being sometimes referred to as the *moment of momentum* about some specified origin point; see also §1.14, figure 1-36). Although the particular algebraic descriptions of twistor theory differ from the conventional ones, there is nothing unconventional about the interpretations I have just given. At least at this stage, twistor theory provides merely a distinctive formalism. It does not introduce any new assumptions about the nature of the physical world (unlike, for example, string theory). It does, however, give us a different slant on things, suggesting that, perhaps, the notion of space-time might be usefully regarded as a secondary quality of the physical world, the geometry of twistor space being regarded as somehow more fundamental. It must also be remarked that the framework of twistor theory has certainly not, as yet, achieved any such exalted status, and its current utility in scattering theory for very high-energy particles (referred to above) rests entirely on the utility of the twistor formalism for the description of processes for which rest masses can be ignored.

It is usual to use coordinates for a twistor $\mathbf{Z}$ (4 complex numbers) for which the first pair of components $Z^0$, $Z^1$ are the two complex components of a quantity $\omega$ referred to as a 2-*spinor* (compare also §1.14) and the second pair $Z^2$, $Z^3$ are the two components of a quantity $\pi$ which is a slightly different type of 2-spinor (different by being of dual, complex-conjugate type), so we can represent the entire twistor as

$$\mathbf{Z} = (\omega, \pi).$$

(In much of the recent literature the notation "$\lambda$" is used for "$\pi$" and "$\mu$" for "$\omega$", which follows the notation I originally used in Penrose [1967a], where I adopted some inappropriate choices of convention – mainly to do with the upper or lower placings of indices – and common practice often follows those unfortunate conventions.) I do not want to go into the precise notion of what a 2-spinor actually is (sometimes called a *Weyl spinor*) here, but we can get something of the idea by referring back to §2.9. The two components (amplitudes) $w$ and $z$, whose ratio $z : w$ defines the spin direction for a spin-$\frac{1}{2}$ particle (see figure 2-18), can be thought of as the two components defining a 2-spinor, and this applies both[3] to $\omega$ and to $\pi$.

For a better geometrical idea of a 2-spinor, however, I refer the reader to figure 4-4, which shows how a (non-zero) 2-spinor may be pictured in space-time terms. Strictly speaking, figure 4-4(b) should be interpreted as within the tangent space of some point of space-time (see figure 1-18(c)), but since our space-time is here flat Minkowski space $\mathbb{M}$, we can regard this picture as referring to the whole of $\mathbb{M}$ instead, taken with respect to some coordinate origin point O. Up to a phase multiplier, the 2-spinor is represented as a future-pointing null vector, called its *flagpole* (the line segment OF in figure 4-4). We can think of the *direction* of the

---

[3] In the standard 2-spinor index notation [Penrose and Rindler 1984], $\omega$ and $\pi$ have the respective index structures $\omega^A$ and $\pi_{A'}$.

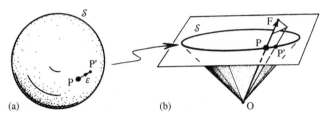

Figure 4-4: The geometrical interpretation of a 2-spinor, where we may think of this as within the tangent space of a space-time point O, or else within the entire Minkowski space $\mathbb{M}$, referred to some coordinate origin O. (a) The Riemann sphere $\mathscr{S}$ represents (b) the future-null "flagpole" directions OF. A tangent vector $\overrightarrow{PP'}$ at a point P of $\mathscr{S}$ represents a "flag plane" direction along OF which, up to sign, represents the phase of the 2-spinor.

flagpole as being given by a point P in the abstract (Riemann) sphere $\mathscr{S}$ of future null directions (figure 4-4(a)). The phase of the 2-spinor is itself (up to an overall sign) represented by a tangent vector $\overrightarrow{PP'}$ to $\mathscr{S}$ at P, where P′ is a neighbouring point to P in $\mathscr{S}$. In space-time terms, this phase is given as a null half-plane bounded by the flagpole, called the *flag plane* (shown in figure 4-4(b)). Although the details of this are not too important for us here, it is useful to have a picture in one's mind that a 2-spinor is a very clear-cut geometrical object (ambiguous only in that our picture does not distinguish a specific 2-spinor from *minus* that 2-spinor).

For the twistor $\mathbf{Z}$, its 2-spinor part $\boldsymbol{\pi}$, up to a phase, describes the particle's energy–momentum 4-vector as the outer product[4]

$$\mathbf{p} = \boldsymbol{\pi}\bar{\boldsymbol{\pi}}$$

(see §1.5), the overbar denoting complex conjugation. If we multiply $\boldsymbol{\pi}$ by a phase factor $e^{i\theta}$ ($\theta$ real), then $\bar{\boldsymbol{\pi}}$ is multiplied by $e^{-i\theta}$, and consequently $\mathbf{p}$ remains unchanged; indeed, $\mathbf{p}$ is the 2-spinor $\boldsymbol{\pi}$'s flagpole. Once $\boldsymbol{\pi}$ is known, the extra data in $\boldsymbol{\omega}$ is equivalent to the particle's relativistic angular momentum (see §1.14), about the coordinate origin, expressed in terms of (symmetrized) products $\boldsymbol{\omega}\bar{\boldsymbol{\pi}}$ and $\boldsymbol{\pi}\bar{\boldsymbol{\omega}}$.

---

[4] A spinor product, expressed by simple juxtaposition, is an *uncontracted* product, so the product $\boldsymbol{\pi}\bar{\boldsymbol{\pi}}$ gives us a vector (actually a *covector*) $p_a = p_{AA'} = \pi_{A'}\bar{\pi}_A$, where each 4-space index is represented (in the abstract-index formalism being used here [Penrose and Rindler 1984]) as a pair of spinor indices, one primed and the other unprimed. Explicitly, the expression for the angular momentum tensor $M^{ab}$ in terms of $\omega^A$ and $\pi_{A'}$, is given, in spinor form [Penrose and Rindler 1986], by $M^{ab} = M^{AA'BB'} = i\omega^{(A}\bar{\pi}^{B)}\varepsilon^{A'B'} - i\bar{\omega}^{(A'}\pi^{B')}\varepsilon^{AB}$, the round brackets denoting symmetrization, the epsilon symbols being skew-symmetrical.

The complex conjugate quantity $\bar{\mathbf{Z}}$, represented accordingly as

$$\bar{\mathbf{Z}} = (\bar{\pi}, \bar{\omega}),$$

is a dual twistor (i.e. element of $\mathbb{T}^*$), which means that it is a natural object for forming scalar products with twistors (§A.4). Thus, if $\mathbf{W}$ is any dual twistor $(\lambda, \mu)$ we can form its scalar product with $\mathbf{Z}$, which is the complex number

$$\mathbf{W} \cdot \mathbf{Z} = \lambda \cdot \omega + \mu \cdot \pi.$$

The *norm* $\|\mathbf{Z}\|$ of the twistor $\mathbf{Z}$ is then the (real) number

$$\|\mathbf{Z}\| = \bar{\mathbf{Z}} \cdot \mathbf{Z} = \bar{\pi} \cdot \omega + \bar{\omega} \cdot \pi$$
$$= 2\hbar s.$$

It turns out that $s$ is the *helicity* of the massless particle described by $\mathbf{Z}$. If $s$ is positive, then the particle has a right-handed spin, whose value is $s$; if $s$ is negative, then the spin is left-handed, with value $|s|$. Thus, a right-handed (circularly polarized) photon would have $s = 1$ and a left-handed one, $s = -1$ (see §2.6). This justifies the pictorial description given in figure 4-2. For a graviton, the right- and left-handed versions would have $s = 2$ and $s = -2$, respectively. For a neutrino and anti-neutrino, if taken to be massless, we would have $s = -1/2$ and $s = +1/2$, respectively.

If $s = 0$, the particle would be spinless, and then the twistor $\mathbf{Z}$, referred to as a *null twistor* ($\bar{\mathbf{Z}} \cdot \mathbf{Z} = 0$), indeed has the geometrical interpretation in Minkowski space $\mathbb{M}$ (or in its compactification $\mathbb{M}^{\#}$, if we allow $\pi = 0$) as a light ray, or *null straight line* $\mathbf{z}$ (null geodesic – see §1.7). This is the world-line of the particle, in accordance with the description given above for the "initial picture" of a null twistor shown in figure 4-1. The light ray $\mathbf{z}$ has the space-time direction of $\mathbf{p}$, where $\mathbf{p}$ also provides an *energy* scaling for $\mathbf{z}$, and this scaling is also determined by the actual twistor $\mathbf{Z}$. The flagpole direction of $\omega$ now also has a direct interpretation, provided that the light ray $\mathbf{z}$ meets the light cone of the coordinate origin O at some finite point Q, as the direction of OQ, the position vector of $\mathbf{y}$ being $\omega\bar{\omega}(i\bar{\omega} \cdot \pi)^{-1}$; see figure 4-5.

This basic correspondence between Minkowski space $\mathbb{M}$ and the twistor space $\mathbb{PN}$ is provided algebraically by what is called the *incidence relation* between a *null* twistor $\mathbf{Z}$ and a space-time point $\mathbf{x}$, which can be expressed as[5]

$$\omega = i\mathbf{x} \cdot \pi,$$

---

[5] The index form of this relation is $\omega^A = ix^{AB'}\pi_{B'}$.

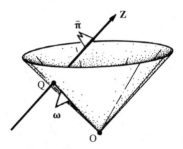

Figure 4-5: The flagpole direction of the ω part of a null twistor $\mathbf{Z} = (\omega, \pi)$. Assuming that the light ray $\mathbf{Z}$ meets the light cone of the coordinate origin O at some finite point Q, then ω's flagpole is in the direction OQ; moreover, ω is itself fixed (given π) by the position vector of Q being $\omega\bar{\omega}(i\bar{\omega} \cdot \pi)^{-1}$.

which, for the reader familiar with matrix notation, stands for

$$\begin{pmatrix} Z^0 \\ Z^1 \end{pmatrix} = -\frac{i}{\sqrt{2}} \begin{pmatrix} t + z & x + iy \\ x - iy & t - z \end{pmatrix} \begin{pmatrix} Z^2 \\ Z^3 \end{pmatrix},$$

where $(t, x, y, z)$ are standard Minkowski space-time coordinates (with $c = 1$) for $\mathbf{x}$. Incidence is interpreted, in $\mathbb{M}$, as the space-time point $\mathbf{x}$ lying on the null line $\mathbf{z}$; in terms of $\mathbb{PN}$, the interpretation of incidence is that the point $\mathbb{P}\mathbf{Z}$ lies on a projective line $\mathbf{X}$, this line being the Riemann sphere representing $\mathbf{x}$ according to our initial picture above, this Riemann sphere being a complex projective straight line in the projective 3-space $\mathbb{PT}$, actually lying in the subspace $\mathbb{PN}$ of $\mathbb{PT}$. See figure 4-1.

When $s \neq 0$ (so the twistor $\mathbf{Z}$ is *non*-null), the incidence relation $\omega = i\mathbf{x} \cdot \pi$ cannot be satisfied by any *real* point $\mathbf{x}$, and there is no unique world-line that is singled out. The particle's location is now, to some degree, *non-local*, as mentioned above [Penrose and Rindler 1986, §§6.2 and 6.3]. However, the incidence relation can be satisfied by *complex* points $\mathbf{x}$ (points of the *complexification* $\mathbb{CM}$ of Minkowski space $\mathbb{M}$), which actually has significance underlying the positive frequency condition satisfied by twistor wave functions, which will concern us shortly.

An important, and somewhat magical, feature of this twistor-space way of looking at physics (related to the above geometry) is the very simple procedure whereby twistor theory provides all solutions to the field equations for mass-less particles of any specified helicity [Penrose 1969b; see also Penrose 1968; Hughston 1979; Penrose and MacCallum 1972; Eastwood et al. 1981; Eastwood

1990]. Certain versions of this formula had, in effect, been found much earlier [see Whittaker 1903; Bateman 1904, 1910]. This comes about naturally when we try to think of how the wave function of a massless particle is to be described in twistor terms. In the conventional physical descriptions, a wave function for a particle (see §§2.5 and 2.6) might be presented as a complex-valued function $\psi(\mathbf{x})$ of (spatial) position $\mathbf{x}$ or, alternatively, as a complex-valued function $\tilde{\psi}(\mathbf{p})$ of 3-momentum $\mathbf{p}$. Twistor theory gives two further ways of representing a massless particle's wave function, namely as a complex-valued function $f(\mathbf{Z})$ of a twistor $\mathbf{Z}$, referred to simply as the particle's *twistor function*, or else as a complex-valued function $\tilde{f}(\mathbf{W})$ of a *dual* twistor $\mathbf{W}$, the particle's dual twistor function. The functions $f$ and $\tilde{f}$ turn out to be necessarily *holomorphic*, i.e. *complex-analytic* (so they do not "involve" the complex-conjugate variables $\bar{\mathbf{Z}}$ or $\bar{\mathbf{W}}$; see §A.10). In §2.13 it was noted that $\mathbf{x}$ and $\mathbf{p}$ are what are called *canonical conjugate* variables; in a corresponding way, $\mathbf{Z}$ and $\bar{\mathbf{Z}}$ are also canonical conjugates of each other.

These twistor functions (and dual twistor functions – but for definiteness let us just concentrate on the twistor functions) have a number of remarkable properties. The most immediate is that for a particle of a definite helicity, its twistor function $f$ is *homogeneous*, which means that, for some number $d$, called the *degree of homogeneity*,

$$f(\lambda \mathbf{Z}) = \lambda^d f(\mathbf{Z}),$$

for any non-zero complex number $\lambda$. The number $d$ is determined by the helicity $s$ by

$$d = -2s - 2.$$

The homogeneity condition tells us that $f$ may actually be viewed as a kind of function on, simply, the *projective* twistor space $\mathbb{PT}$. (Such a function is sometimes referred to as a *twisted function* on $\mathbb{PT}$, the amount of "twist" being specified by $d$.)

The twistor form of a wave function for a massless particle of given helicity $s$ is thus remarkably simple – though there is an important catch, which we shall be coming to shortly. The *simple* aspect of this representation is that the field equations (basically all the appropriate Schrödinger equations) governing the corresponding position-space wave function $\psi(\mathbf{x})$ effectively evaporate almost completely! All we need is a twistor function $f(\mathbf{Z})$ of our twistor variable $\mathbf{Z}$ which is *holomorphic* (i.e. does not involve $\bar{\mathbf{Z}}$; see §A.10) and *homogeneous*.

These field equations are also important in classical physics. For example, when $s = \pm 1$, corresponding to the homogeneities $d = -4$ and $0$, we get general solutions of Maxwell's electromagnetic equations (see §2.6). When $s = \pm 2$

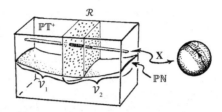

Figure 4-6: The twistor geometry relevant to the contour integration which obtains the positive-frequency field equation (Schrödinger equation) for a free massless particle of any given helicity. The twistor function can be defined on the intersection region $\mathcal{R} = \mathcal{V}_1 \cap \mathcal{V}_2$ of two open sets $\mathcal{V}_1$ and $\mathcal{V}_2$ which together cover $\mathbb{PT}^+$.

(homogeneities $d = -6$ and $+2$) we get general solutions of the *weak-field* (i.e. "linearized") Einstein vacuum equations ($\mathbf{G} = \mathbf{0}$, where "vacuum" means $\mathbf{T} = \mathbf{0}$; see §1.1). In each case the solutions of the field equations arise automatically from the twistor function, according to a simple procedure, familiar in complex-number analysis, known as *contour integration* [see, for example, TRtR, §7.2].

In the quantum context, there is another feature of wave functions (for free massless particles) that we get directly in the twistor formalism. This is the fact that wave functions for free particles have to satisfy an essential condition known as *positive frequency*, which means in effect that the energy has no negative contributions in the wave function (see §4.2). This comes out automatically if we ensure that the twistor function is, in a curious but appropriate sense, *defined on* the top half $\mathbb{PT}^+$ of projective twistor space. Figure 4-6 gives a schematic of the geometry involved. Here, we are thinking of the wave function as being evaluated at a *complex* space-time point $\mathbf{x}$ which is represented by the line, labelled $\mathbf{X}$ in the picture, that is located entirely within $\mathbb{PT}^+$, this line being actually a Riemann sphere, as illustrated in the right-hand part of the picture.

The "curious" appropriate sense, in which $f$ is actually "defined" on $\mathbb{PT}^+$, is illustrated by the dotted region $\mathcal{R}$ within $\mathbb{PT}^+$, this being the actual domain of the function. Thus, $f$ is allowed to have singularities within those parts of $\mathbb{PT}^+$ that lie outside $\mathcal{R}$ (to the right or left of $\mathcal{R}$ in the picture). The line (Riemann sphere) $\mathbf{X}$ meets $\mathcal{R}$ in an annular region, and the contour integral takes place around a loop within this annular region. This gives us the value of the position-space wave function $\psi(\mathbf{x})$ for the (complex) space-time point $\mathbf{x}$, and it automatically satisfies the appropriate field equations *and* energy positivity by virtue of this construction!

What, then is the catch, referred to above? The issue lies in the appropriate meaning underlying the above curious notion of $f$ being "defined on $\mathbb{PT}^+$", whereas actually its domain of definition is the smaller region $\mathcal{R}$. How are we to make mathematical sense of this?

There is, in fact, some sophistication here that I cannot properly go into without inappropriate technicality. But the essential point about $\mathcal{R}$ is that it can be thought of as the region of overlap between two open regions (see §A.5) $\mathcal{V}_1$ and $\mathcal{V}_2$ which together cover $\mathbb{PT}^+$:

$$\mathcal{V}_1 \cap \mathcal{V}_2 = \mathcal{R} \quad \text{and} \quad \mathcal{V}_2 \cup \mathcal{V}_1 = \mathbb{PT}^+;$$

see figure 4-6. (The symbols $\cap$ and $\cup$ denote *intersection* and *union*, respectively; see §A.5.) More generally, we might have considered a covering of $\mathbb{PT}^+$ by a *larger* number of open sets, and then our twistor function would have to be defined in terms of a *collection* of holomorphic functions defined on all the various pairwise intersections among these open sets. From this collection, we would extract a particular quantity, called an element of 1st *cohomology*. It would be this element of 1st cohomology that would really be the quantity that provides the twistor notion of a wave function!

This seems all pretty complicated, and it would certainly get so, were I to explain everything we have to do in detail. However, this complication really expresses an important underlying idea, which, I believe, fundamentally relates to the mysterious *non-locality* that is actually exhibited by the quantum world, as indicated in §2.10. Let me first try to simplify things by shortening the terminology, and referring to an element of 1st cohomology simply as a *1-function*. An ordinary function would then be a 0-function, and we can also have higher-order things called 2-functions (elements of 2nd cohomology, defined in terms of collections of functions defined on *triple* overlaps of open sets of a covering) and so on, with 3-functions, 4-functions, etc. (The type of cohomology I am using here is what is known as *Čech cohomology*; there are other (equivalent, but very different-looking) procedures, such as Dolbeault cohomology [Gunning and Rossi 1965; Wells 1991].)

So how are we to understand what kind of thing a 1-function actually stands for? The clearest way I know of, to explain this idea in simple terms, is to refer the reader to the impossible triangle depicted in figure 4-7. Here, we have the impression of a 3-dimensional structure which could not, in fact, exist in ordinary Euclidean 3-space. Let us imagine that we have a box full of wooden rods and corner pieces, and we are given a list of instructions telling us how these are to be glued together. We suppose that the instructions for fitting pairs of pieces together

Figure 4-7: The impossible triangle (or tribar) provides a good illustration of 1st cohomology. The degree of impossibility in the figure is a non-local quantity, which can, in fact, be precisely quantified as a 1st cohomology element. If we cut across the triangle at any place, its impossibility disappears, illustrating how this quantity cannot be localized in any particular place. A twistor function plays a very similar non-local role and is indeed to be interpreted as an element of (holomorphic) 1st cohomology.

present a picture that, from the vantage point of the viewer, is *locally* consistent, but ambiguous up to an uncertainty concerning the distance from the viewer's eye. Nevertheless, as with figure 4-7, the entire perceived object might not be actually constructible in 3-dimensional space, there being no consistent way for the viewer to assign distances away, for all different parts of the image.

The very non-local impossibility exhibited in this picture gives a good impression of what 1st cohomology is all about and the kind of thing that a 1-function actually expresses. Indeed, given our list of gluing instructions, the procedures of cohomology enable us to construct a precise 1-function that measures a degree of impossibility of the resulting figure, so that whenever this measure comes out as something different from zero we have an impossible object, like that of figure 4-7. Note that by covering up any corner or connecting edge in the picture, we obtain an image of something that could actually be realized in Euclidean 3-space. The impossibility in the picture is thus not a local thing, but is a *global* feature of the picture as a whole. Accordingly, the 1-function describing the measure of this impossibility is indeed a non-local quantity, referring to the structure as a whole and not to any individual part of the structure [Penrose 1991; Penrose and Penrose 1958]. For earlier pictorial representations of such impossible structures, by Maurits C. Escher, Oscar Reutersvärd and others, see Ernst [1986, pp. 125–34] and Seckel [2004].

The twistor wave function of a single particle is, likewise, a non-local entity, namely a 1-function which would be obtainable from local matching functions on overlaps of patches in essentially the same way as with the construction of such an impossible object. Now, rather than this cohomology arising from the rigidity of pieces of wood in ordinary 3-dimensional Euclidean space, the rigidity here is that of holomorphic functions, as is expressed in the process of *analytic continuation*, as pointed out in §A.10 (and also in §3.8). The remarkable "rigidity" exhibited by holomorphic functions seems to provide such a function with a mind of its own about where it wants to go, that it cannot be deflected from. In the current situation, such rigidity may prevent such a function from being defined on the whole of $\mathbb{PT}^+$. A holomorphic 1-function can be viewed as an expression of the kind of obstruction to such globality, and this is actually the non-local nature of a twistor wave function.

In this way, twistor theory reveals, within its formalism, something of the non-local character even of the wave functions of just single particles, which somehow remain as individual entities ("particles") even though the spread of the wave function's extent might be very great – perhaps to vast numbers of light-years, as would be the case when single photons are received from stars in distant galaxies (see §2.6). We must think of the twistor 1-function as indeed being something like an impossible triangle, spread over such vast distances. Finally, finding the particle at one particular location breaks the impossibility, and this would be the case no matter where the photon is finally observed to be located. The 1-function has then done its job, and that particular photon cannot then be found to be located anywhere else.

The situation is actually rather more sophisticated in the case of many-particle wave functions, the twistor description of the wave function for $n$ massless particles being a (holomorphic) $n$-function. It seems likely to me that the puzzles of Bell-inequality violations (see §2.10) in entangled $n$-particle states could be clarified by an examination of these twistor descriptions. But, to my knowledge, this has not yet been seriously attempted [see Penrose 1998b, 2005, 2015a].

Witten's innovative ideas of 2003, introducing the ideas of twistor string theory, avoided these issues of cohomology by employing an ingenious anti-Wick rotation to Minkowski space (compare §1.9), by "rotating" one of the spatial dimensions into a temporal one, thereby providing a flat 4-dimensional "space-time" with 2 timelike dimensions and 2 spacelike ones. Then its (projective) "twistor space" turns out to be a *real* projective 3-space $\mathbb{RP}^3$ rather than the complex $\mathbb{CP}^3$ that is actually $\mathbb{PT}$. This makes life a little easier, at first, since cohomology is avoided and Dirac $\delta$-functions can be used in a way that is more like standard quantum mechanics (see §2.5). However, in my opinion, though I appreciate the utility of this procedure, I feel strongly that it misses much of the potential power inherent in twistor theory when investigating physics at a greater depth.

Witten's original ideas also involved an intriguing suggestion that the Riemann spheres (lines) in $\mathbb{PT}$ that represent points in $\mathbb{CM}$ should be generalized to higher-order Riemann surfaces such as conic sections, cubic, and quartic curves, etc. ("strings", as in §1.6), leading to direct procedures for calculating gluon scatterings. (Ideas of this general nature had also been introduced earlier [Shaw and Hughston 1990], but with quite different kinds of applications in mind.) These ideas stimulated much new interest in the theory underlying gluon scatterings and their calculational derivation, which linked up, most particularly with pioneering work that had been carried out much earlier by Andrew Hodges – almost single-handedly for some 30 years – in developing the theory of twistor diagrams [Hodges and Huggett 1980; Penrose and MacCallum 1972; Hodges 1982, 1985a,b, 1990, 1998, 2006b], the twistor analogue of the Feynman-diagram formalism of standard particle physics (§1.5). Although in more recent years, interest in the string aspect of these new developments appears to have waned somewhat (or moved into the domain of what are called *ambitwistors*, which are the combined twistor/dual twistor representation of complex null geodesics [LeBrun 1985, 1990]), the twistor-related developments have come into their own, and there is much recent new work which simplifies the gluon scatterings further still, so that more and more complicated processes can be calculated. Among the remarkable recent developments have been the useful notion of *momentum twistors* (due originally to Andrew Hodges) and of *amplituhedron*, a concept introduced by Nima Arkani-Hamed, developed from earlier ideas of Hodges, that lies in higher-dimensional versions of twistor space ("Grassmannians"), and which appear to represent a new outlook on the describing of scattering amplitudes in a remarkably comprehensive way [see, for example, Hodges 2006a, 2013a,b; Bullimore et al. 2013; Mason and Skinner 2009; Arkani-Hamed et al. 2010, 2014; Cachazo et al. 2014].

Yet, these fascinating developments are all of the nature of perturbative schemes, where quantities of interest are computed by means of some kind of power series (§§A.10, A.11, 1.5, 1.11, and 3.8). Powerful as such methods are, there are many features that are difficult to access by such means. Most particular among them are the essentially curved-space aspects of gravitation theory. Whereas much can indeed be derived by power-series treatments of problems in general relativity, where perturbational corrections to Newton's flat-space theory can often be carried to great accuracy when gravitational fields are relatively weak, it is a very different story when properties of black holes need to be properly understood. This must be so also for twistor theory where its treatment of nonlinear fields, such as Yang–Mills theory and general relativity by means of perturbative scattering theory, leaves completely untouched some of the potential power that I believe must reside within the twistorial approach to basic physics.

One of the powerful ways in which twistor theory does apply to nonlinear physical theory, although in a way that is fundamentally incomplete so far, is indeed to give a *non*-perturbative treatment of the basic nonlinear physical fields, Einstein's general relativity and Yang–Mills theory, and the interactions of Maxwell theory. This incompleteness, which has been a fundamental and thoroughly frustrating obstruction to the development of twistor theory for nearly 40 years, comes about from the curious lopsidedness of the above twistor-function representations of massless particles, where we find a strange imbalance between the homogeneities for left- and right-handed helicities. This is of little consequence if we stick to the above description of the free linear massless fields (twistor wave functions). But it has, so far, been possible to treat, *non*-perturbatively, the nonlinear interactions of these fields only with regard to their *left-handed* parts.

It has, indeed, been one of the striking benefits of the above twistor formulation of massless fields that the interactions (and self-interactions) of these fields also turn out to have a remarkably concise description – but only in the case of the left-handed part of the field. A "nonlinear graviton" construction, which I found in 1975, produces a *curved* twistor space to represent each left-handed solution of Einstein's equations, describing, in effect, how left-handed gravitons interact with themselves. An extension of this procedure was found by Richard S. Ward about a year later, to handle the left-handed gauge (Maxwell and Yang–Mills) fields of electromagnetic, strong, or weak interactions [Penrose 1976b; Ward 1977, 1980]. But a fundamental problem, referred to as the *googly problem*, had remained unsolved. (A *googly* is a term used in the game of cricket to describe a ball bowled with a right-handed spin, but delivered with the apparent action for delivering a left-handed spin.) The problem had been to find to a corresponding procedure for the *right*-handed gravitational and gauge interactions to that of the above nonlinear graviton, so that the two could be combined to give a complete twistor formulation of the known basic physical interactions.

It should be made clear that if we had used *dual* twistor space, then the relation between the handedness of the spin and the homogeneity degree of the (now dual) twistor function would simply have been reversed. Using the dual twistor space for the opposite helicity cases would not solve the googly problem, because we need a uniform procedure for handling *both* helicities at once – not least because we need to be able to handle massless particles (such as plane-polarized photons; see §2.5) which involve quantum superpositions of both helicities at once. We could have used the dual twistor space instead of twistor space all the way through, of course, but the googly problem remains. This key issue seems not to be fully resolvable within the original framework of twistor theory, using deformed versions of twistor space, despite there having been some hopeful-looking developments in the intervening years [see, for example, Penrose 2000a].

However, a new concept has arisen in the twistor programme over the past few years, which I refer to as *palatial twistor theory* (the name coming from the unusual venue of the key initial ingredient that launched this idea, arising from a brief discussion I had with Michael Atiyah at a pre-lunch gathering in the inspi-rational ambiance of Buckingham Palace) [Penrose 2015a,b], which promises to open up new vistas in the application of twistor ideas. This depends upon an old feature of twistor theory that had played important roles in many of the early developments (although not referred to explicitly above). This is a basic relation between twistor geometry and the ideas of quantum mechanics that resides in the twistor quantization procedure, whereby the twistor variables $\mathbf{Z}$ and $\bar{\mathbf{Z}}$ are taken to be canonical conjugates of one another (referred to above, like the position and momentum variables $\mathbf{x}$ and $\mathbf{p}$ for a particle; see §2.13) as well as being complex conjugates. In the standard canonical quantization procedure, such canonically conjugate variables are replaced by non-commuting operators (§2.13), and the same idea has been applied in the case of twistor theory in various developments over the years [Penrose 1968, 1975b; Penrose and Rindler 1986], where such non-commutativity ($\mathbf{Z}\bar{\mathbf{Z}} \neq \bar{\mathbf{Z}}\mathbf{Z}$) is physically natural, and each of the operators $\mathbf{Z}$ and $\bar{\mathbf{Z}}$ behaves as *differentiation* with respect to the other (see §A.11). What is new about palatial twistors is the incorporation of the algebra of such non-commuting twistor variables into the nonlinear geometric constructions (the nonlinear graviton and Ward gauge-field construction referred to above). The non-commutative algebra does indeed appear to make geometrical sense, in terms of structures not pre-viously investigated (incorporating ideas from the theories of non-commutative geometry and of geometric quantization). See Connes and Berberian [1995] and Woodhouse [1991]. This procedure does indeed appear to provide a formalism broad enough to incorporate both left- and right-handed helicities at the same time, and the potential to describe fully curved space-times in a way that enables the Einstein vacuum equations (with or without $\Lambda$) to be simply encompassed, but whether or not it properly achieves what is really needed remains to be seen.

It may be mentioned that the new appreciation that twistor theory has obtained in recent years, as initiated by the twistor-string ideas of Witten and others, has found its particular utility in very high-energy processes. The particular role for twistors in this context has come about largely from the fact that one may regard particles as being *massless* in these circumstances. Twistor theory is certainly well attuned to the study of massless particles, but it is by no means limited in this way. There are, indeed, various ideas for the incorporation of mass into the general scheme [Penrose 1975b; Perjés 1977, pp. 53–72, 1982, pp. 53–72; Perjés and Sparling 1979; Hughston 1979, 1980; Hodges 1985b; Penrose and Rindler 1986] but these appear not to have played any role in these new developments, as yet. How twistor theory develops in the future with regard to rest mass would appear to be a matter of some considerable interest.

## 4.2. WHITHER QUANTUM FOUNDATIONS?

In §2.13 I have tried to make the case that no matter how well the standard formalism of quantum mechanics is supported by its vast multitude of experimentally confirmed implications – no experiment to date having presented evidence for need of modification – there are, nevertheless, strong arguments for believing that the theory is actually a provisional one and that the linearity, so fundamental to our present understanding of quantum theory, must ultimately be somehow superseded in order that its (mutually inconsistent) **U** and **R** ingredients become merely excellent approximations to some more all-embracing consistent scheme. In fact, as Dirac [1963] himself has eloquently argued:

> Everyone is agreed on the formalism [of quantum physics]. It works so well that nobody can afford to disagree with it. But still the picture that we are to set up behind this formalism is a subject of controversy. I should like to suggest that one not worry too much about this controversy. I feel very strongly that the stage physics has reached at the present day is not the final stage. It is just one stage in the evolution of our picture of nature, and we should expect this process of evolution to continue in the future, as biological evolution continues into the future. The present stage of physical theory is merely a steppingstone toward the better stages we shall have in the future. One can be quite sure that there will be better stages simply because of the difficulties that occur in the physics of today.

If this is to be accepted, we need some kind of indication as to what form the improved rules of quantum theory might actually take. Or at least we need to know the kind of experimental circumstances under which it is to be expected that observational deviations from the predictions of standard quantum theory should begin to become apparent.

I have already made a case, in §2.13, that these circumstances must be those in which *gravity* starts to be significantly involved in quantum superpositions. The idea is that when this level is reached, some sort of nonlinear instability sets in, limiting the duration of such superpositions, and one or another of the alternatives taking part will have become resolved out, after a certain estimated time period. Moreover, I am making the claim that *all* quantum state reductions (**R**) arise in this "**OR**" (objective reduction) way.

A common response to suggestions of this kind is that the gravitational force is so incredibly weak that no presently conceivable laboratory experiment could be expected to be affected by any quantum implications of gravity's presence; still less could the ubiquitous phenomenon of quantum state reduction be the result of such absurdly tiny quantum aspects of gravitational forces or energies, these being far tinier than any of those with which they might be competing in the situations under consideration. A complementary problem would seem to be that, if we are indeed expecting quantum gravity to be relevant to, for example, some simple desk-top quantum experiment, how could we expect our quantum state-reducing process to invoke the comparatively enormous (quantum-gravity) Planck energy $E_P$ – this being an energy scale which is some $10^{15}$ times the energy that the individual particles in the LHC can reach (see §§1.1 and 1.10), and which is about the energy released in the explosion of a sizeable artillery shell. Furthermore, if we are asking for quantum-gravity effects in the fundamental structure of space-time to provide us with something significant in the behaviour of things, we must bear in mind that the scales at which quantum gravity is argued to influence behaviour are the Planck length $l_P$ and the Planck time $t_P$, these being viewed as so ridiculously tiny (see §3.6) that they could hardly be of relevance to ordinary macroscopic physics.

However, I argue very differently here. I am not really suggesting that the tiny gravitational forces involved in a quantum experiment should be the relevant **OR** trigger, nor am I asking that a Planck energy need be conjured up in the process of quantum state reduction. I am asking, instead, that we must look towards a fundamental change in our quantum viewpoint, this change taking very seriously Einstein's picture of gravity as a curved-space-time phenomenon. We should also bear in mind that the Planck length $l_P$ and the Planck time $t_P$

$$l_P = \sqrt{\frac{\gamma\hbar}{c^3}}, \qquad t_P = \sqrt{\frac{\gamma\hbar}{c^5}},$$

are both obtained (as square roots) by multiplying together two quantities – the gravitational constant $\gamma$ and (the reduced) Planck's constant $\hbar$, which are extremely tiny on the scale of ordinary things of our experience, and then dividing by a positive power of a very large quantity, the speed of light. So it is not surprising that these calculations lead us to scales that are tiny almost beyond our comprehension, each being some $10^{-20}$ times smaller than the scale at which the smallest and most rapid processes occur in fundamental particle interactions.

Yet the proposal of *objective* state reduction, which I have argued for in §2.13, and which I now refer to as **OR** (this being essentially similar to the proposal put forward several years earlier by Lajos Diòsi [1984, 1987, 1989], though without my own motivations from general-relativistic principles [Penrose 1993, 1996, 2000b, pp. 266–82]), leads us to a much more reasonable-looking scale of time for the state reduction. Indeed, **OR**'s proposed average lifetime $\tau \approx \hbar/E_G$ for the decay time of a superposition of a stationary object in two separated locations, as given in §2.13, leads us to consider time periods whose calculation involves, in effect, the quotient $\hbar/\gamma$ of these two very small quantities, not their product (and the speed of light does not even come into it), the gravitational (self-)energy quantity $E_G$ (in Newtonian theory) being proportional to $\gamma$. Since there is now no particular reason why $\hbar/E_G$ should be either particularly large or particularly small, we need to look carefully, in any particular instance, to see if this formula leads to time scales that seem plausible for an actual physical process that could underlie a realistic objective quantum state reduction. It may also be remarked that the Planck energy,

$$E_P = \sqrt{\frac{\hbar c^5}{\gamma}},$$

while itself involving this same quotient $\hbar/\gamma$, has a large power of the speed of light in its numerator, which greatly enhances its magnitude.

In any case, in Newtonian theory, gravitational self-energy calculations do always yield expressions that are proportional to $\gamma$, so we see that $\tau$ does indeed scale in proportion to the quotient $\hbar/\gamma$. Owing to the smallness of $\gamma$, the quantity $E_G$ is certainly likely to be exceedingly tiny in the experimental situations I shall be considering (particularly because it is the distribution of mass *displacement* that is relevant here, the total overall mass in this subtracted distribution being zero), so this might lead us to expect a very long time scale for the decay time of a quantum superposition – being in proportion to $\gamma^{-1}$ – which would be in agreement with the fact that in standard quantum mechanics the time of persistence of a quantum superposition should be *infinite* (in accordance with the limit $\gamma \to 0$). But then we must bear in mind that $\hbar$ is also very small in ordinary terms, so it may not be out of the question that the ratio $\hbar/E_G$ can be arranged to come out as a measurable number. Another way to think of this is that, in natural units (Planck units, see §3.6), a second is an extremely long time, namely about $2 \times 10^{43}$, so to get a measurable effect, say of the order of seconds, the gravitational self-energy under consideration need only be something very tiny indeed in natural units.

Another point of relevance here is that our expression $\hbar/E_G$ does not involve the speed of light $c$. This has the simplifying implication that we can consider situations in which the mass movements are very slow. This has various practical advantages, but there are advantages also on the theoretical side, because we do not need to worry about the full intricacies of Einstein's general relativity, and we can content ourselves with a largely Newtonian treatment. Moreover, we can postpone any worries that we might have about the non-local aspects of a *physically real* quantum state reduction "violating causality" (as could be thought to be troublesome in EPR situations like those considered in §2.10), because in Newtonian theory the speed of light, being infinite, provides no bar to the speed of signals, and gravitational influences may be taken as being instantaneous.

I consider the **OR** proposal in a minimalist form (involving the fewest additional assumptions), where we have a roughly equal-amplitude quantum superposition of a pair of states, each of which, on its own, would be stationary. This is the situation considered in §2.13, where I presented a rough argument that there should be an approximate scale of time $\tau$, limiting the probable persistence of such a superposition, after which it would spontaneously resolve itself into one or other of these alternatives, this decay time being given by

$$\tau \approx \frac{\hbar}{E_G},$$

where $E_G$ is the gravitational self-energy of the *difference* between the mass distribution in one of the states in superposition and the mass distribution in the other. If the displacement is merely a rigid translation of one location to the other, then the simpler description of $E_G$ indicated in §2.13 can be given, namely the *interaction energy*, which would be the energy it would cost to make this displacement, considering each of these states to be acted upon only by the *gravitational* field of the other.

The gravitational self-energy of a gravitationally bound system is, quite generally, the energy it would take to disperse the system into its gravitating constituents, taken out to infinity, ignoring all other forces, gravity being treated according to Newtonian theory. For an example, the gravitational self-energy of a uniform sphere of mass $m$ and radius $r$ comes out as $3m^2\gamma/5r$. For the evaluation of $E_G$ in the situation under consideration here, we would calculate it

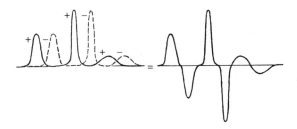

Figure 4-8: $E_G$ is the gravitational self-energy of the mass distribution that results from subtracting the mass expectation distribution of one of the superposed quantum states from that of the other. For each individual state we may well find the mass density highly concentrated in certain regions (e.g. around nuclei), their difference providing a patchwork of positive and of negative mass, leading to a relatively large $E_G$.

from a theoretical mass distribution that comes about by *subtracting* the mass distribution of one of the stationary states from that of the other, so the relevant mass distribution would be positive in some regions and negative in others (see figure 4-8), which is *not* the normal situation under which one calculates gravitational self-energies!

As an example, we could consider the case of the above uniform sphere (radius $r$ and mass $m$), put into a (roughly equal amplitude) *superposition* of two horizontally displaced locations, where their centres are separated by a distance $q$. The (Newtonian) calculation for $E_G$ (self-energy of the *difference* of the two mass distributions) now turns out to give

$$E_G = \begin{cases} \dfrac{m^2\gamma}{r}\left(2\lambda^2 - \dfrac{3\lambda^3}{2} + \dfrac{\lambda^5}{5}\right), & 0 \leqslant \lambda \leqslant 1, \\[2ex] \dfrac{m^2\gamma}{r}\left(\dfrac{6}{5} - \dfrac{1}{2\lambda}\right), & 1 \leqslant \lambda, \end{cases} \qquad \text{where } \lambda = \dfrac{q}{2r}.$$

We see that, as the distance $q$ increases, the value of $E_G$ increases, reaching

$$\frac{7}{10} \times \frac{m^2\gamma}{r}$$

when the two instances of the sphere are in contact ($\lambda = 1$); then with further increase in $q$ the increase in $E_G$ slows, attaining the limiting total value,

$$\frac{6}{5} \times \frac{m^2\gamma}{r},$$

for infinite separation ($\lambda = \infty$). Thus, the major effect on $E_G$ occurs with the separation increasing from coincidence to contact, and only somewhat less can be additionally gained by any further increase in separation.

Of course, any actual material would not really be uniform in its detailed structure, the mass being concentrated mainly in the atomic nuclei. This suggests that the major effect might be achieved by a very tiny displacement, in a quantum superposition of a material body in two different locations, where the nuclei need be displaced merely by their diameters. This could well be so, but there is the important complicating issue that we are really considering *quantum* states, and the nuclei would be expected to be "smeared out" in accordance with such issues as Heisenberg's uncertainty principle (§2.13). Indeed, if that were not so, we might well worry that we should even consider the individual neutrons and protons that are the constituents of those nuclei, or the individual quarks that make up the proton and neutrons. Since quarks, like electrons also, are considered to be *point-like* particles – giving $r = 0$ in the above formula – we seem to obtain $E_G \approx \infty$, whence $\tau \approx 0$, which would seem to lead to the conclusion that almost all superpositions would instantaneously collapse [see also Ghirardi et al. 1990] and therefore there would be no quantum mechanics, according to this proposal!

Yes, we *do* need to take "quantum spreading" into account if we are to take $\tau \approx \hbar/E_G$ seriously. We recall from §2.13 that Heisenberg's uncertainty principle tells us that the more precisely localized a particle's state may be, the more spread out will be its momentum. Accordingly, we cannot expect a localized particle to remain stationary, its stationarity being a requirement for our current considerations. Of course, for the extended bodies that we would be concerned with, in our calculations of $E_G$, we would have to take into account collections of very many particles, all of which would contribute to such a body's stationary state. We would need to work out that body's stationary wave function $\psi$ and then to calculate what is referred to as the *expectation value* of the mass density at each point (which is a standard quantum-mechanical procedure) giving us an expected mass distribution for the whole body. This procedure would be carried out for the body in each of the two locations involved in the quantum superposition, and one of these (expected) mass distributions would need to be subtracted from the other, so that the required gravitational self-energy $E_G$ can be computed. (There is still the technical point, raised in §1.10, that quantum stationary states would, strictly speaking, be spread out over the whole universe – but this can be dealt with by the standard mild cheat of treating the mass centre *classically*. We will return to this issue later in the section.)

We may now ask whether a stronger rational basis can be given for this **OR** proposal than the somewhat tentative considerations presented in §2.13. The idea is that there is an underlying tension between the principles of Einstein's general theory of relativity and those of quantum mechanics, and that this tension will only be properly relieved by a fundamental change in basic principles. My bias here is to place the greater trust in the basic principles of general relativity, and to be more questioning of those that are fundamental to standard quantum mechanics. This emphasis is different from what one finds in most treatments of quantum gravity. Indeed, it is probably the common view among physicists that in such a clash of principles those of general relativity would be more likely to have to be abandoned, being less firmly established experimentally than those of quantum mechanics. I shall try to argue from a largely opposite perspective, taking Einstein's *principle of equivalence* (see §§1.12 and 3.7) to be more basic than the quantum principle of *linear superposition* – largely because it is this very aspect of the quantum formalism that leads us into the paradoxes of applying that theory to macroscopic objects (such as Schrödinger's cat; see §§1.4, 2.5, and §2.11).

The reader may well recall the (Galilei–)Einstein equivalence principle, which asserts that the local effect of a gravitational field is the same as that of an acceleration, or, put another way, it asserts that a local observer freely falling under gravity will feel no gravitational force. An alternative way of expressing this is to say that the gravitational force acting on a body is in proportion to the inertial mass of that body (resistance to its acceleration), a property not shared by any other force of nature. The fact that astronauts in free orbit float freely in their space stations or on their space walks, and do not feel any gravitational pull, is now very familiar to us. It was well appreciated by Galileo (and Newton) – and Einstein used it as the foundational principle of his general theory of relativity.

Let us now imagine a quantum experiment, where the influence of the Earth's gravitational field is being taken into account. We can envisage two different procedures – which I shall call *perspectives* – for accommodating the Earth's field. There is the more straightforward Newtonian perspective, which is simply to treat the Earth's field as providing a downward force $m\mathbf{a}$ on any particle of mass $m$ (the gravitational acceleration vector $\mathbf{a}$ here being assumed constant in both space and time). The Newtonian coordinates are $(\mathbf{x}, t)$, where the 3-vector $\mathbf{x}$ denotes spatial position and $t$ the time. In standard quantum-mechanical language, this perspective would treat the gravitational field by the standard quantum procedure, which would be referred to by the words "adding a gravitational potential term into the Hamiltonian", in accordance with the same procedure that would be adopted for any other physical force. The alternative Einsteinian perspective would be to

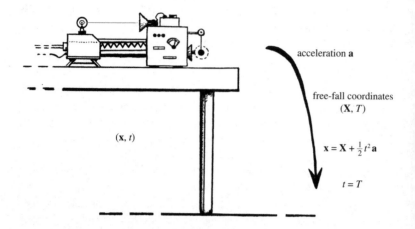

Figure 4-9: A (fanciful) quantum experiment is set up in which the effect of the Earth's gravitational field is involved. The (conventional) Newtonian perspective on the Earth's gravity uses coordinates $(\mathbf{x}, t)$ fixed by the laboratory and treats the Earth's field in the same way as any other force. The Einsteinian perspective uses free-fall coordinates $(\mathbf{X}, T)$ in which the Earth's field vanishes.

imagine that the description is being made with respect to the space and time coordinates $(\mathbf{X}, T)$ used by a freely falling observer so that, according to that observer, the Earth's gravitational field vanishes. The description is then translated back to that of the experimenter in the stationary laboratory (see figure 4-9). The relation between the two sets of coordinates is

$$\mathbf{x} = \mathbf{X} + \tfrac{1}{2}t^2\mathbf{a}, \quad t = T.$$

What we find [Penrose 2009a, 2014a; Greenberger and Overhauser 1979; Beyer and Nitsch 1986; Rosu 1999; Rauch and Werner 2015] is that the wave function $\psi_E$ (§§2.5–2.7) that the Einsteinian perspective gives us is related to the wave function $\psi_N$ of the Newtonian perspective by

$$\psi_E = \exp\left(\mathrm{i}(\tfrac{1}{6}t^3\mathbf{a}\cdot\mathbf{a} - t\bar{\mathbf{x}}\cdot\mathbf{a})\frac{M}{h}\right)\psi_N$$

(with suitable coordinate choices), where $M$ is the total mass of the quantum system being studied and $\bar{\mathbf{x}}$ the Newtonian position vector of the mass centre of that system. The discrepancy is simply a phase factor, so this ought not to lead to any observable differences between the two perspectives (see §2.5) – or should it? In the situation under consideration here, the two descriptions ought indeed to be equivalent, and there is a well-known experiment, performed originally in 1975 [Colella et al. 1975; Colella and Overhauser 1980; Werner 1994; Rauch and Werner 2015], demonstrating agreement between the two perspectives, thereby supporting the notion that quantum mechanics *is* consistent with Einstein's equivalence principle in this context.

However, a curious fact about this phase factor should be noticed, namely that it involves the term

$$\tfrac{1}{6}t^3\mathbf{a}\cdot\mathbf{a}$$

in the exponent (multiplied by the factor $iM/\hbar$), which has the implication that when we try to restrict attention to solutions of the Schrödinger equation that are of positive energy ("physical" solutions, of *positive frequency*; see §4.1) separating them from contributions of negative energy ("unphysical" solutions), we find a discrepancy between the Einsteinian and Newtonian wave functions. Quantum field theory considerations (of relevance also to ordinary quantum mechanics; see Penrose [2014a]) would tell us that the Einsteinian and Newtonian perspectives have provided us with different vacua (see §§1.16 and 3.9), so that the Hilbert spaces that arise from the two perspectives are, in a sense, incompatible with one another, and we cannot consistently add state vectors from one of these Hilbert spaces to those of the other.

In fact, this is just the $c \to \infty$ limit of the *Unruh effect*, described briefly in §3.7, normally considered in the context of black holes, where an accelerating observer, in vacuum, is considered to experience a *temperature* $\hbar a/2\pi kc$ (i.e. just $a/2\pi$ in natural units). The vacuum experienced by the accelerating observer is what is called a *thermal vacuum*, which provides a non-zero ambient temperature, here taking the value $\hbar a/2\pi kc$. In the Newtonian limit $c \to \infty$, under consideration here, this Unruh temperature goes to zero, but the two vacua under consideration here (namely that provided by the Newtonian perspective and that by the Einsteinian perspective) remain *different* because of the nonlinear phase factor given above, which survives when the $c \to \infty$ limit is applied to the Unruh vacuum.[6]

This would not cause any difficulties when, as here, we are considering just a single background gravitational field, such as that of the Earth, and the states being superposed would all have the same vacuum state, no matter whether the Newtonian or Einsteinian perspective were being used throughout. But suppose that we are considering a circumstance in which there is a superposition of two

---

[6] I am grateful to Bernard Kay for confirming for me, in a calculation, this anticipated conclusion.

gravitational fields. This would be the case for a tabletop experiment involving the superposition of a massive object in two different locations (see figure 2-28 in §2.13). The tiny gravitational field of the object itself would be slightly different for each location and, in the quantum state that describes the superposition of the two locations of the object, the superposition of these two gravitational fields must also be considered. Then we *do* have to worry about which perspective we are adopting.

The Earth's gravity also contributes to the total gravitational field in such considerations, but when we calculate the *difference* that we shall see is needed for the computation of $E_G$, we find that the Earth's field cancels out, so that it is only the gravitational field of the *quantum-displaced object* which contributes to $E_G$. There is, however, an element of subtlety in this cancellation which needs to be considered. When the massive object under consideration is displaced in some direction, there must be a compensating displacement of the Earth in the opposite direction, so that the centre of gravity of the Earth–object system remains unchanged. The Earth's displacement is, of course, exceedingly tiny, owing to the vastness of the Earth's mass in comparison with that of the object. But the Earth's very vastness might lead us to question whether even this minute displacement of the Earth might lead to a considerable contribution to $E_G$. Fortunately, an examination of the details of what is involved quickly leads us to the conclusion that the cancellation is indeed effective, and the contribution of the Earth's displacement to $E_G$ can be completely ignored.

But why should we be considering the quantity $E_G$ in any case? If we adopt the Newtonian perspective for treating gravitational fields generally, then there is no difficulty about treating a quantum superposition of the two locations of the object, the gravitational field being regarded in just the same way as would any other field, in accordance with normal quantum-mechanical procedures, and in this perspective linear superposition of gravitational states is allowed, there being only one vacuum arising. However, in my opinion, in view of general relativity's remarkable large-scale observational support, we should take on board the Einsteinian perspective, this being extremely likely to be ultimately more in accordance with nature's ways than the Newtonian perspective. Then we find ourselves driven to the viewpoint that in the superposition of two gravitational fields under consideration, both fields must be treated in accordance with the Einsteinian perspective. This involves us in trying to superpose states belonging to two different vacua – i.e. two incompatible Hilbert spaces – such superpositions being considered to be *not allowed* (see §1.16 and early parts of §3.9).

We need to look at this situation a little more closely, where we are to imagine things on a quite tiny scale, largely between the regions of the superposed body where the nuclei reside, but also external to the body. Although the considerations of the preceding few paragraphs were concerned with a gravitational acceleration field **a** that is spatially *constant*, we may assume that these considerations will still apply locally, at least approximately, in these largely empty regions where there will be a superposition of two different gravitational fields. The point of view I am adopting here is that each of these fields is individually to be treated according to the Einsteinian perspective, so that here the physics has a structure involving an "illegal" superposition of quantum states belonging to two different Hilbert spaces. The state of free fall in one gravitational field would be related to the state of free fall in the other by a phase factor of the type we encountered above, involving a term that is nonlinear in the time $t$ in the exponent $e^{iMQt^3/\hbar}$, for some particular $Q$ like the $\frac{1}{6}\mathbf{a} \cdot \mathbf{a}$ above. But since we are now considering passing from one state of free fall (with acceleration vector $\mathbf{a}_1$) to another state of free fall (acceleration vector $\mathbf{a}_2$), our $Q$ now has the form $\frac{1}{6}(\mathbf{a}_1 - \mathbf{a}_2) \cdot (\mathbf{a}_1 - \mathbf{a}_2)$ rather than simply the $\frac{1}{6}\mathbf{a} \cdot \mathbf{a}$ that we had before, because what is relevant is the *difference* $\mathbf{a}_1 - \mathbf{a}_2$ between the fields of the object in its two different locations, the individual accelerations $\mathbf{a}_1$ and $\mathbf{a}_2$ having only a *relative* significance in relation to the Earth as a reference system.

In fact, each of $\mathbf{a}_1$ and $\mathbf{a}_2$ is now a function of position, but I am supposing that, at least to a good approximation in any small local region, it is this term ($Q$) that is causing us the problem. The superposition of states from such different Hilbert spaces (i.e. with differing vacua) is technically illegal, there now being a local phase factor

$$\exp\left(\frac{iM(\mathbf{a}_1 - \mathbf{a}_2)^2 t^3}{6\hbar}\right)$$

between one of the states and the other in local regions. This tells us that the states belong to incompatible Hilbert spaces, even though the difference $\mathbf{a}_1 - \mathbf{a}_2$ in the two accelerations of free fall would almost certainly be extraordinarily tiny in the kind of experiment under consideration.

Strictly speaking, the notion of alternative vacua is a feature of QFT, rather than of the non-relativistic quantum mechanics under consideration here, but the issue has direct relevance to the latter also. Standard quantum mechanics requires that energies remain positive (i.e. that frequencies remain positive), but this is not normally a problem in ordinary quantum mechanics (for the technical reason that normal quantum dynamics is governed by a positive-definite Hamiltonian, which will preserve this positivity). But the situation arising here is not like that, and we do appear to be forced into violating the condition unless the vacua are kept separate, i.e. that state vectors belonging to one of the Hilbert spaces are not added to (superposed with) those of the other [see Penrose 2014a].

Thus, we appear to be taken outside the normal framework of quantum mechanics, and there does not seem to be an unambiguous way to proceed. What I am proposing, at this stage, is that we follow the same kind of route as in §2.13, namely that we do not directly face the conundrum of the superpositions of different Hilbert-space vacua, but just try to estimate the *error* involved in trying to ignore this problem. As before (in §2.13), the problem term is the quantity $(\mathbf{a}_1 - \mathbf{a}_2)^2$, and the proposal is to take the total of this quantity over the whole of 3-space (i.e. its spatial integral) as the measure of the error in ignoring the problem of the illegal superpositions. The uncertainty involved in ignoring this issue thus leads us again to a quantity $E_G$ as a measure of the energy uncertainty inherent in the system, as before, in §2.13 [Penrose 1996].

In order for us to be able to estimate the length of time that our superposition might persist before mathematical contradictions involved in the illegality of our superposition begin to arise, we may take advantage of Heisenberg's time–energy uncertainty principle $\Delta E \Delta t \geqslant \frac{1}{2}\hbar$ as an estimate for how long the superposition might last, where $\Delta E \approx E_G$ (again just as in §2.13). This is simply the sort of situation that arises with an unstable atomic nucleus, which decays after a certain average time span $\tau$. As in §2.13, we take $\tau$ to be essentially the "$\Delta t$" of Heisenberg's relation, since it is this uncertainty that allows the decay to take place at all, within a finite time. Thus, we always find a fundamental energy uncertainty $\Delta E$ (or, by Einstein's $E = Mc^2$ a mass uncertainty $c^2 \Delta M$), which relates roughly to the decay time $\tau$ via Heisenberg's relation, so $\tau \approx \hbar/2\Delta E$. Hence (ignoring small numerical factors), we again obtain

$$\tau \approx \frac{\hbar}{E_G}$$

as the proposal for the expected lifetime of the superposition, as in §2.13.

Despite the comments above, which point out that this proposal for an *objective* **R** (i.e. **OR**) event which could be taking place at "ordinary" time scales with objects that are not extraordinarily tiny, we can indeed see in it a direct link with the Planck time and Planck length. In figure 4-10, I have tried to sketch the space-time history of such an **OR**-event, where a lump of material is put into a quantum superposition of two separate locations, illustrated as a space-time that gradually bifurcates before the **OR**-event occurs. At the **OR**-event itself, one of the

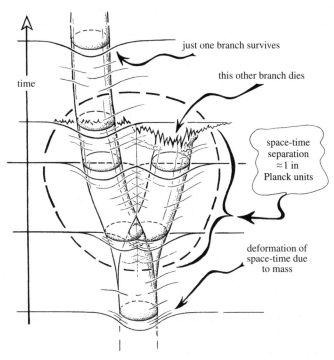

time

just one branch survives

this other branch dies

space-time
separation
≈ 1 in
Planck units

deformation of
space-time due
to mass

Figure 4-10:   A space-time sketch illustrating how a quantum superposition of two different displacements of a massive lump leads to a significant separation of the superposed space-times, deformed separately by the different lump locations. The gravitational **OR** proposal asserts that one of the space-times "dies off" at around the time when the space-time separation between the two components reaches order unity in Planck units.

components of the bifurcation dies off, leaving a single space-time representing the resulting lump location. I have indicated the limited space-time region where the bifurcation occurs, before it is killed off by the **OR** process. The relation with Planck units comes about from the fact that, in this proposal, the 4-volume over which the bifurcation persists is of the order of *unity* in Planck units! Thus, the smaller the *spatial* separation, as the space-time branches, the longer the bifurcation will persist, and the greater the spatial separation, the shorter will be the time of its persistence. (The measure of space-time separation, however, has to be understood in terms of an appropriate symplectic measure on the space of space-times, which is not so easy to grasp, though a rather crude derivation of the $\tau \approx \hbar/E_G$ estimate can be obtained in this way [Penrose 1993, pp. 179–89; see also Hameroff and Penrose 2014].

A natural question to ask is whether there is any observational evidence which either supports or refutes this proposal. It is easy to envisage situations in which $\hbar/E_G$ is either a very long period of time or a very short one. In the case of our Schrödinger's cat of §§1.4 and 2.13, for example, the kind of mass displacement between the cat's locations at the $A$-door and at the $B$-door would be hugely sufficient for $\tau$ to come out as an extremely tiny time (far shorter than the Planck time of $\sim 10^{-43}$ s), so that the spontaneous transition from any superposed cat positions would be essentially instantaneous. On the other hand, in experiments where we have quantum superpositions of individual neutrons in different locations, the value of $\tau$ would be enormous, measured in astronomical terms. The same would apply even to the $C_{60}$ and $C_{70}$ buckyballs (individual molecules with 60 or 70 carbon atoms), which appear to be the largest objects that have been observed, to date, to take part in quantum superpositions of different locations [Arndt et al. 1999], whereas the length of time that these molecules would have actually remained isolated in superposition would have been only a tiny fraction of a second.

In fact, in both kinds of situation, we would need to bear in mind that the quantum state under consideration might well not be isolated from its surroundings, so that there would be a considerable amount of additional material, namely the system's *environment*, whose state would be likely to be entangled with that of the quantum state under consideration. Accordingly, the mass displacement involved in the superposition would have also to take into account the displacements in all of this disturbed environment, and it would be this environmental displacement (involving vast numbers of particles moving in many different directions) which would frequently provide the major contribution to $E_G$. The issue of environmental decoherence features strongly in most of the conventional viewpoints whereby the unitary evolution (**U**) of a quantum system is taken to result in an effective reduction (**R**) of the quantum state, in accordance with the Born rule (§2.6). The idea is that this environment contributes uncontrollably to the quantum system under consideration, and the procedure, as described in §2.13, is to average out all these environmental degrees of freedom whereby the quantum superposed state behaves effectively as though it were a probability mixture of the contributing alternatives. Whereas I have argued in §2.13 that the involvement of a system's messy environment into the quantum state does not actually resolve the measurement paradox of quantum mechanics, it has an important role to play in the **OR** modification of standard quantum theory that I am arguing for here. Once the outside environment begins to get seriously involved in a quantum state, then sufficient mass displacement is rapidly attained through the system's entanglement with this environment, to give a large enough $E_G$ for a very rapid spontaneous reduction of the state to one or other of the alternatives that are involved in the superposition. (This idea borrows much from the earlier "**OR**" proposal of Ghirardi and his associates [Ghirardi et al. 1986].)

At the time of writing, no actual experiment has been carried out that is sensitive enough to confirm or refute this proposal. The environmental contribution to $E_G$ has to be kept very tiny for there to be any hope of observing the relevant effect. There are several projects underway [Kleckner et al. 2008, 2015; Pikovski et al. 2012; Kaltenbaek et al. 2012] which should eventually have something to say on the matter. The only one which I have been at all involved with myself [Marshall et al. 2003; Kleckner et al. 2011] is an experiment being developed under the direction of Dirk Bouwmeester of Leiden and Santa Barbara Universities. In this, a tiny mirror, which is a cube of about 10 micrometres ($10^{-5}$ metres – roughly one tenth of the thickness of a human hair) in size, is to be put into a quantum superposition of two locations, these locations differing from one another by about the diameter of an atomic nucleus. The intention would be to try to maintain this superposition for a period of some seconds or minutes and then return it to its original state to see whether any phase coherence is necessarily lost.

This superposition would be achieved by first dividing the quantum state of a single photon by a beam splitter (see §2.3). One part of the photon's wave function then impacts on the tiny mirror so that the photon's momentum causes it to move slightly (perhaps comparable with the atomic nuclei of the tiny mirror), this being delicately suspended on a cantilever. Since the photon's state is split into two, the mirror's state becomes a superposition of displacement and non-displacement – a tiny Schrödinger's cat. For a visible-light photon, however, a single impact would not be nearly sufficient for what is required, so the same photon is arranged to impact on the tiny mirror again and again, around a million times, by reflecting it back and forth from a fixed (hemispherical) mirror. This multiple impact might perhaps be sufficient to displace the tiny mirror by about the diameter of an atomic nucleus, or perhaps more if this is needed, in a period of seconds or so.

There is some theoretical uncertainty as to how refined one would have to consider the mass distribution to be. Since each of the two components of the superposition is to be taken as individually stationary, there will be a spread in the mass distribution which would presumably depend on the material being used. Stationary solutions of the Schrödinger equation would necessarily involve some spread on the matter distribution, as follows from the Heisenberg uncertainty principle (so $E_G$ should *not* be calculated as though the particles occupied point-like locations – fortunately, as remarked earlier, because that would give an infinite $E_G$). A uniformly spread-out mass distribution would probably not be appropriate either (that being the most *un*favourable case experimentally, giving the smallest possible $E_G$ for a given total mass size/shape and separation). An accurate estimate for $E_G$ would need the stationary Schrödinger equation to be solved, at least approximately, so that the expected mass distribution can be estimated. For such an experiment to be successful, excellent isolation from vibration is required, in addition to the system being kept at a very low temperature in a near-perfect vacuum, and most particularly with mirrors of superb quality.

There is the technical issue, raised earlier in this section (and also in §1.10), about stationary solutions of the Schrödinger equation being necessarily spread out over the entire universe. This can be addressed either by the usual rather ad hoc procedure of taking the mass centre to be located at a fixed place, or else (perhaps preferably) using the Schrödinger–Newton (SN) equation. The latter is a nonlinear extension of the standard Schrödinger equation that takes into account the gravitating effect of the expectation value of the mass distribution provided by the wave function itself into the equation, as a Newtonian gravitational field added into the Hamiltonian [Ruffini and Bonazzola 1969; Diósi 1984; Moroz et al. 1998; Tod and Moroz 1999; Robertshaw and Tod 2006]. The SN equation's main value, so far, with regard to **OR**, has been to provide a proposal for the alternative stationary states into which the system is supposed to reduce, as an action of **OR**.

The cantilever holding the tiny mirror causes it to swing back to its original position in some prearranged time interval of, say, a few seconds or minutes. To ascertain whether or not the state of the tiny mirror has in fact reduced spontaneously during the photon impacts, or whether quantum coherence has been preserved, the photon would be released from this reflecting cavity (of tiny mirror and hemispherical mirror), so that it can retrace its track back to the beam splitter. Meanwhile, the other part of the photon's wave function has to have been marking time, being trapped within another reflecting cavity, made from two stationary mirrors. If, as standard quantum theory maintains, phase coherence is actually preserved between the two separated parts of the photon's wave function, this could be confirmed by arranging for a photon detector to be placed at

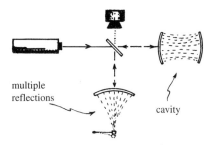

multiple
reflections

cavity

Figure 4-11: A cartoon of the Bouwmeester experiment to test whether or not gravitational **OR** is respected by nature. A single photon emitted by a laser is aimed at a beam splitter, so that the photon's path is split into horizontal and vertical tracks. The horizontal track leads to a cavity, where the photon can be held by continued back-and-forth reflections by mirrors, while the vertical one leads to another cavity in which one mirror is tiny and suspended so that the pressure from the multiple photon reflections would move it slightly. The **OR** proposal asserts that after a measurable time interval the tiny mirror's two superposed locations would spontaneously reduce to one or the other, as opposed to being in a superposition. This could be discerned by the detector at the top by reversing the photon's motion.

an appropriate point on the other side of the beam splitter (see figure 4-11), so that so long as there is no loss of coherence in the system, the returning photon would always (or with an alternative arrangement, never) activate that particular detector.

The current state of this experiment is still somewhat short of critically testing the proposal. Up to a point, its successful performing ought to confirm the predictions of standard quantum mechanics, but at a level considerably exceeding that which has been achieved so far (in terms of mass displaced, between states in superposition). The expectation, however, is that with some further refinements in technique the actual boundaries of standard quantum theory will begin to be seriously probed, and it may well be possible, within a few years, to decide experimentally whether or not proposals like the one I have been putting forward have a basis in observational fact (Weaver et al. 2016; Eerkens et al. 2015; Pepper et al. 2012; see also Kaltenbaek et al. 2016; Li et al. 2011).

To end this section, I should mention some issues relating to this state-reduction proposal that could have importance if the scheme turns out to be experimentally confirmed. It should be clear, from the above descriptions, that the proposal is a genuinely objective one, in the sense that in such an **OR** scheme it is proposed that

**R** takes place out there in the world, and is not something imposed upon the world, in some way, by virtue of a quantum system being actually observed by some sort of conscious entity. In parts of the universe that are completely remote from any conscious observers, **R**-events would occur in exactly the same circumstances, with the same frequencies, and with the same probabilistic outcomes, as they would here, being perhaps peered at by numerous conscious beings. On the other hand, I have, in various places [Penrose 1989, 1994, 1997], promoted the idea that the phenomenon of *consciousness itself* might well be dependent upon such **OR**-events (occurring mainly within neuronal microtubules), each such event providing, in a sense, a moment of "proto-consciousness", the basic element out of which genuine consciousness would be somehow constructed [Hameroff and Penrose 2014].

As part of these investigations, I considered a slight generalization of the **OR**-proposal outlined above, which could apply when there is a quantum superposition of two stationary states whose energies $E_1$ and $E_2$ differ slightly from one another. In standard quantum mechanics, such a superposition would oscillate between these states with a frequency $|E_1 - E_2|/h$ combined with a much higher quantum oscillation of frequency of around $(E_1 + E_2)/2h$. The generalized **OR**-proposal would be that in such a situation, after an average time of about $\tau \approx \hbar/E_{\mathrm{G}}$, the state would reduce spontaneously to a *classical oscillation* of frequency $|E_1 - E_2|/h$ between these two alternatives, where the actual phase of this oscillation would be the "random" choice made by **OR**. However, this cannot be a completely general proposal, as there could be a classical energy barrier, preventing the classical oscillation.

Clearly, all this is extremely far from a coherent mathematical theory of generalized quantum mechanics that would include both **U** and **R** (as well as classical general relativity) as appropriate limits. What suggestions might I be prepared to make about the actual nature of such a theory? Very few, I'm afraid, though I take the view that such a theory would have to represent a major revolution in the framework of quantum mechanics, far from a matter of just tinkering with the current formalism. More specifically, I would be inclined to guess that elements of twistor theory should have a role in this, as it gives some hope that the puzzlingly non-local aspects of quantum entanglement and quantum measurement might perhaps tie in with the non-localities of holomorphic cohomology that the technicalities of twistor theory apparently force upon us (see §4.1). I believe that there is some hope that recent developments in palatial twistor theory, mentioned briefly at the end of §4.1, might perhaps offer some suggestions as to a possible route towards progress [Penrose 2015a,b].

## 4.3. CONFORMAL CRAZY COSMOLOGY?

Apart from the motivating arguments of §§2.13 and 4.2, there are various other reasons to suspect that quantum theory cannot be applied in a standard way to the gravitational field itself, in systems where the role of gravity becomes quantum mechanically significant. One such reason comes from the so-called information paradox of black-hole Hawking evaporation. This is an issue that certainly appears to relate to the possible nature of quantum gravity, and I shall be coming to it shortly. But there is another reason, hovering behind the entire discussion of chapter 3. This is the very strange nature of the Big Bang, as brought out particularly in §§3.4 and 3.6, namely that it was enormously constrained in its gravitational degrees of freedom, and apparently *only* in gravitational freedom.

The conventional view of how one must describe the physics of the Big Bang is that it is the only observed phenomenon (albeit somewhat indirectly observed) where the effects of *quantum gravity* (whatever that theory might actually be) are made manifest. Indeed, the rationale of obtaining a better understanding of the Big Bang has often been presented as an important reason for seriously entering into the frustratingly difficult area of quantum gravity. As a matter of fact, I have, on occasion, used that argument myself for supporting research into quantum gravity (see the preface to *Quantum Gravity* [Isham et al. 1975]).

But can one really expect that any conventional quantum (field) theory applied to the gravitational field could explain the extremely odd structure that the Big Bang must apparently have had, whether or not that momentous event was immediately followed by an inflationary phase. I believe that, for reasons I have tried to present in chapter 3, this *cannot* be the case. We have to explain the extraordinary suppression of gravitational degrees of freedom in the Big Bang. If all those other $10^{10^{124}}$ possibilities were potentially present at the Big Bang event, as the formalism of quantum mechanics would seem to insist, then one would expect that they would all be making their contribution to this initial state. It goes against the normal procedures of QFT that one should be able simply to *decree* that these should be absent. Moreover, it is hard to see how the *pre*-Big Bang schemes of §3.11 could evade these difficulties, as one would expect those gravitational degrees of freedom to have made their huge mark on the geometry *subsequent* to the bounce, as they should surely have been present *before* the bounce.

There is the related feature that any quantum gravity theory of a normal type would be subject to, namely dynamical *time symmetry*, where we are looking at something like the Schrödinger equation (**U**-process), which is time symmetric under the replacement $i \rightarrow -i$, as it would be applied to the time-symmetric equations of Einstein's general relativity. If this quantum theory is supposed to apply to the extremely high-entropy singularities expected to occur in black holes, very possibly of a general BKLM type, then these same space-time singularities (in time-reversed form) ought also to apply to the Big Bang, as allowed by that same "normal type" of quantum theory. But that didn't happen in the Big Bang. Moreover, as I hope I made clear in §3.10, the anthropic argument is next to useless in explaining this kind of enormous restriction on the Big Bang.

Yet the Big Bang was extraordinarily constrained, in a way that we do not at all find in the singularities in black holes. The evidence is indeed strong that in just those places where the effects of quantum gravity "ought" to be making their greatest mark on phenomena, near these singularities, there is an absolutely stupendous *time asymmetry*. This should not be the case if the explanation had simply been the result of a quantum theory of normal type, even if assisted by a generous anthropic contribution. As I have said earlier, there must be another explanation.

My own perspective on this matter is to leave aside quantum theory for the moment, and try to think about the type of *geometry* that must have held near to the Big Bang, and to contrast this with that very wild (very possibly BKLM; see end of §3.2) type of geometry that is expected near a black-hole singularity. The first problem is simply to characterize a condition for gravitational degrees of freedom to be *suppressed* at the Big Bang. For many years (effectively dating back to around 1976), I have expressed this in terms of what I subsequently referred to as the *Weyl curvature hypothesis* [Penrose 1976a, 1987a, 1989, chapter 7, TRtR, §28.8]. The *Weyl conformal tensor*, **C**, measures the type of space-time curvature that is relevant to *conformal* space-time geometry, which, as we have seen in §§3.1, 3.5, 3.7, and 3.9, is the geometry that is defined by the system of light cones (or null cones) in space-time. I would need to enter into a significant amount of tensor calculus in order to write down **C**'s definition in terms of a formula, and this would take us considerably beyond the technical scope of this book. It is therefore fortunate for my discussion that we shall not need this formula here, though certain properties of **C** under conformal change of scale ($\hat{\mathbf{g}} = \Omega^2 \mathbf{g}$) will actually have considerable importance for us shortly.

A particular geometrical role for the tensor **C** is worth noting. This is that the equation **C** = **0**, holding throughout some not too extended simply-connected open space-time region $\mathcal{R}$, asserts that $\mathcal{R}$ (with metric **g**) is *conformally flat*. This means that there exists a real scalar field $\Omega$ (called a *conformal factor*) such that the conformally related space-time metric $\hat{\mathbf{g}} = \Omega^2 \mathbf{g}$ is the flat Minkowski metric in $\mathcal{R}$. (See §§A.6 and A.7 for the intuitive meanings of *simply-connected* and *open*, but these terms do not play any significant role for us here.)

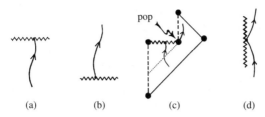

Figure 4-12:  Different types of space-time singularity: (a) future type, encountered only by world-lines that enter it from the past; (b) past type, encountered only by world-lines that emerge from it into the future; (c) in a Hawking-evaporating black hole, the internal singularity has future type, but the ultimate "pop" appears to have past type; (d) a hypothetical naked singularity, which is encountered by world-lines that both come in from the past and leave into the future. According to the cosmic censorship hypothesis, type (d) should not occur in generic classical situations. According to the Weyl curvature hypothesis those of type (b), like the Big Bang, should be enormously constrained by suppression of Weyl curvature.

The full Riemann curvature tensor $\mathbf{R}$ has 20 independent components per point, and it can be split, in effect, into the Einstein tensor $\mathbf{G}$ (see §§1.1 and 3.1) and the Weyl tensor $\mathbf{C}$, each of which has 10 components per point. We recall the Einstein equations $\mathbf{G} = 8\pi\gamma\mathbf{T} + \Lambda\mathbf{g}$. Here $\mathbf{T}$, being the energy tensor of matter, tells us how all the matter degrees of freedom (including those of the electromagnetic field) directly influence the curvature of space-time, through the $\mathbf{G}$ part of the full space-time curvature $\mathbf{R}$, and we also have the contribution $\Lambda\mathbf{g}$ from the cosmological constant. The remaining 10 independent curvature components in $\mathbf{R}$ are those describing the *gravitational* field and are conveniently described by the Weyl tensor $\mathbf{C}$.

The Weyl curvature hypothesis asserts that any space-time singularity of *past* type – i.e. from which timelike curves can emerge into the future, but not enter from the past (figure 4-12(a),(b)) – must have *vanishing Weyl tensor*, in the limit as the singularity is approached inwards from the future along any of these timelike curves. The Big Bang (and any other "bang" singularity of like type that might exist, perhaps such as at the "pop" at the moment of disappearance of a black hole by Hawking evaporation; see figure 4-12(c), and later discussion in this section[7]) must, according to this hypothesis, be free of independent gravitational degrees of

---

[7] Unlike the timelike singularity of figure 4-12(d), the "pop" of figure 4-12(c) is not, in my view, to be regarded as providing a violation of cosmic censorship, as it is more like two separate singularities, namely the BKLM future part, represented by the irregular wiggly line, and a distinct past part representing the "pop". The causal structures of the two are indeed very different, and it makes little sense to identify them.

freedom. The hypothesis says nothing about singularities of future type, or about naked singularities (figure 4-12(d)) for which entering and emerging timelike curves would both be present (classical singularities of that type being supposed absent by strong cosmic censorship; see §§3.4 and 3.10).

A point should be made clear, here, about the Weyl curvature hypothesis. It is presented merely as a geometrical statement that expresses, in a reasonably clear sense, that the gravitational degrees of freedom are hugely suppressed at the Big Bang, or at other past-type space-time singularities (if such exist). It makes no statement about how one might propose to make a definition of an *entropy* in the gravitational field (such as some scalar quantity algebraically constructed from **C** that some people have suggested – not really very appropriately). The huge effect that this hypothesis would have on the *low entropy value* in the Big Bang (§3.6) is simply a direct consequence of this hypothesis, coming from the implied elimination of primordial white holes (or black holes). The actual "low entropy" and calculated improbability value (i.e. $10^{-10^{124}}$) comes from a direct application of the Bekenstein–Hawking formula (§3.6).

There are, however, some technical issues about the precise mathematical interpretation of the Weyl curvature hypothesis, one difficulty arising from the fact that **C** itself, being a tensor quantity, is not really defined *at* a space-time singularity, where such tensorial notions, strictly speaking, are not well defined. Accordingly, the assertion **C** = 0 at such a singularity needs to be phrased in some kind of limiting sense, as the singularity is approached. A trouble with this kind of thing is that there are various inequivalent ways of stating the condition, it being unclear which is most appropriate. In view of such uncertainties, it is fortunate that my Oxford colleague Paul Tod has proposed and made a careful study of an alternative formulation of a mathematical condition on the Big Bang that is not explicitly phrased in terms of **C** at all.

Tod's [2003] proposal is that (as with the FLRW Big Bang singularities; see end of §3.5) our Big Bang can be *conformally* represented as a smooth spacelike 3-surface $\mathscr{B}$, across which the space-time is, in principle, extendible into the past in a conformally smooth way. That is to say, with a suitable conformal factor $\Omega$, we can rescale our *physical* post–Big Bang metric $\check{\mathbf{g}}$ to a new metric **g**

$$\mathbf{g} = \Omega^2 \check{\mathbf{g}}$$

according to which the space-time can now acquire a smooth past boundary $\mathscr{B}$ (where $\Omega = \infty$) across which this *new* metric **g** remains perfectly well defined and smooth. This allows **g** to be continued into a hypothetical "pre–Big Bang" space-time region; see figure 4-13. (I hope the slightly strange notation, whereby

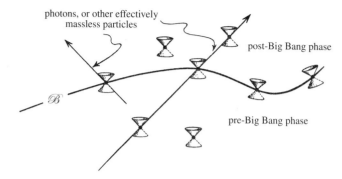

photons, or other effectively
massless particles

post-Big Bang phase

$\mathcal{B}$

pre-Big Bang phase

Figure 4-13: In Tod's proposal for a constraint on the Big Bang (of Weyl-curva-ture-hypothesis type), it is proposed that space-time can be extended into the past, in a conformally smooth way, so that the initial singularity becomes a smooth hypersurface $\mathcal{B}$ across which the conformal space-time extends smoothly to a hypothetical region that precedes the Big Bang. If that preceding region were to be assigned physical reality, massless particles such as photons could pass though $\mathcal{B}$ from before it to after it.

it is "$\breve{g}$" that denotes the actual *physical* metric, will not confuse the reader here; it enables us to refer to quantities defined *at* $\mathcal{B}$ without adornments, which will be helpful for us later.) It should be remarked, however, that Tod's proposal does not tell us that $\mathbf{C} = 0$ at $\mathcal{B}$, but it does tell us that $\mathbf{C}$ must remain *finite* at $\mathcal{B}$ (since the conformal space-time is smooth there), which, nevertheless, is a very strong restriction on the gravitational degrees of freedom at $\mathcal{B}$, certainly ruling out anything like BKLM behaviour.

In Tod's original proposal, this added region prior to the Big Bang was not intended to have any physical "reality"; it was introduced merely as a tidy mathematical artefact which enables a clear-cut formulation of something of the nature of a Weyl curvature hypothesis without having to introduce awkward and arbitrary-looking mathematical limiting conditions. This was very much in the spirit of the way that the *future* asymptotics of general-relativistic space-times had often been studied (an idea I proposed in the 1960s) as a means towards analysing the behaviour of outgoing gravitational radiation [Penrose 1964b, 1965b, 1978; Penrose and Rindler 1986, chapter 9]. In that work, the asymptotic *future* could be looked at in a geometrical way by attaching a smooth conformal boundary to the future of the space-time manifold (see §3.5). In this case, let us call the remote-future physical metric $\hat{g}$ (with apologies again for the notational oddity, including notational changes from that of §3.5 and the change of notation for the *physical* metric, now, from $\breve{\mathbf{g}}$ to $\hat{\mathbf{g}}$, which will be clarified shortly) and we rescale it to a new, conformally related metric $\mathbf{g}$, according to

$$\mathbf{g} = \omega^2 \hat{\mathbf{g}},$$

where now the metric $\mathbf{g}$ extends smoothly across the smooth 3-surface $\mathscr{I}$, where $\omega = 0$. See figure 4-13 again, but now viewed as an extension from the physical space-time at the bottom part of the figure across its future infinity $\mathscr{I}$ (rather than $\mathscr{B}$) into a hypothetical region "to the future of infinity".

Both of these tricks have already been employed extensively in this book, in the conformal diagrams of §3.5, when representing FLRW cosmological models, where, according to the conventions of strict conformal diagrams depicted in figure 3-22 (in §3.5), the big bang $\mathscr{B}$ of each model is represented as a jagged line at its past boundary and the future infinity $\mathscr{I}$ as a smooth line at its future boundary. When rotated about the symmetry axis, according to the conventions of these diagrams, we get smooth 3-dimensional conformal boundaries to the space-times in each case. What is different in the present considerations is that we are thinking of space-time models that are much more general, so that *no* rotational symmetry is expected to be present, there being none of the high symmetry that is assumed for the FLRW models.

How do we know that these tricks are still applicable in these more general circumstances? Here, we find a vast logical distinction between the case for a smooth $\mathscr{B}$ and that for a smooth $\mathscr{I}$. We find that under very broad physical assumptions (assuming the cosmological constant $\Lambda$ is positive, as appears to be observationally the case) the mathematical existence of a smooth conformal future infinity $\mathscr{I}$ is to be generally expected (as is implied by theorems due to Helmut Friedrich [1986]). On the other hand, the existence of a smooth *initial* conformal big-bang 3-surface $\mathscr{B}$ represents an enormous constraint on a cosmological model – as is to be expected in view of Tod's proposal being intended as such a constraint, with the hope of mathematically encoding an improbability even of the scale of that expressed in figures like $10^{-10^{124}}$.

In mathematical terms, the existence of smooth conformal boundaries ($\mathscr{B}$ in the past and $\mathscr{I}$ in the future) is conveniently presented in terms of the theoretical possibility of providing an *extension* of the space-time to the other side of such a boundary 3-surface, but this extension would indeed be taken as just a mathematical trick, introduced simply for convenience in the formulation of conditions that would be somewhat awkward to express in other ways, where now local geometric notions can be brought into play in place of awkward asymptotic limits. Such a viewpoint had been that adopted by theorists, when making use of these ideas of conformal boundaries, both in the case of $\mathscr{I}$ and of $\mathscr{B}$. Yet, we find that the physics itself seems to fit in rather well with these mathematical procedures, and is somewhat suggestive of a rather outrageous (fantastical?) possibility that the actual *physics* of the world might allow for a meaningful extension through such 3-dimensional conformal boundaries, both in the case of $\mathscr{B}$ and in the case of $\mathscr{I}$. This allows us to wonder whether there might have actually been a pre–Big Bang world and also whether there might be another world beyond our future infinity!

The key point is that much of physics – basically that part of physics that does not have to do with mass – seems to be unchanged (i.e. invariant) under the conformal rescalings under consideration here. This happens to be explicitly true for Maxwell's equations for electromagnetism, not only for the equations of the free electromagnetic field, but also for the way that electric charges and currents act as sources for electromagnetism. This is also the case for the (classical) Yang–Mills equations that govern the strong and the weak nuclear forces, these being extensions of the equations of Maxwell where the gauge symmetry group of phase rotations is extended to the larger groups that are needed for weak or strong interactions (see §§1.8 and 1.15).

An important point needs to be made, however, when it comes to the quantum versions of these theories (of particular relevance to the Yang–Mills equations), because conformal anomalies can arise, whereby the quantum theory does not share the full symmetry of the classical theory [Polyakov 1981a,b; Deser 1996]. It may be recalled that this issue had especial relevance to the development of string theory; see §1.6, and also §1.11. Although I do feel that this issue of conformal anomalies would be likely to have considerable importance to the more detailed implications of the ideas that I am describing here, I would argue that they do not in any way invalidate the main scheme.

This conformal invariance is explicitly a property of the field equations for the massless particles that are the carriers of these forces in the case of electromag-netism and of strong interactions – photons and gluons, respectively – though there is a complication in the case of weak interactions, where those carriers are normally taken to be the very massive W and Z particles. One may consider that as we contemplate going back in time towards the Big Bang, temperatures get higher and higher until the rest masses of all particles concerned become totally insignificant (as does the issue of conformal anomalies) in relation to the enor-mously high kinetic energy of the particles' motions. The relevant physics at the Big Bang, being in effect the physics of massless particles, will be conformally invariant physics, and therefore, if we trace things back to the bounding 3-surface

$\mathscr{B}$, the material will basically not notice $\mathscr{B}$ at all. As far as that material is concerned, it should have had a "past" when at $\mathscr{B}$, just like physics everywhere else, and that "past" would be described by what goes on within the proposed *theoretical* extension of the space-time that Tod's proposal demands.

But what kind of universe activity might one expect to have taken part in Tod's hypothetical extended universe region, where we now try to take seriously the possibility of a *pre*-Big Bang physical actuality? The most obvious kind of thing would be some collapsing universe phase, like the ($K > 0$) extended Friedmann model considered in §3.1 (see figure 3-6 or figure 3-8 in §3.1) or many other suggested "bounce" models such as the ekpyrotic proposal described in §3.11. However, all these suffer from the problem raised by the 2nd Law, as referred to in many places in chapter 3, namely that either the 2nd Law followed in the same direction in the pre-bounce phase, and then one has the difficulty of trying to match an extremely messy big crunch (illustrated in figure 3-48) with a smooth Big Bang, or else the 2nd Law operated in the *opposite* direction (away from the bounce in both directions) in which case there is no rationale provided for the existence of a moment (the bounce moment) of such extraordinary improbability, as indicated by the numbers encountered in §3.6 such as $10^{-10^{124}}$.

My proposed idea is a very different one, namely that we examine the opposite end of the time/distance scale and look again at the other conformal mathematical trick just considered, which is the conformal "squashing down" of the remote future, as illustrated in many examples in §3.5 (e.g. in figure 3-25 and figure 3-26(a)) to obtain an extension beyond a smooth 3-surface $\mathscr{I}$ at future infinity. Two points must be made about this. First, the presence of a positive cosmological constant $\Lambda$ entails that $\mathscr{I}$ is a *spacelike* 3-surface [Penrose 1964b; Penrose and Rindler 1986, chapter 9], and second, as remarked earlier, that the continuation across a smooth $\mathscr{I}$ is a *generic* phenomenon as shown explicitly by Friedrich [1998], under certain broad assumptions. As stressed before, this latter point is very much *un*like the generic situation for a big bang, since Tod's proposal, demanding the smooth extension of the space-time through $\mathscr{B}$, represents a (very much desired) enormous *constraint* on the Big Bang.

At least this smooth conformal continuation across $\mathscr{I}$ would be the case if the matter content of the universe in the very remote future were to consist of entirely *massless* ingredients, since this is a presumption behind the above assertion. Is it plausible that, in the extremely remote future, only massless ingredients will remain? There are two main issues to be considered here, one being the nature of the particles that remain, in the very remote future, and the other being black holes.

Let us start with the black holes. These will be growing relentlessly at first, as they swallow more and more material, and then just swallowing the cosmic background radiation when there is nothing else around to eat! But after the cosmic background temperature eventually drops to below each hole's Hawking temperature, the hole would start very slowly to evaporate away until eventually disappearing in a terminal explosion (a relatively very minor one, by astrophysical standards), in an overall time which would take very much longer for the huge supermassive black holes in galactic centres than for the smallest ones of just a few solar masses. After a total time-scale of perhaps some $10^{100}$ years (depending on how large the biggest ones get), all would be gone, according to this picture, which was, in essence, originally put forward by Hawking in 1974 – and which I accept as the most likely prognosis.

What, then, can be said about the particles remaining in the exceedingly remote future? In number, the vast majority will be photons. It is already true that the photon-to-baryon ratio is around $10^9$, the majority of these photons being in the CMB. This number should remain basically constant, despite starlight eventually dying out, and many baryons being swallowed by black holes. There will also be an additional contribution from the Hawking evaporation from supermassive black holes, which will be almost entirely in the form of extremely low-frequency photons.

Yet there remain some massive particles to consider. Some of these, presently regarded as stable, might well eventually decay away, and it is frequently argued that protons may eventually decay. Protons, however, are (positively) electrically charged, and so long as charge conservation remains an exact law of nature, there would have to be a charged remnant. The least massive possibility for such a remnant would have to be a positron, the anti-particle of the electron. It is clear from considerations of horizons, etc. (see figure 4-14 and Penrose [2010, §3.2, figure 3.4]) that there would have to be both electrons and positrons (if not other more massive charged particles) surviving indefinitely. They have nowhere to go, since there are no charged massless particles (as we know from the behaviour of pair-annihilation processes [Bjorken and Drell 1964]). It is worth considering the possibility that charge conservation is not exactly true, but even this unlikely option does us no good, because theoretical considerations inform us that then the photon itself would acquire a mass [Bjorken and Drell 1964]. With regard to uncharged particles remaining, the least massive of the neutrinos would presumably survive, although I understand that it is still within experimentally allowed possibilities that there might conceivably be a neutrino type that is massless; see Fogli et al. [2012].

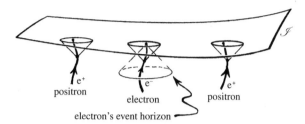

Figure 4-14: This schematic conformal diagram illustrates how, with the spacelike infinity $\mathscr{I}$ demanded by positive $\varLambda$, individual charged particles such as electrons and positrons can eventually become so separated that there is no possibility of their eventually annihilating one another.

The above considerations seem to be telling us that although the conditions for the existence of a smooth (spacelike) future conformal boundary $\mathscr{I}$ are likely to be almost satisfied, there would appear to be occasional massive particles remaining in the ultimate future, somewhat spoiling the purity of the picture. For the scheme that I am about to describe here – *conformal cyclic cosmology* (CCC) – things would really work best if only *massless* particles survive to $\mathscr{I}$. Accordingly, I am conjecturing that in the extremely remote future, rest mass itself eventually dies away, to become zero only in the asymptotic infinite-time limit. This could be at an absurdly slow rate, and there certainly need be no conflict with current observations. We might think of this as something of the nature of an inverse Higgs mechanism, which effectively kicks in only when the ambient temperature reaches some extremely low value. In fact, running values of particle rest masses is a feature of certain particle-physics theories [Chan and Tsou 2007, 2012; Bordes et al. 2015] so it may not be too unreasonable to suppose that all masses eventually run to zero-and "eventually" could be a very long time indeed.

One would expect, according to such theories, that the decay of rest mass would not be at the same rate for all different kinds of particles, so it could not be attributed to an overall decay of the gravitational constant. General relativity requires, for its formulation, a definite notion of time to be determined along any timelike world-line. So long as rest masses are taken as constant, such a time measure would be best given in terms of the prescription of §1.8, where the combination of Einstein's $E = mc^2$ with Planck's $E = h\nu$ tells us that any stable particle whose mass is $m$ behaves fundamentally as a perfect clock of frequency $mc^2/h$. But such a prescription would not work in the very remote future if particle masses decay away at different rates.

CCC does require a positive cosmological constant $\Lambda$ (so that $\mathscr{I}$ will be spacelike). Thus, in a certain sense $\Lambda$ keeps track of scales, so that Einstein's ($\Lambda$) equations remain valid throughout the finite regions of the space-time; however, it is hard to see how $\Lambda$ can be used to build a local clock. An important underlying ingredient of CCC is, indeed, that clocks lose their meaning as $\mathscr{I}$ is approached, so that the ideas of conformal geometry take over, and different physical principles become important, both at $\mathscr{B}$ and at $\mathscr{I}$.

Let us now come to the actual CCC proposal, so that we can see why this may be considered to be desirable. The idea [Penrose 2006, 2008, 2009a,b, 2010, 2014b; Gurzadyan and Penrose 2013] is that our current picture of an ever expanding universe, from its Big Bang origin (but *without* any inflationary phase) to its exponentially expanding infinite future is but one *aeon* in an infinite succession of such aeons, where the $\mathscr{I}$ of each matches conformally smoothly with the $\mathscr{B}$ of the next (see figure 4-15), the resulting conformal 4-manifold being smooth across all the joins. In a sense, this scheme is somewhat like the cyclic/ekpyrotic proposal of Steinhardt–Turok (see §3.11), but without any colliding branes or other input from string/M-theory. It also has points in common with the Veneziano proposal (§3.11), since there is no inflationary phase following each big bang[8] but, in a sense, the exponential expansion in the remote future of each aeon plays a role that supplants the need for any inflation in the succeeding one. Thus, in our own aeon, it would be the inflation-like remote-future expansion of the aeon before ours that provides an explanation for the good reasons favouring inflation. These are issues referred to in §3.9, most notably: (1) the near scale-invariance of the CMB temperature fluctuations, (2) the presence of correlations outside the horizon scale in the CMB, and (3) the early-universe requirement that the local matter density $\rho$ had to be exceptionally close to the critical value $\rho_c$ – all of which can be seen to be very plausible consequences of CCC.

Finally, some important points concerning the viability of CCC need to be addressed. One is the question of how a cyclic scheme such as this can be consistent with the 2nd Law, as it would seem to be the case that cyclicity is incompatible with the 2nd Law. However, a key point here is the fact (already noted in §3.6) that by far the major contribution to the entropy in the universe, even now, is in supermassive black holes in galactic centres, and this contribution will vastly increase in the future. But what will eventually happen to these black holes? It is fully expected that they will ultimately evaporate away by the Hawking process.

---

[8] As in §§3.4 and 3.5, I use the capitalized "Big Bang" for the particular event which initiated our own aeon and "big bang" for other aeons, or for a generic use of the term.

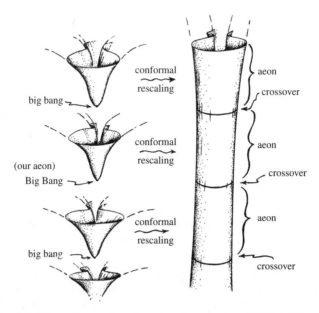

Figure 4-15: The scheme of conformal cyclic cosmology (CCC). In this proposal, the conventional picture (depicted in figure 3-3) of the entire history of our universe (but without its inflationary phase) is but one "aeon" in an unending sequence of generally similar such aeons. The transition from each aeon to the next involves a conformally smooth continuation from the future infinity of each aeon to the big bang of the next (and what would have been the inflationary phase of each aeon is replaced by the ultimate exponential expanding phase of the previous aeon).

I should make the point here that this Hawking evaporation, though in detail dependant upon subtle issues of quantum field theory in curved-space backgrounds, is fully expected merely on the basis of general considerations of the 2nd Law. Here we bear in mind the huge entropy assigned to a black hole (essentially proportional to the square of its mass via the Bekenstein–Hawking formula; see §3.6) leads to a clear expectation of the black hole's Hawking temperature (essentially inversely proportional to its mass; §3.7), leading to the fact that it should eventually lose mass and evaporate away [Bekenstein 1972, 1973; Bardeen et al. 1973; Hawking 1975, 1976a,b]. I am not disputing any of this behaviour, which is effectively *driven* by the 2nd Law. However, an important conclusion that Hawking arrived at in his early considerations was that there must be information loss in the dynamics of black holes – or, as I would prefer to put it, a loss of dynamical

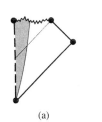
pop

(a)                    (b)

Figure 4-16: The undulations in the jagged line, denoting a black-hole singularity, in these conformal diagrams is to indicate that it is of a generic (perhaps BKLM) nature, yet remaining spacelike in accordance with strong cosmic censorship: (a) generic classical collapse to a black hole, (b) collapse to black hole followed by final disappearance due to Hawking evaporation. Shaded areas denote matter distributions. Compare figures 3-19 and 3-29.

degrees of freedom inside a black hole, and this introduces a fundamentally new ingredient into the discussion.

In my opinion, this loss of degrees of freedom *is* a clear implication of the space-time geometry of collapse to a black hole, as revealed in the conformal diagrams representing such a collapse, despite the fact that many physicists argue for the contrary view. In the strict conformal diagram figure 3-29(a) of §3.5, the original Oppenheimer–Snyder spherically symmetrical collapse picture was depicted, and we clearly see how all material bodies, once they have crossed the horizon, are inevitably forced to destruction at the singularity with no hope of signalling details of their internal structure to the outside world, so long as normal ideas of classical causality are followed. Moreover, so long as strong cosmic censorship is maintained (see §§3.4, 3.10, and Penrose [1998a, TRtR, §28.8]), the overall picture of a *generic* collapse will not be very different, and I have tried to give an impression of this in the conformal diagram of figure 4-16(a), where we may imagine that the somewhat irregular jagged line at the top represents something like a BKLM singularity. Again, all material bodies that have crossed the horizon will inevitably meet destruction at the singularity. In figure 4-16(b), I have modified this picture to depict the case of a Hawking-evaporated black hole, and we see that the situation with regard to the material falling in is no different. If we try to imagine that the situation will actually be very different when local quantum effects are taken into consideration, we should bear in mind the time scales that may be involved. A body falling into a supermassive black hole might take weeks or even years to reach the singularity after falling through the horizon, and it is hard to imagine that a classical description will not be amply adequate to describe its progress towards its inevitable fate. If quantum entanglements are argued somehow to transfer the information in that body, as it nears the singularity, to outside the horizon (as some theorists appear to expect), the horizon perhaps being light-weeks or even light-years away, then we will have a very serious conflict with the no-signalling restrictions of quantum entanglement (see §§2.10 and 2.12).

At this point, I have to address the issue of *firewalls*, which some theorists have argued for as an alternative to black-hole horizons [Almheiri et al. 2013; see also Susskind et al. 1993; Stephens et al. 1994]. According to this proposal, arguments based on general principles of quantum field theory (related to those that argue for the Hawking temperature) are invoked to demonstrate that these principles lead to the conclusion that an observer who attempts to fall through a black-hole horizon would instead find a firewall, where enormous temperatures would be encountered, resulting in the destruction of the poor observer. To me, this provides yet another argument to demonstrate that the basic principles of current quantum mechanics (most particularly unitarity **U**) cannot be generally true in a gravitational context. From the point of view of general relativity, the local physics at a black hole's horizon should not differ from the local physics elsewhere. Indeed, the horizon itself does not even have a local definition, for its actual location depends upon how much material is going to fall into the hole in the future. After all, despite the admittedly hugely numerous confirmations of current quantum-mechanical theory in small-scale phenomena, with regard to large-scale phenomena, it is $\Lambda$ general relativity that enjoys unrefuted success.

Nevertheless, most physicists who seriously consider this issue appear to be highly disturbed by this prospect of information loss, and it has become to be known as the *black-hole information paradox*. It is referred to as a paradox because it implies a gross violation of the fundamental quantum-mechanical principle of *unitarity* **U**, which would deeply undermine the total quantum *faith*! As it will be clear to the reader who has persevered to this point, I am not a supporter of the contention that **U** must be true at all levels and that, indeed, its violation (which in any case has to take place in most circumstances during measurement) will occur when gravitation gets involved. With black holes, gravitation is indeed profoundly involved, and I have no trouble, myself, with the prospect of **U**-violation in black-hole quantum dynamics. In any case, I have long regarded the black-hole information issue as providing a strong contribution to the argument that the **U**-violation that necessarily takes place in an objective **R**-process must be gravitationally based, and could well be related to this so-called black-hole information paradox [Penrose 1981, TRtR, §30.9]. Accordingly, I here take the strong view (unpopular with many physicists, including Hawking himself, after 2004 [Hawking 2005]) that information loss *does* take place at a black hole's singularity. Thus, *phase-space volume* must be dramatically reduced as a result of this process, between the initial formation of the black hole and its final disappearance through Hawking evaporation.

How does this help with the 2nd Law in CCC? The argument depends upon a careful consideration of the definition of entropy. We recall from §3.3 that the Boltzmann definition is given in terms of the logarithm of a phase-space volume $V$

$$S = k \log V,$$

where $V$ is defined as encompassing all states that resemble the one under consideration, with regard to all the relevant macroscopic parameters. Now when a black hole is present in the situation under consideration, the question arises as to whether or not to count all those degrees of freedom that describe things that have fallen into the hole. These will be directed into the singularity and at some stage will be destroyed – removed from consideration of all the processes external to the hole.

When the hole has finally evaporated away, then one could consider that, at that moment, all these swallowed degrees of freedom would be completely removed from consideration. Alternatively, one might choose not to take into consideration such degrees of freedom at any stage of the black hole's existence, after they have fallen through the hole's event horizon. As a further possible interpretation, one might consider that the loss is gradual, spread out over the hole's extremely long lifetime. This really makes little difference, however, since we are concerned only with the overall information loss, throughout the black hole's history.

We recall (§3.3) that the logarithm in Boltzmann's formula allows us to write the total entropy for the system $S_{\text{tot}}$, where the swallowed degrees of freedom *are* taken into consideration, as a sum

$$S_{\text{tot}} = S_{\text{ext}} + k \log V_{\text{swal}}.$$

Here $S_{\text{ext}}$ is the entropy calculated using a phase space in which the swallowed degrees of freedom are *not* taken into account and $V_{\text{swal}}$ is the phase-space volume of all the swallowed degrees of freedom. The entropy $S_{\text{swal}} = k \log V_{\text{swal}}$ is taken out of useful consideration for the system when the black hole has finally evaporated away, so it makes physical sense to switch one's entropy definition from $S_{\text{tot}}$ to $S_{\text{ext}}$ once the hole has gone.

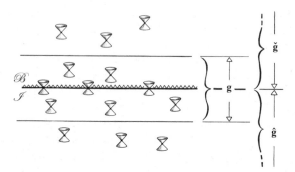

Figure 4-17: The crossover 3-surface connects the previous aeon to the subsequent one, being both the future infinity $\mathscr{I}$ of the previous one and the big bang $\mathscr{B}$ of the subsequent one ($\mathscr{I} = \mathscr{B}$). The metric **g** is to be completely smooth over an open "bandage" region containing the crossover surface, and is conformal to the Einstein physical metric $\hat{\mathbf{g}}$ prior to crossover ($\mathbf{g} = \omega^2\hat{\mathbf{g}}$) and to the Einstein physical metric $\check{\mathbf{g}}$ subsequent to crossover ($\mathbf{g} = \Omega^2\check{\mathbf{g}}$). The $\omega$-field is taken to be smooth throughout the bandage region, vanishing at crossover, and the reciprocal hypothesis $\Omega = -\omega^{-1}$ is taken to hold over this region.

We see from this that, according to CCC, there is no violation of the 2nd Law – and a good deal of the behaviour of black holes and their evaporation may, indeed, be considered to be driven by the 2nd Law. However, because of the loss of degrees of freedom within a black hole, the 2nd Law is, in a sense, transcended. By the time all the black holes have completely evaporated away in an aeon (after some $10^{100}$ years since its big bang), the entropy definition that would initially be employed as appropriate would have become inappropriate after that period of time, and a new definition, providing a far smaller entropy value, would have become relevant some while before the crossover into the next aeon.

In order to see why this has the effect, in the next aeon, of suppression of gravitational degrees of freedom, it is necessary to look a little into the equations that govern the transition from aeon to aeon. Using the notation introduced earlier in this section, we have a picture like that exhibited in figure 4-17. Here $\hat{\mathbf{g}}$ is the Einstein physical metric in the remote future of the *earlier* aeon, just prior to crossover, and $\check{\mathbf{g}}$ is Einstein's physical metric just following the big bang of the *subsequent* aeon. We may recall that the smoothness of the geometry in the neighbourhood of $\mathscr{I}$ was to be expressed in terms of a metric **g**, defined locally in a narrow region containing $\mathscr{I}$, with respect to which the geometry of $\mathscr{I}$ becomes that of an ordinary spacelike 3-surface, $\mathbf{g}$ being conformally related to the pre-$\mathscr{I}$ physical metric $\hat{\mathbf{g}}$ via $\mathbf{g} = \omega^2 \hat{\mathbf{g}}$.

Similarly, the smoothness at $\mathscr{B}$ is expressed in terms of some metric which we again call $\mathbf{g}$, defined locally in a narrow region containing $\mathscr{B}$, with respect to which the geometry of $\mathscr{B}$ becomes that of an ordinary spacelike 3-surface, $\mathbf{g}$ being conformally related to the *post-$\mathscr{B}$* physical metric $\check{\mathbf{g}}$ via $\mathbf{g} = \Omega^2 \check{\mathbf{g}}$. What CCC proposes is that these two "$\mathbf{g}$" metrics can be chosen to match smoothly (thence called the *bandage* metric) covering the crossover region, containing the $\mathscr{I}$ of the earlier aeon, now identified with the $\mathscr{B}$ of the subsequent one. This gives us

$$\omega^2 \hat{\mathbf{g}} = \mathbf{g} = \Omega^2 \check{\mathbf{g}},$$

where, in addition, I adopt the reciprocal hypothesis that $\Omega$ be the inverse of $\omega$, but with a minus sign

$$\Omega = -\omega^{-1},$$

where $\omega$ smoothly moves from negative to positive as we move from the earlier to the later aeon, with $\omega = 0$ *at* the crossover 3-surface ($\mathscr{I} = \mathscr{B}$). See figure 4-17. This allows both $\Omega$ and $\omega$ to be positive in the regions where they are relevant as conformal factors.

We need rather more than this to provide a unique propagation from the previous to the succeeding aeon, and there are still some issues about how best to ensure this uniqueness (apparently involving some symmetry breaking related to what is involved with the standard Higgs mechanism for the reappearance of mass following crossover). All this is beyond the technical scope of this book, but it should be mentioned here that this entire procedure goes very much against the common viewpoint that some quantum-gravity theory would be needed to understand the detailed nature of the Big Bang. Here we just have classical differential equations, and this is potentially far more predictive, especially in view of the fact that there is no really accepted theory of quantum gravity! The reason that one is *not* forced into the arena of quantum gravity, according to my own perspectives, is that the enormous space-time curvatures (i.e. very tiny Planck-scale radii of curvature) that are encountered at $\mathscr{B}$ are all in the form of Einstein curvature $\mathbf{G}$ (equivalent to Ricci curvature; see §1.1) and those do not measure gravity. The gravitational degrees of freedom are not in $\mathbf{G}$ but in $\mathbf{C}$, and $\mathbf{C}$ remains perfectly finite in the neighbourhood of crossover, according to CCC, so quantum gravity need not be important there.

Despite uncertainties concerning the exact form of CCC's detailed equations, one thing can be clearly stated concerning the propagation of gravitational degrees of freedom at the crossover. The issue is a somewhat curious and subtle one, but the essentials can be stated fairly directly as follows. The Weyl tensor $\mathbf{C}$, since it describes conformal curvature, must indeed be a conformally invariant entity, but there is another quantity that I shall call $\mathbf{K}$, which can be taken to be equal to $\mathbf{C}$ in the metric $\hat{\mathbf{g}}$ of the earlier aeon, so I write this:

$$\hat{\mathbf{K}} = \hat{\mathbf{C}}.$$

But these two tensors have different conformally invariant interpretations. Whereas the interpretation of $\mathbf{C}$ (in whatever metric) is indeed Weyl's conformal curvature, that of $\mathbf{K}$ is the graviton field, which satisfies a conformally invariant wave equation (in fact, the same as the one that is described by twistor theory in §4.1, given by spin 2, i.e. $|s| = 2\hbar$, that is $d = +2$ or $-6$). What is curious is that the conformal invariance of this wave equation demands different conformal weights for $\mathbf{K}$ and for $\mathbf{C}$, so that if the above displayed equation holds, we find that in the $\mathbf{g}$ metric

$$\mathbf{K} = \Omega \mathbf{C}.$$

Because of the conformal invariance of $\mathbf{K}$'s wave equation, $\mathbf{K}$ propagates to a finite value at $\mathscr{I}$, from which we immediately deduce that $\mathbf{C}$ must vanish there (since $\Omega$ becomes infinite), and since the conformal geometry must match across $\mathscr{I} = \mathscr{B}$, we find that in fact $\mathbf{C}$ vanishes at the big bang of the subsequent aeon also. Thus, CCC gives us a clear satisfaction of the Weyl curvature hypothesis in the original form $\mathbf{C} = 0$, rather than just the finite form of $\mathbf{C}$ that we directly get from Tod's proposal applied to a single aeon.

We have classical differential equations carrying all the information reaching the $\mathscr{I}$ from the earlier aeon into the $\mathscr{B}$ of the subsequent aeon. The information in gravitational waves reaches $\mathscr{I}$ in the form of $\mathbf{K}$, and this information propagates into the subsequent aeon in the guise of $\Omega$, itself. What we find (driven, in effect, by the reciprocal hypothesis) is that the conformal factor $\Omega$ has to acquire a "reality" as a new scalar field in the subsequent aeon, this scalar field dominating the matter that emerges in the big bang of the subsequent aeon. I conjecture that this $\Omega$-field is, in fact, the initial form of dark matter in the subsequent aeon – where we recall from §3.4 that this mysterious substance currently provides around 85% of the material content of the universe. The $\Omega$-field indeed has to be interpreted as some kind of energy-bearing matter in the subsequent aeon; it *has* to be there, contributing to the energy tensor of that aeon, according to the equations of CCC (which become simply the Einstein $\Lambda$-equations for the subsequent aeon). It must provide an additional contribution to that of all the massless fields (such as electromagnetism) that propagate through from the earlier aeon – except for gravity. It is this $\Omega$-field that picks up the information in **K** from the previous aeon, so the **K** information is not lost, but it emerges in the subsequent aeon as disturbances in $\Omega$ and not as gravitational degrees of freedom [Gurzadyan and Penrose 2013].

The idea would be that at the time when the Higgs mechanism takes hold in the subsequent aeon, the $\Omega$-field would acquire a mass and it would then become the dark matter that is required for consistency with astrophysical observation (§3.4). There would need to be some close connection between $\Omega$ and the Higgs field. Moreover, this dark matter would need to decay away completely into other particles through the course of the subsequent aeon, so that there would not be a build-up from aeon to aeon.

Finally, there is the question of observational tests of CCC. In fact the whole scheme is fairly tight, so there should be many areas where CCC says something genuinely testable through observation. At the time of writing, I have concentrated on just two features of the scheme. In the first of these I consider encounters between supermassive black holes in the aeon previous to ours. Throughout each aeon's history, such encounters must be very frequent overall. (For example, in our own aeon, our Milky Way galaxy is on a collision course with the Andromeda galaxy, due in about $10^9$ years, and there is a reasonably strong chance that our own $\sim 4 \times 10^6$ solar mass black hole will spiral into Andromeda's $\sim 10^8$ solar mass black hole as a result of this.) Such encounters would result in enormous, almost impulsive, bursts of gravitational wave energy which, according to CCC, would result in initially impulsive disturbances in the initial dark matter distribution of the succeeding aeon. Events like this in the previous aeon to ours, would lead to (often concentric) *circular* signals in our CMB that ought to be discernable [Penrose 2010; Gurzadyan and Penrose 2013]. In fact, there does appear to be a significant signal that such activity is actually present in the CMB, seen in both the WMAP and Planck satellite data (see §3.1), observed in analyses carried out by two independent groups [Gurzadyan and Penrose 2013, 2016; Meissner et al. 2013]. This would appear to provide some distinctive evidence in favour of a previous – and surprisingly inhomogeneous – universe aeon, in accordance with the CCC proposal. If this is a genuine interpretation of the data, then one appears to conclude that there is a considerable inhomogeneity in the supermassive black-hole distribution in the aeon previous to ours. Though this was not anticipated by the CCC scheme, it can certainly be readily accommodated within it. It is much harder to see how such inhomogeneity could arise from the conventional inflationary picture, where the CMB temperature fluctuations would have a random quantum origin.

There is a second observational consequence of CCC, which was brought to my attention by Paul Tod early in 2014, namely that CCC provides a possible source for *primordial magnetic fields*. An apparent need for magnetic fields to be present in the early Big Bang (irrespective of CCC) comes from the fact that magnetic fields are actually observed in the large *voids* that inhabit some vast regions of intergalactic space [see Ananthaswamy 2006]. The conventional explanation for the existence of galactic and intergalactic magnetic fields is that these come about from galactic dynamical processes involving *plasma* (separate protons and electrons cohabiting large regions of space), which serve to stretch and strengthen previously present magnetic fields within and between galaxies. However, such processes are not available where there are no galaxies, which is the case for voids, so that the observed presence of magnetic fields in voids presents something of a mystery. It thus appears that such magnetic fields must be *primordial*, i.e. present already in the early Big Bang itself.

According to Tod's suggestion, such fields could indeed have come through unscathed into our early Big Bang from the regions where there had been galactic clusters in the *previous* aeon. Magnetic fields are, after all, subject to the equations of Maxwell, which, as has been mentioned above, are *conformally invariant*, and therefore such fields are able to pass from the remote future of one aeon into the early origins of the next. They would thus appear as *primordial* magnetic fields in our own aeon.

Such a primordial magnetic field could provide a possible source for the so-called $B$-modes in photon polarization in the CMB that appear to have been observed by the BICEP2 team, and widely reported on 17 March 2014 [Ade et al. 2014] as a "smoking gun" of inflation! At the time of writing, some doubts have been raised as to the significance of these observations, it being argued that the role of intergalactic dust had not been adequately taken into consideration [Mortonson and Seljak 2014]. Nevertheless, CCC provides an alternative source for such $B$-modes, and it will be interesting to see which type of explanation turns out to fit the facts best. As a final comment, in this connection, the presence of such previous-aeon galactic clusters might well reveal themselves, according to CCC, through supermassive black-hole encounters, so the two observational implications of CCC mentioned here can be related to one another. All this leads to intriguing issues involving further observational tests, and it will be fascinating to see how CCC fares with regard to such expectations.

## 4.4. A PERSONAL CODA

When I was being interviewed by a Dutch journalist a few years ago, at one point he asked me whether I considered myself to be a "maverick". I think that, in my response, I took this word in a sense that was a little different from what he had intended (and my *Concise Oxford Dictionary* now does seem to bear out his own version). I rather took the view that a maverick is someone who not only went against conventional thinking, but who, to some degree, might deliberately do so for the sake of standing away from the crowd. I replied to my interviewer that I did not see myself in such a light at all, and that in most respects, with regard to the basic physical theories underlying our current pictures of the workings of the world, I am inclined to be quite conservative, and much more accepting, in my opinions, of conventional wisdom than most of those others I know who may try to make headway in pushing at the boundaries of scientific understanding.

Let us take, as a good example of what I mean, Einstein's general theory of relativity (with cosmological constant $\Lambda$); for I am, indeed, completely happy with it as a beautiful classical theory of gravity and of space-time, to be thoroughly trusted so long as we are not encroaching too enormously closely on those singularities where curvatures go wild and Einstein's theory, as such, may well reach its limits. I am certainly much happier with the consequences of general relativity than was even Einstein himself, in his later years, at least. If Einstein's theory tells us that there must be weird objects, essentially made of just empty space, that can swallow entire stars, then so be it – but Einstein himself refused to take this concept that we now refer to as a *black hole* on board, and he tried to argue that such ultimate gravitational collapses would surely not take place. He evidently believed that his own general theory needed fundamental change even at the classical level; for he spent much of his later years of life (while in Princeton) trying to modify his own magnificent theory of general relativity in various (often mathematically unattractive) ways in attempts to bring electromagnetism into the scope of these modifications, while largely ignoring other physical fields.

Of course, as I have argued in §4.3, I am happy to extend Einstein's general relativity in unusual directions, where a strict adherence to it would tell us that the Big Bang must be the beginning, and extensions to before that remarkable event would represent something not within Einstein's great theory. Yet, I would point out that my extension to it is extremely mild, merely allowing a slight broadening of its concepts so that it may have a somewhat greater scope of applicability than before. Indeed, CCC agrees exactly with $\Lambda$ general relativity, just as Einstein put it forward in 1917, and is completely in accordance with that theory as it is found in all old cosmology books (albeit with matter sources in the very early universe that would not be found there). Moreover, CCC accepts Einstein's $\Lambda$ exactly as he gave it, and not introducing some mysterious "dark energy" or "false vacuum" or "quintessence", subject to equations that might allow wild deviations from classical Einstein theory.

Even when it comes to quantum mechanics, where in §2.13, I express scepticism about the complete quantum faith that so many physicists seem to adhere to, I am fully accepting of almost all of its very peculiar implications, such as the non-locality exhibited by EPR (Einstein–Podolsky–Rosen) effects. My acceptance begins to falter only where Einstein's space-time curvature can be shown to be in conflict with quantum principles. Accordingly, I am happy with all those experiments that continue to support the strangeness of quantum theory, since they remain, as of now, significantly below the level where the tensions with general relativity would be expected to make their marks.

When it comes to the fashionable aspects of spatial higher dimensionality (and to a somewhat lesser degree supersymmetry), I am again very conservative in my rejection of these ideas, but I have a confession to make. I have presented, here, my objections to extra space dimensions almost entirely from the point of view of the difficulties presented by the enormously excessive functional freedom in those extra dimensions. I do believe that those objections are indeed valid, and I have never seen them properly addressed by the supra-dimensionalists. But those are not my *real* deep-down objections to supra-dimensionality!

What are these objections then? On some other occasions when I have been interviewed, or otherwise questioned by friends or acquaintances, I have been asked what my reasons might be for objecting to higher-dimensional theories – to which I might well respond that I have a public reason and a private reason. The public objection would indeed be one based largely on the problems raised by the excessive functional freedom, but what about the private one? For that, I need to put the development of my own ideas in historical perspective.

My own early attempts to develop concepts which combine space-time notions with quantum principles started while I was a mathematics graduate student in the early 1950s, and later a Research Fellow at St John's College at Cambridge, where I was often greatly stimulated by long discussions with my friend and mentor Dennis Sciama, and also others such as Felix Pirani, and I had been inspired by some superb lectures, particularly those of Hermann Bondi and Paul Dirac. In addition to this, I had, from my undergraduate days at University College London, been smitten by the power and the magic of complex analysis and geometry, and had become convinced that this magic must also lie deep in the fundamental workings of the world. I had seen that within the 2-component spinor formalism (a topic that Dirac had made clear to me in his lectures) there is not only a close link between 3-dimensional spatial geometry and quantum-mechanical amplitudes, but also a somewhat different one between the Lorentz group and the Riemann sphere (see §4.1). Both these relationships demanded the particular space-time dimensionality that we directly see about us, but I was not able to uncover a key relationship between them that twistor theory subsequently revealed, until about half a decade later (in 1963).

To me, this was the culmination of many years of searching, and although there were other key motivations that drove these ideas in this particular direction [Penrose 1987c], the essential "Lorentzian" combination of a 3-dimensionality of space together with a 1-dimensionality of time were thoroughly imbedded in the entire venture. Moreover, many of the later developments (such as the twistor representation of the wave functions of massless fields, indicated in §4.1) seemed to confirm the value of these motivational strands. So when I heard that string theory – to which I had initially been distinctly attracted, partly because of its early use of Riemann surfaces – had moved itself in the direction of requiring all those extra spatial dimensions, I was horrified, and far from being tempted by the romantic attractions of a higher-dimensional universe. I found it impossible to believe that nature would have rejected all those beautiful connections with Lorentzian 4-space – and I still do.

One might, of course, regard my dogged adherence to Lorentzian 4-space as another example of my inner conservatism when it comes to basic science. Indeed, when physicists get their ideas right, I see no reason to change them. It is when they are not quite right, or perhaps far from being quite right, that I have my worries. Of course, there may well need to be some fundamental changes, even when the theory works so beautifully. Newtonian mechanics is a case in point, and I believe that the same must be the case for quantum theory. But that does not detract from the firm place that both these magnificent theories hold in the development of fundamental science. It took nearly two centuries before it became clear that the Newtonian particulate universe needed modification by the inclusion of Maxwellian continuous fields, and then another half-century before the changes inflicted by relativity and quantum theory began to come into play. It will be interesting to see whether quantum theory can remain unscathed for that long.

Let me end by making a few final comments about the role of fashion in its frequent grip on scientific ideas. I very much admire and benefit from the way that modern technology, mainly by way of the internet, allows immediate access to so much of the broadening body of scientific knowledge. Yet I fear that this very breadth may itself lead to a tightening of the grip of fashion. There is so much out there which is now so accessible that it is extremely difficult to know which things among that multitude contain new ideas to which attention should be paid. How does one make judgements as to what may be important and what owes its prominence merely to its popularity? How does one wade through the multitude that is there largely because it *is* a multitude, rather than because it contains ideas, either new or old, that have genuine substance, coherence, and truth? It is a difficult question and I can supply no clear answer.

However, the role of fashion in science is clearly not a new one, and I have indicated something of this role in the science of the past in §1.1. The formation of independent coherent judgements in ways that are not unduly influenced by fashion is a difficult balance to achieve. Personally, I had the huge advantage of growing up with a highly talented and inspirational father, Lionel, a biological scientist specializing in human genetics, who had very broad interests and skills: in mathematics, art, and music, and also a talent for writing – although I fear that his personal skills would sometimes reveal limitations, when it came to managing relationships with his family, despite the enjoyment and edification that we all obtained, in sharing with him his many interests and original insights. The general intellectual level in the family was distinctive, and I learnt much also from my highly precocious older brother, Oliver, particularly in the area of physics.

Lionel clearly had a very independent mind, and if he felt that some generally accepted line of thinking was wrong, he would not refrain from pointing this out. I remember particularly when a colleague of his had used a famous family tree on the cover of his book. This distinctive family had been regarded as exhibiting a classic example of Y-chromosome inheritance, a medical condition that would be directly handed down from each affected father to every one of his sons, and apparently had been so for several generations, with no females in the family ever being involved. The condition was a severe disturbance of the skin (*ichthyosis hystrix gravior*) and someone suffering from it might be referred to as a *porcupine man*. Lionel told his colleague that he didn't believe the family tree, because he simply could not accept that this particular condition was of the right kind for Y-chromosome inheritance. Moreover, the men had been exhibited in circuses, in the eighteenth century, and Lionel thought it likely that the circus owners would be very tempted to promote a story like the direct father-to-sons inheritance. His colleague expressed extreme doubt about Lionel's scepticism, so Lionel took it upon himself to demonstrate his case, by making many excursions, accompanied by my mother Margaret, to examine the relevant old church registers, to see what the porcupine men's family tree was *really* like. After several weeks, he triumphantly produced a quite different and more plausible family tree, showing that the condition could not be an example of Y-chromosome inheritance after all, but could be directly explained as a straightforward dominant condition.

I always felt that Lionel had a powerful instinctive feeling for what was likely to be the truth (although he was not *always* right). His instinctive feelings were not necessarily in areas of science, and one thing that he did feel strongly about was the Shakespearian authorship issue. He had been fairly well persuaded by a book by Thomas Looney [1920] that the true author of the Shakespeare plays was Edward de Vere, the 17th Earl of Oxford, and Lionel even went so far as to try to test this authorship by a statistical analysis of de Vere's acknowledged writings, as compared with that of the plays (somewhat inconclusively, as it turned out). Most of Lionel's colleagues thought that this belief was going several steps too far. For my own part, I found the case against the normally accepted authorship very strong (as it seemed to me most unlikely that the author of these great works would have owned no books and left no evidence of his handwriting, save a few illiterate-looking signatures, but I had no real views as to who the true author might be). It is interesting to see that a strong case has recently been made for de Vere's authorship in a book by Mark Anderson [2005]. No matter how difficult it may be to shift a *scientific* viewpoint that has become generally established, it would seem that to do the same in the *literary* world – particularly for such a firmly entrenched dogma with huge commercial interests – would be simply *vast* by comparison!

# A

## Mathematical Appendix

### A.1. ITERATED EXPONENTS

In this section, I want to say something about raising a number to some *power*. That, of course, means multiplying the number by itself that many times. Thus, the notation

$$a^b,$$

where $a$ and $b$ are positive integers (positive whole numbers), means $a$ multiplied by itself a total of $b$ times (so $a^1 = a$, $a^2 = a \times a$, and $a^3 = a \times a \times a$, etc.); thus $2^3 = 8$, $2^4 = 16$, $2^5 = 32$, $3^2 = 9$, $3^3 = 27$, $4^2 = 16$, $5^2 = 25$, $10^5 = 100000$, etc. We can also extend this, without difficulty, to the case where $a$ need not be positive, nor need $b$ be positive if $a \neq 0$ (e.g. $a^{-2} = 1/a^2$), and the notion also applies when neither $a$ nor $b$ need be an integer (e.g. they can be real, or even *complex* numbers that we shall be coming to later, in §§A.9 and A.10). (There are, however, issues of multi-valuedness that can then arise [see, for example, TRtR, §5.4].) One minor point, concerning the terminology that I am adopting fairly consistently in this book, is that, rather than using terms like "billion", "trillion", or "quadrillion", which are somewhat uninformative (partly owing to certain ambiguities of usage that are still relevant) and very limited when it comes to some of the extremely large numbers that we encounter at various places in this book (particularly in chapter 3), I will use exponent notation like $10^{12}$ pretty systematically for numbers larger than a million ($10^6$).

This is all fairly straightforward, but we might be interested in performing such an operation at a second level, so that we would be considering some quantity

$$a^{b^c}.$$

I should make clear what this notation means. It does *not* mean $(a^b)^c$, in other words $a^b$ raised to the power $c$, for the good practical reason that we could perfectly well write that quantity without repeated exponents, as $a^{bc}$ (i.e. $a$ raised to the power $b \times c$). What the displayed expression $a^{b^c}$ is actually intended to mean is the (usually very much larger) quantity

$$a^{(b^c)},$$

in other words, $a$ raised to the power $b^c$. Thus, $2^{2^3} = 2^8 = 256$, which is *not* $(2^2)^3 = 64$.

Now I want to make a somewhat basic point about such quantities, namely that for reasonably large numbers $a$, $b$, $c$, we find that $a^{b^c}$ depends rather little on the value of $a$, whereas $c$ is all-important. (For more intriguing information on these issues, see Littlewood [1953] and Bollobás [1986, pp. 102–3].) We can see this rather clearly if we rewrite $a^{b^c}$ in terms of logarithms. Now, being a mathematician and a little bit of a purist, I tend to use *natural* logarithms, so when I write "log", I indeed mean the natural logarithm "$\log_e$" (although this is frequently denoted by "ln"). If you prefer the ordinary "logs to the base 10" to these, please jump to the paragraph following the next. But for fellow purists, the natural logarithm is just the *inverse* of the standard *exponential* function. This means that the real number

$$y = \log x$$

(for a positive real number $x$) is defined through the equivalent (inverse) equation

$$e^y = x,$$

where $e^y$ is the standard exponential function, sometimes written "exp $y$", defined by the infinite series

$$\exp y = e^y = 1 + \frac{y}{1!} + \frac{y^2}{2!} + \frac{y^3}{3!} + \frac{y^4}{4!} + \cdots,$$

where

$$n! = 1 \times 2 \times 3 \times 4 \times 5 \times \cdots \times n$$

(see figure A-1). Putting $y = 1$ in the above series, we find

$$e = e^1 = 2.7182818284590452 \cdots.$$

We shall come to this series again in §A.7.

It should be noted that (rather remarkably) the notation "$e^y$" is consistent with what we had before, namely that if $y$ is a positive integer, then $e^y$ is indeed the

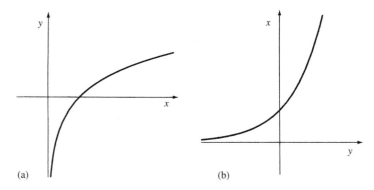

Figure A-1: (a) The logarithm function $y = \log x$ is the inverse function of (b) the exponential function $x = e^y$ (using non-standard conventions for axes). Note that to pass to an inverse function we interchange the $x$-axis with the $y$-axis, i.e. reflect in the diagonal $y = x$.

number "e" multiplied by itself $y$ times. Moreover, the addition-to-multiplication law satisfied by exponents holds, i.e.

$$e^{y+z} = e^y e^z.$$

It follows from this, since "log" is the inverse function to "exp", that we have the multiplication-to-addition law satisfied by logarithms

$$\log(ab) = \log a + \log b$$

(which is equivalent to what we had above if we put $a = e^y$ and $b = e^z$). We also have

$$a^b = e^{b \log a}$$

(because $e^{\log a} = a$, so that $e^{b \log a} = (e^{\log a})^b = a^b$) from which it follows that

$$a^{b^c} = e^{e^{c \log b + \log \log a}}$$

(because $e^{c \log b + \log \log a} = e^{c \log b} e^{\log \log a} = b^c \log a$). Since the function $\log x$ grows very slowly, for large $x$, and the function $\log \log x$ grows far more slowly even than that, we tend to find that, for reasonably large values of $a, b, c$, it is $c$ that is the most relevant in determining the size of $c \log b + \log \log a$, and therefore the size of $a^{b^c}$, and that the value of $a$ is likely to make very little difference at all.

Perhaps to the lay reader it is easier to see what happens by phrasing things in terms of logarithms to the base 10 instead, i.e. in terms of "$\log_{10}$". (This has the advantage, in popular expositions, that I don't have to explain what "e" is!) I shall use the notation "Log", here, for "$\log_{10}$". Accordingly, the real number $u = \mathrm{Log}\,x$ (for any positive real number $x$) is defined through the equivalent inverse equation

$$10^u = x$$

and we have

$$a^b = 10^{b\,\mathrm{Log}\,a}$$

from which it follows (just as above) that

$$a^{b^c} = 10^{10^{c\,\mathrm{Log}\,b + \mathrm{Log}\,\mathrm{Log}\,a}}.$$

We can now very simply illustrate the slowness of the growth $\mathrm{Log}\,x$, by noting

$$\mathrm{Log}\,1 = 0, \quad \mathrm{Log}\,10 = 1, \quad \mathrm{Log}\,100 = 2,$$
$$\mathrm{Log}\,1000 = 3, \quad \mathrm{Log}\,10000 = 4, \quad \text{etc.}$$

The extreme slowness of growth of $\mathrm{Log}\,\mathrm{Log}\,x$ is illustrated, accordingly, by noting

$$\mathrm{Log}\,\mathrm{Log}\,10 = 0, \quad \mathrm{Log}\,\mathrm{Log}\,10000000000 = \mathrm{Log}\,\mathrm{Log}\,10^{10} = 1,$$
$$\mathrm{Log}\,\mathrm{Log}(\text{one googol}) = \mathrm{Log}\,\mathrm{Log}\,10^{100} = 2, \quad \mathrm{Log}\,\mathrm{Log}\,10^{1000} = 3, \quad \text{etc.,}$$

where we recall that $10^{1000}$ would be written, without exponents, by "one" followed by a *thousand* "zeros", and where a *googol* is the number written as "one" followed by a hundred "zeros".

In chapter 3, we shall come across some very large numbers like $10^{10^{124}}$. This particular number (roughly estimating how "special" the universe was at the time of the Big Bang), and the arguments provided there would lead one to consider only the smaller number $e^{10^{124}}$. However, by the above, we find

$$e^{10^{124}} = 10^{10^{124 + \mathrm{Log}\,\mathrm{Log}\,e}}.$$

The quantity $\mathrm{Log}\,\mathrm{Log}\,e$ turns out to have the value $-0.362$, roughly, so we see that changing e to 10 on the left-hand side corresponds to the mere replacement of 124 by roughly 123.638 in the uppermost exponent which, to the nearest integer, is still 124. In fact the estimated number "124" in this expression is not known to a very great accuracy, and a more "correct" figure might well be perhaps 125 or

123. Although in many earlier writings, I had actually used the figure $e^{10^{123}}$ for this degree of specialness, a figure pointed out to me by Don Page in around 1980, this was at a time when the prevalence of dark matter was not well appreciated; see §3.4. A larger figure $e^{10^{124}}$ (or $e^{10^{125}}$) takes the dark matter into account. Thus, replacing e by 10 in this is really neither here nor there! In this case, $b$ is not very large ($b = 10$), so the term Log Log e still makes a bit of difference, but not that much since 124 is still a lot larger than Log Log e.

Another feature of such large numbers is that if we multiply or divide a pair of them, where the top exponents differ by even a rather small amount, then the number with the larger top exponent is likely to swamp the other number completely, so we can more or less totally ignore the smaller number's actual presence in such a multiplication or division sum! To see this, we first observe that

$$10^{10^x} \times 10^{10^y} = 10^{10^x + 10^y} \quad \text{and} \quad 10^{10^x} \div 10^{10^y} = 10^{10^x - 10^y}.$$

Then we take note of the fact that if $x > y$, then the (lower) exponent of 10 in the product is $10^x + 10^y = 1000\ldots001000\ldots00$ and the (lower) exponent in the quotient is $10^x - 10^y = 1000\ldots000999\ldots99$. Here, for the product, the first block "000...00" has $x - y - 1$ zeros in it and the second block "000...00" has $y$ zeros. For the quotient, the first block has $x - y$ zeros in it and the second block has $y - 1$ nines. Clearly, if $x$ is significantly larger than $y$, then the "1" in the middle of the first number, or the 9s at the end, make practically no difference to it (where, of course, we must be a little careful about what we mean by "no difference", as subtracting one from the other would still be an enormously large number!). If $x - y$ is even as little as 2, the lower exponent changes by no more than 1%, and it changes by very much less if $x - y$ is much more than 2, so we can indeed ignore the $10^{10^y}$ in the product $10^{10^x} \times 10^{10^y}$. Similarly, in the division sum, the smaller number $10^{10^y}$ is again swamped, and can also normally be totally ignored in $10^{10^x} \div 10^{10^y}$. We shall be seeing this sort of thing playing a role in §3.5.

## A.2. FUNCTIONAL FREEDOM OF FIELDS

More important to the considerations of chapter 1, in particular, are numbers of the form $a^{b^c}$ in the "limit" when $a$ and $b$ become *infinite*, and I shall write such a quantity as $\infty^{\infty^n}$. We may ask what does this really mean? And what is the importance of such quantities to physics? To respond to the first question it is best that I answer the second question first. And for that, it should be borne in mind

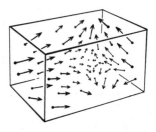

Figure A-2: A magnetic field in ordinary 3-space is a good example of a physical (vector) field.

that much of physics is described in terms of what physicists call *fields*. So what is the physicist's notion of a field?

A good way to get a feeling for a physicist's field is to think of a magnetic field. At each point of space, there is a *direction* of the magnetic field (defined by two angles specifying, say, the east–west and up–down inclination) and also an *intensity* of the field (one further number), giving three in all. Or, in a more straightforward way, we can just think of the three real-number components of the *vector* quantity that gives the full measure of the magnetic field at the chosen point. See figure A-2. (A magnetic field is an example of a *vector field*, a concept described more completely in §A.7.) So how many possible magnetic fields throughout space are there? Clearly, there are infinitely many. But infinity is a very crude measure of the number, and I wish to refine this considerably.

It will be helpful, first, to imagine a toy model of the situation, in which the continuum $\mathbb{R}$ of all possible real numbers is replaced by a finite system **R**, consisting of just $N$ elements, where $N$ is an extremely large positive integer – so we are imagining that instead of the entire continuum, we approximate this by a discrete set of very densely packed points (all in a line). The three real numbers that describe our magnetic field at some point P would now be thought of as just three elements of **R**, so there would be just $N$ possibilities for the first of these numbers, another $N$ possibilities for the second and another $N$ for the third, giving

$$N \times N \times N = N^3$$

in all, so there are $N^3$ possible different magnetic fields, in this toy model, at any given point P in space. But we want to know how many possible fields there are, where the field may vary arbitrarily from point to point in space. In this model, each dimension of the continuum of space-time is also to be described by the finite set **R**, so each of the three spatial coordinates (normally real numbers denoted by $x$, $y$, $z$) is now to be thought of as an element of **R**, so we also have $N^3$ different points in space in our toy model. At any particular point P, there are $N^3$ possible magnetic fields. At two different points P, Q, there are $N^3$ possible fields at each of P and Q, and therefore $N^3 \times N^3 = (N^3)^2 = N^6$ possibilities for the field values at the two points taken together (assuming the field values at different points are independent of each other); at three different points, there are $(N^3)^3 = N^9$ possibilities; at four points there are $(N^3)^4 = N^{12}$, and so on. Thus, at all $N^3$ different points together, we have

$$(N^3)^{N^3} = N^{3N^3}$$

possibilities in all, for magnetic fields throughout space, in this toy model.

There is a bit of confusion in this example because the number $N^3$ appears in two different guises, where the first "3" is the number of components that a magnetic field has at each point and the second "3" refers to the number of dimensions of space. Other types of field can have different numbers of components. For example, the temperature or the density of a material at a point would each have just a single component, whereas *tensorial* quantities such as the strain in a material would have more components per point. We could consider a field of some $c$-component field quantity, in place of our 3-component magnetic field, and then our toy model would give us a total of

$$(N^c)^{N^3} = N^{cN^3}$$

different possibilities for such a field. We could also consider a space of a number of dimensions $d$ that differs from the 3 that we are familiar with. Then, in our toy model, the space being now $d$-dimensional, the number of possible $c$-component fields would be

$$(N^c)^{N^d} = N^{cN^d}.$$

Of course, we are more properly interested in real physics, rather than such a toy model, where the number $N$ is to be taken as infinite – although we should bear in mind that we do not really know the actual mathematical structure of the true physics of nature, so that the term *real physics* here refers to the particular mathematical models that are used in our current highly successful theories. For these successful theories, $N$ is indeed infinite, so we substitute $N = \infty$ into the above formula, to obtain

$$\infty^{c\infty^d}$$

for the number of different possible $c$-component fields in a $d$-dimensional space.

In the particular physical situation that I used to start this excursion, namely the number of possible different magnetic field configurations there could be throughout the whole of space, we have $c = d = 3$, so we get the answer

$$\infty^{3\infty^3}.$$

However, we must bear in mind that this was based (in the toy model) on the assumption that the field values at different points are *independent* of one another. In the present context, if we consider magnetic fields throughout space, this is, in a relevant sense, untrue. For there is a restriction that is satisfied by magnetic fields referred to as a *constraint equation* (recognized by the experts as "div **B** $= 0$" in this case, where **B** is the magnetic field vector – an example of a *differential* equation, as discussed briefly in §A.11). This expresses the fact that there are no such things as separate north or south magnetic poles, where these hypothetical entities would act as independent "sources" for the magnetic field, and their absence, as our physical understanding goes at the moment, is indeed a physical fact (though see §3.9). This constraint has the implication that there is a restriction on any magnetic field, which interrelates the field values at different points in space. This has the more specific implication that the field values are not all independent throughout 3-dimensional space but one of the 3 (and it is up to us which) is, in effect, fixed by the other 2 together with what that component does throughout some 2-dimensional subregion $S$ of space. The upshot of this is that the $3\infty^3$ in the exponent should really be "$2\infty^3 + \infty^2$", but since we may regard the "$\infty^2$" correction to the exponent to be completely swamped by the much larger "$2\infty^3$", we can basically forget about it and write the *functional freedom* of magnetic fields in ordinary 3-space (subject to this constraint) to be

$$\infty^{2\infty^3}.$$

In a refinement of this notation, taking into account the work of Cartan [see Bryant et al. 1991; Cartan 1945, particularly §§68 and 69 on pp. 75–76 of the original edition], we can indeed assign meaning to expressions like $\infty^{2\infty^3 + \infty^2}$, where the exponent may be thought of as a polynomial in "$\infty$", with the coefficients being non-negative integers. In this example, we have two free functions of three variables together with one free function of two variables. I shall not need to make use of this refinement of the notation in this book, however.

Clearly, several points of clarification must be made concerning this useful notation (which appears to have been first used by the distinguished and highly original American physicist John A. Wheeler [1960; Penrose 2003, pp. 185–201,

TRtR, §16.7]). The first point I should make is that these infinite numbers do not refer to the ordinary (Cantor) sense of *cardinality* that describes the sizes of general infinite sets. Some readers may well have some familiarity with Cantor's amazing theory of infinite numbers. If you have not yourself come across this theory, do not worry. I mention Cantor's theory here only so as to make the contrast with what we are doing here, which is different. But if you happen to know about Cantor's theory, the following remarks may be helpful, so you can see how these differences come about.

In Cantor's system of infinite numbers – the ones called *cardinal* numbers – the (cardinal) number of the set $\mathbb{Z}$ of all different integers is denoted by $\aleph_0$ ("aleph nought" or "aleph null"), so the *number* of different integers is indeed $\aleph_0$. The number of different *real numbers* is then $2^{\aleph_0}$, which is usually written $C \ (= 2^{\aleph_0})$. (We can represent real numbers as unending strings of binary digits, e.g. 10010111.0100011..., which is, roughly speaking, given by an $\aleph_0$'s worth of binary alternatives, which are $2^{\aleph_0}$ in number.) However, this is not refined enough for what we need here. For example, if we try to think of the size of a $d$-dimensional space as $N^d$, as $N$ increases to $\aleph_0$, then in Cantor's scheme we always just get $\aleph_0$ again, no matter how large $d$ may be. Indeed, in Cantor's notation, $(\aleph_0)^d = \aleph_0$, for any positive integer $d$. In the case $d = 2$, this is just the fact that the system of *pairs* of integers $(r, s)$ can be counted with just a single integer $t$, as illustrated in figure A-3, and this expresses $(\aleph_0)^2 = \aleph_0$. This extends to any string of $d$ integers, simply by repeating this process, showing that $(\aleph_0)^d = \aleph_0$. But in any case, this cannot be what is meant by the "$\infty$" in our expressions above, as we are thinking of our finite set $\mathbf{R}$, of $N$ elements, as being a model for the *continuum*, which has $2^{\aleph_0} = C$ elements in Cantor's theory. (It is not unreasonable to think of $C$ as arising as a limit of $2^N$ as $N \to \infty$, as we can think of a real number between 0 and 1 as expressed as a binary expansion (e.g. 0.1101000101110010...). If we stop the expansion at $N$ digits, then we have $2^N$ possibilities. Letting $N \to \infty$, we obtain the full continuum of possible real numbers between 0 and 1, with minor redundancies.) But this, by itself, does not help us, since we still just get $C^d = C$, for any positive integer $d$, in Cantor's theory. (See Gardner [2006] and Lévy [1979] for more information about Cantor's theory.)

Cantor's theory of (cardinal) infinities is really concerned just with *sets*, which are not thought of as being structured as some kind of continuous space. For our purposes here, we do need to take into account continuity (or smoothness) aspects of the spaces that we are concerned with. For example, the points of the 1-dimensional line $\mathbb{R}$ are just as numerous, in Cantor's sense, as the points of the 2-dimensional plane $\mathbb{R}^2$ (coordinatized by the pairs $x$, $y$ of real numbers) – as

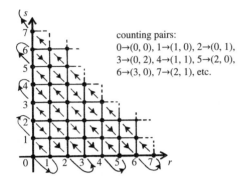

counting pairs:
$0 \to (0, 0)$, $1 \to (1, 0)$, $2 \to (0, 1)$,
$3 \to (0, 2)$, $4 \to (1, 1)$, $5 \to (2, 0)$,
$6 \to (3, 0)$, $7 \to (2, 1)$, etc.

Figure A-3: Cantor procedure of counting pairs $(r, s)$ of natural numbers by a single natural number.

noted in the paragraph above. However, when we think of the points of the real line $\mathbb{R}$ or of the real plane $\mathbb{R}^2$, respectively, as organized into a *continuous* line or a *continuous* plane, the latter must indeed be thought of as a much "larger" entity, in the limit when the finite $N$-element set **R** becomes the continuous $\mathbb{R}$. This is illustrated by the fact that the counting procedure for pairs, shown in figure A-3, cannot be made to be continuous. (Though "continuous" in the limited sense that "close" elements of our counting sequence indeed always give us "close" pairs $(r, s)$, it is not true in the necessary technical *reverse* sense that close *pairs* always give close members of the counting sequence.)

In Wheeler's notation, the size of our continuous line $\mathbb{R}$ is described as $\infty^1$ ($=\infty$), and the size of our continuous plane $\mathbb{R}^2$ as $\infty^2$ ($>\infty$). Similarly, the size of the 3-dimensional space $\mathbb{R}^3$ (of triples of real numbers $x$, $y$, $z$) is $\infty^3$ ($>\infty^2$), etc. The space of smoothly varying magnetic fields on Euclidean 3-space ($\mathbb{R}^3$) is infinite dimensional, but it still has a size, which can be expressed in Wheeler's notation as $\infty^{2\infty^3}$, as noted above (this being when the constraint div **B** $= 0$ is taken into account; otherwise, with div **B** $= 0$ *not* being assumed, we would get $\infty^{3\infty^3}$).

The key point of all this, which is what I make much use of in chapter 1, is that whereas we have (in this "continuous" sense)

$$\infty^{a\infty^d} > \infty^{b\infty^d} \quad \text{if } a > b,$$

we also have

$$\infty^{a\infty^c} \gg \infty^{b\infty^d} \quad \text{if } c > d,$$

the latter holding whatever the relation between the positive numbers $a$ and $b$, where I am using the sign "$\gg$" to indicate the enormous degree to which the left-hand side exceeds the right-hand side. Thus, as was the case with the finite integers used in §A.1, it is the size of *top* exponent which is by far the most important when considering these scales of size. We interpret this as the fact that whereas in a given $d$-dimensional space we get more freely (but continuously) varying fields if the number of components is greater, we find that for spaces of *differing dimension* it is this space-dimension difference which is all important, and any difference in the number of components that the fields may have per point is completely swamped by this. In §A.8 we shall be able to come to a better appreciation of the underlying reasons behind this basic fact.

The phrase "degrees of freedom" is frequently used in the context of physical situations and, indeed, I frequently make use of that terminology in this book. It should be emphasized, however, that this is not the same as "functional freedom". Basically, if we have a physical field with $n$ degrees of freedom we are likely to be referring to something with the functional freedom of

$$\infty^{n\infty^3}$$

since the "number" of degrees of freedom has to do with a parameter number *per point* of 3-space. Thus, in the case of the functional freedom $\infty^{2\infty^3}$ for the magnetic fields, as given above, we have 2 degrees of freedom which is certainly more than the freedom $\infty^{1\infty^3}$ in a 1-component scalar field, but a scalar field in a 5-dimensional space-time would have the *far* greater *functional* freedom $\infty^{1\infty^4}$ than the $\infty^{2\infty^3}$ of our magnetic field in ordinary 3-space (or 4-dimensional space-time).

## A.3. VECTOR SPACES

For a more complete understanding of these matters, it is important to have a better idea of how higher-dimensional spaces are treated mathematically. In §A.5 we shall consider the general notion of a *manifold*, which is a space that can be of any (finite) number of dimensions, but which can also, in an appropriate sense, be *curved*. But before entering into a discussion of such curved-space geometries, it will be useful, for various reasons, to consider the underlying algebraic structure of higher-dimensional *flat* spaces. Euclid himself considered geometries of 2 or 3 dimensions, but perceived no rationale for the consideration of geometries whose dimension might be greater than this, nor is there evidence

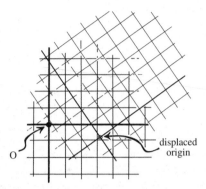

Figure A-4: The choice of coordinates for a space can be very arbitrary, even for ordinary Cartesian rectilinear coordinates for Euclidean 2-space, as with the two such systems illustrated here.

that he even contemplated such possibilities. However, with the introduction of coordinate methods, due basically to Descartes (though others such as Oresme in the fourteenth century and even Apollonius of Perga, in the third century BCE, appear to have had ideas of this kind many years earlier), it became evident that the algebraic formalism employed for 2 or 3 dimensions could be generalized to higher dimensions, even if the utility of such higher-dimensional spaces was far from evident. Whereas 3-dimensional Euclidean space could now be studied using coordinate procedures, a point in 3-space being represented by a *triple* of real numbers $(x, y, z)$, one could readily generalize this to a coordinate $n$-tuple $(x_1, x_2, x_3, \ldots, x_n)$, representing a point in some kind of $n$-dimensional space. Of course, the particular representation of points in terms of $n$-tuples of real numbers, in this way, involves a great deal of arbitrariness, the particular labelling of points being very dependent upon the choice of coordinate *axes* that are being used, and also upon the *origin* point O from which these axes are directed – as we see already when Cartesian coordinates are used to describe the points of a Euclidean plane (figure A-4). But if we allow ourselves to assign a particular status to the point O, then we can take the geometry *relative* to O to be well represented by a particular algebraic structure known as a *vector space*.

A vector space consists of a collection of algebraic elements **u, v, w, x,** ..., called *vectors*, these labelling the individual points in the space, and also numbers called *scalars* $a, b, c, d, \ldots$ that can be used to measure distances (or negatives of distances). The scalars are usually taken to be just ordinary real numbers, i.e.

elements of $\mathbb{R}$, but we find, most particularly in chapter 2, that for the proper understanding of quantum mechanics we shall also be interested in situations where the scalars are *complex* numbers (elements of $\mathbb{C}$, see §A.9). Whether real or complex, the scalars satisfy the rules of ordinary algebra, where pairs of scalars can be combined using operations of addition "$+$", multiplication "$\times$", and their respective inverses subtraction "$-$" and division "$\div$" (although the "$\times$" symbol is normally omitted and "$\div$" often replaced by the *solidus* "$/$"), where division does not act on 0. We have the familiar algebraic rules for scalars

$$a + b = b + a, \quad (a + b) + c = a + (b + c), \quad a + 0 = a,$$
$$(a + b) - c = a + (b - c), \quad a - a = 0, \quad a \times b = b \times a,$$
$$(a \times b) \times c = a \times (b \times c), \quad a \times 1 = a, \quad (a \times b) \div c = a \times (b \div c),$$
$$a \div a = 1, \quad a \times (b + c) = (a \times b) + (a \times c),$$
$$(a + b) \div c = (a \div c) + (b \div c),$$

$a, b, c$ being arbitrary scalars (though we require $c \neq 0$ when acted upon by $\div$) and 0, 1 being *particular* scalars. We write $-a$ for $0 - a$ and $a^{-1}$ for $1 \div a$, and usually write $ab$ for $a \times b$, etc. (These are the abstract rules defining the kind of system that mathematicians refer to as a commutative *field*, $\mathbb{R}$ and $\mathbb{C}$ being particular examples. This is not to be confused with the physicist's notion of *field* described in §A.2.)

The vectors are subject to two operations, *addition* $\mathbf{u} + \mathbf{v}$ and *multiplication by a scalar* $a\mathbf{u}$, these being subject to:

$$\mathbf{u} + \mathbf{v} = \mathbf{v} + \mathbf{u}, \quad \mathbf{u} + (\mathbf{v} + \mathbf{w}) = (\mathbf{u} + \mathbf{v}) + \mathbf{w},$$
$$a(\mathbf{u} + \mathbf{v}) = a\mathbf{u} + a\mathbf{v}, \quad (a + b)\mathbf{u} = a\mathbf{u} + b\mathbf{u}, \quad a(b\mathbf{u}) = (ab)\mathbf{u},$$
$$1\mathbf{u} = \mathbf{u}, \quad 0\mathbf{u} = \mathbf{0},$$

where "$\mathbf{0}$" is a *particular* zero vector and we can write $-\mathbf{v}$ for $(-1)\mathbf{v}$ and $\mathbf{u} - \mathbf{v}$ for $\mathbf{u} + (-\mathbf{v})$. For ordinary Euclidean geometry of 2 or 3 dimensions, it is easy to understand the geometric interpretation of these basic vector operations. We need to fix an origin point O, which we regard as being labelled by the zero vector $\mathbf{0}$, and we think of any other vector $\mathbf{v}$ as labelling some point V, of the space, where we may think of the $\mathbf{v}$ as representing the parallel displacement – i.e. *translation* – of the whole space which takes O to V, and this can be represented diagrammatically as the oriented line segment $\overrightarrow{OV}$, depicted in diagrams as this segment with an arrow from O to V (figure A-5).

Here the scalars are real numbers, and multiplying a vector by a positive real scalar $a$ keeps its direction fixed but scales it up (or down) by the factor $a$.

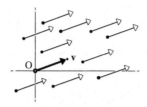

Figure A-5: An ($n$-dimensional) real vector space can be understood in terms of the family translational motions of Euclidean ($n$-dimensional) space. A vector **v** can itself be represented by a directed line segment $\overrightarrow{OV}$, where O is a chosen origin point and V a point of the space, but we may also think of **v** as representing the entire vector field that describes the translational motions that displaces O to V.

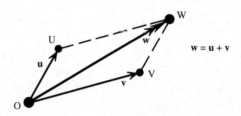

Figure A-6: The parallelogram law for the addition of vectors: $\mathbf{u} + \mathbf{v} = \mathbf{w}$ is expressed in the fact that OUWV is a parallelogram (possibly degenerate).

Multiplying a vector by a negative real scalar does the same, the direction of the resulting vector now being reversed. The sum **w** $(= \mathbf{u} + \mathbf{v})$ of two vectors **u** and **v** is represented as the composition of the two displacements effected by **u** and by **v**, which, in terms of diagrams, is specified by the point W which completes OUWV to a parallelogram (see figure A-6), where in the degenerate case when O, U, V are in line, W is located so that the directed distance OW is the sum of those given by OU and OV. If we wish to describe the condition that the three points U, V, and W are *collinear* (i.e. all in one line), we can express this in terms of the corresponding vectors **u**, **v**, and **w** as

$$a\mathbf{u} + b\mathbf{v} + c\mathbf{w} = \mathbf{0},$$

where $a + b + c = 0$ for some non-zero scalars $a, b, c$, or, equivalently, $\mathbf{w} = r\mathbf{u} + (1 - r)\mathbf{v}$ for some non-zero scalar $r$ (where $r = -a/c$).

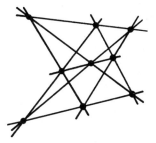

Figure A-7: The ancient theorem of Pappus can be proved by vector methods.

This algebraic description of Euclidean space is very abstract, but it does allow theorems of Euclidean geometry to be reduced to routine calculation, although if the calculation is applied in a direct (and unsubtle) way, it can become very complicated even for relatively simple-looking geometrical theorems. As an example, one could take the fourth-century theorem of Pappus (figure A-7), which asserts that if two sets of three collinear points A, B, C and D, E, F in a plane are *cross-joined* to form three other points X, Y, Z, so that X is the intersection of the lines AE and BD, Y the intersection of AF and CD, and Z the intersection of BF and CE, then X, Y, and Z are also collinear. This can be proved by such direct calculational means, though it is somewhat complicated if simplifying (short-cut) procedures are not adopted.

This particular theorem has the advantage that it depends solely on the notion of collinearity. Euclidean geometry depends also on having a notion of *distance*, and this also can be incorporated into the vector algebra via a notion referred to as an *inner product* (or scalar product) between pairs of vectors $\mathbf{u}$, $\mathbf{v}$, yielding a scalar number which I shall write (in accordance with the quantum-mechanical literature) as $\langle \mathbf{u}|\mathbf{v}\rangle$ although many other notations are frequently used, such as $(\mathbf{u}, \mathbf{v})$ and $\mathbf{u} \cdot \mathbf{v}$. We shall come to the geometrical interpretation of $\langle \mathbf{u}|\mathbf{v}\rangle$ shortly, but let us see its algebraic properties first:

$$\langle \mathbf{u} \mid \mathbf{v}+\mathbf{w}\rangle = \langle \mathbf{u}|\mathbf{v}\rangle + \langle \mathbf{u}|\mathbf{w}\rangle, \quad \langle \mathbf{u}+\mathbf{v} \mid \mathbf{w}\rangle = \langle \mathbf{u}|\mathbf{v}\rangle + \langle \mathbf{w}|\mathbf{v}\rangle,$$
$$\langle \mathbf{u}|a\mathbf{v}\rangle = a\langle \mathbf{u}|\mathbf{v}\rangle$$

and in many types of vector spaces (such as when the scalars are real numbers)

$$\langle \mathbf{u}|\mathbf{v}\rangle = \langle \mathbf{v}|\mathbf{u}\rangle \quad \text{and} \quad \langle a\mathbf{u}|\mathbf{v}\rangle = a\langle \mathbf{u}|\mathbf{v}\rangle,$$

where we normally also demand

$$\langle \mathbf{u} | \mathbf{u} \rangle \geqslant 0,$$

where

$$\langle \mathbf{u} | \mathbf{u} \rangle = 0 \quad \text{only when } \mathbf{u} = \mathbf{0}.$$

In the case of *complex* scalars (see §A.9), these last two relations are often modified, to give what we call a *Hermitian* inner product, for which $\langle \mathbf{u} | \mathbf{v} \rangle = \overline{\langle \mathbf{v} | \mathbf{u} \rangle}$, as is required for quantum mechanics (in the way described in §2.8; see §A.9 for the meaning of the overbar). It then follows that $\langle a\mathbf{u} | \mathbf{v} \rangle = \bar{a} \langle \mathbf{u} | \mathbf{v} \rangle$.

The geometrical notion of *distance* can now be expressed in terms of this inner product. The distance from the origin O to the point U defined by the vector $\mathbf{u}$ is a scalar $u$ such that

$$u^2 = \langle \mathbf{u} | \mathbf{u} \rangle$$

and since in most types of vector space $\langle \mathbf{u} | \mathbf{u} \rangle$ is a positive real number (unless $\mathbf{u} = \mathbf{0}$), we can then define $u$ as its positive square root

$$u = \sqrt{\langle \mathbf{u} | \mathbf{u} \rangle}.$$

In §§2.5 and 2.8, the notation

$$\| \mathbf{u} \| = \langle \mathbf{u} | \mathbf{u} \rangle$$

is used for what I call the *norm* of $\mathbf{u}$ and $u = \sqrt{\langle \mathbf{u} | \mathbf{u} \rangle}$ as the *length* of $\mathbf{u}$ (although some authors would refer to $\sqrt{\langle \mathbf{u} | \mathbf{u} \rangle}$ as the norm of $\mathbf{u}$), where the lightface italic version of the boldface letter representing a vector will represent its length (e.g. "$v$" represents the length of $\mathbf{v}$, etc.). In accordance with this, in ordinary Euclidean geometry, the interpretation of $\langle \mathbf{u} | \mathbf{v} \rangle$ itself is

$$\langle \mathbf{u} | \mathbf{v} \rangle = uv \cos \theta,$$

where $\theta$ is the angle[1] between the lines OU and OV (where we note that $\theta = 0$, with $\cos 0 = 1$, when U = V). The distance between the two points U and V would be the length of $\mathbf{u} - \mathbf{v}$, i.e. the square root of

$$\| \mathbf{u} - \mathbf{v} \| = \langle \mathbf{u} - \mathbf{v} \mid \mathbf{u} - \mathbf{v} \rangle.$$

---

[1] In basic trigonometry, "$\cos \theta$", namely the *cosine* of the angle $\theta$, is defined from a Euclidean right-angled triangle ABC, with angle $\theta$ at A and right angle at B, as the *ratio* AB/AC. The quantity $\sin \theta =$ BC/AC is the *sine* of $\theta$ and the quantity $\tan \theta = $ BC/AB is its *tangent*. I denote the *inverses* of these functions by $\cos^{-1}$, $\sin^{-1}$, and $\tan^{-1}$, respectively (so $\cos(\cos^{-1} X) = X$, etc.).

We say that the vectors $\mathbf{u}$ and $\mathbf{v}$ are *orthogonal*, written $\mathbf{u} \perp \mathbf{v}$, if their scalar product vanishes:

$$\mathbf{u} \perp \mathbf{v} \quad \text{means} \quad \langle \mathbf{u} | \mathbf{v} \rangle = 0.$$

From the above, we see that this corresponds to $\cos \theta = 0$, so that the angle $\theta$ is a *right angle* and the lines OU and OV are *perpendicular*.

## A.4. VECTOR BASES, COORDINATES, AND DUALS

A (finite) *basis* of a vector space is a set of vectors $\varepsilon_1, \varepsilon_2, \varepsilon_3, \dots, \varepsilon_n$ with the property that every vector $\mathbf{v}$ of the space can be expressed as a *linear combination*

$$\mathbf{v} = v_1 \varepsilon_1 + v_2 \varepsilon_2 + v_3 \varepsilon_3 + \cdots + v_n \varepsilon_n$$

of the members of this set, which is to say that the vectors $\varepsilon_1, \varepsilon_2, \varepsilon_3, \dots, \varepsilon_n$ *span* the entire vector space – and also, for a basis, we need the fact that the vectors in the set are *linearly independent*, so that we need *all* the $\varepsilon$s in order to span the space. This latter condition is equivalent to saying that $\mathbf{0}$ $(=\mathbf{v})$ can be represented as such an expression only when all the coefficients $v_1, v_2, v_3, \dots, v_n$ are zero – or, equivalently, that the above representation of any $\mathbf{v}$ is *unique*. For any particular vector $\mathbf{v}$, the coefficients $v_1, v_2, v_3, \dots, v_n$ in the above expression are the *coordinates* of $\mathbf{v}$ with respect to this basis, and often referred to as $\mathbf{v}$'s *components* in this basis (where, grammatically, $\mathbf{v}$'s "components" should actually be the quantities $v_1\varepsilon_1$, $v_2\varepsilon_2$, etc., but the conventional terminology refers just to the scalars $v_1$, $v_2$, $v_3$, etc., as being the *components*). The number of elements in the set of basis vectors is the *dimension* of the vector space, this number being independent of the particular choice of basis, for a given vector space. In the case of 2-dimensional Euclidean space, any two non-zero and non-proportional vectors will do as a basis (i.e. any vectors $\mathbf{u}$ and $\mathbf{v}$ labelling points U and V that are not on a line through O). For 3-dimensional Euclidean space, any $\mathbf{u}, \mathbf{v}, \mathbf{w}$ that are linearly independent will do (the corresponding points U, V, W not all in one plane with O). The directions of the basis vectors at O provide, in each case, possible choices of coordinate axes, so the representation of a point P in terms of a basis $(\mathbf{u}, \mathbf{v}, \mathbf{w})$ would be given by

$$\mathbf{p} = x\mathbf{u} + y\mathbf{v} + z\mathbf{w},$$

where the *coordinates* of P would be $(x, y, z)$. Thus, from this algebraic perspective, it is a matter of no great difficulty to generalize from 2 or 3 dimensions to $n$ dimensions, for any positive integer $n$.

For a general basis, there is no demand that the coordinate axes be perpendicular to one another, but for standard *Cartesian* coordinates (so-called, though Descartes himself did not insist his coordinate axes be perpendicular), we do ask that they be orthogonal:

$$\mathbf{u} \perp \mathbf{v}, \quad \mathbf{u} \perp \mathbf{w}, \quad \mathbf{v} \perp \mathbf{w}.$$

Moreover, in a geometrical context, it is usual that the measure of distance is the same, and accurately represented, in the directions of all the axes. This amounts to the *normalization* condition that the coordinate basis vectors $\mathbf{u}$, $\mathbf{v}$, $\mathbf{w}$ are, additionally, all *unit vectors* (i.e. all of unit length):

$$\|\mathbf{u}\| = \|\mathbf{v}\| = \|\mathbf{w}\| = 1.$$

Such a basis is said to be *orthonormal*.

In $n$ dimensions, a set of $n$ non-zero vectors $\boldsymbol{\varepsilon}_1, \boldsymbol{\varepsilon}_2, \boldsymbol{\varepsilon}_3, \ldots, \boldsymbol{\varepsilon}_n$ constitutes an orthogonal basis if they are mutually orthogonal,

$$\boldsymbol{\varepsilon}_j \perp \boldsymbol{\varepsilon}_k, \quad \text{whenever } j \neq k \text{ (with } j, k = 1, 2, 3, \ldots, n),$$

and an orthonormal basis if, in addition, they are all unit vectors,

$$\|\boldsymbol{\varepsilon}_i\| = 1 \quad \text{for all } i = 1, 2, 3, \ldots, n.$$

These two conditions are often combined together in the form

$$\langle \boldsymbol{\varepsilon}_i | \boldsymbol{\varepsilon}_j \rangle = \delta_{ij},$$

where the *Kronecker delta* symbol, defined by

$$\delta_{ij} = \begin{cases} 1 & \text{if } i = j, \\ 0 & \text{if } i \neq j, \end{cases}$$

is being used here. One can easily show from this (the scalars being real numbers) that the Cartesian coordinate form of the inner product of $\mathbf{u}$ with $\mathbf{v}$ and the distance $|UV|$ between U and V are, respectively,

$$\langle \mathbf{u} | \mathbf{v} \rangle = u_1 v_1 + u_2 v_2 + \cdots + u_n v_n$$

and

$$|UV| = |\mathbf{u} - \mathbf{v}| = \sqrt{(u_1 - v_1)^2 + (u_2 - v_2)^2 + \cdots + (u_n - v_n)^2}.$$

To end this section, let us consider one final notion that applies immediately to any (finite-dimensional) vector space $\mathbf{V}$, namely that of its *dual* vector space which is another vector space $\mathbf{V}^*$, of the same dimension as $\mathbf{V}$, which is closely related to it, and often identified with it, but which should really be thought of as a separate space. An element $\mathbf{p}$ of $\mathbf{V}^*$ is what is called a *linear map* (or linear *function*) of $\mathbf{V}$ to the system of scalars, which is to say that $\mathbf{p}$ is a function of elements of $\mathbf{V}$, which is a scalar written $\mathbf{p}(\mathbf{v})$, where $\mathbf{v}$ is any vector belonging to $\mathbf{V}$, this function being *linear* in the sense that

$$\mathbf{p}(\mathbf{u} + \mathbf{v}) = \mathbf{p}(\mathbf{u}) + \mathbf{p}(\mathbf{v}) \quad \text{and} \quad \mathbf{p}(a\mathbf{u}) = a\mathbf{p}(\mathbf{u}).$$

The space of all such $\mathbf{p}$s is again indeed a vector space, which we call $\mathbf{V}^*$, where we define its basic operations of addition $\mathbf{p} + \mathbf{q}$ and scalar multiplication $a\mathbf{p}$ by

$$(\mathbf{p} + \mathbf{q})(\mathbf{u}) = \mathbf{p}(\mathbf{u}) + \mathbf{q}(\mathbf{u}) \quad \text{and} \quad (a\mathbf{p})(\mathbf{u}) = a\mathbf{p}(\mathbf{u})$$

for all $\mathbf{u}$ belonging to $\mathbf{V}$. One can verify that these rules do define $\mathbf{V}^*$ as a vector space, of the same dimension as $\mathbf{V}$, and that associated with any basis $(\boldsymbol{\varepsilon}_1, \ldots, \boldsymbol{\varepsilon}_n)$ for $\mathbf{V}$, there is a *dual basis* $(\boldsymbol{\varrho}_1, \ldots, \boldsymbol{\varrho}_n)$ for $\mathbf{V}^*$, where

$$\boldsymbol{\varrho}_i(\boldsymbol{\varepsilon}_j) = \delta_{ij}.$$

If we repeat this operation of "dualization", to obtain the vector $n$-space $\mathbf{V}^{**}$, we find that we get back to $\mathbf{V}$ again, where $\mathbf{V}^{**}$ is naturally identified with the original space $\mathbf{V}$, so we can write

$$\mathbf{V}^{**} = \mathbf{V},$$

the action of an element $\mathbf{u}$ of $\mathbf{V}$, in its role as $\mathbf{V}^{**}$, being simply defined by $\mathbf{u}(\mathbf{p}) = \mathbf{p}(\mathbf{u})$.

How do we interpret the elements of the dual space $\mathbf{V}^*$ in a geometrical or physical way? Let us think, again, in terms of our Euclidean 3-space ($n = 3$). Recall that, relative to a chosen origin point O, an element $\mathbf{u}$ of the vector space $\mathbf{V}$ may be thought of as representing some other point U in our Euclidean space (or as the translational motion of it which displaces O to U). An element $\mathbf{p}$ of $\mathbf{V}^*$, sometimes referred to as a *covector*, will be associated, instead, with a *plane* P through the origin O, containing all points U for which $\mathbf{p}(\mathbf{u}) = 0$ (figure A-8). The

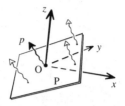

Figure A-8:  For an $n$-dimensional vector space $\mathbf{V}$, any non-zero element $\mathbf{p}$ of its dual space $\mathbf{V}^*$ (referred to as a *covector*) can be interpreted in $\mathbf{V}$ as a hyperplane through the origin O, with a kind of "strength" (quantum mechanically, a frequency) attached. The case $n = 3$ is illustrated, where the covector $\mathbf{p}$ is depicted as a 2-plane P, in relation to coordinate axes $\mathbf{x}$, $\mathbf{y}$, $\mathbf{z}$ at the vector-space origin O.

plane P characterizes the covector completely up to proportionality, but it does not distinguish $\mathbf{p}$ from $a\mathbf{p}$, where $a$ is any non-zero scalar. However, in physical terms we may think of the scale of $\mathbf{p}$ as some kind of *strength* associated with the plane P. We may regard this strength as providing it with some kind of *momentum* directed away from the plane P. In §2.2, we find that in quantum mechanics, this momentum attaches a "frequency of oscillation" away from this plane, which we may associate with the reciprocal wavelength of a plane-wave disturbance moving away from P.

This picture does not make use of the metric "length" structure inherent in a *Euclidean* 3-space. But, with the help of the inner product $\langle \cdots | \cdots \rangle$ that such a structure provides us with, we may then "identify" the vector space $\mathbf{V}$ with its dual $\mathbf{V}^*$, whereby the covector $\mathbf{v}^*$ associated with a vector $\mathbf{v}$ would be the "operator" $\langle \mathbf{v} | \rangle$, whose action on an arbitrary vector $\mathbf{u}$ would be the scalar $\langle \mathbf{v}|\mathbf{u}\rangle$. In terms of the geometry of our Euclidean 3-space, the plane associated with the dual vector $\mathbf{v}^*$ would be the plane through O that is *perpendicular* to OV.

These descriptions apply also to vector spaces of *any* (finite) dimension $n$, where instead of the 2-plane description of a covector in 3-space we would have an $(n-1)$-plane description of a covector in $n$-space, passing through the origin O. Such a higher-dimensional plane, taken to be of just one dimension less than that of the ambient space, is often referred to as a *hyperplane*. Again, for a full description of a covector, rather than of just a covector up to proportionality, we need to assign a "strength" to the hyperplane, which can again be thought of as a kind of momentum or "frequency" (reciprocal wavelength) directed away from the hyperplane.

The preceding discussion was given in terms of *finite*-dimensional vector spaces. However, vector spaces of *infinite* dimension can also be considered.

Such spaces, where a basis would have to have an infinite number of elements, occur in quantum mechanics. Most of what has been said above still holds, but the main difference arises when we try to consider the notion of a *dual* vector space, where it is usual to place a restriction on the linear maps that are considered to compose the dual space $\mathbf{V}^*$, in order to ensure that the relation $\mathbf{V}^{**} = \mathbf{V}$ continues to hold.

## A.5. MATHEMATICS OF MANIFOLDS

Let us now move on to the more general notion of a *manifold*, which need not be flat, like Euclidean space, but curved in various ways, and may perhaps have a different topology from Euclidean space. Manifolds are of fundamental importance to modern physics. This is partly because Einstein's general theory of relativity describes gravity in terms of a curved space-time manifold. But perhaps even more importantly, many other concepts in physics are best understood in terms of manifolds, such as the configuration spaces and phase spaces that we shall come to in §A.6. These often have very large dimensions and sometimes have complicated topology.

So what is a manifold? Basically, it is just a smooth space of some finite number $n$ of dimensions, and we may refer to it as an *n-manifold*. Now, what does the adjective "smooth" mean in this context? In order to be mathematically precise, we would need to address this issue properly in terms of higher-dimensional *calculus*. In this book, I have chosen not to enter into any serious discussion of the mathematical formalism of calculus (beyond the brief remarks made at the end of §A.11), but some sort of intuitive feeling for the basic concepts involved will indeed be needed.

So what is to be meant by a "smooth $n$-dimensional space"? Consider any point P of the space. If the space is to be *smooth* at P, then we can imagine scaling up our picture of it, in the neighbourhood of P, by larger and larger factors, stretching it out away from P, but keeping P centrally placed. If our space is smooth at P, then in the *limit* of stretching it will look like a *flat* $n$-dimensional space. See figure A-9, where the depicted cone vertex P would *not* be a point of smoothness, for example. For a globally smooth manifold, this limiting "stretched-out" space, though flat, shouldn't be thought of as actually being a Euclidean $n$-space exactly, because it need not possess the *metric* structure (i.e. a notion of *distance*) possessed by a Euclidean space. However, it must in the limit possess the structure of a *vector space*, as described in §§A.3 and A.4, where the origin would be the ultimate

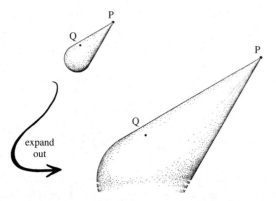

Figure A-9: The manifold depicted at the top would not be smooth at the point P, because no matter how much it is magnified we do not get a flat limiting space. However, it is smooth at Q because on magnification its curvature gets less and less the more it is magnified, and the limiting space is flat there.

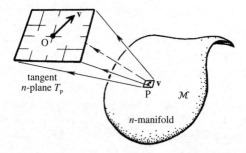

Figure A-10: A tangent vector **v** at a point P of a (smooth) manifold $\mathcal{M}$ would be an element of the *tangent space* $T_P$ at P. We may think of the vector space $T_P$ as the immediate neighbourhood of P infinitely stretched out. O marks $T_P$'s origin.

location of the point P itself, upon which we are focusing our attention. (Think of expanding out a Google map, indefinitely, from its selected point.)

This limiting vector space is referred to as the *tangent space* at the point P, often denoted by $T_P$. The various elements of $T_P$ would themselves be referred to as *tangent vectors* at P. (See figure A-10.) For a good intuitive picture of the geometrical meaning of a tangent vector, think of a tiny arrow based at P, but pointing away from P along the manifold. The various *directions* along the

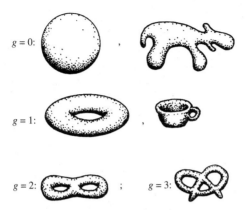

$g = 0$:

$g = 1$:

$g = 2$: ; $g = 3$:

Figure A-11: Examples of 2-spaces with various topologies. The quantity $g$ is the *genus* of the surface (number of "handles"). (Compare figure 1-44.)

manifold at P would be given by the various non-zero vectors in $T_P$ (up to scalar proportionality). To be a globally smooth *n-manifold* our space would have to be smooth at each of its points, having a well-defined tangent $n$-space defined at each.

Sometimes, a manifold may be endowed with more structure than just the *smoothness* that is provided by the presence of the local tangent spaces. A *Riemannian* manifold, for example, possesses a local *length* measure that would be assigned by taking the tangent spaces to be *Euclidean* vector spaces – by providing each $T_P$ with an inner product $\langle \cdots | \cdots \rangle$, as described in §A.3. There are other kinds of local structure that have importance in physics, such as the *symplectic* structures that apply to the *phase spaces* that we shall encounter later. For the usual phase spaces, it turns out that there is a kind of inner product $[\cdots | \cdots]$, for tangent vectors at a point, for which the *anti*-symmetrical $[\mathbf{u}|\mathbf{v}] = -[\mathbf{v}|\mathbf{u}]$ holds, as opposed to the symmetrical $\langle \mathbf{u}|\mathbf{v} \rangle = \langle \mathbf{v}|\mathbf{u} \rangle$ that holds for a Riemannian manifold.

On a global scale, a manifold may have a simple topology, like $n$-dimensional Euclidean space, or it may have a much more elaborate topology, such as the 2-dimensional examples indicated in figure A-11 and figure 1-44 in §1.16. But in each case, whatever the overall topology might be, an $n$-manifold is, in the sense described earlier, everywhere *locally* like a flat $n$-dimensional vector space, and it does not need to have a local notion of distance or angle, like Euclidean $n$-space $\mathbb{E}^n$. Recall that we can assign *coordinates* to designate the different points of a vector space, as described in §A.4. We wish to consider the issue of assigning coordinates to an $n$-manifold generally. In the case of Euclidean $n$-space, we can regard it as being modelled, as a whole, by the space $\mathbb{R}^n$ of $n$-tuples of real numbers $(x_1, x_2, \ldots, x_n)$, as some particular Cartesian coordinates as in figure A-4, but any such representation is far from unique. A manifold, generally, can also be described in terms of coordinates, but there is an even greater arbitrariness in this coordinatization than the vector-space coordinatization of a Euclidean space. There is also an issue of whether such coordinates can apply to the manifold *globally*, or only in local regions. We shall need to consider all these matters.

Let us return to our above assignment of coordinates $(x_1, x_2, \ldots, x_n)$ to a Euclidean space $\mathbb{E}^n$. If we do this via the assignment of coordinates to a vector space, as just described, we need to note that $\mathbb{E}^n$ has no particular point O singled out as "the origin", which, for a vector space, would be assigned the coordinates $(0, 0, \ldots, 0)$. This is clearly arbitrary, and it adds to the arbitrariness already present of choosing a particular *basis* for the vector space. In terms of coordinates, this arbitrariness in choice of origin can be expressed in the freedom of "translating"[2] a given coordinate system, say $\mathfrak{C}$, into another one $\mathfrak{A}$, by adding to each component $x_i$, in $\mathfrak{C}$'s description, a fixed number $A_i$ (normally different for each value of $i$) so that if a point P is represented in $\mathfrak{C}$ by the $n$-tuple $(x_1, x_2, \ldots, x_n)$ then P will be represented in $\mathfrak{A}$ by the coordinate $n$-tuple $(X_1, X_2, \ldots, X_n)$, where

$$X_i = x_i + A_i \quad (i = 1, 2, \ldots, n),$$

the origin O of $\mathfrak{C}$ now being represented in $\mathfrak{A}$ by the $n$-tuple $(A_1, A_2, \ldots, A_n)$.

This is only a very simple transformation of coordinates, merely giving us another coordinate system of the same "linear" type as before. Changing the vector-space basis would also only gives us another coordinate system of this same particular type. Frequently, in the study of structures within Euclidean geometry, more general systems of coordinates referred to as *curvilinear coordinates* are used. One of the most familiar of these is the system of *polar* coordinates for the Euclidean plane (figure A-12(a)), where the standard Cartesian $(x, y)$ would be replaced by $(r, \theta)$, where

$$y = r \sin \theta, \qquad x = r \cos \theta,$$

---

[2] The word "translate" serves double duty here, as in addition to its colloquial meaning of changing from one system of description to another, there is its mathematical meaning, which is to displace without rotation.

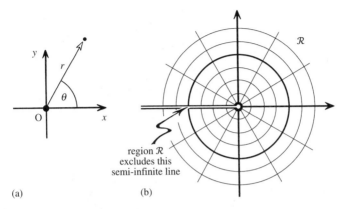

Figure A-12: The "curved" system of polar coordinates $(r, \theta)$. (a) Relation to standard Cartesian $(x, y)$. (b) To provide a proper coordinate chart $\mathcal{R}$, some line out from the centre needs to be excluded, here the half-line $\theta = \pm\pi$.

and conversely

$$r = \sqrt{x^2 + y^2}, \qquad \theta = \tan^{-1} \frac{y}{x}.$$

As its name implies, the coordinate lines of a curvilinear coordinate system need not be straight (or flat planes, etc., in higher-dimensional situations), and we see from figure A-12(b) that, whereas the $\theta = $ const. lines are straight, those given by $r = $ const. are curved, being circular. The example of polar coordinates also illustrates another common feature of curvilinear coordinates, namely that they frequently do not cover the entire space in a smooth one-to-one way. The central point $(0, 0)$ in the $(x, y)$ system is not properly represented in the $(r, \theta)$ system ($\theta$ having no unique value there), and also if we circle around this point, we find that $\theta$ jumps by $2\pi$ (i.e. $360°$). Our polar coordinates do, however, appropriately designate the points of the region $\mathcal{R}$ of the plane which excludes the central point O (given when $r = 0$) and the half-line going off in the opposite direction from O from that given by $\theta = 0$, and which would have been given ambiguously by $\theta = \pm\pi$, i.e. $\theta = \pm180°$; see figure A-12(b). (It should be noted that here I am considering the polar coordinate $\theta$ as running from $-180°$ to $+180°$, whereas the range 0 to $360°$ is frequently used.)

This region $\mathcal{R}$ provides an example of what would be called an *open subset* of the Euclidean plane $\mathbb{E}^2$. In an intuitive way we can think of the notion of "open", as applied to a subset $\mathcal{R}$ of an $n$-manifold $\mathcal{M}$, as a region within $\mathcal{M}$ that has the full

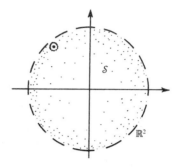

Figure A-13: A portion $\delta$ of a manifold is referred to as an *open set* if it is a subset for which each of its points is contained within a coordinate ball that lies entirely within $\delta$. This is illustrated here, in the 2-dimensional case, $\delta$ being the subset of $\mathbb{R}^2$ given by $x^2 + y^2 < 1$, where we can see that any chosen point within $\delta$ lies within a small circular disc that is completely contained within $\delta$. The region $x^2 + y^2 \leqslant 1$ would not satisfy this condition because it fails for points chosen on its boundary (which is now part of the set).

dimensionality $n$ of $\mathcal{M}$ and which does not include any boundary or "edge" that $\mathcal{R}$ might have. (In the case of polar coordinates for the plane, such an "edge" would be the excluded part of the $x$-axis given by non-positive $x$.) Another example of an open subset of $\mathbb{E}^2$ would be the region – a *disc* (or "2-ball") – lying entirely within the unit circle (i.e. given by $x^2 + y^2 < 1$). On the other hand, neither the unit circle itself ($x^2 + y^2 = 1$) nor the region consisting of this disc *together with* its unit-circle boundary (i.e. the *closed* unit disc $x^2 + y^2 \leqslant 1$) would be open. The corresponding statements apply also in higher dimensions so, in $\mathbb{E}^3$, the "closed" region $x^2 + y^2 + z^2 \leqslant 1$ would *not* be open but the 3-ball $x^2 + y^2 + z^2 < 1$ would be, etc. A little more technically, an open region $\mathcal{R}$ of an $n$-manifold $\mathcal{M}$ can be defined by the property that any point $p$ of $\mathcal{R}$ is centred within a small enough coordinate $n$-ball lying entirely within $\mathcal{M}$. This is illustrated in figure A-13 in the 2-dimensional case of an open disc, where each point of the disc, no matter how close it is to the boundary, lies within a smaller circular region lying entirely within the disc.

In general, for topological reasons, we may find that a global assignment of coordinates to an entire manifold $\mathcal{M}$, using a single coordinate system $\mathfrak{C}$, is not possible, where it turns out that every attempt at coordinatization breaks down somewhere (such as at the north and south poles and along the international dateline, in the case of latitude and longitude coordinates for the spherical Earth). In such situations, to assign coordinates to $\mathcal{M}$, we would not do this with a single

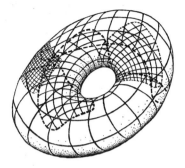

Figure A-14:  The picture illustrates an open covering of a space – here a 2-dimensional torus – by open coordinate regions of $\mathbb{R}^2$ (the sets $\mathscr{R}_1, \mathscr{R}_2, \mathscr{R}_3, \ldots$ of the text).

coordinate system, but we would need to cover the entire $\mathcal{M}$ with a patchwork of overlapping open regions $\mathscr{R}_1, \mathscr{R}_2, \mathscr{R}_3, \ldots$ (see figure A-14) called an *open covering* of $\mathcal{M}$, where we would assign a coordinate system $\mathfrak{C}_i$ to each respective $\mathscr{R}_i$ ($i = 1, 2, 3, \ldots$). Within each overlap between different pairs of open sets of the covering, i.e. within each non-vacuous intersection

$$\mathscr{R}_i \cap \mathscr{R}_j$$

(the symbol "$\cap$" denoting *intersection*), we would have two different coordinate systems, namely $\mathfrak{C}_i$ and $\mathfrak{C}_j$, and we would need to specify the transformation (like the transformation between the Cartesian $(x, y)$ and polar $(r, \theta)$ systems of coordinates considered above; see figure A-12(a)). Piecing together coordinate patches in this way, we can construct spaces with complicated geometry or topology, as illustrated in the 2-dimensional case in figure A-11 and figure 1-44(a) of §1.16.

We must bear in mind that the coordinates are to be regarded merely as *auxiliaries*, introduced as a convenience in order that properties of a manifold may be investigated in detail. The coordinates normally have no specific meaning in themselves, and, in particular, the notion of Euclidean distance between points in terms of these coordinates would have no relevance. (We recall from §A.4 the Cartesian-coordinate formula for Euclidean distance between points $(X, Y, Z)$ and $(x, y, z)$ in $\mathbb{E}^3$: $\sqrt{(X - x)^2 + (Y - y)^2 + (Z - z)^2}$.) Instead, we would be interested in properties of the manifold that are *independent* of whichever coordinate system(s) we happen to choose (and in polar coordinates in the plane, for

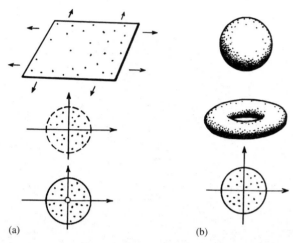

Figuro A-15: (a) Various examples of non-compact 2-manifolds: the entire Euclidean plane, the open unit disc, the closed unit disc with the origin removed. (b) Various examples of compact 2-manifolds: the sphere $S^2$, the torus $S^1 \times S^1$, the closed unit disc.

example, the distance formula would look quite different). This issue has particular importance in Einstein's general theory of relativity, where space-time is taken to be a 4-manifold, where no particular choice of space and time coordinates is taken to have any absolute status. This is referred to as the *principle of general covariance* in general relativity theory (see §§1.2, 1.7, and 2.13).

A manifold may be what is called *compact*, which means, basically, that it is closed up on itself, like a closed curve (dimension $n = 1$) or the closed surfaces depicted in figure A-15(a), or the closed topological surface shown in figure 1-44(a) of §1.16 (dimension $n = 2$). Or it may be *non-compact*, like Euclidean $n$-space or the surface with holes shown in figure 1-44(b). The distinction between non-compact and compact surfaces is illustrated in figure A-15, where we may think of non-compact spaces as "going off to infinity" or "having punctures", like the "holes" in figure 1-44(b) (where the *boundary curves* of the three holes are not taken to be part of the manifold). A bit more technically, a compact manifold has the property that any infinite sequence of points within it has a *limit point*, which means a point P in the manifold such that every open set containing P contains infinitely many members of the sequence (see figure A-16). (For more details of such matters, and technical issues that I have skated over, see Tu [2010] and Lee [2003].)

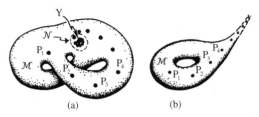

Figure A-16:  A characterization of compactness for a manifold $\mathcal{M}$: (a) in a compact $\mathcal{M}$, every infinite sequence of points $P_1, P_2, P_3, \ldots$ has an accumulation point $y$ in $\mathcal{M}$; (b) in a non-compact $\mathcal{M}$, some infinite sequence of points $P_1, P_2, P_3, \ldots$ has no accumulation point in $\mathcal{M}$. (An accumulation point $y$ has the property that every open set $\mathcal{N}$ containing Y also contains infinitely many of the $P_i$.)

Sometimes we consider regions within a manifold that have *boundaries*, such regions are not being quite manifolds in the sense given here, but can be more general spaces referred to as *manifolds with boundary* (such as the surface depicted in figure 1-44(b) of §1.16, but where the boundaries of the holes are now taken to be part of this manifold-with-boundary). Such spaces can easily be compact without "closing up on themselves" (figure A-15(b)). A manifold may be *connected* – which means (in commonplace terms) that it consists only of one piece – or else *disconnected*. A 0-manifold consists of a single point if it is connected, or of a finite set of two or more separate points if disconnected. Often the term "closed"[3] is used to describe a manifold which is compact (and without any boundary).

## A.6. MANIFOLDS IN PHYSICS

In physics, the most obvious use of a manifold is the flat 3-manifold of ordinary Euclidean 3-space. However, according to Einstein's general theory of relativity (see §1.7), we must now think in terms of spaces that might be *curved*. The magnetic fields considered in §A.2, for example, when considered in a curved 3-space, would be examples of *vector fields*, such as that depicted in figure A-17. The *space-times* of general relativity, moreover, are curved 4-manifolds, and we often have to consider fields in space-time (such as electromagnetic fields) that are of a more complicated nature than just vector fields.

---

[3] This is one of the more confusing pieces of mathematical terminology, since it conflicts with the topological notion of *closed set* that we considered above. *Any* manifold constitutes a closed set, in the topological sense (i.e. complementary to the notion of *open*, described above: a closed set contains all its limit points [Tu 2010; Lee 2003]), whether or not it is closed in the sense of a manifold.

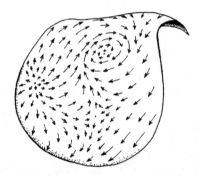

Figure A-17: A smooth vector field on a manifold. The three marked points with no arrow are places where the vector field becomes zero.

Yet, in ordinary (non-string-theoretic) physics, we are frequently interested in manifolds of dimension greater than 3 or 4 (where we use a 3-manifold to describe ordinary space and a 4-manifold for space-time), and we may ask why, other than for the amusements of pure mathematics, we should be concerned with manifolds of such large dimension, or with manifolds whose topology might be other than Euclidean. It should be made clear that manifolds whose dimensions are much greater than 4 and which may possess complicated topologies do play many crucial roles in conventional physical theory. This is quite irrespective of the requirements of many modern physical proposals (such as string theory, as discussed in chapter 1) which demand more than 3 spatial dimensions. Among the simplest and most important examples of high-dimensional manifolds are *configuration* spaces and *phase* spaces. Let us briefly consider these two.

A configuration space is a mathematical space – a manifold $C$ – each of whose points represents a complete description of the locations of all the individual parts of some physical system under consideration (see figure A-18). A simple example would be the 6-dimensional configuration space each of whose points represents the location (including its spatial orientation) of some rigid body B in ordinary Euclidean 3-space (figure A-19). We need 3 coordinates to fix, say, B's centre of gravity (mass centre) G, and 3 more to fix B's spatial orientation, giving 6 in all.

The 6-space $C$ is *non-compact* because $G$ can be at any location in the infinite Euclidean 3-space; moreover $C$ also has a non-trivial (and interesting) topology. It is what is called "non-simply connected" because there are closed curves in $C$ which cannot be continuously deformed to a point [Tu 2010; Lee 2003]. Such a curve is that representing a rotation of B continuously through 360°. Curiously, the curve representing the *repeat* of this process, i.e. a continuous rotation through

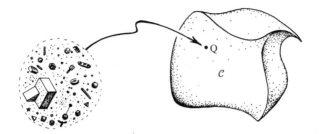

Figure A-18:  A point Q of the configuration space $\mathcal{C}$ represents the location (and orientation, for an asymmetrical shape) of every member of the entire system under consideration.

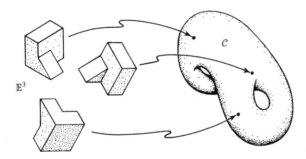

Figure A-19:  The configuration space of a single irregular-shaped rigid body in Euclidean 3-space $\mathbb{R}^3$ is a non-compact, curved, topologically non-trivial 6-manifold.

720°, *can* be continuously deformed to a point [see, for example, TRtR, §11.3], illustrating what is referred to as *topological torsion* [Tu 2010; Lee 2003].

Configuration spaces of very much larger dimension are frequently considered in physics, such as occurs with the case a gas, where one might be concerned with the detailed location of all the molecules in the gas. If there are $N$ molecules (considered as individual point particles without internal structure), then the configuration space would have $3N$ dimensions. Of course, $N$ might be very large indeed, but nevertheless the general mathematical framework for studying manifolds, built up from our intuitions about 1, 2, and 3 dimensions, turns out to be extremely powerful for the analysis of such complicated systems.

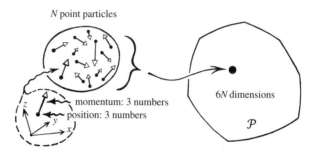

Figure A-20: While the configuration space $\mathcal{C}$ of $N$ structureless classical point particles is a $3N$-manifold, the phase space $\mathcal{P}$ also takes into consideration the 3 momentum degrees of freedom, so $\mathcal{P}$ is $6N$-dimensional.

Phase space $\mathcal{P}$ is a very similar notion to that of configuration space, but where now the *motions* of the individual ingredients have to be taken into account. In the second example of a configuration space considered above, where each individual point of the $3N$-manifold $\mathcal{C}$ represents the complete set of locations of all the molecules in a gas, the corresponding phase space would be a $6N$-manifold $\mathcal{P}$, where the motion of each particle would also be represented. We could imagine doing this by taking the 3 components of the velocity (determining the velocity vector) of each particle, but for technical reasons it turns out to be more appropriate to take the 3 components of *momentum* of each particle. The momentum vector of a particle (at least for the situations of relevance to us here) is simply the velocity vector multiplied by (i.e. scaled-up by) that particle's *mass*. This vector gives us 3 more components per particle than we had before, so we have 6 components in all for each particle, and the phase space $\mathcal{P}$ for our system of $N$ structureless particles will indeed have $6N$ dimensions (figure A-20).

If the particles do have some internal structure, then things get more complicated. Recall, above, that for a rigid body, the configuration space already has 6 dimensions, since 3 numbers defining the angular orientation of the body must be taken into account. To describe the angular *motions* of the body (taken about its mass centre), we need – in addition to the 3 components of the momentum given by the motion of its mass centre – the 3 further components of *angular momentum* about the mass centre to be incorporated into the phase space, giving us a 12-dimensional phase-space manifold $\mathcal{P}$. For $N$ particles, each structured as a rigid body, we would therefore require a phase space of $12N$ dimensions. As a general rule, the phase space $\mathcal{P}$ of a physical system will indeed have *twice* the number of dimensions that would be the case for its configuration space $\mathcal{C}$.

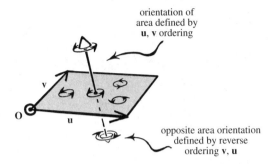

orientation of
area defined by
**u**, **v** ordering

opposite area orientation
defined by reverse
ordering **v**, **u**

Figure A-21: The 2-plane element determined by vectors **u**, **v** in the tangent space at a point of a manifold has an orientation that depends on the order in which **u** and **v** are taken. In a 3-dimensional ambient space, we may think of this orientation in terms of a direction out from the plane, to one side or the other, but it is better to think in terms of a sense of "twist" around the 2-plane element, since this applies also in a higher-dimensional ambient space. If the ambient space is a symplectic manifold, then the area that the symplectic structure assigns to the 2-plane has a sign that depends upon the 2-plane's orientation.

Phase spaces have many beautiful mathematical properties – being what mathematicians call *symplectic* manifolds – of especial relevance to dynamical behaviour. As mentioned in §A.5, each tangent space of such a manifold $\mathcal{P}$ possesses an *antisymmetric* "inner product" $[\mathbf{u}, \mathbf{v}] = -[\mathbf{v}, \mathbf{u}]$, determined by what is referred to as a *symplectic form*. We note that this does not help us to provide a measure of magnitude to a tangent vector, since it directly implies that $[\mathbf{u}, \mathbf{u}] = 0$, for any tangent vector **u** whatever. However, the symplectic form does provide a measure of *area* to any *2-dimensional* surface element, where $[\mathbf{u}, \mathbf{v}]$ would be the element of area for the surface element spanned by the two vectors **u** and **v**. Because of the antisymmetry, this is an *oriented* area, in the sense that if we reverse the order of **u** and **v** (which amounts to describing the area in the opposite sense; see figure A-21) it changes sign. Having this area measure on an infinitesimal scale, we can build (technically, *integrate*) it up to provide a measure of area for any (say, compact, to ensure finiteness) 2-dimensional surface (see figure A-15(c)). We can further develop this area notion, by taking *products* of such things, to provide a measure of "volume" for any even-dimensional (say, compact) surface region lying within $\mathcal{P}$. This applies to the entire space $\mathcal{P}$, since it is necessarily even-dimensional, and to any full-dimensional region within $\mathcal{P}$ (where in each case *finiteness* would be ensured by *compactness*). This volume measure is referred to as the *Liouville measure*.

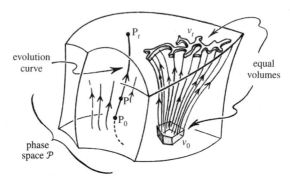

Figure A-22: The dynamical evolution of a classical system is described, in terms of phase space $\mathcal{P}$, by its evolution curve. Each point P in $\mathcal{P}$ represents the instant-aneous locations and motions of all constituents of the system, and the dynamical equations then determine the system's evolution, providing the evolution curve from P, reaching some point $P_t$, describing the system at a later time $t$. The deter-minism of the dynamical equations tells us that there is a unique evolution curve through each P, in the future direction, and the same holds into the past, extending back until it reaches some initial point $P_0$, representing the system's initial state. The symplectic structure of $\mathcal{P}$ provides a volume (Liouville measure) to any com-pact region in $\mathcal{V}$, and Liouville's theorem tells us that this volume is preserved by propagation along the evolution curves, no matter how convoluted this region may become.

Although the detailed mathematical properties of symplectic manifolds will not be of particular relevance to this book, it is worth pointing out two particular features of this geometry here. These concern the curves in $\mathcal{P}$, known as *evolution curves*, which represent the possible evolutions, as time progresses, of the phys-ical system under consideration, this evolution being taken to be in accordance with the dynamical equations governing the system (which can be simply the dynamics of classical Newtonian theory, or else the more sophisticated dynamics of relativity theory or of many other physical proposals). This dynamics is taken to be *deterministic* in the way that normal classical physical systems are taken to behave whereby, for a system composed of particles, the behaviour is completely determined by the locations and the momenta of all the constituent particles at any chosen time $t$. If there are dynamical continuous fields (such as electromagnetism) present, then we expect a similar type of deterministic evolution. Accordingly, in terms of the phase space $\mathcal{P}$, each evolution curve $c$, representing a possible entire evolution of the system, is completely fixed by any chosen point on $c$. The entire family of evolution curves form what mathematicians term a *foliation* of $\mathcal{P}$, where there is exactly one evolution curve through any chosen point of $\mathcal{P}$; see figure A-22.

The first of these features that the symplectic nature of $\mathcal{P}$ provides us with is that the precise locations of all these evolution curve $\mathcal{P}$ are completely fixed once one knows the value of the *energy* of the system for every point of $\mathcal{P}$ (this energy function being referred to as the *Hamiltonian functional*) – though this remarkable and important role of energy does not play any direct part in our considerations here. The second feature, which *is* of importance for us here, however, is that the natural Liouville measure that phase spaces have defined on them (determined by the symplectic structure) is *preserved* throughout the time evolution according to the given dynamical laws. This is a striking fact, known as *Liouville's theorem*. For a $2n$-dimensional phase space $\mathcal{P}$, this volume provides a real-number measure of *size* $L_n(\mathcal{V})$ assigned to any (compact) $2n$-dimensional subregions $\mathcal{V}$ of $\mathcal{P}$. As the time parameter $t$ increases, whereby the points of $\mathcal{P}$ move along their evolution curves, the entire region $\mathcal{V}$ will move, within $\mathcal{P}$, in such a way that its $2n$-volume $L_n(\mathcal{V})$ always remains the same. This will have particular implications in relation to chapter 3.

## A.7. BUNDLES

An important mathematical notion, which is a key ingredient of our modern understanding of the kinds of structures that can reside on manifolds, or of the forces of nature, is what is referred to as a *fibre bundle*, or simply a *bundle* for short [Steenrod 1951; TRtR, chapter 15]. We can think of this as a way of bringing the notion of a field, in the physicist's sense, within the general geometrical framework of manifolds as described in §A.4. This will also enable us to understand more clearly the issue of functional freedom, introduced in §A.2.

For our present point of view, we may think of a bundle $\mathcal{B}$, as an $(r + d)$-manifold that is smoothly built up out of a continuous family of copies of a smaller-dimensional $r$-manifold $\mathcal{F}$, these being referred to as the *fibres* of $\mathcal{B}$. The structure of this family is itself to have the form of another manifold $\mathcal{M}$, a $d$-manifold referred to as the *base space*, so each point of the base space $\mathcal{M}$ corresponds to a particular instance of the manifold $\mathcal{F}$ within the entire family that makes up $\mathcal{B}$. Thus, loosely speaking, we may think as our bundle $\mathcal{B}$, in the following way:

$\mathcal{B}$ is a continuous $\mathcal{M}$'s worth of $\mathcal{F}$s.

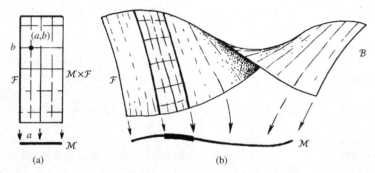

Figure A-23: The picture illustrates the idea of a fibre bundle. The *total space* $\mathcal{B}$ is a manifold that may be thought of as a "continuous $\mathcal{M}$'s worth of $\mathcal{F}$s", where the manifold $\mathcal{M}$ is called the *base space* and $\mathcal{F}$, the *fibre*. There is a *projection* $\pi$ (indicated by the arrows) that maps each instance of $\mathcal{F}$ in $\mathcal{B}$ down to a point of $\mathcal{M}$, and we regard that particular instance of $\mathcal{F}$ in $\mathcal{B}$ to be the fibre "above" that point of $\mathcal{M}$. (a) Above any small enough open subset of $\mathcal{M}$, will be a region of $\mathcal{B}$ that is the *product space* of that subset with $\mathcal{F}$ (see figure A-25), but (b) in its entirety $\mathcal{B}$ need not be such a produce, because of some kind of "twist" in its global structure.

We refer to $\mathcal{B}$ as an $\mathcal{F}$ *bundle over* $\mathcal{M}$, where the entire bundle $\mathcal{B}$ is itself a manifold whose dimension is the sum of the dimensions of $\mathcal{M}$ and of $\mathcal{F}$. The description of $\mathcal{B}$ as being an $\mathcal{M}$'s worth of $\mathcal{F}$s is understood, more technically, as there being a *projection* $\pi$ which maps $\mathcal{B}$ down to $\mathcal{M}$ where the *inverse image* of any point in $\mathcal{M}$ (i.e. the entire part of $\mathcal{B}$ that $\pi$ maps down to that particular point) is one of the copies of $\mathcal{F}$ which comprises $\mathcal{B}$. What this means is that the projection $\pi$ smoothly squashes down each entire $\mathcal{F}$, of which $\mathcal{B}$ is composed, to a single point of $\mathcal{M}$ (see figure A-23). The base space $\mathcal{M}$ and fibre space $\mathcal{F}$ thus combine together in this way to give us what we call the *total space* $\mathcal{B}$ of the bundle.

We want everything in this description to be *continuous*, so in particular this projection must be a continuous map (i.e. free from jumps); but here I shall also demand that all our maps and spaces are *smooth* (preferably, technically what is called $C^\infty$ [see, for example, TRtR, §6.3]), so that the ideas of calculus can be applied, as far as needed. In this book, I have not assumed that the reader need be familiar with the actual formalism of calculus (some of the basic notions being given in §A.11), but some intuitive feeling for the notions of differentiation, integration, and of tangent vectors, etc., is indeed helpful (as touched upon in §A.5). The notion that *differentiation* is concerned with rates of change and slopes

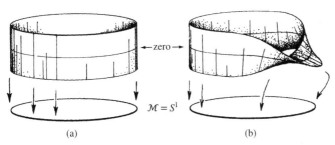

(a)                                            (b)

Figure A-24: The two possible bundles where the fibre $\mathcal{F}$ is a line segment and the base space $\mathcal{M}$ is a circle $S^1$ are (a) the cylinder and (b) the Möbius band.

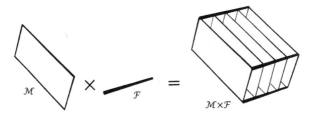

Figure A-25: The product space $\mathcal{M} \times \mathcal{F}$ of manifolds $\mathcal{M}$ and $\mathcal{F}$ is a particular type of bundle of $\mathcal{F}n$ over $\mathcal{M}$ known as a *trivial bundle*, consisting of pairs $(a, b)$, where $a$ is a point of $\mathcal{M}$ and $b$ is a point of $\mathcal{F}$. It can also be read as a trivial bundle of $\mathcal{M}$ over $\mathcal{F}$.

of curves, etc., and that *integration* is concerned with areas and volumes, etc., and some rough acquaintance with these notions will be of value in many places (see figure A-44 in §A.11).

Two simple examples of bundles are those illustrated in figure A-24, where in this case the base space $\mathcal{B}$ is a *circle* and the fibre space $\mathcal{F}$ is a *line segment*. The two topologically distinct possibilities are the *cylinder* (figure A-24(a)) and the *Möbius band* (figure A-24(b)). The cylinder is an example of what is referred to as a *product space*, or *trivial* bundle, where the product $\mathcal{M} \times \mathcal{F}$ of two spaces $\mathcal{M}$ and $\mathcal{F}$ is to be thought of as the space of *pairs* $(a, b)$, where $a$ is a point of $\mathcal{M}$ and $b$ is a point of $\mathcal{F}$ (see figure A-25). We may observe that this product notion is consistent with that applied to pairs of positive integers. For the number of pairs $(a, b)$, when $a$ runs over integers $1, 2, 3, \ldots, A$ and $b$ runs over $1, 2, 3, \ldots, B$ is indeed simply the product $AB$.

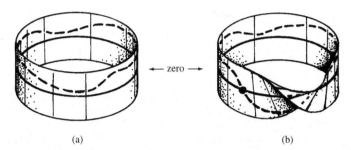

(a)                                                    (b)

Figure A-26: The broken lines are examples of cross-sections of the bundles of figure A-24. One way of distinguishing two bundles from each other is (a) the cylinder has many nowhere-vanishing cross-sections, such as that illustrated, whereas (b) for the Möbius band every cross-section has a zero (crosses the zero line), as illustrated here.

The more general case of what is sometimes referred to as a *twisted product* is instantiated by the Möbius band. This illustrates the fact that a bundle is always *locally* a product space – in the sense that if we take any point $a$ of the base space $\mathcal{M}$, then there is a small enough open region (see §A.5) of $\mathcal{M}_a$, in $\mathcal{M}$, containing $a$, for which the *part* $\mathcal{B}_a$ of the bundle $\mathcal{B}$ that lies *above* $\mathcal{M}_a$ (i.e. that part of $\mathcal{B}$ that $\pi$ projects down to $\mathcal{M}_a$) can itself be expressed as a product

$$\mathcal{B}_a = \mathcal{M}_a \times \mathcal{F}.$$

This *local* product structure always holds for a bundle, even though the whole bundle may not be able to be expressed (continuously) in this way, as indeed it cannot for the case of the Möbius band (figure A-24(b)).

This clear topological distinction between the cylinder and the Möbius band can be understood in terms of what are called *cross-sections* of bundles. A cross-section of a bundle $\mathcal{B}$ is a submanifold $\mathcal{X}$ of $\mathcal{B}$ (i.e. a smaller manifold $\mathcal{X}$ smoothly contained within $\mathcal{B}$) which intersects every fibre in exactly one point. (It is sometimes helpful to think of a cross-section as the image of some map from the base space $\mathcal{M}$ back into the bundle $\mathcal{B}$, with the above property, since $\mathcal{X}$ will always be topologically identical with $\mathcal{B}$.) As is the case with all *product* spaces (when $\mathcal{F}$ contains more than one point), there will be cross-sections that do not intersect one another (e.g. take $(a_1, b)$ and $(a_2, b)$, where $a_1$ and $a_2$ are distinct elements of $\mathcal{F}$ and $b$ runs over the whole of $\mathcal{M}$). We see this illustrated in the cases of the cylinder in figure A-26(a). But with the Möbius band, *every*

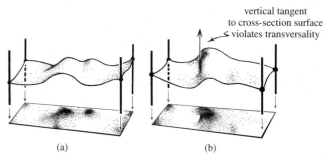

Figure A-27:   The transversality condition for a cross-section illustrated. (In this local picture the base space is a flat plane and the fibres are vertical lines.) (a) Here transversality is satisfied, the undulations in the cross-section never attain a vertical direction in its slope. (b) Though smooth, this cross-section attains a vertical tangent direction and so is not transversal (the field it represents having an infinite derivative there).

pair of cross-sections must intersect (as one may readily persuade oneself; see figure A-26(b)). This illustrates the topological non-triviality of the Möbius band.

From the physical point of view, cross-sections of bundles are important because they give a nice geometrical picture of what a *physical* field is, where we think of $\mathcal{M}$ as being space or space-time. Recall the magnetic fields considered in §A.2. We can think of such a field as a cross-section of the bundle whose base space is ordinary Euclidean 3-space, where the fibre above any point P is the 3-dimensional vector space of possible magnetic fields at P. We shall be coming to this again shortly, in §A.8. Here we are concerned with the notion of *smooth* cross-sections. This smoothness means that not only are all the spaces and maps under consideration smooth, but also we must insist that any such cross-section $\mathcal{X}$ is everywhere *transversal* to the fibres – in the sense that there is no tangent direction to $\mathcal{X}$, at a point $P_0$ of its intersection with a fibre $\mathcal{F}_0$, which coincides with a tangent direction to $\mathcal{F}_0$. See figure A-27 for examples illustrating satisfaction or violation of transversality.

It is important to realize that in order for a bundle to be *non-trivial* (i.e. not a product), it is necessary that the fibre space $\mathcal{F}$ has an exact *symmetry* of some kind. In the case of the Möbius band, it is the symmetry of being able to flip the line (i.e. the fibre $\mathcal{F}$) over end-to-end without changing its nature that allows the construction of this non-trivial example. This applies quite generally, and a space $\mathcal{F}$ without any symmetries at all would not allow the construction of a non-trivial bundle with $\mathcal{F}$ as fibre. This fact is important for us also when we consider the *gauge theories* that underlie the modern theories of the forces of nature (see §1.8), which depend upon a notion, referred to as a *gauge connection*, whose non-triviality depends crucially on the fibres $\mathcal{F}$ possessing a non-trivial (continuous) symmetry, so that neighbouring fibres in a bundle can be related to one another in slightly different alternative ways, depending upon which choice of "connection" is under consideration.

One piece of terminology is useful to take note of here. In any bundle $\mathcal{B}$, with base space $\mathcal{M}$ and fibre $\mathcal{F}$, we may say that $\mathcal{M}$ is a *factor space* of $\mathcal{B}$. This, of course, applies in the trivial case of a product bundle, where each of $\mathcal{M}$ and $\mathcal{F}$ is a factor space of $\mathcal{M} \times \mathcal{F}$. It must be contrasted with the very different situation, where we say that a space $\mathcal{M}$ is a *subspace* of another space $\mathcal{S}$ if it can be smoothly identified as some region within $\mathcal{S}$, and we may write this as

$$\mathcal{M} \hookrightarrow \mathcal{S}.$$

The obvious distinction between these two very different (but curiously often confused) concepts has importance in string theory; see §§1.10, 1.11, and 1.15, and figure 1-32 in §1.10.

One particular class of bundle that has great importance in physics, and also in pure mathematics, is the class of *vector bundles*, for which the fibre space $\mathcal{F}$ is a *vector space* (see §A.3). Examples of vector bundles would be those relevant to the magnetic fields considered in §A.2, as we shall be seeing in §A.8, whose possible values at any one point constitute a vector space. The same would be true of electric fields or of many other kinds of fields of interest in physics where, at any point, we can add such fields together or multiply them by a real scalar number to get another such field for possible consideration. Another class of example would be the *phase spaces* considered in §A.6. In this latter case, we are concerned with the kind of vector bundle referred to as the *cotangent bundle* $T^*(\mathcal{C})$ of a configuration space $\mathcal{C}$, which turns out to be automatically a symplectic manifold, as mentioned in §A.6.

How is a cotangent bundle defined? The *tangent* bundle $T(\mathcal{M})$ of an $n$-manifold $\mathcal{M}$ is the vector bundle whose base space is $\mathcal{M}$ and whose fibre over each point of $\mathcal{M}$ is the *tangent space* at that point (see §A.5). Each tangent space is an $n$-dimensional vector space, so the total space $T(\mathcal{M})$ is a $2n$-manifold (see figure A-28(a)). The *cotangent* bundle $T^*(\mathcal{M})$ of $\mathcal{M}$ is constructed in just the same way, except that the fibre at each point of $\mathcal{M}$ is now the *cotangent* space (the *dual* of the tangent space; see §A.4) at that point (figure A-28(b)). When $\mathcal{M}$ is the configuration space $\mathcal{C}$ of some (classical) physical system, then the cotangent vectors can be identified with the system of *momenta*, whence the identification of the

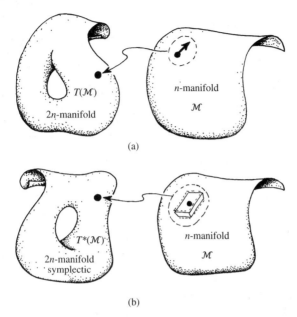

(a)

(b)

Figure A-28: (a) Each point of the $2n$-dimensional tangent bundle $T(\mathcal{M})$ of an $n$-manifold $\mathcal{M}$ represents a point of $\mathcal{M}$ together with a tangent vector to $\mathcal{M}$ at that point. (b) Each point of the symplectic $2n$-dimensional cotangent bundle $T^*(\mathcal{M})$ of $\mathcal{M}$ represents a point of $\mathcal{M}$ together with a tangent covector to $\mathcal{M}$ there.

cotangent bundle $T^*(\mathcal{C})$ as the system's *phase space* (§A.6). Thus, an ordinary phase space would, indeed, be the total space of a (generally non-trivial) vector bundle over the corresponding configuration space, where the fibres provide all the various momenta, and the projection $\pi$ is the map which simply "forgets" all the momenta.

Other examples of bundles, occurring naturally in physics, are those basic to the formalism of quantum mechanics, as depicted in figure 2-16(b) of §2.8, where the complex $n$-dimensional vector space known as a *Hilbert space* $\mathcal{H}^n$ (excluding its origin $\mathbf{O}$) is seen to be a bundle over the *projective* Hilbert space $\mathbb{P}\mathcal{H}^n$, each fibre being a copy of the Wessel plane (§A.10) with its origin removed. Moreover, the $(2n - 1)$-sphere $S^{2n-1}$ of *normalized* Hilbert-space vectors is a circle-bundle ($S^1$-bundle) over $\mathbb{P}\mathcal{H}^n$. Other examples of physically important bundles occur with the *gauge* theories of physical interactions as referred to above. Most particularly, as described in §1.8, the bundle describing the (Weyl) gauge

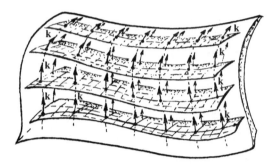

Figure A-29:  On a manifold $\mathcal{M}$ with a metric structure (such as a space-time, in general relativity) there may be a Killing vector field **k**, which expresses (perhaps only locally) a continuous symmetry of $\mathcal{M}$. If $\mathcal{M}$ is a space-time and **k** is timelike, then $\mathcal{M}$ is called *stationary*. If, in addition, **k** is orthogonal to a family of metrically identical spacelike 3-surfaces $\mathcal{S}$, as shown here, then $\mathcal{M}$ is called *static*, but it is not normally appropriate to think of $\mathcal{M}$ as having the structure of a bundle (either one way or the other), since the time scales may vary along $\mathcal{S}$.

theory of electromagnetism is seen to be, in effect, the Kaluza–Klein 5-dimensional "space-time", where the 5th dimension of that theory is a circle along which there is a *symmetry*, and the whole 5-manifold takes the form of a circle-bundle over the 4-manifold of ordinary space-time. See figure 1-12 in §1.6. The symmetry direction is given by what is called a *Killing vector* (field), along which the metric structure of the manifold remains unchanged.

Related to this is the notion of a *stationary* space-time, which has a global Killing vector **k** that is everywhere *timelike*, remaining unchanged along the time direction given by **k**. If **k** is orthogonal to a family of spacelike 3-surfaces, then we say that the space-time is *static*, as shown in figure A-29. However, the bundle structure provided by the timelike curves along the **k** directions may be viewed as somewhat unnatural because these time curves are really mostly inequivalent to one another, having different time scales.

As mentioned earlier, for a bundle to be *non-trivial* the fibre space $\mathcal{F}$ must possess some kind of *symmetry* (like the end-to-end flip needed for the Möbius bundle). The different symmetry operations that can be applied to some given structure constitute what is referred to mathematically as a *group*. Technically, a group, in abstract terms, is a system of operations $a, b, c, d$, etc., that can be applied sequentially, the action of successive operations being written (as with ordinary multiplication) by simple juxtaposition ($ab$, etc.). These operations always satisfy

$(ab)c = a(bc)$, and there is an identity element $e$ for which $ae = a = ea$ for all $a$, and each element $a$ has an inverse $a^{-1}$ such that $a^{-1}a = e$. Specific names are given to various groups commonly employed in physical theories, such as $O(n)$, $SO(n)$, $U(n)$, etc., where, in particular, $SO(3)$ is the group of rotations of an ordinary sphere in Euclidean 3-space, without reflections allowed, and $O(3)$ is the same but *with* reflections allowed; $U(n)$ is the symmetry group of an $n$-dimensional Hilbert space, as described in §2.8, so in particular $U(1)$ (actually the same as $SO(2)$) is the *unimodular group* of phase rotations in the Wessel plane, i.e. multiplication by $e^{i\theta}$ ($\theta$ real).

## A.8. FUNCTIONAL FREEDOM VIA BUNDLES

The concept of a vector bundle has a particular interest for us here, because it gives us insights into the matter of functional freedom that we considered in a rather intuitive way in §A.2. To understand this, we must return to the issue of why, in physics, we are especially concerned with (smooth) cross-sections of bundles. The answer, as briefly addressed above, is that *physical fields* can be interpreted as such cross-sections, their smoothness (including transversality) expressing the smoothness of the field in question. Here, we are to take the base space $\mathcal{M}$ to be *physical space* (normally thought of as a 3-manifold) or physical *space-time* (normally a 4-manifold). The transversality condition expresses the fact that the *derivative* (gradient, or rate of change, either spatially or temporally) of the field in question is always finite.

For a specific example to illustrate this, we could consider a scalar field defined throughout space $\mathcal{M}$. Then $\mathcal{F}$ would simply be a copy of the continuum $\mathbb{R}$ of real numbers, since a *scalar* field (in this context) is simply a smooth assignment of a real number (the field's *strength*) to each point of $\mathcal{M}$. Accordingly, we take our bundle to be simply the trivial bundle

$$\mathcal{B} = \mathcal{M} \times \mathbb{R},$$

with no "twist" involved. A cross-section $\mathcal{X}$ of $\mathcal{B}$ would provide us with a real number at each point of $\mathcal{M}$ in a smooth way, which is what a scalar field *is*. A simple picture of what is going on here is obtained if we just think of an ordinary graph of a function, where $\mathcal{M}$ is now taken to be also 1-dimensional, being just another copy of $\mathbb{R}$ (see figure A-30(a)). The graph itself is the cross-section. Transversality demands that the slope of the curve is never vertical. A vertical slope would tell us that the function has an infinite derivative there, which is not

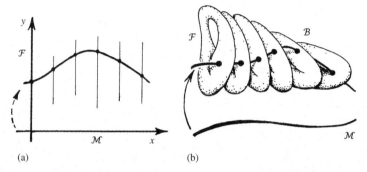

Figure A-30: (a) The ordinary graph of a function $y = f(x)$ provides an elementary illustration of a cross-section of a bundle describing a physical field. Here, the fibres are vertical lines crossing the graph, some of which are drawn in, and the horizontal axis is the manifold $\mathcal{M}$ in this elementary case. Transversality asserts that the slope of the curve never becomes vertical. (b) This illustrates the general case, where $\mathcal{M}$, and also the fibre $\mathcal{F}$ of possible field values at a point of $\mathcal{M}$, can be general manifolds. Any particular field configuration would be represented as a cross-section of the bundle (satisfying transversality).

allowed for a smooth field. This is a very special case of a field represented as a cross-section of a bundle. The different possible "values" that a field could take might not even be a linear space, but might constitute some complicated manifold with a non-trivial topology, as suggested in figure A-30(b), where the space-time itself might be a more complicated space.

For a slightly more complicated example than that of figure A-30(a), let us consider the magnetic fields of §A.2. Here we are thinking of ordinary 3-space, so $\mathcal{M}$ is a 3-manifold (Euclidean 3-space) and $\mathcal{F}$ is the 3-space of possible magnetic fields at a point (again a 3-space, because we need 3 components to define a magnetic field at each point). We can think of $\mathcal{F}$ as being identified with $\mathbb{R}^3$ (the space of triples $(B_1, B_2, B_3)$ of real numbers, these being the 3 components of the magnetic field; see §A.2), and our bundle $\mathcal{B}$ can be regarded as simply being the "trivial" product $\mathcal{M} \times \mathbb{R}^3$. Now, since we have a magnetic *field*, rather than merely the field value at a particular point, we are looking at a smooth *cross-section* of our bundle $\mathcal{B}$ (see figure A-31). A magnetic field is an example of a *vector field*, where at each point of the base space (here $\mathcal{M}$) we have a vector assigned, in a smooth way. In general, a vector field is simply a smooth cross-section of some vector bundle, but that term is most frequently used when the vector bundle in question is the *tangent bundle* of the space in question. See figure A-17 in §A.6.

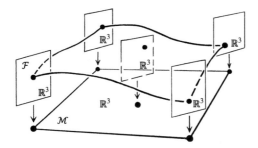

Figure A-31: The picture is intended to suggest how a magnetic field in flat 3-space ($\mathbb{R}^3$) can be represented as a cross-section of the trivial $\mathbb{R}^3$-bundle over $\mathbb{R}^3$ (i.e. $\mathbb{R}^3 \times \mathbb{R}^3$), where we must think of all the planes as actually being $\mathbb{R}^3$s.

If $\mathcal{M}$ were the kind of *curved* 3-space that occurs in general relativity, the identification $\mathcal{B} = \mathcal{M} \times \mathbb{R}^3$ would not really be appropriate because, in general, there would be no natural identification between tangent spaces at different points of $\mathcal{M}$. In many higher-dimensional situations, the tangent bundle $\mathcal{B}$ of a $d$-dimensional $\mathcal{M}$ would not even be *topologically* the same as $\mathcal{M} \times \mathbb{R}^d$ (though the case $d = 3$ is a curious exception). Such global issues will not be of great importance in the present context, however, because even in general relativity our considerations here will be entirely local in the space (or space-time), and, for that, a "trivial" local structure $\mathcal{M} \times \mathbb{R}^d$ is adequate.

A virtue of looking at things from this point of view is that the issue of *functional freedom* becomes particularly apparent. Suppose we have an $n$-component field defined on a $d$-dimensional manifold $\mathcal{M}$. We are then concerned with a (smooth) cross-section $\mathcal{X}$ of the $(d+n)$-dimensional bundle $\mathcal{B}$. We shall be concerned only with *local* behaviour in $\mathcal{M}$, so we may as well assume that we are working with the *trivial* bundle $\mathcal{B} = \mathcal{M} \times \mathbb{R}^n$. If the field is chosen completely freely, then the manifold $\mathcal{X}$ will be a freely chosen $d$-dimensional submanifold of the $(d + n)$-manifold $\mathcal{B}$. ($\mathcal{X}$ is a $d$-manifold because, as noted in §A.4, it is topologically identical with $\mathcal{M}$.) Strictly, however, it is not quite true that $\mathcal{X}$ is completely freely chosen because, first, we need to ensure that the transversality condition is everywhere satisfied and, second, that $\mathcal{X}$ does not "twist around" in some way, so as to meet the fibres more than once. These provisos are not important, however, for considerations of functional freedom, because, at a *local* level, a generically chosen $d$-manifold within the $(d + n)$-manifold $\mathcal{B}$ will indeed be transversal and will meet each $n$-dimensional fibre $\mathcal{F}$, in its vicinity, just once. The amount of

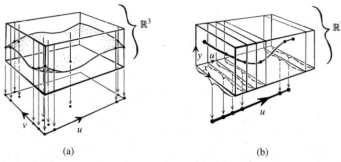

Figure A-32: (a) The functional freedom $\infty^{\infty^2}$ of choosing 1-component fields in 2-space is that of choosing 2-surfaces in $\mathbb{R}^3$, the latter being viewed as an $\mathbb{R}^1$ bundle over $\mathbb{R}^2$ (the $(u, v)$-plane). This may be contrasted with (b) the functional freedom $\infty^{2\infty}$ of choosing 2-component fields (the $(x, y)$-planes) in 1-space ($u$-coordinate), which is that of choosing 1-surfaces (i.e. curves) in $\mathbb{R}^3$, the latter being viewed as an $\mathbb{R}^2$ bundle over $\mathbb{R}^1$.

(local) freedom in choosing an $n$-component field in a given $d$-manifold is simply the (local) freedom in choosing a $d$-manifold $\mathcal{X}$ in an ambient $(d+n)$-manifold $\mathcal{B}$.

Now, the key issue is that it is the value of $d$ that is all important, and it does not matter so much how big $n$ (or $d + n$) is. So how do we "see" this? How do we get a feeling for "how many" $d$-manifolds there are in a $(d + n)$-manifold?

It is a good idea to consider the cases $d = 1$ and $d = 2$, in other words curves and surfaces, in an ambient 3-manifold (which can be ordinary Euclidean 3-space), because then we can easily visualize what is happening (figure A-32). When $d = 2$, we are looking at ordinary scalar fields in 2-space, so the base space can be taken (locally) as being $\mathbb{R}^2$ and the fibre as $\mathbb{R}^1 (= \mathbb{R})$, so our cross-sections are just *surfaces* (2-surfaces) in $\mathbb{R}^3$ (Euclidean 3-space). The functional freedom – i.e. the "number" of possible freely chosen scalar fields – is given by the freedom in choosing 2-surfaces in 3-space (figure A-32(a)). In the case $d = 1$, however, it is the *base* space which is locally $\mathbb{R}^1$ and the *fibre* which is $\mathbb{R}^2$, so the cross-sections are now just *curves* in $\mathbb{R}^3$ (figure A-32(b)).

We now ask: why are there so many more surfaces in $\mathbb{R}^3$ than there are curves in $\mathbb{R}^3$? In other words (interpreting this issue in terms of cross-sections of bundles, i.e. in terms of the functional freedom in fields), why is $\infty^{\infty^2} \gg \infty^{2\infty}$ (or $\infty^{1\infty^2} \gg \infty^{2\infty^1}$), in the notation of §A.2? First, I should explain the "2" in $\infty^{2\infty}$. If we want to describe our curve, we can do this by simply looking at one component at a time of the $\mathbb{R}^2$ fibre $\mathcal{F}$. This amounts to considering the *projection* of

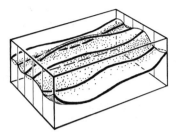

Figure A-33: This picture illustrates why $\infty^{\infty^2} \gg \infty^{k\infty}$ no matter how large the positive integer $k$ may be. Through $k$ curves (here $k = 6$), well separated and smooth (since we are considering a local situation, where the curves do not wind around and keep coming back), we can always find many surfaces to pass through them, so there must be more surfaces in $\mathbb{R}^3$ than any finite number $k$ of curves in $\mathbb{R}^3$.

our curve in two different directions, these being the two coordinate directions, thereby giving us *two* curves, one in each of two planes (i.e. in the $(x, u)$-plane and in the $(y, u)$-plane, where $x$ and $y$ are the fibre coordinates and $u$ is the base-space coordinate). This *pair* of plane curves is equivalent to the original space curve. The freedom for each plane curve is $\infty^\infty$ (one smooth real-valued function of a single real variable) so for the pair of curves, we have the freedom $\infty^\infty \times \infty^\infty = \infty^{2\infty}$.

To see why the (local) freedom $\infty^{\infty^2}$ of 2-surfaces in $\mathbb{R}^3$ is much greater than this – and greater, indeed, than the freedom in any finite number $k$ of plane curves (locally) – we can consider $k$ parallel plane sections of the 2-surface (which we may now picture as the 2-surface of figure A-32(a) sliced vertically by $k$ planes given by $k$ different constant values of the coordinate $v$ in the $\mathbb{R}^2$ base-space of this figure). Each of these $k$ curves has a (local) functional freedom $\infty^\infty$, so the total freedom of these $k$ curves is $(\infty^\infty)^k = \infty^{k\infty}$. (Clearly, a family of $k$ curves can be thought of as just a *single* curve, if we allow it to be a disconnected one. This is one reason why these considerations are to apply only *locally*. A local patch of a disconnected curve would be just like a local patch of a single connected curve, which has less freedom than $k$ separate local patches of curves.) Evidently, no matter how large a finite number $k$ is, there will be vastly more freedom in filling in the 2-surface between these $k$ sections; see figure A-33. This illustrates the fact that $\infty^{\infty^2} \gg \infty^{k\infty}$, no matter how large the finite number $k$ may be.

Although I have illustrated this argument for $\infty^{r^{\infty^d}} \gg \infty^{s^{\infty^f}}$, only in the case $r = 1$, $d = 2$, $s = 1$ (generalized to $s = k$), and $f = 1$, the general case can be demonstrated by using just the same line of reasoning, even though such very direct visualization is not available to us. Basically, we just to need to generalize our curve in $\mathbb{R}^3$ to an $f$-manifold in $\mathbb{R}^{f+k}$ and our surface in $\mathbb{R}^3$ to a $d$-manifold in $\mathbb{R}^{d+r}$, where the former case represents cross-sections of a $k$-bundle over an $f$-manifold, and the latter, cross-sections of an $r$-bundle over a $d$-manifold. So long as $d > f$, there are vastly more of the latter than there are of the former, no matter how big or small $r$ and $s$ might be.

Up to this point I have been considering *freely chosen* fields (or cross-sections), but we recall from §A.2 that with actual magnetic fields in 3-space, there is the *constraint* (div $\mathbf{B} = 0$). This tells us that our (thus constrained) magnetic fields are represented not as just *arbitrary* smooth cross-sections of $\mathscr{B}$, but as cross-sections that are subject to this relation. As mentioned in §A.2, this has the effect of telling us that one of the 3 components of the magnetic field, say $B_3$, would be determined by the other two, $B_1$ and $B_2$, together with the information of what $B_3$ is doing on a 2-dimensional submanifold $\mathscr{S}$ of the 3-manifold $\mathscr{M}$. As far as functional freedom is concerned, we need not pay too much attention to what is happening over $\mathscr{S}$ (since the 2-dimensional $\mathscr{S}$ provides us with a smaller functional freedom than does the remaining 3-dimensional $\mathscr{M}$), so our major functional freedom is the $\infty^{2\infty^3}$ provided by freely chosen 3-manifolds in the 5-space $\mathscr{B} = \mathscr{M} \times \mathbb{R}^2$.

There is one further constraint issue of importance here, and this is what occurs when we take $\mathscr{M}$ to be 4-dimensional *space-time*, rather than just 3-dimensional space. The normal situation in physics is that we have *field equations* which provide us with a *deterministic evolution* of the physical fields throughout space-time, once sufficient data have been specified at one particular time. In relativity theory – especially in Einstein's general relativity – we prefer *not* to refer to time as though it were given to us globally throughout the universe in some absolute sense, but we tend to describe things in terms merely of some arbitrarily specified time coordinate $t$. In this case, some initial value of $t$, say $t = 0$, would provide us with a *spacelike* (as it is usually called; see §1.7) initial 3-surface $\mathscr{N}$, and the appropriate fields specified on $\mathscr{N}$ would then normally determine the fields throughout the 4-dimensional space-time uniquely, by virtue of the field equations. (There are situations in general relativity in which what are called *Cauchy horizons* may arise, in which departures from strict uniqueness can arise, but this issue is not important for the "local" matters of relevance here.) There are frequently constraints holding for the fields within the initial 3-surface also, but in any case, in normal physics, one is dealing with the functional freedom pertinent to the 3-surface $\mathscr{N}$, which is $\infty^{N\infty^3}$, for some positive integer $N$, the "3" coming from the dimension of the initial 3-surface $\mathscr{N}$. If, in some proposed theory, such as string theory (see §1.9), the functional freedom seems to be of the form $\infty^{N'\infty^d}$, where $d > 3$, then we shall need a very good explanation why this excessive freedom is supposed not to show up in physical behaviour.


## A.9. COMPLEX NUMBERS

The mathematical considerations of §§A.2–A.8 have been aimed primarily towards considerations of *classical* physics, where physical fields, point particles, and space-time itself, are described in terms of the real-number system $\mathbb{R}$ (where coordinates and field strengths, etc., are normally taken to be real numbers). However, when quantum mechanics was introduced in the first quarter of the twentieth century, it was found to depend fundamentally on the more extended system, $\mathbb{C}$, of *complex* numbers. Accordingly, as strongly indicated in §§1.4 and 2.5, these complex numbers are now found to underlie the behaviour of the actual physical world at its tiniest known scales.

What are complex numbers? These are numbers which involve the seemingly impossible process of taking the square root of a negative quantity. Recall that the square root of a number $a$ is a number $b$ satisfying $b^2 = a$, so the square root of 4 is 2, the square root of 9 is 3, of 16 is 4, of 25 is 5, and the square root of 2 is $1.414213562\ldots$, and so on. We can allow that the *negatives* of these respective square roots $(-2, -3, -4, -5, -1.414213562\ldots,$ etc.) would also qualify as "square roots" (since $(-b)^2 = b^2$). But if $a$ is itself negative, we have a problem, because whether $b$ is positive or negative, its square is always positive, so it is hard to see how we are ever going to get a negative number simply by squaring something. We may regard the basic problem to be finding a square root of $-1$, because if we had a number "i", as we shall call it, which satisfies $i^2 = -1$, then 2i should satisfy $(2i)^2 = -4$, $(3i)^2 = -9$, $(4i)^2 = -16$, etc., and generally $(ib)^2 = -b^2$. Of course, as we have just seen, whatever such an "i" is, it cannot be an ordinary real number, and it is often referred to as an *imaginary number*, as are all the real multiples of i, such as 2i or 3i, or $-i$, $-2i$, etc.

However, the terminology is misleading, for it suggests that there is some greater "reality" to these so-called real numbers than there is to the so-called imaginary numbers. This impression comes about, I suppose, because there is the feeling that distance measures and time measures are, in some sense "really" such real-number quantities. But we do not know this. We know that these real numbers are indeed very good for describing distances and times, but we do not know that this description holds good at absolutely *all* scales of distance or time.

We have no actual understanding of the nature of a physical continuum at a scale of, say, one googolth (see §A.1) of a metre or of a second, for example. The so-called real numbers are *mathematical* constructions, which are, nevertheless, immensely valuable for the formulation of the physical laws of classical physics.

Yet, real numbers may also be regarded as "real" in the *Platonic* sense – the same Platonic sense as any other consistent mathematical structure – if we are to adopt the common standpoint among mathematicians whereby mathematical consistency is the sole criterion for such Platonic "existence". However, the so-called imaginary numbers form just as consistent a mathematical structure as do the so-called real numbers, so, in this same Platonic sense, they are also just as "real". A separate (and, indeed, *open*) question is the extent to which either of these number systems precisely models the actual world.

The *complex* numbers – elements of the number system $\mathbb{C}$ – are the numbers formed by adding together such (so-called) real and imaginary numbers, i.e. they are numbers of the form $a + ib$, where $a$ and $b$ are elements of the real-number system $\mathbb{R}$. These numbers seem to have been first encountered by the very remarkable Italian physician and mathematician Gerolamo Cardano in 1545, and their algebra was described in detail by another deeply insightful Italian, the engineer Raphaello Bombelli, in 1572. (See, for example, Wykes [1969]; imaginary numbers on their own appear, however, to have been considered much earlier, e.g. by Heron of Alexandria in the first century CE.) Many magical properties of complex numbers were revealed in subsequent years, and their purely mathematical utility is today unquestioned. They had found many applications, also, to physical problems, such as in the theory of electrical circuits and in hydrodynamics. But right up until the early twentieth century they had been regarded as purely mathematical constructs, or aids to calculation, without their having any *direct* realization in the physical world.

But now, with the coming of quantum mechanics, things have changed dramatically, $\mathbb{C}$ finding a central place in the mathematical formulation of that theory which appears to be just as direct as are the various roles of $\mathbb{R}$ in classical physics. The basic quantum-mechanical physical role of $\mathbb{C}$, as introduced in §§1.4 and 2.5–2.9, depends upon various remarkable mathematical properties of complex numbers. Recall, from above, that such a number is one of the form $x + iy$, where $x$ and $y$ are real numbers (elements of $\mathbb{R}$) and the quantity i satisfies

$$i^2 = -1.$$

The ordinary rules of algebra that apply to real-number quantities also apply equally well to complex numbers. For this, the operations of addition and multiplication of complex numbers are defined in terms of real-number operations by

$$(x + \mathrm{i}y) + (u + \mathrm{i}v) = (x + u) + \mathrm{i}(y + v),$$
$$(x + \mathrm{i}y) \times (u + \mathrm{i}v) = (xu - yv) + \mathrm{i}(xv + yu),$$

where $x$, $y$, $u$, and $v$ are real. Also, the reverse operations of subtraction and division of complex numbers are determined (except division by zero) via the operation of forming the negative of, or the inverse of, a complex number by

$$-(x+\mathrm{i}y) = (-x)+\mathrm{i}(-y) \quad \text{and} \quad (x+\mathrm{i}y)^{-1} = \frac{x}{x^2 + y^2} - \mathrm{i}\frac{y}{x^2 + y^2} = \frac{x - \mathrm{i}y}{x^2 + y^2},$$

$x$ and $y$ being real (and not both zero, in the latter case). It is usual, however, to write a complex number using just a *single* symbol, so we might for example simply write $z$ for $x + \mathrm{i}y$ and $w$ for $u + \mathrm{i}v$ in the above:

$$z = x + \mathrm{i}y \quad \text{and} \quad w = u + \mathrm{i}v,$$

and then directly write their sum as $z + w$, their product as $zw$, and the negative and the inverse of $z$ simply as $-z$ and $z^{-1}$, respectively. The difference and quotient of complex numbers are now simply defined by $z - w = z + (-w)$ and $z \div w = z \times (w^{-1})$, with the definitions of $-w$ and $w^{-1}$ being given as for $z$ above.

We find that we can manipulate complex numbers just in the same way as reals, but in many respects, the rules become much more systematic than they were with reals. An important illustration of this is the so-called fundamental theorem of algebra, which tells us that any polynomial in a single variable $z$

$$a_0 + a_1 z + a_2 z^2 + a_3 z^3 + \cdots + a_{n-1} z^{n-1} + a_n z^n$$

can always be factorized into a product of $n$ linear terms. As an example of what this means, we can consider the simple quadratic polynomials $1 - z^2$ and $1 + z^2$. The factorization of the first of these may well be familiar to the reader, and uses only real-number coefficients, but we need complex numbers for the second:

$$1 - z^2 = (1 + z)(1 - z), \qquad 1 + z^2 = (1 + \mathrm{i}z)(1 - \mathrm{i}z).$$

This particular example just begins to illustrate how complex numbers make algebra more systematic, but it only uses the rule $\mathrm{i}^2 = -1$ in a very immediate way and the magic of complex numbers is not yet brought out. However, we begin to see something of this magic in the full theorem (where, in what follows, we can assume that the final coefficient $a_n$ is non-zero, whence we can then divide through by its value, so we may as well take $a_n = 1$), which tells us that we can factorize *any* (real or complex) polynomial

$$a_0 + a_1 z + a_2 z^2 + \cdots + a_{n-1} z^{n-1} + z^n = (z - b_1)(z - b_2)(z - b_3) \cdots (z - b_n)$$

in terms of complex numbers $b_1, b_2, b_3, \ldots, b_n$. Observe that if $z$ takes any of the values $b_1, \ldots, b_n$, then the polynomial *vanishes* (since the right-hand side does). The magic is that the simple procedure of adjoining the single number "i" to the system of reals, in order to be able to solve the very specific simple little equation $1 + z^2 = 0$, we find that we get, completely *free*, solutions to *all* non-trivial one-variable polynomial equations!

Generalizing in another direction, we find that all equations of the form $z^\alpha = \beta$ can be solved, where $\alpha$ and $\beta$ are given non-zero complex numbers. We get this all *free*, where we just started from the very particular case $\alpha = 2$, $\beta = -1$ (i.e. $z^2 = -1$). We shall see some other aspects of complex-number magic in the next section. (For further examples of this magic, see Nahin [1998] and TRtR [chapters 3, 4, 6, and 9].)

## A.10. COMPLEX GEOMETRY

A standard representation of complex numbers (first explicitly described by the Norwegian/Danish surveyor and mathematician Caspar Wessel, in a report written in 1787 and published in a detailed paper in 1799) is to label the points of a Euclidean plane with them, where the single complex number $z = x + iy$ labels the point with Cartesian coordinates $(x, y)$ (figure A-34). In honour of Wessel's priority, I shall refer to this plane as the *Wessel plane* here, despite the common terms *Argand plane* and *Gauss plane*, referring to some much later publications describing this geometry (in 1806 and 1831, respectively). Gauss is on record as asserting that he had thought of this idea many years prior to his publication (though not at the age of 10, when Wessel's report was written). The records do not appear to say when the idea first occurred to Wessel or to Argand [see Crowe 1967]. The sum and the product of two complex numbers each has a simple geometrical characterization. The *sum* of complex numbers $w$ and $z$ is given by the now familiar *parallelogram* law (figure A-35(a), and compare §A.3, figure A-6), whereby the line from 0 to $w + z$ is a diagonal of the parallelogram formed by those two points and the two original points $w$ and $z$; the *product* is given by a *similar triangle* law

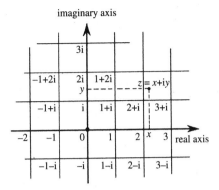

Figure A-34: The Wessel plane (complex plane) represents $z = x + iy$ as $(x, y)$ in a standard Cartesian representation.

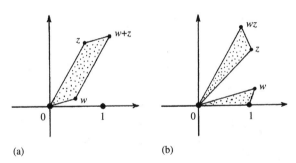

(a)                              (b)

Figure A-35: The geometric realization, in the Wessel plane, of (a) addition, in terms of the parallelogram law, and (b) multiplication, in terms of the similar triangles law.

(figure A-35(b)), according to which the triangle formed by the respective points $0, 1, w$ is similar (without reflection) to that formed by $0, z, wz$. (There are also various degenerate cases of these, where the parallelogram or triangle becomes squashed down to a line, which need to be described appropriately.)

The geometry of the Wessel plane clarifies many issues which may not, at first sight, have anything to do with complex numbers. One important example relates to the convergence of power series. A power series is an expression

$$a_0 + a_1 z + a_2 z^2 + a_3 z^3 + a_4 z^4 + \cdots,$$

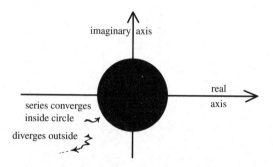

Figure A-36: For any power series in complex numbers $A_0 + A_1 Z + A_2 z^2 + A_3 z^3 + A_4 z^4 + \cdots$, there will be a circle, centred at the origin the Wessel plane and called the *circle of convergence*, for which the series converges for any $z$ strictly within the circle (open black region) and diverges for any $z$ strictly outside the circle (open white region). This allows that the radius of convergence (the circle's radius) may be zero (series never convergent except for $z - 0$) or infinite (convergent for all $z$).

where $a_0, a_1, a_2, \ldots$ are complex constants, in which (unlike a polynomial) the terms continue indefinitely. (In fact, polynomials would come under the heading of power series when we take all the $a_r$ to be zero beyond a certain $r$ value.) For a given value of $z$, we may find that the sum of the terms *converges* to some specific complex number or that it *diverges*, i.e. does not converge. (This is interpreted as the sum of a successively increasing number of terms – the *partial sums* $\Sigma_r$ of the series – which may, or may not, converge to some specific complex value $S$. Technically, *convergence to S* means that for any given positive number $\varepsilon$, no matter how small, there will be some value of $r$ for which the difference $|S - \Sigma_q|$ is less than $\varepsilon$ for all $x$, whenever $q$ is greater than $r$.)

A remarkable role for Wessel's complex plane now emerges: if the series converges for some (non-zero) values of $z$ and diverges for others, then there is a *circle* (called the circle of convergence) centred at the origin in the Wessel plane with the property that for every complex number strictly inside the circle the series converges, and for every complex number strictly outside the circle, the series diverges to infinity; see figure A-36. What the series does for points actually *on* the circle is, however, a more delicate question.

This remarkable result resolves various issues that are otherwise somewhat puzzling, such as why the series $1 - x^2 + x^4 - x^6 + x^8 - \cdots$, for a real variable $x$, should start to diverge just at the points when $x$ gets greater than 1 or less than

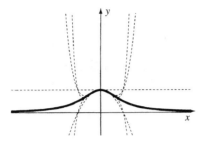

Figure A-37: The real-variable function $y = f(x) = 1/(1+x^2)$, shown here plotted with the dark continuous line, is represented within the interval $-1 < x < 1$ by the infinite series $1 - x^2 + x^4 - x^6 + x^8 - x^{10} + \cdots$, but the series diverges for $|x| > 1$. The partial sums $y = 1$, $y = 1 - x^2$, $y = 1 - x^2 + x^4$, $y = 1 - x^2 + x^4 - x^6$, and $y = 1 - x^2 + x^4 - x^6 + x^8$ are plotted using broken lines, this indicating the points of divergence. From the viewpoint of real variables alone, there seems to be no reason why the function should suddenly start to diverge just at the places where $|x|$ exceeds unity since the curve $y = f(x)$ exhibits no particular feature there, being as smooth as one could wish just at the places where divergence begins.

$-1$, whereas the algebraic expression for the *sum* of the series (for $-1 < x < 1$), which happens to be $1/(1 + x^2)$, does nothing particular at the values $x = \pm 1$ (see figure A-37). The problem arises at the *complex* value $z = \text{i}$ (or $z = -\text{i}$), where the function $1/(1 + z^2)$ becomes infinite, and we infer that the circle of convergence must pass through the points $z = \pm\text{i}$. That circle also passes through $z = \pm 1$, so we must expect divergence for real values of $x$ just outside the circle, i.e. where $|x| > 1$ (see figure A-38).

There is a further point that I would like to make with regard to divergent series, like the one just considered. We may ask if there is any sense in assigning the answer "$1/(1 + x^2)$" to the series when $x$ is greater than 1. In particular, taking $x = 2$, we would get

$$1 - 4 + 16 - 64 + 256 - \cdots = \tfrac{1}{5},$$

which is, of course, an absurdity if we simply try to add up the terms one by one, if only because all the terms on the left are integers whereas we have a *fraction* on the right. Yet, there appears to be something "correct" in the answer $\tfrac{1}{5}$, since if we call the "sum" of the series $\Sigma$ and add $4\Sigma$ to it we seem to get

$$\Sigma + 4\Sigma = 1 - 4 + 16 - 64 + 256 - 1024 + \cdots$$
$$+ 4 - 16 + 64 - 256 + 1024 - \cdots = 1,$$

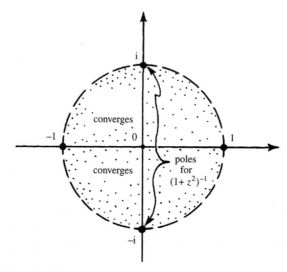

Figure A-38: In the Wessel plane, we see what the trouble with $f(x) = 1/(1+x^2)$ is. In its complex form $f(z) = 1/(1+z^2)$, where $z = x+iy$, we see that the function becomes infinite at the "poles" $z = \pm i$, and the circle of convergence cannot extend beyond those points. Hence the real series for $f(x)$ must also diverge for $|x| > 1$.

so $5\Sigma = 1$, and we indeed arrive at $\Sigma = \frac{1}{5}$. Using similar types of argument, one can "prove" the even more remarkable-looking equation (derived by Leonhard Euler in the eighteenth century)

$$1 + 2 + 3 + 4 + 5 + 6 + \cdots = -\tfrac{1}{12},$$

which, curiously enough, has a significant role to play in string theory (see §3.8 and equation (1.3.32) in Polchinski [1998]).

Logically, one would be considered to have "cheated" in doing term-by-term subtractions of badly divergent series to obtain these answers; yet there is some deep underlying truth to them, as can be revealed through a procedure called *analytic continuation*. This can be sometimes used to justify such manipulations of divergent series, allowing the range of a function, validly defined by the series in one region of the Wessel plane, to extend to other regions where the original series diverges. It should be noted that, as part of this procedure, we need to expand functions about points other than the origin, which means considering series of the form $a_0 + a_1(z - Q) + a_2(z - Q)^2 + a_3(z - Q)^3 + \cdots$ to represent a function expanded about a point $z = Q$. For an example, see figure 3-36 in §3.8.

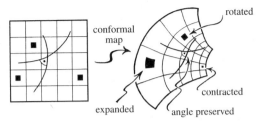

Figure A-39: A holomorphic map of one portion of the Wessel plane to another is characterized by the fact that it is conformal and non-reflective. Geometrically, "conformal" means that the angles between intersecting curves are preserved by the map; equivalently infinitesimal shapes are preserved: they may become larger, smaller, or rotated, but not altered in shape, in the limit of small sizes.

The process of analytic continuation exhibits a remarkable kind of *rigidity* possessed by holomorphic functions. They cannot be "bent" in arbitrary ways, as can smooth real-valued functions. The fully detailed nature of a holomorphic function in any small local region constrains what it can do a long way off. In a strange sense, a holomorphic function seems to have a mind of its own, which it cannot be deflected from. This feature has important roles for us in §§3.8 and 4.1.

It is often useful to think in terms of *transformations* of the Wessel plane. Two of the simplest are given by adding a fixed complex number $A$ to the coordinate $z$ in the plane, or by multiplying the coordinate $z$ by a fixed complex number $B$

$$z \mapsto A + z \quad \text{or} \quad z \mapsto Bz$$

corresponding, respectively, to performing a *translation* of the plane (rigid motion without rotation) or a rotation and/or uniform expansion/contraction. These are transformations of the plane which preserve shapes (without reflecting them) but not necessarily sizes.

The transformations (maps) that are given by functions – called *holomorphic* functions – which are built up from $z$ by sums, products, together with constant complex numbers, and the taking of limits, so that they can be described in terms of power series, have the characteristic that they are, geometrically, what are called *conformal* (and non-reflective). These maps have the property that infinitesimally small shapes are preserved in the transformation (although they can be rotated and/or isotropically expanded or contracted); another way of stating the conformal property is that *angles* between curves are preserved in such a transformation. See figure A-39. The notions of conformal geometry have considerable importance also in higher dimensions; see §§1.15, 3.1, 3.5, 4.1, and 4.3.

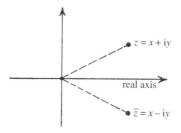

Figure A-40: The operation of complex conjugation ($z \mapsto \bar{z}$), i.e. of reflection in the real axis of the Wessel plane, is not holomorphic. Though it is clearly conformal, it reverses the orientation of the Wessel plane.

An example of a *non*-holomorphic function of a complex number $z$ is the quantity $\bar{z}$, defined by

$$\bar{z} = x - \mathrm{i}y,$$

where $z = x + \mathrm{i}y$ with $x$ and $y$ real. The map $z \mapsto \bar{z}$ is conformal in the sense that small angles are preserved, but it does not count as holomorphic because there is a reversal of orientation, the map being defined by a *reflection* of the Wessel plane in the real axis (see figure A-40). This is an example of an *anti*-holomorphic function, which is the complex conjugate of a holomorphic function (see §1.9). Though also conformal, anti-holomorphic functions reverse orientation, in the sense that they effect a reflection in the local structure. It is as well that we do not count $\bar{z}$ as holomorphic, for if we did, then the whole point would be lost, because, for example, the real and imaginary parts of $z$ would have to count as holomorphic since $x = \frac{1}{2}(z + \bar{z})$ and $y = \frac{1}{2}(z - \bar{z})$. Moreover, this would also apply to the quantity $|z|$, called the *modulus* of $z$, given by

$$|z| = \sqrt{z\bar{z}} = \sqrt{x^2 + y^2}.$$

We note that (by the Pythagorean theorem) $|z|$ is simply the distance from the origin 0 of the point $z$ in the Wessel plane. Clearly, the map $z \mapsto z\bar{z}$ is very far from conformal, since it squashes the whole plane down to the non-negative part of the real axis, so it is certainly not holomorphic. A useful point of view is to think of a holomorphic function of $z$ as one which "does not involve $\bar{z}$". Thus, $z^2$ is holomorphic; $z\bar{z}$ is not.

Holomorphic functions are central to complex-number analysis. They are the analogues of the *smooth* functions of real-number analysis. But with complex

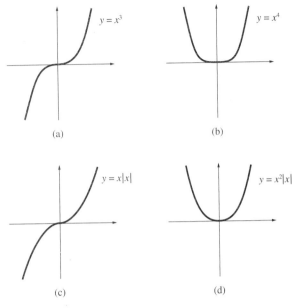

Figure A-41: Real-number functions can have various degrees of smoothness. The curves (a) $y = x^3$ and (b) $y = x^4$ have infinitely many degrees of smoothness, being referred to as *analytic* ($C^\omega$, meaning extendable to smooth complex-number functions). On the other hand, the curve (c) $y = x|x|$, which is $x^2$ when $x$ is positive and $-x^2$ when $x$ is negative, has only 1 degree of smoothness (i.e. $C^1$) and curve (d) $y = x^2|x|$, which is $x^3$ when $x \geqslant 0$ and $-x^3$ when $x < 0$, has 2 degrees of smoothness ($C^2$), despite their superficial similarity to the previous two curves.

analysis there is a remarkable piece of magic that is not at all shared by its real-number counterpart. Real-number functions can have all sorts of degrees of smoothness. For example, the function $x \times |x|$, which is $x^2$ when $x$ is positive and $-x^2$ when $x$ is negative, has only 1 degree of smoothness (technically $C^1$), whereas the function $x^3$, whose graph looks superficially similar, has infinitely many degrees of smoothness (technically $C^\infty$ or $C^\omega$). As another example, $x^2 \times |x|$ ($x^3$ when $x \geqslant 0$ and $-x^3$ when $x < 0$) has 2 degrees of smoothness (technically $C^2$), whereas the very similar-looking $x^4$ has infinitely many, etc. (See figure A-41.) However, with complex functions, everything is much simpler, because even the lowest degree of complex smoothness ($C^1$) implies the highest also ($C^\infty$) and, moreover, it implies the property of expandability as a power series ($C^\omega$), so

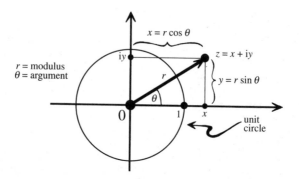

Figure A-42: The relation between polar and Cartesian coordinates in the Wessel plane, expressed by the formula $z = re^{i\theta} = r\cos\theta + ir\sin\theta$. The quantity $r$ is called the modulus and $\theta$ the argument of the complex number $z$.

every complex-smooth function is automatically holomorphic. For more details, see Rudin [1986] and TRtR [chapters 6 and 7].

One particular holomorphic function of great interest, namely the *exponential function* $e^z$ (commonly written "exp $z$"), is one that we already encountered for real numbers in §A.1, defined by the series

$$e^z = 1 + \frac{z}{1!} + \frac{z^2}{2!} + \frac{z^3}{3!} + \frac{z^4}{4!} + \cdots$$

($n! = 1 \times 2 \times 3 \times \cdots \times n$). This series actually converges for *all* values of $z$ (so its circle of convergence has become infinite). If $z$ lies on the *unit circle* in the Wessel plane – namely the circle of unit radius centred at the origin 0 (see figure A-42) – then we have the magical (Cotes–De Moivre–Euler) formula

$$e^{i\theta} = \cos\theta + i\sin\theta,$$

where $\theta$ is the angle (measured in the anti-clockwise sense) that the radius out to $z$ makes with the positive real axis. We may also take note of the extension of this formula to points $z$ that need not lie on the unit circle of Wessel's plane:

$$z = re^{i\theta} = r\cos\theta + ir\sin\theta,$$

where the modulus of $z$ is $r = |z|$, as noted above, and $\theta$ is referred to as the *argument* of $z$; see figure A-42.

The whole theory of real manifolds (briefly indicated in §A.5) extends also to *complex* manifolds, where the real-number coordinates of real manifolds are replaced by complex-number coordinates. However, we always have the option of regarding a complex number $z = x + iy$ as being represented as a pair $(x, y)$ of real numbers. From this perspective, we can re-express a complex $n$-manifold as a real $2n$-manifold (with a certain kind of structure, called a *complex structure*, which comes from the holomorphic properties of the complex coordinates). We may note, from this, that a real manifold that can be reinterpreted, in this way, as a complex manifold, must necessarily be even dimensional. But this condition is, in itself, very far from being sufficient for a real $2n$-dimensional real manifold to be able to be assigned a complex structure, in order that such a reinterpretation might be possible. Especially for reasonably large values of $n$, this possibility is a very rare privilege.

In the case of a 1-complex-dimensional manifold, these matters are a lot easier to understand. For a *complex curve*, in real-number terms, we get certain types of real 2-surfaces, these being known as *Riemann surfaces*. In real-number terms, a Riemann surface is an ordinary real surface endowed with a *conformal* structure (which means, as indicated above, that the notion of *angle* between curves on the surface is determined) and an *orientation* (which simply means that the notion of a local "anti-clockwise rotation" can be maintained consistently over the whole surface; see figure A-21). Riemann surfaces can have different kinds of topology, and some examples are illustrated in figure A-13 of §A.5. These play a key role in string theory (§1.6). Usually, Riemann surfaces are taken to be *closed*, i.e. compact without boundary, but such surfaces can also be considered to have *holes*, or *punctures*, in them (figure 1-44), these playing a role in *string theory* (§1.6).

Of particular significance to us is the simplest of the Riemann surfaces, namely the one with the topology of an ordinary sphere, called the *Riemann sphere*, which, in §2.7, plays a special role in relation to quantum-mechanical spin. We can easily construct the Riemann sphere by simply adjoining a single point (which we can label as "∞") to the entire Wessel plane. To see that the *entire* Riemann sphere can be regarded as a genuine (1-complex-dimensional) complex manifold, we can cover this sphere with two coordinate patches, where one patch is the original Wessel plane coordinatized by $z$, and where the other is another copy of the Wessel plane, coordinatized by $w(= z^{-1})$. This now includes our new point "$z = \infty$", simply as the $w$-origin ($w = 0$), but it now leaves out the $z$-origin. These two Wessel planes are thus patched together via $z = w^{-1}$ to give us the entire Riemann sphere (figure A-43).

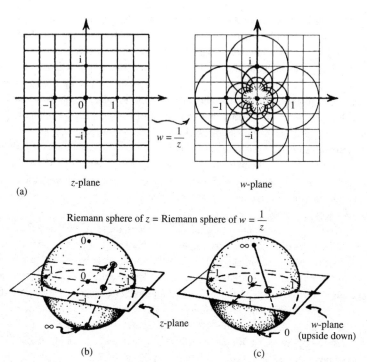

z-plane      w-plane

(a)

Riemann sphere of $z$ = Riemann sphere of $w = \dfrac{1}{z}$

(b)      (c)

Figure A-43: The Riemann sphere is a manifold that can be assembled from two coordinate patches, each of which is a copy of the Wessel plane, here the $z$-plane and the $w$-plane related by $w = z^{-1}$. (a) How the lines of constant real and imaginary parts of $z$, in the $z$-plane, appear when mapped to the $w$-plane. (b) Stereographic projection from the Riemann sphere's south pole gives the $z$-plane. (c) Stereographic projection from the Riemann sphere's north pole gives the $w$-plane, shown upside down.

## A.11. HARMONIC ANALYSIS

A powerful procedure that is frequently adopted by physicists in order to address those equations that arise in physical problems is that of harmonic analysis. These equations are normally *differential* equations (frequently of a type referred to as *partial* differential equations, one such being the "div $\mathbf{B} = 0$" referred to in §A.2). Differential equations are part of the subject of *calculus*, and since I have deliberately refrained from entering into a detailed discussion of this topic,

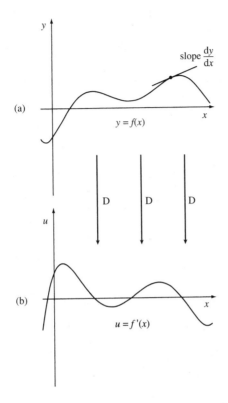

Figure A-44: The operation of differentiation, here denoted by D, replaces a function $f(x)$ by a new function $f'(x)$, where the value of $f'(x)$ at each $x$ is the slope of $f(x)$ at $x$. The inverse operation of *integration*, concerned with areas under the lower curve, would be described by the *reversed* arrow directions.

I provide here only a rough intuitive idea of the basic algebraic properties of differential operators.

What is the operation of differentiation? For a function $f(x)$ of *one* variable, this differentiation operation, which let us denote by D, when acting on the function $f$, replaces $f$ by a new function $f'$, called the *derivative* of $f$, whose value $f'(x)$ at $x$ is the *slope* of the original function $f$ at $x$, and we could write $Df = f'$ (see figure A-44). We can also consider the *second* derivative $f''$, of $f$, whose value $f''(x)$ at $x$ measures the slope of $f'$ at $x$. This turns out to be a measure of how much the original function $f$ "bends" at $x$ (and would measure *acceleration*, if $x$ were a measure of time). We can write

$$f'' = D(Df) = D^2 f$$

and continue this to obtain the $k$th derivative $D^k f$ of $f$, for any positive integer $k$. The *reverse* operation to D (sometimes written $D^{-1}$, or more usually by use of the "integral sign" $\int$) leads to the *integral calculus of areas and volumes*.

When there are more variables, $u, v, \ldots$, which could be (local) coordinates for an $n$-dimensional space, the notion of derivative can apply to each coordinate separately. We can write $D_u$ for the derivative with respect to $u$ (referred to as a *partial* derivative, where all the other variables are kept constant) and $D_v$ for the derivative with respect to $v$, etc. Again these may be raised to various powers (i.e iterated various numbers of times, and also added together in various combinations. As a good illustration, there is a particular much-studied differential operator referred to as the *Laplacian* (first employed by the highly esteemed French mathematician Pierre-Simon de Laplace, in the late eighteenth century, and published in his classic work *Mécanique Céleste* [Laplace 1829–39]). The Laplacian is normally denoted by $\nabla^2$ (or by $\Delta$), where in a 3-dimensional Euclidean space with Cartesian coordinates $u, v, w$, we have

$$\nabla^2 = D_u^2 + D_v^2 + D_w^2,$$

which tells us that when acting on some function $f$ (of the three variables $u, v, w$), the quantity $\nabla^2 f$ denotes the sum of the second derivatives of $f$ with respect to $u$, with respect to $v$, and with respect to $w$, i.e.

$$\nabla^2 f = D_u^2 f + D_v^2 f + D_w^2 f.$$

Equations involving $\nabla^2$ have vastly many applications in both physics and mathematics, starting with Laplace's own $\nabla^2 \varphi = 0$, which he used to describe the Newtonian gravitational field in terms of a scalar quantity known as a *potential function* $\varphi$ for the gravitational field. (The vector describing the strength and direction of the gravitational field would have, as its three components, $-D_u \varphi$, $-D_v \varphi$, and $-D_w \varphi$.) Another important example arises in the case of 2-dimensional Euclidean space, with Cartesian coordinates $x$ and $y$ (so now $\nabla^2 = D_x^2 + D_y^2$), where we consider this plane as the Wessel plane for the complex number $z = x + iy$. Then we find that any holomorphic function $\psi$ of $z$ (see §A.10) has real and imaginary parts $f$ and $g$

$$\psi = f + ig,$$

each of which satisfies Laplace's equation:

$$\nabla^2 f = 0, \qquad \nabla^2 g = 0.$$

Laplace's equation is an example of a differential equation that is *linear*, which means that if we have any two solutions, say $\nabla^2 \phi = 0$ and $\nabla^2 \chi = 0$, then any linear combination,

$$\lambda = A\phi + B\chi,$$

where $A$ and $B$ are constants, will also be a solution:

$$\nabla^2 \lambda = 0.$$

Although linearity is unusual for differential equations generally, we find that linear equations do play foundational roles in theoretical physics. The example of Newtonian gravitational theory, as expressed in terms of Laplace's potential $\varphi$, above, has already been referred to. Other important examples of linear differential equations are Maxwell's equations for the electromagnetic field (§§1.2, 1.6, 1.8, 2.6, and 4.1) and the basic Schrödinger equation of quantum mechanics (§§2.4–2.7 and 2.11).

In the case of such linear equations, harmonic analysis can provide us with a very powerful method of solution. The name "harmonic" comes from music, where musical tones can be analysed in terms of various particular "pure tones". For example, a violin string can vibrate in different ways. The *fundamental* tone has a particular frequency $v$, where the entire string oscillates in its simplest way (without any nodes), but it can also vibrate with various *harmonics*, with frequencies $2v$, $3v$, $4v$, $5v$, etc., and where the shape of the vibrating string (with 1 node, 2 nodes, 3 nodes 4 nodes, etc.) matches that of the wave form of the pure harmonic tone being produced (see figure A-45). The basic differential equation governing the vibration of the string is linear, so the general vibrational state can be built up in terms of linear combinations of these *modes*, these being the individual pure tones of vibration (i.e. the fundamental, and then with all the harmonics). To represent the general solution of the differential equation for the string, we need merely specify a sequence of numbers, each number representing, in an appropriate sense, the magnitude of the contribution of each mode. Any wave form, provided that it is periodic with the frequency of the fundamental tone, can be uniquely expressed as a sum of sinusoidal components in this way (where the term *sinusoidal* refers to the sine curve shape that is exhibited in the function $y = \sin x$, as in figure A-46). This representation of a periodic function

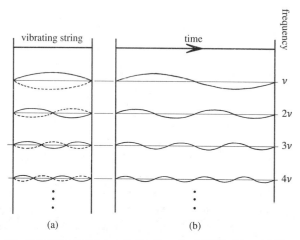

Figure A-45:  The different modes of vibration of a (violin) string. (a) The shape of each vibration mode of the string itself. (b) The temporal behaviour of the vibration, where the frequency is some integer multiple of the fundamental frequency $\nu$.

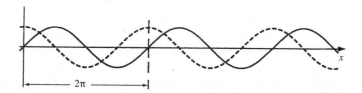

Figure A-46:  The continuous curve is the graph of the function $\sin x$; the dotted curve is $\cos x$.

into such harmonics is referred to as *Fourier analysis*, in honour of the French mathematician Joseph Fourier, who first studied this representation of periodic wave forms in terms of sinusoidal harmonics. Later in this section, we shall see an elegant way that such a representation can come about.

This general kind of procedure applies to linear differential equations generally, where the individual modes are particular simple solutions of the equation that are readily obtained, in terms of which all other solutions can be expressed by taking linear combinations of the modes (usually infinite linear combinations). Let us see this in the particular case of the Laplace equation in 2-dimensional Euclidean space. This is an especially simple case, where we can appeal directly to the algebra and analysis of complex numbers, where the required modes can be written down directly by this means. The reader should not be misled by this; in more general situations things cannot be done so rapidly. Nevertheless, the basic points I wish to make can be nicely demonstrated in this case by appealing to the complex-number descriptions.

As noted above, we can think of each solution of the Laplace equation $\nabla^2 f = 0$, in two dimensions, to be the real part $f$ of a holomorphic function $\psi$ (or, equivalently the imaginary part – it doesn't matter which we choose, because the imaginary part of $\psi$ is simply the real part of the slightly different holomorphic function $-i\psi$). We can express the general solution of our differential equation $\nabla^2 f = 0$ as linear combinations of basic *modes* (the analogues of the various harmonics of the violin string) and to find out what these modes are, we can pass to the corresponding holomorphic quantities $\psi$. Since these are holomorphic functions of the complex number $z$, they can be expressed as *power series* in

$$\psi = a_0 + a_1 z + a_1 z^2 + a_1 z^3 + \cdots,$$

where $z = x + iy$, and taking the real part all the way through, we get the expression for $f$ in terms of $x$ and $y$. The individual modes would be the various terms in the power series, i.e. the real and imaginary parts of individual powers

$$z^k = (x + iy)^k$$

(multiplied by some appropriate constant number, depending on $k$ – and we now need the imaginary part as well as the real part because the coefficients are complex). Thus, the mode, in terms of $x$ and $y$, would be the real and imaginary parts of this expression (e.g. $x^3 - 3xy^2$ and $3x^2y - y^3$, in the case $k = 3$).

To be more precise about this, we need to know what region of the plane we are interested in. Let us first suppose that this region is the entire Wessel plane, so we are interested in solutions of Laplace's equation that cover this entire plane. In terms of the holomorphic function $\psi$, we shall need a power series with an *infinite* radius of convergence, a particular example being the exponential function $e^z$. In this case, the coefficients $1/k!$ tend rapidly to zero as $k$ tends to infinity, thereby ensuring the convergence, for all $z$, of the power series (example A)

$$e^z = 1 + \frac{z}{1!} + \frac{z^2}{2!} + \frac{z^3}{3!} + \frac{z^4}{4!} + \cdots,$$

that we encountered in §A.10 (and §A.1). On the other hand, another series considered in §A.10 (example B)

$$(1 + z^2)^{-1} = 1 - z^2 + z^4 - z^6 + z^8 - \cdots,$$

while converging within the unit circle $|z| = 1$ diverges outside it. An intermediate case (example C) is

$$\left(1 + \frac{z^2}{4}\right)^{-1} = 1 - \frac{z^2}{4} + \frac{z^4}{16} - \frac{z^6}{64} + \frac{z^8}{256} - \cdots ,$$

convergent within the circle $|z| = 2$.

Thus, whereas we can choose to represent our solutions of the Laplace equation simply by the sequences of coefficients, i.e. $(1, 1, 1/2, 1/6, 1/24, \dots)$ for example A, $(1, 0, -1, 0, 1, 0, -1, 0, 1, \dots)$ for example B, and $(1, 0, -1/4, 0, 1/16, 0, -1/64, 0, 1/256, \dots)$ for example C, we do need to be careful to examine the way in which these numbers behave as the sequences proceed to infinity in order to know whether any particular such sequence of numbers actually represents a solution of our differential equation throughout our intended region of definition. An extreme example of this issue arises if our region of definition happens to be the Riemann sphere (see §A.10), obtained by adjoining a single additional point "∞" to the Wessel plane. There is a theorem that the only holomorphic functions that exist globally on the Riemann sphere are in fact *constants*, so the sequences of numbers representing solutions of the Laplace equation on the Riemann sphere are *all* of the form $(K, 0, 0, 0, 0, 0, 0 \dots)$!

These examples also illustrate another aspect of harmonic analysis. Often, we are interested in specifying *boundary values* for solutions of differential equations. For example, we might wish to find a solution to Laplace's equation $\nabla^2 f = 0$ in $n$-dimensional Euclidean space, that holds both on and within a unit $(n-1)$-sphere $\mathscr{S}$. It is, in fact, a theorem [see Evans 2010; Strauss 1992] that if we specify $f$ to be any arbitrarily chosen real-valued function on $\mathscr{S}$ (taken to be smooth, let us say), then there will be a unique solution of $\nabla^2 f = 0$ *within* $\mathscr{S}$ that attains the given values *on* $\mathscr{S}$. Now we can ask what happens to each of the individual modes, in a harmonic decomposition of the solutions of the Laplace equation.

Again, it is instructive to examine the case $n = 2$ first, and to take $\mathscr{S}$ to be the unit circle in the Wessel plane, and we look for solutions of the Laplace equation on the unit disc. If we consider the *mode* defined by a particular power $z^k$, we see, using the polar representation of $z$ given in §A.10, namely

$$z = r e^{i\theta} = r \cos \theta + i r \sin \theta,$$

that on the unit circle $\mathscr{S}$ ($r = 1$) we have

$$z^k = e^{ik\theta} = \cos k\theta + i \sin k\theta.$$

Around the unit circle, for each such mode, the real and imaginary parts of $z$ vary in a sinusoidal way, just like the $k$th harmonic that could be produced by the violin string considered earlier (i.e. $\cos k\theta$ and $\sin k\theta$, where the coordinate $\theta$ now plays the role of time, and is allowed to wind indefinitely around the circle as the time increases (figure A-45). For a *general* solution of the Laplace equation on this unit disc, the value of $f$, as a function of the angular coordinate $\theta$, is arbitrary, so long as it has the periodicity that is determined by the circle, namely a period of $2\pi$. (Of course, any other periodicity could be considered also, in the same way, simply by scaling the length of the circle up or down, as desired.) This is just the Fourier decomposition of a periodic function mentioned earlier, in connection with the vibrating violin string.

In the above, I considered the value of $f$ around the boundary circle $\mathcal{S}$ to vary as a smooth function, but the procedure works considerably more generally than this. For example, even in the case of example B above, the boundary function is far from smooth, having singularities at the two places $\theta = \pm\pi/2$, corresponding to $\pm i$, in the Wessel plane. On the other hand, in example C (or indeed, example A) if we look only at the part of the solution on the unit disc, we find a completely smooth behaviour of $f$ on the bounding unit circle $\mathcal{S}$. The minimal requirements for $f$ on the boundary will not be of concern for us here, however.

In higher dimensions ($n > 2$) the same type of analysis can be applied. Solutions of the Laplace equation in the interior of the hypersphere $\mathcal{S}$ – an $(n - 1)$-dimensional sphere – can be split into harmonics, which, just as in the 2-dimensional case, correspond to different powers of the radial coordinate $r$. In the case $n = 3$, $\mathcal{S}$ is an ordinary 2-sphere, and although the simple description in terms of complex functions will not work, we can still consider our "modes" to be distinguished from one another by their dependence on the power $k$ to which the radial coordinate $r$ is raised. It is usual to adopt coordinates $\theta$, $\phi$, on each sphere, centred at the origin ($r = R$, with $R$ constant) called *spherical polar* coordinates, closely related to the latitude and longitude coordinates on the Earth. The details are not important for us, but pictured in figure A-47.

The usual modes have the form

$$r^k Y_{k,m}(\theta, \phi),$$

where the $Y_{k,m}(\theta, \phi)$ are *spherical harmonics* (introduced by Laplace in 1782) these being particular explicit functions of $\theta$ and $\phi$, whose detailed forms will not concern us [Riley et al. 2006]. The "$k$" value (usually denoted by $\ell$, in standard notation) runs over all the natural numbers $k = 0, 1, 2, 3, 4, 5, \ldots$, and $m$, also an integer, is allowed to be negative, with $|m| \leqslant k$. Accordingly, the allowable

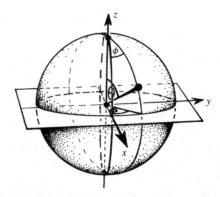

Figure A-47: The conventional spherical polar angles $\theta$ and $\phi$, for $S^2$ embedded in the standard way in $\mathbb{R}^3$.

values for $(k, m)$ would be

$$(0, 0), \ (1, -1), \ (1, 0), \ (1, 1), \ (2, -2), \ (2, -1),$$
$$(2, 0), \ (2, 1), \ (2, 2), \ (3, -3), \ (3, -2), \ \ldots.$$

To specify a particular solution of Laplace's equation within the solid ball contained within $\mathcal{S}$ (i.e. $1 \geqslant r \geqslant 0$), we would need to know the contribution of each of these modes, an infinite sequence of real numbers

$$f_{0,0}, \ f_{1,-1}, \ f_{1,0}, \ f_{1,1}, \ f_{2,-2}, \ f_{2,-1}, \ f_{2,0}, \ f_{2,1}, \ f_{2,2}, \ f_{3,-3}, \ f_{3,-2}, \ \ldots,$$

telling us exactly the contribution of each. This sequence of numbers specifies $f$ on the bounding sphere $\mathcal{S}$, or equivalently the corresponding solution to Laplace's equation in the interior of $\mathcal{S}$. (Continuity/smoothness issues concerning $f$ on $\mathcal{S}$, would be reflected in complicated questions about how the sequence $f_{k,m}$ proceeds to infinity.)

A particular point I wish to raise here is that, powerful as such methods are with regard to studying individual solutions, notably with regard to numerical computation, there is an important matter that becomes obscured, namely that of *functional freedom*, which was our particular concern in §§A.2 and A.8, and plays a key role for the discussion given in chapter 1. In specifying solutions to differential equations in this way, as with Laplace's equation or other more complicated systems, harmonic analysis eventually provides us with a solution in the form of an infinite sequence of numbers. The very dimension of the space on which the solution is defined, let alone its size or shape, has frequently become

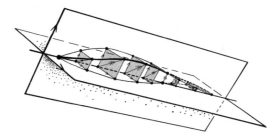

Figure A-48: The small vibrations of a string in three dimensions can be resolved into the vibrations in two orthogonal planes, where at each point of the string the vector of displacement is resolved into two components in these planes at right angles.

hidden within some complicated asymptotic property of that sequence, there being a tendency to lose sight of the issue of functional freedom completely.

Even in the simplest situation of a vibrating string – the violin string considered earlier in this section – we find that a mere mode analysis can mislead us with regard to functional freedom, if we are not careful. Let us consider two different situations, in one of which the string is allowed to vibrate only in one plane, such as with a violin string carefully stroked with a bow. In the other situation, such as when the string is plucked, the vibrations can involve displacements of the string in both of the two dimensions outwards from the string's direction. (I here ignore displacements that are in the direction *along* the string, which could be activated by stroking the fingers along its length.) The modes of vibration of the string can be separated into those in two perpendicular planes through the direction of the string, and all other vibrations can be considered to be composed of these (see figure A-48). Since the two planes are on an equal footing with each other, we simply get exactly the same modes in each plane, with exactly the same frequencies of vibration. Thus, the only difference between the modes for the bowed string (vibrations restricted to one plane) and the plucked string (unrestricted vibrations) are that, in the latter case, each mode occurs twice over. The functional freedom in the first case is $\infty^{2\infty^1}$ and in the second case, the significantly larger $\infty^{4\infty^1}$. The "2" and "4" come from the magnitude and velocity of the displacement outwards, at each point of the string, there being twice as many such quantities needed in the second case. The upper "1" comes from the 1-dimensionality of the string, and this would be a larger number $n$ if the string were replaced by an "$n$-brane" (see §1.15), an entity that plays an important role in currently fashionable activity of string theory. We see the importance of the functional-freedom issue in chapter 1.

In relation to this, it is useful also to consider the vibrations of a 2-dimensional surface such as a *drum*. It is quite usual to analyse such things in terms of a mode analysis, where the general ways in which the drum might vibrate could be expressed in terms of the different contributions to this motion from each of the separate modes, which could be expressed as an infinite sequence of numbers, say $p_0, p_1, p_2, p_3, \ldots$, giving the amount of the contribution from each mode. At first glance, this might not seem so different from representing the vibrations of a bowed violin string by a similar-looking sequence $q_0, q_1, q_2, q_3, \ldots$, where these represent the contributions from the different modes of vibration of the string. Yet for the drum-surface displacements the functional freedom would be *enormously* larger, namely $\infty^{2\infty^2}$, as opposed to the above $\infty^{2\infty^1}$ that we had for the string. We can get something of a feeling for this difference if we consider the drum surface to be a *square*, given by Cartesian coordinates $(x, y)$, where each of $x$ and $y$ are restricted to lie between 0 and 1. Then we could (rather unconventionally) try to represent the displacements of the drum surface in terms of "modes" that are the products $F_{ij}(x, y) = g_i(x)h_j(y)$ of the modes $g_i(x)$ along the $x$-axis with modes $h_j(y)$ along the $y$-axis. The mode analysis would then describe the general displacements of the drum surface in terms of a sequence of numbers $f_{0,0}, f_{0,1}, f_{1,0}, f_{0,2}, f_{1,1}, f_{2,0}, f_{0,3}, f_{2,1}, f_{1,2}$, etc., these providing the magnitude of the contribution from each $F_{ij}(x, y)$. There is nothing wrong with this sort of thing, but it does not directly bring out the vast difference between the functional freedom $\infty^{\infty^2}$ in the 2-dimensional drum displacements and the much smaller functional freedom $\infty^{\infty^1}$ in each of the separate 1-dimensional displacements in $x$ and $y$ coordinates respectively (or of the $\infty^{2\infty^1}$ that would represent displacements in $x$ together with displacements in $y$, this basically giving the far smaller freedom in the "product displacements" of the form $g(x)h(y)$).

# References

Abbott, B. P., et al. (LIGO Scientific Collaboration) 2016 Observation of gravitational waves from a binary black hole merger. arXiv:1602.03837.

Ade, P. A. R., et al. (BICEP2 Collaboration) 2014 Detection of B-mode polarization at degree angular scales by BICEP2. *Physical Review Letters* **112**:241101.

Aharonov, Y., Albert, D. Z., and Vaidman, L. 1988 How the result of a measurement of a component of the spin of a spin-1/2 particle can turn out to be 100. *Physical Review Letters* **60**:1351–54.

Albrecht, A., and Steinhardt, P. J. 1982 Cosmology for grand unified theories with radiatively induced symmetry breaking. *Physical Review Letters* **48**:1220–23.

Alexakis, S. 2012 *The Decomposition of Global Conformal Invariants.* Annals of Mathematics Studies 182. Princeton University Press.

Almheiri, A., Marolf, D., Polchinski, J., and Sully, J. 2013 Black holes: complementarity or firewalls? *Journal of High Energy Physics* **2013**(2):1–20.

Anderson, M. 2005 *"Shakespeare" by Another Name: The Life of Edward de Vere, Earl of Oxford, the Man Who Was Shakespeare.* New York: Gotham Books.

Ananthaswamy, A. 2006 North of the Big Bang. *New Scientist* (2 September), pp. 28–31.

Antusch, S., and Nolde, D. 2014 BICEP2 implications for single-field slow-roll inflation revisited. *Journal of Cosmology and Astroparticle Physics* **5**:035.

Arkani-Hamed, N., Dimopoulos, S., and Dvali, G. 1998 The hierarchy problem and new dimensions at a millimetre. *Physics Letters* B **429**(3):263–72.

Arkani-Hamed, N., Cachazo, F., Cheung, C., and Kaplan, J. 2010 The S-matrix in twistor space. *Journal of High Energy Physics* **2**:1–48.

Arkani-Hamed, N., Hodges, A., and Trnka, J. 2015 Positive amplitudes in the amplituhedron. *Journal of High Energy Physics* **8**:1–25.

Arndt, M., Nairz, O, Voss-Andreae, J., Keller, C., van der Zouw, G., and Zeilinger, A. 1999 Wave–particle duality of $C_{60}$. *Nature* **401**:680–82.

Ashok, S., and Douglas, M. 2004 Counting flux vacua. *Journal of High Energy Physics* **0401**:060.

Ashtekar, A., Baez, J. C., Corichi, A., and Krasnov, K. 1998 Quantum geometry and black hole entropy. *Physical Review Letters* **80**(5):904–7.

Ashtekar, A., Baez, J. C., and Krasnov, K. 2000 Quantum geometry of isolated horizons and black hole entropy. *Advances in Theoretical and Mathematical Physics* **4**:1–95.

Ashtekar, A., Pawlowski, T., and Singh, P. 2006 Quantum nature of the Big Bang. *Physical Review Letters* **96**:141301.

Aspect, A., Grangier, P., and Roger, G. 1982 Experimental realization of Einstein–Podolsky–Rosen–Bohm *Gedankenexperiment*: a new violation of Bell's inequalities. *Physical Review Letters* **48**:91–94.

Bardeen, J. M., Carter, B., and Hawking, S. W. 1973 The four laws of black hole mechanics. *Communications in Mathematical Physics* **31**(2):161–70.

Barrow, J. D., and Tipler, F. J. 1986 *The Anthropic Cosmological Principle*. Oxford University Press.

Bateman, H. 1904 The solution of partial differential equations by means of definite integrals. *Proceedings of the London Mathematical Society* (2) **1**:451–58.

——. 1910 The transformation of the electrodynamical equations. *Proceedings of the London Mathematical Society* (2) **8**:223–64.

Becker, K., Becker, M., and Schwarz, J. 2006 *String Theory and M-Theory: A Modern Introduction*. Cambridge University Press.

Bedingham, D., and Halliwell, J. 2014 Classical limit of the quantum Zeno effect by environmental decoherence. *Physical Review* A **89**:042116.

Bekenstein, J. 1972 Black holes and the second law. *Lettere al Nuovo Cimento* **4**:737–40.

——. 1973 Black holes and entropy. *Physical Review* D **7**:2333–46.

Belinskiĭ, V. A., Khalatnikov, I. M., and Lifshitz, E. M. 1970 Oscillatory approach to a singular point in the relativistic cosmology. *Uspekhi Fizicheskikh Nauk* **102**:463–500. (English translation in *Advances in Physics* **19**:525–73.)

Belinskiĭ, V. A., Lifshitz, E. M., and Khalatnikov, I. M. 1972 Construction of a general cosmological solution of the Einstein equation with a time singularity. *Soviet Physics JETP* **35**:838–41.

Bell, J. S. 1964 On the Einstein–Podolsky–Rosen paradox. *Physics* **1**:195–200. (Reprinted in Wheeler and Zurek [1983, pp. 403–8].)

——. 1981 Bertlmann's socks and the nature of reality. *Journal de Physique* **42**, C2(3), p. 41.

——. 2004 *Speakable and Unspeakable in Quantum Mechanics: Collected Papers on Quantum Philosophy*, 2nd edn (with a new introduction by A. Aspect). Cambridge University Press.

Bennett, C. H., Brassard, G., Crepeau, C., Jozsa, R. O., Peres, A., and Wootters, W. K. 1993 Teleporting an unknown quantum state via classical and Einstein–Podolsky–Rosen channels. *Physical Review Letters* **70**:1895–99.

Besse, A. 1987 *Einstein Manifolds*. Springer.

Beyer, H., and Nitsch, J. 1986 The non-relativistic COW experiment in the uniformly accelerated reference frame. *Physics Letters* B **182**:211–15.

Bisnovatyi-Kogan, G. S. 2006 Checking the variability of the gravitational constant with binary pulsars. *International Journal of Modern Physics* D **15**:1047–52.

Bjorken, J., and Drell, S. 1964 *Relativistic Quantum Mechanics*. McGraw-Hill.

Blau, S. K., and Guth, A. H. 1987 Inflationary cosmology. In *300 Years of Gravitation* (ed. S. W. Hawking and W. Israel). Cambridge University Press.

Bloch, F. 1932 Zur Theorie des Austauschproblems und der Remanenzerscheinung der Ferromagnetika. *Zeitschrift für Physik* **74**(5):295–335.

Bohm, D. 1951 The paradox of Einstein, Rosen, and Podolsky. In *Quantum Theory*, ch. 22, § 15–19, pp. 611–23. Englewood Cliffs, NJ: Prentice-Hall. (Reprinted in Wheeler and Zurek [1983, pp. 356–68].)

——. 1952 A suggested interpretation of the quantum theory in terms of "hidden" variables, I and II. *Physical Review* **85**:166–93. (Reprinted in Wheeler and Zurek [1983, pp. 41–68].)

Bohm, D., and Hiley, B. J. 1993 *The Undivided Universe: An Ontological Interpretation of Quantum Theory*. Abingdon and New York: Routledge.

Bojowald, M. 2007 What happened before the Big Bang? *Nature Physics* **3**:523–25.

——. 2011 *Canonical Gravity and Applications: Cosmology, Black Holes, and Quantum Gravity*. Cambridge University Press.

Bollobás, B. (ed.) 1986 *Littlewood's Miscellany*. Cambridge University Press.

Boltzmann, L. 1895 On certain questions of the theory of gases. *Nature* **51**:413–15.

Bordes, J., Chan, H.-M., and Tsou, S. T. 2015 A first test of the framed standard model against experiment. *International Journal of Modern Physics* A **27**:1230002.

Börner, G. 1988 *The Early Universe* Springer.

Bouwmeester, D., Pan, J. W., Mattle, K., Eibl, M., Weinfurter, H., and Zeilinger, A. 1997 Experimental teleportation. *Nature* **390**:575–79.

Boyer, R. H., and Lindquist, R. W. 1967 Maximal analytic extension of the Kerr metric. *Journal of Mathematical Physics* **8**:265–81.

Breuil, C., Conrad, B., Diamond, F., and Taylor, R. 2001 On the modularity of elliptic curves over Q: wild 3-adic exercises. *Journal of the American Mathematical Society* **14**:843–939.

Bryant, R. L., Chern, S.-S., Gardner, R. B., Goldschmidt, H. L., and Griffiths, P. A. 1991 *Exterior Differential Systems*. MSRI Publication 18. Springer.

Bullimore, M., Mason, L., and Skinner, D. 2010 MHV diagrams in momentum twistor space. *Journal of High Energy Physics* **12**:1–33.

Buonanno, A., Meissner, K. A., Ungarelli, C., and Veneziano, G. 1998a Classical inhomogeneities in string cosmology. *Physical Review* D **57**:2543.

——. 1998b Quantum inhomogeneities in string cosmology. *Journal of High Energy Physics* **9801**:004.

Byrnes, C. T., Choi, K.-Y., and Hall, L. M. H. 2008 Conditions for large non-Gaussianity in two-field slow-roll inflation. *Journal of Cosmology and Astroparticle Physics* **10**:008.

Cachazo, F., Mason, L., and Skinner, D. 2014 Gravity in twistor space and its Grassmannian formulation. In *Symmetry, Integrability and Geometry: Methods and Applications* (SIGMA) **10**:051 (28 pages).

Candelas, P., de la Ossa, X. C., Green, P. S., and Parkes, L. 1991 A pair of Calabi–Yau manifolds as an exactly soluble superconformal theory. *Nuclear Physics* B **359**:21.

Cardoso, T. R., and de Castro, A. S. 2005 The blackbody radiation in a $D$-dimensional universe. *Revista Brasileira de Ensino de Física* **27**:559–63.

Cartan, É. 1945 *Les Systèmes Différentiels Extérieurs et leurs Applications Géométriques*. Paris: Hermann.

472    References

Carter, B. 1966 Complete analytic extension of the symmetry axis of Kerr's solution of Einstein's equations. *Physical Review* **141**:1242–47.

——. 1970 An axisymmetric black hole has only two degrees of freedom. *Physical Review Letters* **26**:331–33.

——. 1983 The anthropic principle and its implications for biological evolution. *Philosophical Transactions of the Royal Society of London* A **310**:347–63.

Cartwright, N. 1997 Why physics? In *The Large, the Small and the Human Mind* (ed. R. Penrose). Cambridge University Press.

Chan, H.-M., and Tsou, S. T. 1980 U(3) monopoles as fundamental constituents. CERN-TH-2995 (10 pages).

——. 1998 *Some Elementary Gauge Theory Concepts*. World Scientific Notes in Physics. Singapore: World Scientific.

——. 2007 A model behind the standard model. *European Physical Journal* C **52**:635–63.

——. 2012 *International Journal of Modern Physics* A **27**:1230002.

Chandrasekhar, S. 1931 The maximum mass of ideal white dwarfs. *Astrophysics Journal* **74**:81–82.

——. 1934 Stellar configurations with degenerate cores. *The Observatory* **57**:373–77.

Christodoulou, D. 2009 *The Formation of Black Holes in General Relativity*. Monographs in Mathematics, European Mathematical Society.

Clarke, C. J. S. 1993 *The Analysis of Space-Time Singularities*. Cambridge Lecture Notes in Physics. Cambridge University Press.

Coleman, S. 1977 Fate of the false vacuum: semiclassical theory. *Physical Review* D **15**:2929–36.

Coleman, S., and De Luccia, F. 1980 Gravitational effects on and of vacuum delay. *Physical Review* D **21**:3305–15.

Colella, R., and Overhauser, A. W. 1980 Neutrons, gravity and quantum mechanics. *American Scientist* **68**:70.

Colella, R., Overhauser, A. W., and Werner, S. A. 1975 Observation of gravitationally induced quantum interference. *Physical Review Letters* **34**:1472–74.

Connes, A., and Berberian, S. K. 1995 *Noncommutative Geometry*. Academic Press.

Conway, J., and Kochen, S. 2002 The geometry of the quantum paradoxes. In *Quantum [Un]speakables: From Bell to Quantum Information* (ed. R. A. Bertlmann and A. Zeilinger), chapter 18. Springer.

Corry, L., Renn, J., and Stachel, J. 1997 Belated decision in the Hilbert–Einstein priority dispute. *Science* **278**:1270–73.

Crowe, M. J. 1967 *A History of Vector Analysis: The Evolution of the Idea of a Vectorial System* Toronto: University of Notre Dame Press. (Reprinted with additions and corrections, 1985, New York: Dover.)

Cubrovic, M., Zaanen, J., and Schalm, K. 2009 String theory, quantum phase transitions, and the emergent Fermi liquid. *Science* **325**:329–444.

Davies, P. C. W. 1975 Scalar production in Schwarzschild and Rindler metrics. *Journal of Physics* A **8**:609.

Davies, P. C. W., and Betts, D. S. 1994 *Quantum Mechanics* (2nd edn). CRC Press.

de Broglie, L. 1956 *Tentative d'Interpretation Causale et Nonlineaire de la Mechanique Ondulatoire*. Paris: Gauthier–Villars.

Deser, S. 1996 Conformal anomalies – recent progress. *Helvetica Physica Acta* **69**:570–81.

Deutsch, D. 1998 *Fabric of Reality: Towards a Theory of Everything*. Penguin.

de Sitter, W. 1917a On the curvature of space. *Proceedings of Koninklijke Nederlandse Akademie van Wetenschappen* **20**:229–43.

——. 1917b On the relativity of inertia. Remarks concerning Einstein's latest hypothesis. *Proceedings of Koninklijke Nederlandse Akademie van Wetenschappen* **19**:1217–25.

DeWitt, B. S., and Graham, N. (eds) 1973 *The Many Worlds Interpretation of Quantum Mechanics*. Princeton University Press.

Dicke, R. H. 1961 Dirac's cosmology and Mach's principle. *Nature* **192**:440–41.

Dieudonné, J. 1981 *History of Functional Analysis*. North-Holland.

Diósi, L. 1984 Gravitation and quantum-mechanical localization of macro-objects *Physics Letters* **105A**, 199–202.

——. 1987 A universal master equation for the gravitational violation of quantum mechanics. *Physics Letters* **120A**, 377–81.

——. 1989 Models for universal reduction of macroscopic quantum fluctuations *Physical Review* A **40**:1165–74.

Dirac, P. A. M. 1930 (1st edn) 1947 (3rd edn) *The Principles of Quantum Mechanics*. Oxford University Press and Clarendon Press.

——. 1933 The Lagrangian in quantum mechanics. *Physikalische Zeitschrift der Sowjetunion* **3**:64–72.

——. 1937 The cosmological constants. *Nature* **139**:323.

——. 1938 A new basis for cosmology. *Proceedings of the Royal Society of London* A **165**:199–208.

——. 1963 The evolution of the physicist's picture of nature. (Conference on the foundations of quantum physics at Xavier University in 1962.) *Scientific American* **208**:45–53.

Douglas, M. 2003 The statistics of string/M theory vacua. *Journal of High Energy Physics* **0305**:46.

Eastwood, M. G. 1990 The Penrose transform. In *Twistors in Mathematics and Physics*, LMS Lecture Note Series 156 (ed. T. N. Bailey and R. J. Baston). Cambridge University Press.

Eastwood M. G., Penrose, R., and Wells Jr, R. O. 1981 Cohomology and massless fields. *Communications in Mathematical Physics* **78**:305–51.

Eddington, A. S. 1924 A comparison of Whitehead's and Einstein's formulas. *Nature* **113**:192.

——. 1935 Meeting of the Royal Astronomical Society, Friday, January 11, 1935. *The Observatory* **58**(February 1935):33–41.

Eerkens, H. J., Buters, F. M., Weaver, M. J., Pepper, B., Welker, G., Heeck, K., Sonin, P., de Man, S., and Bouwmeester, D. 2015 Optical side-band cooling of a low frequency optomechanical system. *Optics Express* **23**(6):8014-20 (doi: 10.1364/OE.23.008014).

Ehlers, J. 1991 The Newtonian limit of general relativity. In *Classical Mechanics and Relativity: Relationship and Consistency* (International Conference in memory of Carlo Cataneo, Elba, 1989). Monographs and Textbooks in Physical Science, Lecture Notes 20 (ed. G. Ferrarese). Napoli: Bibliopolis.

Einstein, A. 1931 Zum kosmologischen Problem der allgemeinen Relativitätstheorie. *Sitzungsberichte der Königlich Preuss ischen Akademie der Wissenschaften*, pp. 235–37.

———. 1939 On a stationary system with spherical symmetry consisting of many gravitating masses. *Annals of Mathematics* Second Series **40**:922–36 (doi: 10.2307/1968902).

Einstein, A., and Rosen, N. 1935 The particle problem in the general theory of relativity. *Physical Review* (2) **48**:73–77.

Einstein, A., Podolsky, B., and Rosen, N. 1935 Can quantum-mechanical description of physical reality be considered complete? *Physical Review* **47**:777–80. (Reprinted in Wheeler and Zurek [1983, pp. 138–41].)

Eremenkno, A., and Ostrovskii, I. 2007 On the pits effect of Littlewood and Offord. *Bulletin of the London Mathematical Society* **39**:929–39.

Ernst, B. 1986 Escher's impossible figure prints in a new context. In *M. C. Escher: Art and Science* (ed. H. S. M. Coxeter, M. Emmer, R. Penrose and M. L. Teuber). Amsterdam: Elsevier.

Evans, L. C. 2010 *Partial Differential Equations*, 2nd edn (Graduate Studies in Mathematics). American Mathematical Society.

Everett, H. 1957 "Relative state" formulation of quantum mechanics. *Review of Modern Physics* **29**:454–62. (Reprinted in Wheeler and Zurek [1983, pp. 315–323].)

Feeney, S. M., Johnson, M. C., Mortlock, D. J., and Peiris, H. V. 2011a First observational tests of eternal inflation: analysis methods and WMAP 7-year results. *Physical Review* D **84**:043507.

———. 2011b First observational tests of eternal inflation. *Physical Review Letters* **107**: 071301.

Feynman, R. 1985 *QED: The Strange Theory of Light and Matter*, p. 7. Princeton University Press.

Feynman, R. P., Hibbs, A. R., and Styer, D. F. 2010 *Quantum Mechanics and Path Integrals* (emended edition). Dover Books on Physics.

Fickler, R., Lapkiewicz, R., Plick, W. N., Krenn, M., Schaeff, C. Ramelow, S., and Zeilinger, A. 2012 Quantum entanglement of high angular momenta. *Science 2* **338**:640–43.

Finkelstein, D. 1958 Past–future asymmetry of the gravitational field of a point particle. *Physical Review* **110**:965–67.

Fogli, G. L., Lisi, E., Marrone, A., Montanino, D., Palazzo, A., and Rotunno, A. M. 2012 Global analysis of neutrino masses, mixings, and phases: entering the era of leptonic CP violation searches. *Physical Review* D **86**:013012.

Ford, I. 2013 *Statistical Physics: An Entropic Approach*. Wiley.

Forward, R. L. 1980 *Dragon's Egg*. Del Ray Books.

———. 1985 *Starquake*. Del Ray Books.

Francesco, P., Mathieu, P., and Senechal, D. 1997 *Conformal Field Theory.* Springer.

Fredholm, I. 1903 Sur une classe d'équations fonctionnelles. *Acta Mathematica* **27**:365–90.

Friedrich, H. 1986 On the existence of $n$-geodesically complete or future complete solutions of Einstein's field equations with smooth asymptotic structure. *Communications in Mathematical Physics* **107**:587–609.

——. 1998 Einstein's equation and conformal structure. In *The Geometric Universe: Science, Geometry, and the Work of Roger Penrose* (ed. S. A. Huggett, L. J. Mason, K. P. Tod, S. T. Tsou and N. M. J. Woodhouse). Oxford University Press.

Friedrichs, K. 1927 Eine invariante Formulierung des Newtonschen Gravitationsgesetzes und des Grenzüberganges vom Einsteinschen zum Newtonschen Gesetz. *Mathematische Annalen* **98**:566–75.

Fulling, S. A. 1973 Nonuniqueness of canonical field quantization in Riemannian spacetime. *Physical Review* D **7**:2850.

Gamow, G. 1970 *My World Line: An Informal Autobiography.* Viking Adult.

Gardner, M. 2006 *Aha! Gotcha. Aha! Insight. A Two Volume Collection.* The Mathematical Association of America.

Gasperini, M., and Veneziano, G. 1993 Pre-Big Bang in string cosmology. *Astroparticle Physics* **1**:317–39.

——. 2003 The pre-Big Bang scenario in string cosmology. *Physics Reports* **373**:1–212.

Geroch, R., Kronheimer E. H., and Penrose, R. 1972 Ideal points in space-time. *Proceedings of the Royal Society of London* A **347**:545–67.

Ghirardi, G. C., Rimini, A., and Weber, T. 1986 Unified dynamics for microscopic and macroscopic systems. *Physical Review* D **34**:470–91.

Ghirardi, G. C., Grassi, R., and Rimini, A. 1990 Continuous-spontaneous-reduction model involving gravity. *Physical Review* A **42**:1057–64.

Gibbons, G. W., and Hawking, S. W. 1976 Cosmological event horizons, thermodynamics, and particle creation. *Physical Review* D **15**:2738–51.

Gibbons, G. W., and Perry, M. J. 1978 Black holes and thermal Green functions. *Proceedings of the Royal Society of London* A **358**:467–94.

Gingerich, O. 2004 *The Book Nobody Read: Chasing the Revolutions of Nicolaus Copernicus.* Heinemann.

Givental, A. 1996 Equivariant Gromov–Witten invariants. *International Mathematics Research Notices* **1996**:613–63.

Goddard, P., and Thorn, C. 1972 Compatibility of the dual Pomeron with unitarity and the absence of ghosts in the dual resonance model. *Physics Letters* B **40**(2):235–38.

Goenner, H. (ed.) 1999 *The Expanding Worlds of General Relativity.* Birkhäuser.

Green, M., and Schwarz, J. 1984 Anomaly cancellations in supersymmetric $D = 10$ gauge theory and superstring theory. *Physics Letters* B **149**:117–22.

Greenberger, D. M., and Overhauser, A. W. 1979 Coherence effects in neutron diffraction and gravity experiments. *Review of Modern Physics* **51**:43–78.

Greenberger, D. M., Horne, M. A., and Zeilinger, A. 1989 Going beyond Bell's theorem. In *Bell's Theorem, Quantum Theory, and Conceptions of the Universe* (ed. M. Kafatos), pp. 3–76. Dordrecht: Kluwer Academic.

Greene, B. 1999 *The Elegant Universe: Superstrings, Hidden Dimensions and the Quest for the Ultimate Theory*. London: Jonathan Cape.

Greytak, T. J., Kleppner, D., Fried, D. G., Killian, T. C., Willmann, L., Landhuis, D., and Moss, S. C. 2000 Bose–Einstein condensation in atomic hydrogen. *Physica* B **280**:20–26.

Gross, D., and Periwal, V. 1988 String perturbation theory diverges. *Physical Review Letters* **60**:2105–8.

Guillemin, V., and Pollack, A. 1974 *Differential Topology*. Prentice Hall.

Gunning, R. C., and Rossi, R. 1965 *Analytic Functions of Several Complex Variables*. Prentice Hall.

Gurzadyan, V. G., and Penrose, R. 2013 On CCC-predicted concentric low-variance circles in the CMB sky. *European Physical Journal Plus* **128**:1–17.

———. 2016 CCC and the Fermi paradox. *European Physical Journal Plus* **131**:11.

Guth, A. H. 1997 *The Inflationary Universe*. London: Jonathan Cape.

———. 2007 Eternal inflation and its implications. *Journal of Physics* A **40**:6811–26.

Hameroff, S., and Penrose, R. 2014 Consciousness in the universe: a review of the "Orch **OR**" theory. *Physics of Life Reviews* **11**(1):39–78.

Hanbury Brown, R., and Twiss, R. Q. 1954 Correlation between photons in two coherent beams of light. *Nature* **177**:27–32.

———. 1956a A test of a new type of stellar interferometer on Sirius. *Nature* **178**:1046–53.

———. 1956b The question of correlation between photons in coherent light rays. *Nature* **178**:1447–51.

Hanneke, D., Fogwell Hoogerheide, S., and Gabrielse, G. 2011 Cavity control of a single-electron quantum cyclotron: measuring the electron magnetic moment. *Physical Review* A **83**:052122.

Hardy, L. 1993 Nonlocality for two particles without inequalities for almost all entangled states. *Physical Review Letters* **71**:1665.

Harrison, E. R. 1970 Fluctuations at the threshold of classical cosmology. *Physical Review* D **1**:2726.

Hartle, J. B. 2003 *Gravity: An Introduction to Einstein's General Relativity*. Addison Wesley.

Hartle, J. B., and Hawking, S. W. 1983 Wave function of the universe. *Physical Review* D **28**:2960–75.

Hartle, J., Hawking, S. W., and Thomas, H. 2011 Local observation in eternal inflation. *Physical Review Letters* **106**:141302.

Hawking, S. W. 1965 Occurrence of singularities in open universes. *Physical Review Letters* **15**:689–90.

———. 1966a The occurrence of singularities in cosmology. *Proceedings of the Royal Society of London* A **294**:511–21.

Hawking, S. W. 1966b The occurrence of singularities in cosmology. II. *Proceedings of the Royal Society of London* A **295**:490–93.

——. 1967 The occurrence of singularities in cosmology. III. Causality and singularities. *Proceedings of the Royal Society of London* A **300**:187–201.

——. 1974 Black hole explosions? *Nature* **248**:30–31.

——. 1975 Particle creation by black holes. *Communications in Mathematical Physics* **43**:199–220.

——. 1976a Black holes and thermodynamics. *Physical Review* D **13**(2):191–97.

——. 1976b Breakdown of predictability in gravitational collapse. *Physical Review* D **14**:2460–73.

——. 2005 Information loss in black holes. *Physical Review* D **72**:084013-6.

Hawking, S. W., and Ellis, G. F. R. 1973 *The Large-Scale Structure of Space-Time.* Cambridge University Press.

Hawking, S. W., and Penrose, R. 1970 The singularities of gravitational collapse and cosmology. *Proceedings of the Royal Society of London* A **314**:529–48.

Heisenberg, W. 1971 *Physics and Beyond*, pp. 73–76. Harper and Row.

Hellings, R. W., et al. 1983 Experimental test of the variability of *G* using Viking Lander ranging data. *Physical Review Letters* **51**:1609–12.

Hilbert, D. 1912 *Grundzüge einer allgemeinen theorie der linearen integralgleichungen.* Leipzig: B. G. Teubner.

Hodges, A. P. 1982 Twistor diagrams. *Physica* A **114**:157–75.

——. 1985a A twistor approach to the regularization of divergences. *Proceedings of the Royal Society of London* A **397**:341–74.

——. 1985b Mass eigenstates in twistor theory. *Proceedings of the Royal Society of London* A **397**:375–96.

——. 1990 Twistor diagrams and Feynman diagrams. In *Twistors in Mathematics and Physics*, LMS Lecture Note Series 156 (ed. T. N. Bailey and R. J. Baston). Cambridge University Press.

——. 1998 The twistor diagram programme. In *The Geometric Universe; Science, Geometry, and the Work of Roger Penrose* (ed. S. A. Huggett, L. J. Mason, K. P. Tod, S. T. Tsou, and N. M. J. Woodhouse). Oxford University Press.

——. 2006a Scattering amplitudes for eight gauge fields. arXiv:hep-th/0603101v1.

——. 2006b Twistor diagrams for all tree amplitudes in gauge theory: a helicity-independent formalism. arXiv:hep-th/0512336v2.

——. 2013a Eliminating spurious poles from gauge-theoretic amplitudes. *Journal of High Energy Physics* **5**:135.

——. 2013b Particle physics: theory with a twistor. *Nature Physics* **9**:205–6.

Hodges, A. P., and Huggett, S. 1980 Twistor diagrams. *Surveys in High Energy Physics* **1**:333–53.

Hodgkinson, I. J., and Wu, Q. H. 1998 *Birefringent Thin Films and Polarizing Elements.* World Scientific.

Hoyle, F. 1950 *The Nature of the Universe.* Basil Blackwell.

Hoyle, F. 1957 *The Black Cloud*. William Heinemann.

Huggett, S. A., and Tod, K. P. 1985 *An Introduction to Twistor Theory*. LMS Student Texts 4. Cambridge University Press.

Hughston, L. P. 1979 *Twistors and Particles*. Lecture Notes in Physics 97. Springer.

———. 1980 The twistor particle programme. *Surveys in High Energy Physics* 1:313–32.

Isham, C. J., Penrose, R., and Sciama, D. W. (eds) 1975 *Quantum Gravity: An Oxford Symposium*. Oxford University Press.

Jackiw, R., and Rebbi, C. 1976 Vacuum periodicity in a Yang–Mills quantum theory. *Physical Review Letters* **37**:172–75.

Jackson, J. D. 1999 *Classical Electrodynamics*, p. 206. Wiley.

Jaffe, R. L. 2005 Casimir effect and the quantum vacuum. *Physical Review* D **72**:021301.

Jenkins, D., and Kirsebom, O. 2013 The secret of life. *Physics World* February, pp. 21–26.

Jones, V. F. R. 1985 A polynomial invariant for knots via von Neumann algebra. *Bulletin of the American Mathematical Society* **12**:103–11.

Kaku, M. 2000 *Strings, Conformal Fields, and M-Theory*. Springer.

Kaltenbaek, R., Hechenblaiker, G., Kiesel, N., Romero-Isart, O., Schwab, K. C., Johann, U., and Aspelmeyer, M. 2012 Macroscopic quantum resonators (MAQRO). *Experimental Astronomy* **34**:123–64.

Kaltenbaek, R., et al. 2016 Macroscopic quantum resonators (MAQRO): 2015 update. *EPJ Quantum Technology* **3**:5 (doi 10.1140/epjqt/s40507-016-0043-7).

Kane, G. L., and Shifman, M. (eds) 2000 *The Supersymmetric World: The Beginnings of the Theory*. World Scientific.

Kerr, R. P. 1963 Gravitational field of a spinning mass as an example of algebraically special metrics. *Physical Review Letters* **11**:237–38.

Ketterle, W. 2002 Nobel lecture: when atoms behave as waves: Bose–Einstein condensation and the atom laser. *Reviews of Modern Physics* **74**:1131–51.

Khoury, J., Ovrut, B. A., Steinhardt, P. J., and Turok, N. 2001 The ekpyrotic universe: colliding branes and the origin of the hot big bang. *Physical Review* D **64**:123522.

———. 2002a Density perturbations in the ekpyrotic scenario. *Physical Review* D **66**:046005 (arXiv:hepth/0109050).

Khoury, J., Ovrut, B. A., Seiberg, N., Steinhardt, P. J., and Turok, N. 2002b From big crunch to big bang. *Physical Review* D **65**:086007 (arXiv:hep-th/0108187).

Kleckner, D., Pikovski, I., Jeffrey, E., Ament, L., Eliel, E., van den Brink, J., and Bouwmeester, D. 2008 Creating and verifying a quantum superposition in a micro-optomechanical system. *New Journal of Physics* **10**:095020.

Kleckner, D., Pepper, B., Jeffrey, E., Sonin, P., Thon, S. M., and Bouwmeester, D. 2011 Optomechanical trampoline resonators. *Optics Express* **19**:19708–16.

Kochen, S., and Specker, E. P. 1967 The problem of hidden variables in quantum mechanics. *Journal of Mathematics and Mechanics* **17**:59–88.

Kraagh, H. 2010 An anthropic myth: Fred Hoyle's carbon-12 resonance level. *Archive for History of Exact Sciences* **64**:721–51.

Kramer, M. (and 14 others) 2006 Tests of general relativity from timing the double pulsar. *Science* **314**:97–102.

Kruskal, M. D. 1960 Maximal extension of Schwarzschild metric. *Physical Review* **119**:1743–45.

Lamoreaux, S. K. 1997 Demonstration of the Casimir force in the 0.6 to 6 μm range. *Physical Review Letters* **78**:5–8.

Landau, L. 1932 On the theory of stars. *Physikalische Zeitschrift der Sowjetunion* **1**:285–88.

Langacker, P., and Pi, S.-Y. 1980 Magnetic Monopoles in Grand Unified Theories. *Physical Review Letters* **45**:1-4.

Laplace, P.-S. 1829–39 *Mécanique Céleste* (translated with a commentary by N. Bowditch). Boston, MA: Hilliard, Gray, Little, and Wilkins.

LeBrun, C. R. 1985 Ambi-twistors and Einstein's equations. *Classical and Quantum Gravity* **2**:555–63.

——. 1990 Twistors, ambitwistors, and conformal gravity. In *Twistors in Mathematics and Physics*, LMS Lecture Note Series 156 (ed. T. N. Bailey and R. J. Baston). Cambridge University Press.

Lee, J. M. 2003 *Introduction to Smooth Manifolds*. Springer.

Lemaître, G. 1933 L'universe en expansion. *Annales de la Société scientifique de Bruxelles* A **53**:51–85 (cf. p. 82).

Levi-Città, T. 1917 Realtà fisica di alcuni spazî normali del Bianchi. *Rendiconti Reale Accademia Dei Lincei* **26**:519–31.

Levin, J. 2012 In space, do all roads lead to home? *Plus Magazine*, Cambridge.

Lévy, A. 1979 *Basic Set Theory*. Springer. (Reprinted by Dover in 2003.)

Li, T., Kheifets, S., and Raizen, M. G. 2011 Millikelvin cooling of an optically trapped microsphere in vacuum. *Nature Physics* **7**:527–30 (doi: 10.1038/NPHYS1952).

Liddle, A. R., and Leach, S. M. 2003 Constraining slow-roll inflation with WMAP and 2dF. *Physical Review* D **68**:123508.

Liddle, A. R., and Lyth, D. H. 2000 *Cosmological Inflation and Large-Scale Structure*. Cambridge University Press.

Lifshitz, E. M., and Khalatnikov, I. M. 1963 Investigations in relativistic cosmology. *Advances in Physics* **12**:185–249.

Lighthill, M. J. 1958 *An Introduction to Fourier Analysis and Generalised Functions*. Cambridge Monographs on Mechanics. Cambridge University Press.

Linde, A. D. 1982 A new inflationary universe scenario: a possible solution of the horizon, flatness, homogeneity, isotropy and primordial monopole problems. *Physics Letters* B **108**:389–93.

——. 1983 Chaotic inflation. *Physics Letters* B **129**:177–81.

——. 1986 Eternal chaotic inflation. *Modern Physics Letters* A **1**:81–85.

——. 2004 Inflation, quantum cosmology and the anthropic principle. In *Science and Ultimate Reality: Quantum Theory, Cosmology, and Complexity* (ed. J. D. Barrow, P. C. W. Davies, and C. L. Harper), pp. 426–58. Cambridge University Press.

Littlewood, J. E. 1953 *A Mathematician's Miscellany*. Methuen.

Littlewood, J. E., and Offord, A. C. 1948 On the distribution of zeros and $a$-values of a random integral function. *Annals of Mathematics* Second Series **49**:885–952. Errata **50**:990–91.

Looney, J. T. 1920 *"Shakespeare" Identified in Edward de Vere, Seventeenth Earl of Oxford*. London: C. Palmer; New York: Frederick A. Stokes Company.

Luminet, J.-P., Weeks, J. R., Riazuelo, A., Lehoucq, R., and Uzan, J.-P. 2003 Dodecahedral space topology as an explanation for weak wide-angle temperature correlations in the cosmic microwave background. *Nature* **425**:593–95.

Lyth, D. H., and Liddle, A. R. 2009 *The Primordial Density Perturbation*. Cambridge University Press.

Ma, X. 2009 Experimental violation of a Bell inequality with two different degrees of freedom of entangled particle pairs. *Physical Review* A **79**:042101-1–042101-5.

Majorana, E. 1932 Atomi orientati in campo magnetico variabile. *Nuovo Cimento* **9**:43–50.

Maldacena, J. M. 1998 The large $N$ limit of superconformal field theories and supergravity. *Advances in Theoretical and Mathematical Physics* **2**:231–52.

Marshall, W., Simon, C., Penrose, R., and Bouwmeester, D. 2003 Towards quantum superpositions of a mirror. *Physical Review Letters* **91**:13–16; 130401.

Martin, J., Motohashi, H., and Suyama, T. 2013 Ultra slow-roll inflation and the non-Gaussianity consistency relation *Physical Review* D **87**:023514.

Mason, L., and Skinner, D. 2013 Dual superconformal invariance, momentum twistors and Grassmannians. *Journal of High Energy Physics* **5**:1–23.

Meissner, K. A., Nurowski, P., and Ruszczycki, B. 2013 Structures in the microwave background radiation. *Proceedings of the Royal Society of London* A **469**:20130116.

Mermin, N. D. 1990 Simple unified form for the major no-hidden-variables theorems. *Physical Review Letters* **65**:3373–76.

Michell, J. 1783 On the means of discovering the distance, magnitude, &c. of the fixed stars, in consequence of the diminution of the velocity of their light. *Philosophical Transactions of the Royal Society of London* **74**:35.

Mie, G. 1908 Beiträge zur Optik trüber Medien, speziell kolloidaler Metallösungen. *Annalen der Physik* **330**:377–445.

——. 1912a Grundlagen einter Theorie der Materie. *Annalen der Physik* **342**:511–34.

——. 1912b Grundlagen einter Theorie der Materie. *Annalen der Physik* **344**:1–40.

——. 1913 Grundlagen einter Theorie der Materie. *Annalen der Physik* **345**:1–66.

Miranda, R. 1995 *Algebraic Curves and Riemann Surfaces*. American Mathematical Society.

Misner, C. W. 1969 Mixmaster universe. *Physical Review Letters* **22**:1071–74.

Moroz, I. M., Penrose, R., and Tod, K. P. 1998 Spherically-symmetric solutions of the Schrödinger–Newton equations. *Classical and Quantum Gravity* **15**:2733–42.

Mortonson, M. J., and Seljak, U. 2014 A joint analysis of Planck and BICEP2 modes including dust polarization uncertainty. *Journal of Cosmology and Astroparticle Physics* **2014**:035.

Mott, N. F., and Massey, H. S. W. 1965 Magnetic moment of the electron. In *The Theory of Atomic Collisions*, 3rd edn, pp. 214–19. Oxford: Clarendon Press. (Reprinted in Wheeler and Zurek [1983, pp. 701–6].)

Muckhanov, V. 2005 *Physical Foundations of Cosmology*. Cambridge University Press.

Nahin, P. J. 1998 *An Imaginary Tale: The Story of Root*($-1$). Princeton University Press.

Nair, V. 1988 A current algebra for some gauge theory amplitudes. *Physics Letters* B **214**:215–18.

Needham, T. R. 1997 *Visual Complex Analysis*. Oxford University Press.

Nelson, W., and Wilson-Ewing, E. 2011 Pre-big-bang cosmology and circles in the cosmic microwave background. *Physical Review* D **84**:0435081.

Newton, I. 1730 *Opticks*. (Dover, 1952.)

Olive, K. A., et al. (Particle Data Group) 2014 *Chinese Physics* C **38**:090001 (hppt://pdg.lbl.gov).

Oppenheimer, J. R., and Snyder, H. 1939 On continued gravitational contraction. *Physical Review* **56**:455–59.

Painlevé, P. 1921 La mécanique classique et la théorie de la relativité. *Comptes Rendus de l'Académie des Sciences (Paris)* **173**:677–80.

Pais, A. 1991 *Niels Bohr's Times*, p. 299. Oxford: Clarendon Press.

——. 2005 *Subtle Is the Lord: The Science and the Life of Albert Einstein* (new edition with a foreword by R. Penrose). Oxford University Press.

Parke, S., and Taylor, T. 1986 Amplitude for $n$-gluon scatterings. *Physical Review Letters* **56**:2459.

Peebles, P. J. E. 1980 *The Large-Scale Structure of the Universe*. Princeton University Press.

Penrose, L. S., and Penrose, R. 1958 Impossible objects: a special type of visual illusion. *British Journal of Psychology* **49**:31–33.

Penrose, R. 1959 The apparent shape of a relativistically moving sphere. *Proceedings of the Cambridge Philosophical Society* **55**:137–39.

——. 1963 Asymptotic properties of fields and space-times. *Physical Review Letters* **10**:66–68.

——. 1964a The light cone at infinity. In *Conférence Internationale sur les Téories Relativistes de la Gravitation* (ed. L. Infeld). Paris: Gauthier Villars; Warsaw: PWN.

——. 1964b Conformal approach to infinity. In *Relativity, Groups and Topology: The 1963 Les Houches Lectures* (ed. B. S. DeWitt and C. M. DeWitt). New York: Gordon and Breach.

——. 1965a Gravitational collapse and space-time singularities. *Physical Review Letters* **14**:57–59.

——. 1965b Zero rest-mass fields including gravitation: asymptotic behaviour. *Proceedings of the Royal Society of London* A **284**:159–203.

——. 1967a Twistor algebra. *Journal of Mathematical Physics* **82**:345–66.

——. 1967b Conserved quantities and conformal structure in general relativity. In *Relativity Theory and Astrophysics*. Lectures in Applied Mathematics 8 (ed. J. Ehlers). American Mathematical Society.

Penrose, R. 1968 Twistor quantization and curved space-time. *International Journal of Theoretical Physics* **1**:61–99.

——. 1969a Gravitational collapse: the role of general relativity. *Rivista del Nuovo Cimento* Serie I **1**(Numero speciale):252–76. (Reprinted in 2002 in *General Relativity and Gravity* **34**:1141–65.)

——. 1969b Solutions of the zero rest-mass equations. *Journal of Mathematical Physics* **10**:38–39.

——. 1972 *Techniques of Differential Topology in Relativity*. CBMS Regional Conference Series in Applied Mathematics 7. SIAM.

——. 1975a Gravitational collapse: a review. (Physics and astrophysics of neutron stars and black holes.) *Proceedings of the International School of Physics "Enrico Fermi" Course* **LXV**:566–82.

——. 1975b Twistors and particles: an outline. In *Quantum Theory and the Structures of Time and Space* (ed. L. Castell, M. Drieschner and C. F. von Weizsäcker). Munich: Carl Hanser.

——. 1976a The space-time singularities of cosmology and in black holes. *IAU Symposium Proceedings Series*, volume 13: *Cosmology*.

——. 1976b Non-linear gravitons and curved twistor theory. *General Relativity and Gravity* **7**:31–52.

——. 1978 Singularities of space-time. In *Theoretical Principles in Astrophysics and Relativity* (ed. N. R. Liebowitz, W. H. Reid, and P. O. Vandervoort). Chicago University Press.

——. 1980 A brief introduction to twistors. *Surveys in High-Energy Physics* **1**(4):267–88.

——. 1981 Time-asymmetry and quantum gravity. In *Quantum Gravity 2: A Second Oxford Symposium* (ed. D. W. Sciama, R. Penrose, and C. J. Isham), pp. 244–72. Oxford University Press.

——. 1987a Singularities and time-asymmetry. In *General Relativity: An Einstein Centenary Survey* (ed. S. W. Hawking and W. Israel). Cambridge University Press.

——. 1987b Newton, quantum theory and reality. In *300 Years of Gravity* (ed. S. W. Hawking and W. Israel). Cambridge University Press.

——. 1987c On the origins of twistor theory. In *Gravitation and Geometry: A Volume in Honour of I. Robinson* (ed. W. Rindler and A. Trautman). Naples: Bibliopolis.

——. 1989 *The Emperor's New Mind: Concerning Computers, Minds, and the Laws of Physics*. Oxford University Press.

——. 1990 Difficulties with inflationary cosmology. In *Proceedings of the 14th Texas Symposium on Relativistic Astrophysics* (ed. E. Fenves). New York Academy of Sciences.

——. 1991 On the cohomology of impossible figures. *Structural Topology* **17**:11–16.

——. 1993 Gravity and quantum mechanics. In *General Relativity and Gravitation 13. Part 1: Plenary Lectures 1992* (ed. R. J. Gleiser, C. N. Kozameh, and O. M. Moreschi). Institute of Physics.

——. 1994 *Shadows of the Mind: An Approach to the Missing Science of Consciousness*. Oxford University Press.

Penrose, R. 1996 On gravity's role in quantum state reduction. *General Relativity and Gravity* **28**:581–600.

——. 1997 *The Large, the Small and the Human Mind*. Cambridge University Press.

——. 1998a The question of cosmic censorship. In *Black Holes and Relativistic Stars* (ed. R. M. Wald). University of Chicago Press.

——. 1998b Quantum computation, entanglement and state-reduction. *Philosophical Transactions of the Royal Society of London* A **356**:1927–39.

——. 2000a On extracting the googly information. *Twistor Newsletter* **45**:1–24. (Reprinted in *Roger Penrose, Collected Works*, volume 6 (1997–2003), chapter 289, pp. 463–87. Oxford University Press.

——. 2000b Wavefunction collapse as a real gravitational effect. In *Mathematical Physics 2000* (ed. A. Fokas, T. W. B. Kibble, A. Grigouriou, and B. Zegarlinski). Imperial College Press.

——. 2002 John Bell, state reduction, and quanglement. In *Quantum [Un]speakables: From Bell to Quantum Information* (ed. R. A. Bertlmann and A. Zeilinger), pp. 319–31. Springer.

——. 2003 On the instability of extra space dimensions. In *The Future of Theoretical Physics and Cosmology; Celebrating Stephen Hawking's 60th Birthday* (ed. G. W. Gibbons, E. P. S. Shellard, and S. J. Rankin), pp. 185–201. Cambridge University Press.

——. 2004 *The Road to Reality: A Complete Guide to the Laws of the Universe*. London: Jonathan Cape. (Referred to as TRtR in the text.)

——. 2005 The twistor approach to space-time structures. In *100 Years of Relativity; Space-time Structure: Einstein and Beyond* (ed. A. Ashtekar). World Scientific.

——. 2006 Before the Big Bang: an outrageous new perspective and its implications for particle physics. In *EPAC 2006 – Proceedings, Edinburgh, Scotland* (ed. C. R. Prior), pp. 2759–62. European Physical Society Accelerator Group (EPS-AG).

——. 2008 Causality, quantum theory and cosmology. In *On Space and Time* (ed. S. Majid), pp. 141–95. Cambridge University Press.

——. 2009a Black holes, quantum theory and cosmology (Fourth International Workshop DICE 2008). *Journal of Physics Conference Series* **174**:012001.

——. 2009b The basic ideas of conformal cyclic cosmology. In *Death and Anti-Death*, volume 6: *Thirty Years After Kurt Gödel (1906–1978)* (ed. C. Tandy), chapter 7, pp. 223–42. Stanford, CA: Ria University Press.

——. 2010 *Cycles of Time: An Extraordinary New View of the Universe*. London: Bodley Head.

——. 2014a On the gravitization of quantum mechanics. 1. Quantum state reduction. *Foundations of Physics* **44**:557–75.

——. 2014b On the gravitization of quantum mechanics. 2. Conformal cyclic cosmology. *Foundations of Physics* **44**:873–90.

——. 2015a Towards an objective physics of Bell non-locality: palatial twistor theory. In *Quantum Nonlocality and Reality – 50 Years of Bell's Theorem* (ed. S. Gao and M. Bell). Cambridge University Press.

Penrose, R. 2015b Palatial twistor theory and the twistor googly problem. *Philosophical Transactions of the Royal Society of London* **373**:20140250.

Penrose, R., and MacCallum, M. A. H. 1972 Twistor theory: an approach to the quantization of fields and space-time. *Physics Reports* C **6**:241–315.

Penrose, R., and Rindler, W. 1984 *Spinors and Space-Time*, volume 1: *Two-Spinor Calculus and Relativistic Fields*. Cambridge University Press.

——. 1986 *Spinors and Space-Time*, volume 2: *Spinor and Twistor Methods in Space-Time Geometry*. Cambridge University Press.

Pepper, B., Ghobadi, R., Jeffrey, E., Simon, C., and Bouwmeester, D. 2012 Optomechanical superpositions via nested interferometry. *Physical Review Letters* **109**:023601 (doi: 10.1103/PhysRevLett.109.023601).

Peres, A. 1991 Two simple proofs of the Kochen–Specker theorem. *Journal of Physics* A **24**:L175–78.

Perez, A., Sahlmann, H., and Sudarsky, D. 2006 On the quantum origin of the seeds of cosmic structure. *Classical and Quantum Gravity* **23**:2317–54.

Perjés, Z. 1977 Perspectives of Penrose theory in particle physics. *Reports on Mathematical Physics* **12**:193–211.

——. 1982 Introduction to twistor particle theory. In *Twistor Geometry and Non-Linear Systems* (ed. H. D. Doebner and T. D. Palev), pp. 53–72. Springer.

Perjés, Z., and Sparling, G. A. J. 1979 The twistor structure of hadrons. In *Advances in Twistor Theory* (ed. L. P. Hughston and R. S. Ward). Pitman.

Perlmutter, S., Schmidt, B. P., and Riess, A. G. 1998 Cosmology from type Ia supernovae. *Bulletin of the American Astronomical Society* **29**.

Perlmutter, S. (and 9 others) 1999 Measurements of $\Omega$ and $\Lambda$ from 42 high-redshift supernovae. *Astrophysical Journal* **517**:565–86.

Pikovski, I., Vanner, M. R., Aspelmeyer, M., Kim, M. S., and Brukner, C. 2012 Probing Planck-scale physics with quantum optics. *Nature Physics* **8**:393–97.

Piner, B. G. 2006 Technical report: the fastest relativistic jets from quasars and active galactic nuclei. *Synchrotron Radiation News* **19**:36–42.

Planck, M. 1901 Über das Gesetz der Energieverteilung im Normalspektrum. *Annalen der Physik* **4**:553.

Polchinski, J. 1994 What is string theory? Series of Lectures from the 1994 Les Houches Summer School (arXiv:hep-th/9411028).

——. 1998 *String Theory*, volume I: *An Introduction to the Bosonic String*. Cambridge University Press.

——. 1999 Quantum gravity at the Planck length. *International Journal of Modern Physics* A **14**:2633–58.

——. 2001 *String Theory*, volume 1: *Superstring Theory and Beyond*. Cambridge University Press.

Polchinski, J. 2004 Monopoles, duality, and string theory. *International Journal of Modern Physics* A **19**:145–54.

Polyakov, A. M. 1981a Quantum geometry of bosonic strings. *Physics Letters* B **103**:207–10.

Polyakov, A. M. 1981b Quantum geometry of fermionic strings. *Physics Letters B* **103**:211–13.

Popper, K. 1963 *Conjectures and Refutations: The Growth of Scientific Knowledge.* Routledge.

Ramallo, A. V. 2013 Introduction to the AdS/CFT correspondence. *Journal of High Energy Physics* **1306**:092.

Rauch, H., and Werner, S. A. 2015 *Neutron Interferometry: Lessons in Experimental Quantum Mechanics, Wave–Particle Duality, and Entanglement,* 2nd edn. Oxford University Press.

Rees, M. J. 2000 *Just Six Numbers: The Deep Forces That Shape the Universe.* Basic Books.

Riess, A. G. (and 19 others) 1998 Observational evidence from supernovae for an accelerating universe and a cosmological constant. *Astronomical Journal* **116**:1009–38.

Riley, K. F., Hobson, M. P., and Bence, S. J. 2006 *Mathematical Methods for Physics and Engineering: A Comprehensive Guide,* 3rd edn. Cambridge University Press.

Rindler, W. 1956 Visual horizons in world-models. *Monthly Notices of the Royal Astronomical Society* **116**:662–77.

———. 2001 *Relativity: Special, General, and Cosmological.* Oxford University Press.

Ritchie, N. M. W., Story J. G., and Hulet, R. G. 1991 Realization of a measurement of "weak value". *Physical Review Letters* **66**:1107–10.

Robertshaw, O., and Tod, K. P. 2006 Lie point symmetries and an approximate solution for the Schrödinger–Newton equations. *Nonlinearity* **19**:1507–14.

Roseveare, N. T. 1982 *Mercury's Perihelion from Le Verrier to Einstein.* Oxford: Clarendon Press.

Rosu, H. C. 1999 Classical and quantum inertia: a matter of principle. *Gravitation and Cosmology* **5**(2):81–91.

Rovelli, C. 2004 *Quantum Gravity.* Cambridge University Press.

Rowe, M. A., Kielpinski, D., Meyer, V., Sackett, C. A., Itano, W. M., Monroe, C., and Wineland, D. J. 2001 Experimental violation of a Bell's inequality with efficient detection. *Nature* **409**:791–94.

Rudin, W. 1986 *Real and Complex Analysis.* McGraw-Hill Education.

Ruffini, R., and Bonazzola, S. 1969 Systems of self-gravitating particles in general relativity and the concept of an equation of state. *Physical Review* **187**(5):1767–83.

Saunders, S., Barratt, J., Kent, A., and Wallace, D. (eds) 2012 *Many Worlds? Everett, Quantum Theory, and Reality.* Oxford University Press.

Schoen, R., and Yau, S.-T. 1983 The existence of a black hole due to condensation of matter. *Communications in Mathematical Physics* **90**:575–79.

Schrödinger, E. 1935 Die gegenwärtige Situation in der Quantenmechanik. *Naturwissenschaftenp* **23**:807–12, 823–28, 844–49. (Translation by J. T. Trimmer 1980 in *Proceedings of the American Philosophical Society* **124**:323–38.) Reprinted in Wheeler and Zurek [1983].

———. 1956 *Expanding Universes.* Cambridge University Press.

Schrödinger, E. 2012 *What Is Life?* with *Mind and Matter and Autobiographical Sketches* (foreword by R. Penrose). Cambridge University Press.

Schrödinger, E., and Born, M. 1935 Discussion of probability relations between separated systems. *Mathematical Proceedings of the Cambridge Philosophical Society* **31**:555–63.

Schwarzschild, K. 1900 Ueber das zulaessige Kruemmungsmaass des Raumes. *Vierteljahrsschrift der Astronomischen Gesellschaft* **35**:337–47. (English translation by J. M. Stewart and M. E. Stewart in 1998 *Classical and Quantum Gravity* **15**:2539–44.)

Sciama, D. W. 1959 *The Unity of the Universe*. Garden City, NY: Doubleday.

——. 1969 *The Physical Foundations of General Relativity* (Science Study Series). Garden City, NY: Doubleday.

Seckel A. 2004 *Masters of Deception. Escher, Dalí & the Artists of Optical Illusion*. Sterling.

Shankaranarayanan, S. 2003 Temperature and entropy of Schwarzschild–de Sitter spacetime. *Physical Review* D **67**:08026.

Shaw, W. T., and Hughston, L. P. 1990 Twistors and strings. In *Twistors in Mathematics and Physics*, LMS Lecture Note Series 156 (ed. T. N. Bailey and R. J. Baston). Cambridge University Press.

Skyrme, T. H. R. 1961 A non-linear field theory. *Proceedings of the Royal Society of London* A **260**:127–38.

Smolin, L. 2006 *The Trouble with Physics: The Rise of String Theory, the Fall of Science, and What Comes Next*. Houghton Miffin Harcourt.

Sobel, D. 2011 *A More Perfect Heaven: How Copernicus Revolutionised the Cosmos*. Bloomsbury.

Stachel, J. (ed.) 1995 *Einstein's Miraculous Year: Five Papers that Changed the Face of Physics*. Princeton University Press.

Stapp, H. P. 1979 Whieheadian approach to quantum theory and the generalized Bell theorem. *Foundations of Physics* **9**:1–25.

Starkman, G. D., Copi, C. J., Huterer, D., and Schwarz, D. 2012 The oddly quiet universe: how the CMB challenges cosmology's standard model. *Romanian Journal of Physics* **57**:979–91 (http://arxiv.org/PS_cache/arxiv/pdf/1201/1201.2459v1.pdf).

Steenrod, N. E. 1951 *The Topology of Fibre Bundles*. Princeton University Press.

Stein, E. M., Shakarchi, R. 2003 *Fourier Analysis: An Introduction*. Princeton University Press.

Steinhardt, P. J., and Turok, N. 2002 Cosmic evolution in a cyclic universe. *Physical Review* D **65**:126003.

——. 2007 *Endless Universe: Beyond the Big Bang*. Garden City, NY: Doubleday.

Stephens, C. R., 't Hooft, G., and Whiting, B. F. 1994 Black hole evaporation without information loss. *Classical and Quantum Gravity* **11**:621.

Strauss, W. A. 1992 *Partial Differential Equations: An Introduction*. Wiley.

Streater, R. F., and Wightman, A. S. 2000 *PCT, Spin Statistics, and All That*, 5th edn. Princeton University Press.

Strominger, A., and Vafa, C. 1996 Microscopic origin of the Bekenstein–Hawking entropy. *Physics Letters* B **379**:99–104.

Susskind, L. 1994 The world as a hologram. *Journal of Mathematical Physics* **36**(11): 6377–96.

Susskind, L., and Witten, E. 1998 The holographic bound in anti–de Sitter space. http://arxiv.org/pdf/hep-th/9805114.pdf

Susskind, L., Thorlacius, L., and Uglum, J. 1993 The stretched horizon and black hole complementarity. *Physical Review* D **48**:3743.

Synge, J. L. 1921 A system of space-time coordinates. *Nature* **108**:275.

——. 1950 The gravitational field of a particle. *Proceedings of the Royal Irish Academy* A **53**:83–114.

——. 1956 *Relativity: The Special Theory*. North-Holland.

Szekeres, G. 1960 On the singularities of a Riemannian manifold. *Publicationes Mathematicae Debrecen* **7**:285–301.

't Hooft, G. 1980a Naturalness, chiral symmetry, and spontaneous chiral symmetry breaking. *NATO Advanced Study Institute Series* **59**:135–57.

't Hooft, G. 1980b Confinement and topology in non-abelian gauge theories. Lectures given at the Schladming Winterschool, 20–29 February. *Acta Physica Austriaca Supplement* **22**:531–86.

't Hooft, G. 1993 Dimensional reduction in quantum gravity. In *Salamfestschrift: A Collection of Talks* (ed. A. Ali, J. Ellis, and S. Randjbar-Daemi). World Scientific.

Teller, E. 1948 On the change of physical constants. *Physical Review* **73**:801–2.

Thomson, M. 2013 *Modern Particle Physics*. Cambridge University Press.

Tod, K. P. 2003 Isotropic cosmological singularities: other matter models. *Classical and Quantum Gravity* **20**:521–34.

——. 2012 Penrose's circle in the CMB and test of inflation. *General Relativity and Gravity* **44**:2933–38.

Tod, K. P., and Moroz, I. M. 1999 An analytic approach to the Schrödinger–Newton equations. *Nonlinearity* **12**:201–16.

Tolman, R. C. 1934 *Relativity, Thermodynamics, and Cosmology*. Oxford: Clarendon Press.

Tombesi, F., et al. 2012 Comparison of ejection events in the jet and accretion disc outflows in 3C 111. *Monthly Notices of the Royal Astronomical Society* **424**:754–61.

Trautman, A. 1970 Fibre bundles associated with space-time. *Reports on Mathematical Physics* (Torun) **1**:29–62.

Tsou, S. T., and Chan, H. M. 1993 *Some Elementary Gauge Theory Concepts*, Lecture Notes in Physics, volume 47. World Scientific.

Tu, L. W. 2010 *An Introduction to Manifolds*. Springer.

Unruh, W. G. 1976 Notes on black hole evaporation. *Physical Review* D **14**:870.

Unruh, W. G., and Wald, R. M. 1982 Entropy bounds, acceleration radiation, and the generalized second law. *Physical Review* D **27**:2271.

Veneziano, G. 1991 *Physics Letters* B **265**:287.

Veneziano, G. 1998 A simple/short introduction to pre-Big-Bang physics/cosmology. arXiv:hep-th/9802057v2.

Vilenkin, A. 2004 Eternal inflation and chaotic terminology. arXiv:gr-qc/0409055.

von Klitzing, K. 1983 Quantized Hall effect. *Journal of Magnetism and Magnetic Materials* **31–34**:525–29.

von Klitzing, K., Dorda, G., and Pepper, M. 1980 New method for high-accuracy determination of the fine-structure constant based on quantized Hall resistance. *Physical Review Letters* **45**:494–97.

von Neumann, J. 1927 Wahrscheinlichkeitstheoretischer Aufbau der Quantenmechanik. *Göttinger Nachrichten* **1**:245–72.

———. 1932 Measurement and reversibility *and* The measuring process. In *Mathematische Grundlagen der Quantenmechanik*, chapters V and VI. Springer. (Translation by R. T. Beyer 1955: *Mathematical Foundations of Quantum Mechanics*, pp. 347–445. Princeton University Press. Reprinted in Wheeler and Zurek [1983, pp. 549–647].)

Wald, R. M. 1984 *General Relativity*. University of Chicago Press.

Wali, K. C. 2010 Chandra: a biographical portrait. *Physics Today* **63**:38–43.

Wallace, D. 2012 *The Emergent Multiverse: Quantum Theory According to the Everett Interpretation*. Oxford University Press.

Ward, R. S. 1977 On self-dual gauge fields. *Physics Letters* A **61**:81–82.

———. 1980 Self-dual space-times with cosmological constant. *Communications in Mathematical Physics* **78**:1–17.

Ward, R. S., and Wells Jr, R. O. 1989 *Twistor Geometry and Field Theory*. Cambridge University Press.

Weaver, M. J., Pepper, B., Luna, F., Buters, F. M., Eerkens, H. J., Welker, G., Perock, B., Heeck, K., de Man, S., and Bouwmeester, D. 2016 Nested trampoline resonators for optomechanics. *Applied Physics Letters* **108**:033501 (doi: 10.1063/1.4939828).

Weinberg, S. 1972 *Gravitation and Cosmology: Principles and Applications of the General Theory of Relativity*. Wiley.

Wells Jr, R. O. 1991 *Differential Analysis on Complex Manifolds*. Prentice Hall.

Wen, X.-G., and Witten, E. 1985 Electric and magnetic charges in superstring models. *Nuclear Physics* B **261**:651–77.

Werner, S. A 1994 Gravitational, rotational and topological quantum phase shifts in neutron interferometry. *Classical and Quantum Gravity* A **11**:207–26.

Wesson, P. (ed.) 1980 *Gravity, Particles, and Astrophysics: A Review of Modern Theories of Gravity and G-Variability, and Their Relation to Elementary Particle Physics and Astrophysics*. Springer.

Weyl, H. 1918 Gravitation und Electrizität. *Sitzungsberichte der Königlich Preuss ischen Akademie der Wissenschaften*, pp. 465–80.

Weyl, H. 1927 *Philosophie der Mathematik und Naturwissenschaft*. Oldernburg.

Wheeler, J. A. 1960 Neutrinos, gravitation and geometry. In *Rendiconti della Scuola Internazionale di Fisica Enrico Fermi XI Corso*, July 1959. Bologna: Zanichelli. (Reprinted 1982.)

Wheeler, J. A., and Zurek, W. H. (eds) 1983 *Quantum Theory and Measurement*. Princeton University Press.

Whittaker, E. T. 1903 On the partial differential equations of mathematical physics. *Mathematische Annalen* **57**:333–55.

Will, C. 1993 *Was Einstein Right?*, 2nd edn. Basic Books.

Witten, E. 1989 Quantum field theory and the Jones polynomial. *Communications in Mathematical Physics* **121**:351–99.

——. 1998 Anti–de Sitter space and holography. *Advances in Theoretical and Mathematical Physics* **2**:253–91.

——. 2004 Perturbative gauge theory as a string theory in twistor space. *Communications in Mathematical Physics* **252**:189–258.

Woodhouse, N. M. J. 1991 *Geometric Quantization*, 2nd edn. Oxford: Clarendon Press.

Wykes, A. 1969 *Doctor Cardano. Physician Extraordinary*. Frederick Muller.

Xiao, S. M., Herbst, T., Scheldt, T., Wang, D., Kropatschek, S., Naylor, W., Wittmann, B., Mech, A., Kofler, J., Anisimova, E., Makarov, V., Jennewein, Y., Ursin, R., and Zeilinger, A. 2012 Quantum teleportation over 143 kilometres using active feed-forward. *Nature Letters* **489**:269–73.

Zaffaroni, A. 2000 Introduction to the AdS–CFT correspondence. *Classical and Quantum Gravity* **17**:3571–97.

Zee, A. 2003 (1st edn) 2010 (2nd edn) *Quantum Field Theory in a Nutshell*. Princeton University Press.

Zeilinger, A. 2010 *Dance of the Photons*. New York: Farrar, Straus, and Giroux.

Zel'dovich, B. 1972 A hypothesis, unifying the structure and entropy of the universe. *Monthly Notices of the Royal Astronomical Society* **160**:1P.

Zimba, J., and Penrose, R. 1993 On Bell non-locality without probabilities: more curious geometry. *Studies in History and Philosophy of Society* **24**:697–720.

# Index